Archaeomalacology:
Shells in the Archaeological Record

Edited by

Katherine Szabó
Catherine Dupont
Vesna Dimitrijević
Luis Gómez Gastélum
Nathalie Serrand

BAR International Series 2666
2014

Published in 2016 by
BAR Publishing, Oxford

BAR International Series 2666

Archaeomalacology: Shells in the Archaeological Record

ISBN 978 1 4073 1308 5

BAR Publishing is the trading name of British Archaeological Reports (Oxford) Ltd. British Archaeological Reports was first incorporated in 1974 to publish the BAR Series, International and British. In 1992 Hadrian Books Ltd became part of the BAR group. This volume was originally published by Archaeopress in conjunction with British Archaeological Reports (Oxford) Ltd / Hadrian Books Ltd, the Series principal publisher, in 2014. This present volume is published by BAR Publishing, 2016.

Printed in England

BAR
PUBLISHING

BAR titles are available from:

BAR Publishing
122 Banbury Rd, Oxford, OX2 7BP, UK
EMAIL info@barpublishing.com
PHONE +44 (0)1865 310431
FAX +44 (0)1865 316916
www.barpublishing.com

Editors of the volume

Katherine Szabó
Associate Professor
Centre for Archaeological Science
School of Earth and Environmental Sciences
University of Wollongong
NSW 2522
Australia
kat@uow.edu.au

Catherine Dupont
Researcher
CNRS UMR 6566 CReAAH « Centre de Recherche en Archéologie, Archéosciences, Histoire »
CNRS, Université de Rennes 1, Rennes 2, Nantes, Le Mans et Ministère de la Culture
Bâtiment 24-25, Université de Rennes I – Campus Beaulieu 74205CS, 35042 Rennes Cx
France
catherine.dupont@univ-rennes1.fr

Vesna Dimitrijević
Professor
Laboratory for Bioarchaeology, Department of Archaeology
Faculty of Philosophy, University of Belgrade
Čika Ljubina 18-20, 11000 Belgrade
Serbia
vdimitri@f.bg.ac.rs

Luis Gómez Gastélum
Professor
Departamento de Ciencias Sociales
Centro Universitario de Tonalá
Universidad de Guadalajara
Mexico
gastelum@cencar.udg.mx

Nathalie Serrand
Researcher
In charge of archaeological research and field operations
Inrap: Institut national de recherches archéologiques préventives
Route de Dolé
97113 Gourbeyre
Guadeloupe
nathalie.serrand@inrap.fr

With the technical assistance of Francis Bertin (CNRS, CReAAH, UMR6566, Université de Rennes I – Campus Beaulieu 74205CS, 35042 Rennes Cx France) for the page setting. Assistance with English language editing was provided by Luke Gliganic.

Contents

Preface

This publication is one of the volumes of the proceedings of the 11th International Conference of the International Council for Archaeozoology (ICAZ), which was held in Paris (France) 23rd-28th August 2010. ICAZ was founded in the early 1970s and ever since has acted as the main international organisation for the study of animal remains from archaeological sites. The International Conferences of ICAZ are held every four years, with the Paris meeting – the largest ever – following those in Hungary (Budapest), the Netherlands (Groningen), Poland (Szczecin), England (London), France (Bordeaux), USA (Washington, DC), Germany (Constance), Canada (Victoria), England (Durham) and Mexico (Mexico City). The next meeting is scheduled be held in Argentina in 2014. The Paris conference – attended by some 720 delegates from 56 countries – was organised as one general and thirty thematic sessions, which attracted, in addition to archaeozoologists (zooarchaeologists), scholars from related disciplines such as bone chemistry, genetics, morphometry anthropology, archaeobotany, and mainstream archaeology. This conference was also marked by the involvement in the international archaeozoological community of increasing numbers of individuals from countries of Latin America and of South and East Asia.

As nearly 800 papers were presented at the Paris conference in the form of either oral or poster presentations, it was not possible to organize a comprehensive publication of the proceedings. It was left up to the session organizers to decide if the proceedings of their session would be published and to choose the form such a publication would take. A comprehensive list of publication plan of the 11th ICAZ International Conference is regularly updated and posted on the ICAZ web site.

The conference organizers would like to take this opportunity to thank the Muséum national d'Histoire naturelle, the Université Pierre et Marie Curie, the Centre national de la Recherche scientifique and the ICAZ Executive Committee for their support during the organization of the conference, and all session organisers – some of them being now book editors – for all their hard work. The conference would not have met with such success without the help of the Alpha Visa Congrès Company, which was in charge of conference management. Further financial help came from the following sources: La Région Île-de-France, the Bioarch European network (French CNRS; Natural History Museum Brussels; Universities of Durham, Aberdeen, Basel and Munich), the LeCHE Marie Curie International Training Network (granted by the European Council), the Institute of Ecology and Environment of the CNRS, the Institut National de Recherche en Archéologie Préventive (INRAP), the European-Chinese Cooperation project (ERA-NET Co-Reach), the Centre National Interprofessionnel de l'Économie Laitière (CNIEL) and its Observatory for Food Habits (OCHA), the Ville de Paris, the Société des Amis du Muséum, the French Embassies in Beijing and Moscow, the laboratory "Archaeozoology-Archaeobotany" (UMR7209, CNRS-MNHN), the School of Forensics of Lancaster, English Heritage and private donors.

Jean-Denis Vigne, Christine Lefèvre and Marylène Patou-Mathis
Organizers of the 11th ICAZ International Conference

Introduction

"Archaeomalacology: Shells in the Archaeological Record" is the Proceedings of the ICAZ Archaeomalacology Working group, who met at the 11th Conference of the International Council of Archaeozoology.

Session overview
Katherine Szabó

The study of invertebrate remains has traditionally been overshadowed by analysis of animal bones in the archaeological record. Yet evidence for human collection and modification of shells in particular has great temporal depth and is to be found in all corners of the world. This full-day session seeks to highlight the potential of molluscan remains to investigate issues of resource procurement and use, the social context of shell gathering and utilisation, and the various ways in which archaeomalacology can assist in current issues of environmental change and management. Three sub-sessions examine the use of shell as a raw material, the status and nature of shellfish as a food resource, and the potential of shell to probe questions of environmental and site transformations through time.

Acquisition and use of shell raw materials in prehistory
Vesna Dimitrijević, Catherine Dupont, Sándor Gulyás, Nathalie Serrand

Shells in general, and in particular species well-known across the world, such as spiny oysters (Spondylus spp.), tusk shells (Dentalium spp.), conchs (Strombus spp.), cowries (Trivia spp./Cypraea spp.) or mud snails (Cyclope neritea), were valued raw materials throughout prehistory as well as in more recent historic times. Common ways of shell acquisition involved harvesting coastal areas, including diving trips by those groups who lived by the sea, or developing exchange networks for those inhabiting inland areas. There is also the recorded collection of fossil shells. Shell products sometimes might have signalled a specific local character of a particular group, or were similarly valued across a wider region, thus illustrating bonds, rather than differentiation between diverse social groups.

This sub-session focuses on patterns of acquisition and use of shell raw materials as well as on the production sequences of shell items in time and space. Discussions of these patterns also relate to the question of economy, understood both as subsistence and industry. Specific themes of interest include the exploitation of shells as raw materials in relation to their dietary functions, or choices made to use particular shells along with or as opposed to other raw materials.

Shell middens and shells as a food resource
Luis Gómez Gastélum

In this sub-session we seek to integrate archaeological, historical, anthropological and biological studies that give insight into relationships between human and mollusc communities with a specific focus upon gathering of molluscs for subsistence. While many studies over the years have seen molluscs as a marginal resource of little overall importance, the testimony of shell middens as well as widespread information from historical and anthropological sources clearly tell us that shellfish-gathering was a regular and important activity. This importance is not limited to economics or dietary structure, but also forms a part of social and ritual practices. Shellfish gathering and midden creation has a larger social context.

This sub-session provides a venue to explore the relationships between human groups and molluscan resources and especially encourages the combination of information derived from multiple disciplines, as well as studies that seek to contextualise shell-gathering in a wider socio-economic context.

Shells as indicators of palaeoenvironment, site formation and transformation
Katherine Szabó

Shells, whether deliberately introduced to a site by humans or whether introduced through other natural processes, can give insight into aspects of the ancient environment as well as the formation and transformation of archaeological deposits. In stratified sites, changes in the composition and structure of shell assemblages through time can shed light on environmental change, whilst the comparison of various sites within an archaeological landscape can help us understand mosaic landscapes and the spread of different resources. The comparison of archaeological and modern shell assemblages can also provide an invaluable perspective to issues of resource management, modern environmental change and degradation, and the impact of invasive species.

Molluscan remains can also be of great assistance in untangling patterns of site formation and transformation, through studies of taphonomic alteration, bioturbation, or fine-grained investigations of naturally- and culturally-introduced shells within deposits.

This sub-session aims to investigate the potential of archaeological shell to answer questions not directly related to subsistence or material culture and especially welcomes contributions which mobilise the study of archaeological shell in relation to modern resource management and environmental change.

1 - THE USE OF MARINE MOLLUSC SHELLS AT THE NEOLITHIC SITE SHKARAT MSAIED, JORDAN

Aiysha ABU-LABAN
Department of Cross-Cultural and Regional Studies, University of Copenhagen, Snorresgade 17-19, 2300 Copenhagen S, Denmark,
aiysha@hum.ku.dk

Abstract : *Marine shells from the Mediterranean Sea and the Red Sea were widely used as ornaments by early Neolithic hunter-gatherers as well as sedentary groups in the southern Levant. A study of marine shells from the Neolithic site Shkarat Msaied (9,500±90 to 8,880±80 BP) shows that the diversity of species is high. Most of the shells were used as beads, and rather simple modification techniques were applied. The variation of bead types is quite limited when compared to other sites in the region. From an intra-site spatial analysis of artefacts at Shkarat Msaied it is suggested that shell artefacts are like stone and bone tools stored in communal buildings. Because of the high frequency of marine shells in the building designated to the interment of the dead, they are strongly associated with mortuary rituals.*

Keywords : *Marine molluscs, Shell beads, Neolithic, Exchange, Southern Levant*

Introduction

During the Middle Pre-Pottery Neolithic period (MPPNB) (c. 9,300-8,300 BP) (Kuijt and Goring-Morris 2002:366), marine shells were collected and used, mainly as ornaments, by hunter-gatherer groups in the southern Levant (Goring-Morris 2005:97; Kirkbride 1967:9). Many scholars consider their presence in the material cultural repertoire an *ipso facto* proof of exchange between groups (Bar-Yosef and Belfer-Cohen 1989; Kuijt and Goring-Morris 2002). When found at inland sites far from their sources, marine shells are assigned a high value and typically categorised as prestige items (Banning 1998:215; Watkins 2008:160). The study of marine mollusc shells from the Neolithic site Shkarat Msaied provides an insight into the use of shells among hunter-gatherer societies in a semi-arid inland region (Abu-Laban 2010). Excavation of the site Shkarat Msaied was carried out by a team from the University of Copenhagen between 1999 and 2005, and again in 2010.[1]

The site Shkarat Msaied

Shkarat Msaied is situated c. 13 km north of Petra (fig. 1-1), in the Nemelleh region; a semi-arid steppe area c. 980 m above sea level between two mountain peaks (Gebel 1988:8; Jensen *et al.* 2005:115). The eastern area of the site is dominated by a sandstone plateau, and the west by plains of drainage systems, which run into Wadi Araba (Gebel 1988:71, 83). The vegetation today is characterised by forests of oak and pistachio, as well as shrubs and other minor plants. These plant species generally grow in the area of the Arabian Plateau and surround the site from the east. In the west, the *siqs* are wooded with pistachio, oak and juniper (Gebel 1988:81; Jensen *et al.* 2005:115).

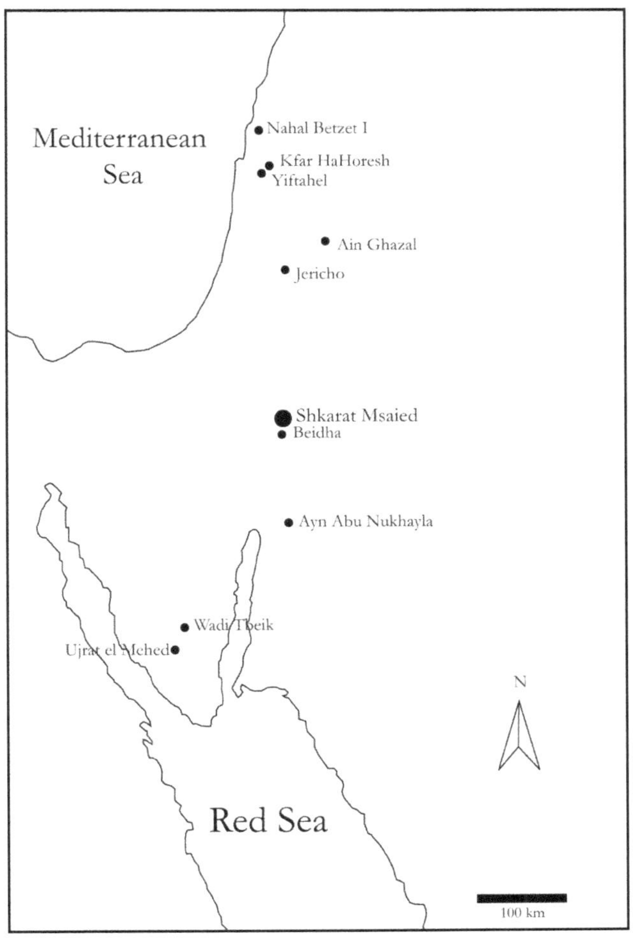

Figure 1-1. Map of MPPNB sites in the southern Levant mentioned in the article.

[1] The study does not encompass the shell assemblage from the last season in 2010.

Shkarat Msaied was mainly occupied during the MP-PNB period – seven uncalibrated conventional 14C dates give a range from 9,500±90 to 8,880±80 BP (Hermansen *et al.* 2006:3). The inhabitants comprised a small group of hunter-gatherers who also practised cultivation and possibly the herding of goats (Jensen *et al.* 2005:117-119). The site was only occupied in certain seasons of the year; most probably in spring when wild plants could be harvested and cereals processed (Jensen forthcoming). Shkarat Msaied is estimated to have covered an area of c. 1000 m², of which 600 m² have been exposed (Jensen *et al.* 2005:115). The site consists mainly of circular buildings arranged in clusters and with open spaces between some of them (fig. 1-2). The northern part of the settlement shows evidence of domestic use, whereas the southern part bears evidence of activities of a more communal character. The open spaces functioned as passages, and in certain areas there is evidence of activities such as food processing and tool production.

From a regional perspective and compared to the larger sedentary settlements (4-5 ha) in the Mediterranean zone and the smaller camps (maximum 300 m² in size) occupying the desert regions (Kuijt and Goring-Morris 2002), Shkarat Msaied is a medium sized settlement.

The marine mollusc shells – species and aspects of exchange between groups

Although the amount of marine shells found at the site is low – 379 (NISP), the diversity of species is quite high, with 26 different species identified (table 1-1). More than half of the shell assemblage (57%) ultimately derives from the Red Sea, and less than one fifth (18%) from the Mediterranean. It was impossible to determine the species of the rest of the shell assemblage because the specimens were either too fragmented or worn. In fact, 26% of the shell assemblage was categorised as fragments, i.e. where less than half of the shell is present. Because of the distances over which the shells were transported –100 km to the Red

[2] Wood charcoal samples (Lab no. Aar-9335, Aar-9336, Aar-9337, Wk-15159, Wk-15160, Wk-26490, Wk-26491)

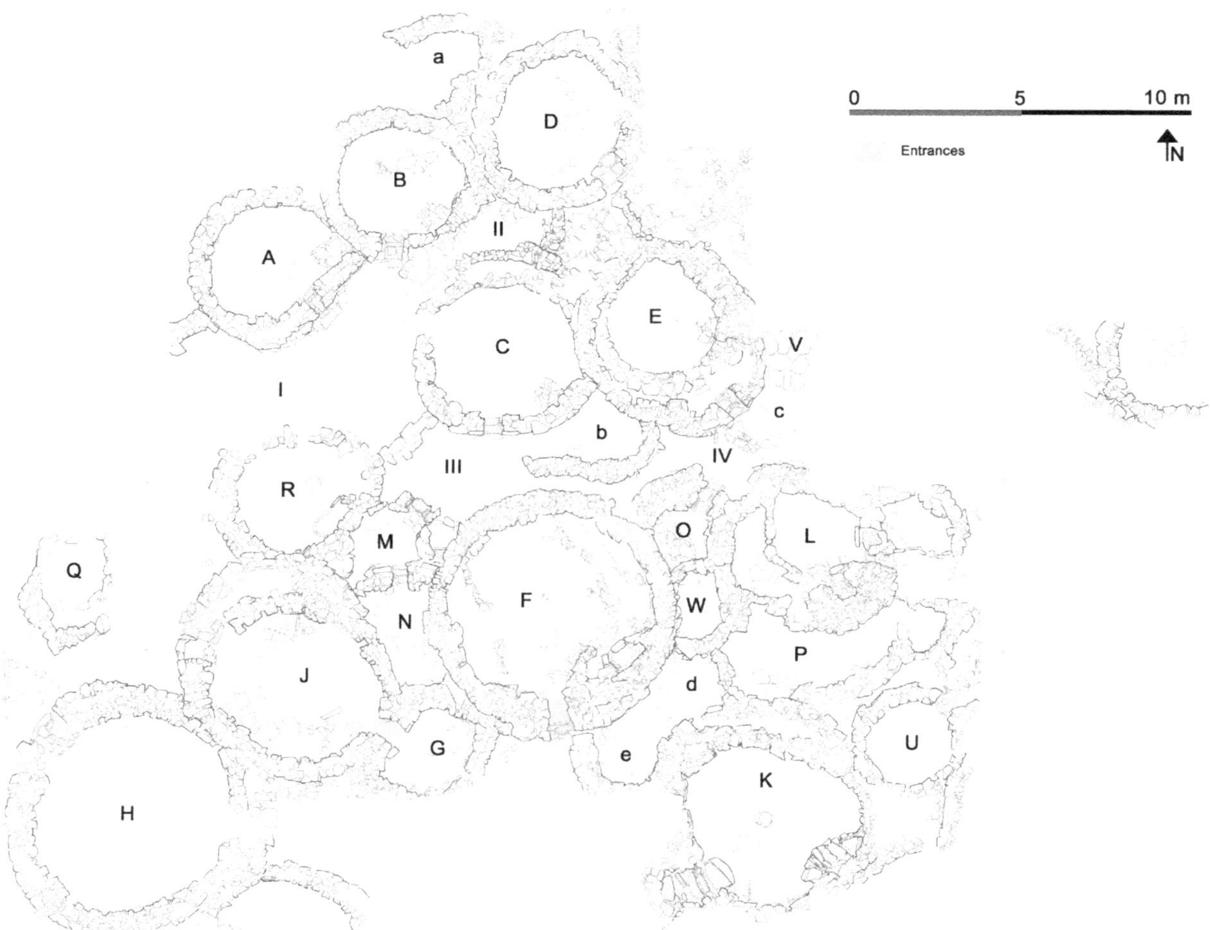

Figure 1-2. Site plan of Shkarat Msaied (Courtesy of Moritz Kinzel).

Sea and 160 km to the Mediterranean Sea – the possibility that the settlers procured the shells directly has been discounted. Thus, the marine shells at the site are considered to have been introduced as a direct consequence of exchange with neighbouring groups.

Comparing the distribution of the shells to that of other sites in the southern Levant, there is a patterned relationship between the quantity found in a given site and its location relative to the sources. The more distant a given material group is from the source, the lower the quantity. Such a distribution pattern produces a fall-off curve from which the process of exchange is explained as down-the-line exchange (Fig. 1-3.a and b). This kind of exchange is characterised as not being hierarchically organised, as there is no sign of a central place at which a certain material group was collected and/or produced. Exchange is consequently viewed as being exclusively reciprocal (Earle 1999:614; Renfrew 1977:73). This explanation model is one-dimensional as mechanisms of exchange are based solely on measurable factors such as geographical distance and statistical densities from excavations.

The most abundant species at Shkarat Msaied are cowries (Cypraeidae) including species in the genera *Cypraea, Erosaria, Lyncina* and *Monetaria* (148 NISP), of which the Red Sea *Erosaria nebrites* (93 NISP) is the most common. They are followed by species of *Nerita* (57 NISP).

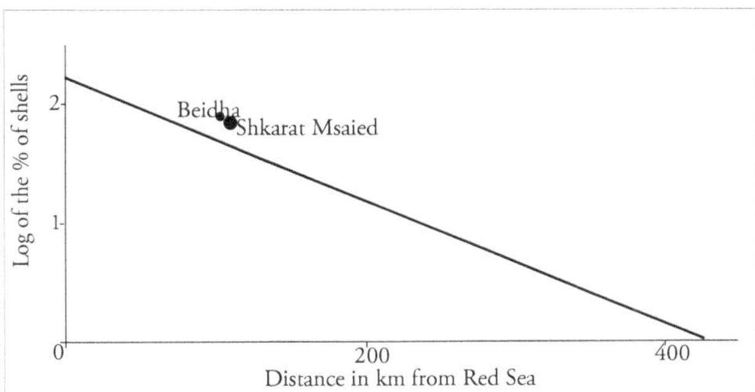

Figure 1-3a. Fall-off curve of Red Sea shell species in the southern Levant.

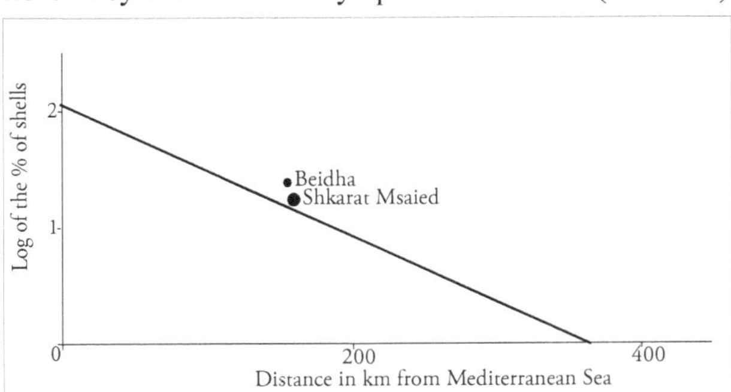

Figure 1-3b. Fall-off curve of Mediterranean shell species in the southern Levant.

These two taxa are also among the dominant marine mollusc categories in other contemporary settlements such as Beidha (Reese unpublished manuscript) and Ayn Abu Nukhayla (fig. 1-4), as well as in smaller encampments in the Sinai such as Ujrat el Mehed and Wadi Tbeik. There is, however, an element of sub-regional variation, as the Sinai sites and Ayn Abu Nukhayla have *Conus* and *Dentalium* shells as the dominant genera (Bar-Yosef Mayer 1999:72; Spatz 2008:77). These two taxa are found in low quantities at Shkarat Msaied and Beidha. On the other hand, the Mediterranean bivalve species *Acanthocardia tuberculata* is well represented followed in quantity by the Red Sea species *Glycymeris livida* both at Shkarat Msaied and Beidha. Moreover, *Nassarius* shells are quite abundant at Ayn Abu Nukhayla (Spatz 2008:76), although the distance of this site to the Mediterranean Sea is higher as compared to Shkarat Msaied and Beidha, where only 6 and 8 *Nassarius* specimens were found respectively.

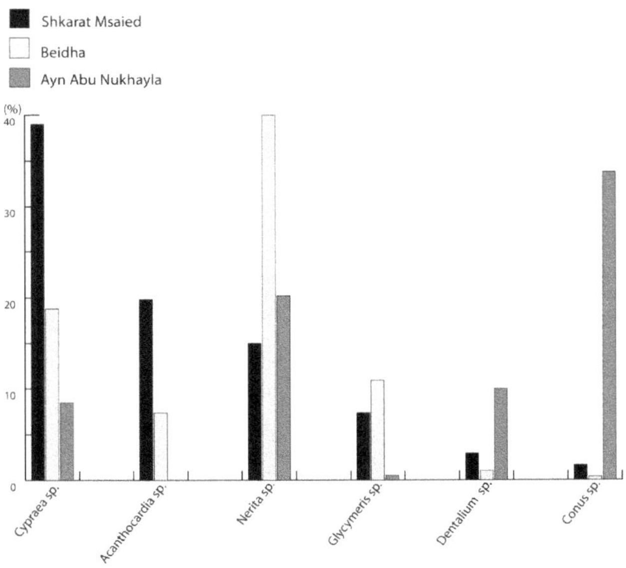

Figure 1-4. Comparison of selected species' quantity (in %) in the sites Shkarat Msaied, Beidha and Ayn Abu Nukhayla.

Cowrie (="*Cypraea*") shells seem to have been a 'trend' in the southern part of the southern Levant only, as they are not so common in the Mediterranean region at sites such as Nahal Betzet I, Kfar HaHoresh and Yiftahel (Spatz 2008). In this region however, *Cypraea* shells occur in mortuary contexts as grave goods and as eye inlays on plastered skulls as has been observed at Jericho (Kenyon 1957:124). The bivalve species *Acanthocardia* and *Cerastoderma* are among the dominant species constituting together between 30-85% of the marine assemblage at the aforementioned sites situated in the Mediterranean region (Spatz 2008).
Although we are not dealing with an organised exchange of shells, there are indications that each group

had a preference for certain species. It is postulated that the shells were primarily selected for their morphological features or shape. Shkarat Msaied and Beidha, only 7 km apart, have unsurprisingly similar preferences for shell species.

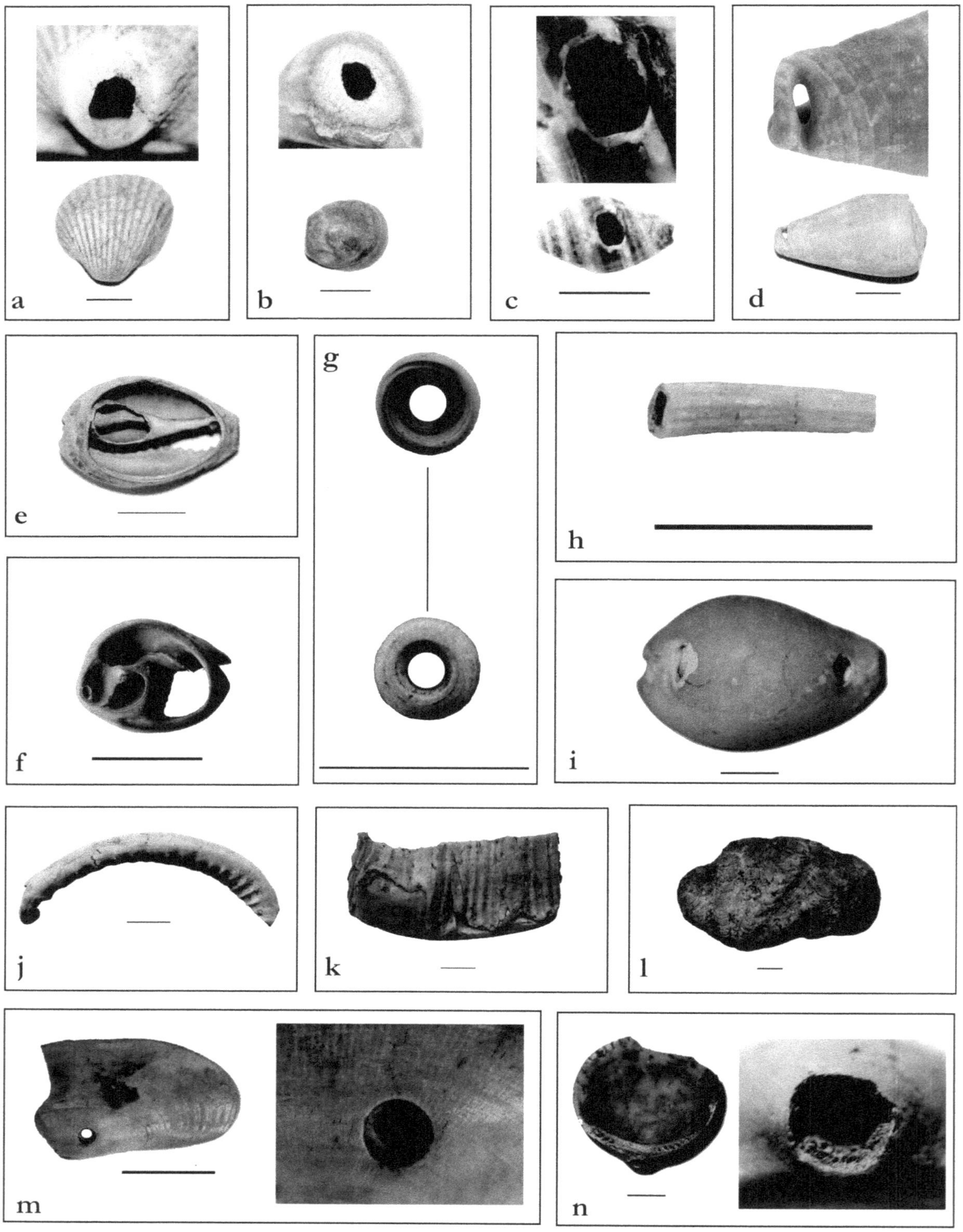

Figure 1-5. Modified shells in Shkarat Msaied. a. *Cerastoderma* sp. (61217), b. *Nerita* sp. (3715c), c. *Engina mendicaria* (251), d. *Conus* sp. (52506b), e. *Erosaria* sp. (71318c), f. *Nassarius* sp. (61328), g. Gastropoda indet. (61912), h. *Dentalium* sp. (1972a), i. *Phalium granulatum undulatum* (3550), j. *Cypraea* sp. (3119). Possibly modified shell: k. Gastropoda indet.(1340) scraper? l. *Tridacna maxima* (52211) possible container. Naturally perforated shells: m. *Donax* sp. (71002), n. *Glycymeris* sp. (52503a). Scale=1cm.

	Species	Natural Habitat	NISP	Modified [1]	Unmodified	Fragment
Bivalves	*Acanthocardia* sp.	Mediterranean Sea	21	0	2	19
	Acanthocardia tuberculata	Mediterranean Sea	54	40	13	1
	Cerastoderma glaucum	Mediterranean Sea	5	3	2	0
	Chama sp.	?	1	0	1	0
	Donax trunculus	Mediterranean Sea	2	2	0	0
	Glycymeris livida	Red Sea	28	16	7	6
	Pinctada sp.	Red Sea	8	0	0	8
	Tridacna maxima	Red Sea	3	1	2	0
Gastropods	*Canarium erythrinus*	Red Sea	1	1	0	0
	Canarium fusiformis	Red Sea	1	1	0	0
	Clanculus pharaonius	Red Sea	2	2	0	0
	Clypeomorus bifasciata	Red Sea	1	1	0	0
	Conomurex fasciatus	Red Sea	1	1	0	0
	Conus mediterraneus	Mediterranean Sea	1	1	0	0
	Conus tessulatus	Red Sea	1	1	0	0
	Conus sp.	?	4	4	0	0
	Cypraea sp.	?	36	35	0	1
	Engina mendicaria	Red Sea	3	3	0	0
	Erosaria macandrewi	Red Sea	1	1	0	0
	Erosaria nebrites	Red Sea	93	89	4	0
	Lamibs truncata sebae	Red Sea	2	0	1	1
	Lyncina carneola	Red Sea	2	2	0	0
	Lyncina lynx	Red Sea	2	2	0	0
	Monetaria annulus	Red Sea	16	15	1	0
	Nassarius gibbosulus	Mediterranean Sea	6	6	0	0
	Nerita sanguinolenta	Red Sea	57	31	26	0
	Phalium granulatum undulatum	Mediterranean Sea	3	3	0	0
	Polinices mammilla	Red Sea	3	3	0	0
	Thais savignyi	Red Sea	1	1	0	0
	Gastropoda indet.	?	3	3	0	0
	Mollusca indet.	?	5	0	0	5
	Scaphopoda indet.	?	11	11	0	0

[1] Includes fragments of shells with signs of modification

Table 1-1. Marine mollusc species at Shkarat Msaied.

Producing shell beads

More than two thirds of the shells (71%) at Shkarat Msaied have undergone deliberate modification. Most of the shells (96%) can be defined as simple beads, as they only have one hole (Bar-Yosef Mayer 1999:30). The hole is produced at the umbo for all bivalves and at the apex for most of the gastropods (fig. 1-5a and 5b). In some gastropod shells the hole was produced on the body area, either at the body whorl or near the aperture (fig. 1-5c and 5d). Studies show that the preferred technique to produce holes on the umbo and apex was grinding. Gouging or direct percussion was used typically to produce a hole on the body of the shell (Francis 1989). I conducted a small scale experiment on a group of shells, which were ground on a flat piece of sandstone. Sandstone is abundant at the site and used extensively as a building material and for producing tools. Preliminary results show that producing a hole by grinding the umbonal area on bivalves takes less than three minutes, whereas producing a hole on the body area of *Clanculus pharaonius* shells can take up to 10 minutes.

For *Cypraea* and *Nassarius* shells, the dorsum is typically removed by hammering and/or grinding (Francis 1989:29). The edges of some of the cowrie shells seem to have been polished to give a smooth and even surface (fig. 1-5e and 5f).

Sawing is more time consuming than the aforementioned techniques and has only been applied to a few specimens such as a *Conus* shell (fig. 1-5d), and for the production of disc beads (fig. 1-5g). The disc beads are made from gastropod shells and seem to involve sawing off the spire from the rest of the shell as well as removing the top of the spire. A sharp edged flint blade could be used as a sawing device, and the process can take up to two hours (Francis 1989:28). The *Dentalium* shells might also have been sawn or simply broken (fig. 1-5h). Only one cowrie shell bead had the dorsum intact as holes were produced on each side of it; possibly by gouging (fig. 1-5i).

Besides exploiting shells for the use as beads, it is possible that the three *Phalium granulatum undulatum* fragments found at the site were originally used as bangles or bracelets (Bar-Yosef Mayer 1999:31) (fig. -5j). Some shells might have had more utilitarian functions. A fragment of a shell could have been used as a scraper (fig. 1-5k), and two complete unmodified *Tridacna maxima* shells could have been used as some kind of container (fig. 1-5l). One of the *Tridacna* shells was found in situ on a large working sandstone slab in building D.

There are few examples of shells which have natural perforations. Three shells have holes which are characterised as having fine regular striations. The holes have a tapered diameter which is wider at the outer surface (fig. 1-5m). Such holes can only be produced by carnivorous gastropods (personal communication Kenneth Thomas 2008). Otherwise, most natural holes are generally observed in bivalve shells at the umbo area and have an uneven diameter (fig. 1-5n). Although it would require the use of wear analysis to be certain, naturally perforated shells were probably also used as ornaments (Anderson 2001:132; Claassen 1998:40).

From the study of the modified shells, investment of time as well as the level of skill required for the production of most of the shell beads at Shkarat Msaied proved to be on a low scale. In general, the modification of the shells did not require specialised tools. At present it is impossible to say whether the shells were brought to the site pre-modified or were worked on the site. There is, however, no direct evidence of a shell bead workshop at Shkarat Msaied. First of all, there is no concentration of unmodified shells that could be used as raw material (Bar-Yosef Mayer 1999:65) in any specific area, although umodified shells constitute 18% of the entire shell assemblage. Secondly, there is a lack of evidence of shells directly associated with tools that could be related to shell bead production. In the neighbouring site of Beidha a whole sequence of shell bead production – such as blanks and finished beads – were found together on a large sandstone slab with tools related to bead production in one building (Byrd 2005:117).

Figure 1-6. Two cowrie dorsum fragments (52616 and 2900).

It is difficult to detect any signs of such a *chaîne opératoire* of shell bead production at Shkarat Msaied. As this study has shown, most of the techniques applied to produce the beads would not leave any significant debris. The only evidence we have of shell debris are two dorsum fragments from cowrie shells (fig. 1-6).

For the time being, it can be asserted that the inhabitants of Shkarat Msaied had the technological know-how for producing beads. Remains of what could either be a dump or workshop area for the production of green stone disc beads were found in Area I and below enclosure 'a' (Jensen 2008). The process of producing these greenstone beads would have required more time and a higher level of skill compared to the simple shell beads. Studies and experiments show that special tools such as borers are required to produce disc beads (Jensen 2008; Wright, Critchley and Garrard 2008).[3] It is tempting to suggest that if the shell beads were produced on site, this was conducted on a household level, as opposed to the production of stone beads, which was an organised endeavour conducted on a collective level.

Compared to other sites in the region, the variety of bead types at Shkarat Msaied seems restricted, as it only comprises simple beads and disc beads. At other sites such as Ayn Abu Nukhayla, the assemblage also includes bead pendants (Bar-Yosef Mayer 1999:70; Spatz 2008:81).

Intra-site distribution of shells and exchange within a group

In order to shed more light on the use of marine shells at Shkarat Msaied, an intra-site spatial analysis was conducted comparing marine shell distributions to the distributions of other material categories; bone and stone tools.[4] Shells, bone and stone tools are all distributed across the whole site, and are found both in domestic and communal spaces (fig. 1-2). In the domestic buildings (A, B, C, D, E and R), the maximum number of shells found was 25 specimens. They were either absent or found in low quantities in the passage areas (III and IV) and enclosures (b and c), as well as in the small rooms (M, N, G, O and W). Ground stone artefacts were, however, quite abundant there and this implies that shell bead production and/or storage was not part of the function of these spaces.

The analysis shows that the highest concentration of shells as well as stone and bone tools is to be found in building P and enclosures 'd' and 'e'. These facilities are interpreted as storage and possibly also for general production. These areas are not directly accessible from any of the domestic buildings, thus they cannot be linked to any private use and should be interpreted as communal. The very high number of finds in areas P, 'd' and 'e' compared to the buildings designated a domestic function gives an indication that production and distribution of not only tools but also ornaments was controlled on a collective or group level. Shell ornaments, even if they were used for personal expression (Sciama 1998), were like utilitarian objects controlled collectively.

Looking further at the communal buildings F, K and J, all bear evidence of special purpose activities. Building F featured the highest quantity of shells recovered (45 NISP), whereas buildings K and J yielded no more than 8 and 11 specimens respectively. Building J housed activities related to food production, as suggested by grinding tools found *in situ* on the floor. In building K, more than 20 stone objects were placed deliberately on one of the stair cases and on the floor. Some of the stone objects showed remnants of red ochre, and their context implies a ritual activity. Here, only a few shells were found, suggesting that the rituals carried out in this building did not involve marine shells.

Building F was used for mortuary rituals, as it housed all burials.[5] The burials were both primary and secondary, and all bodies were interred below the floors of the building. Finding a high frequency of marine shells in this building strongly indicates that marine shells were used in mortuary ceremonies. Only three shells were found in direct burial context, however, they were in a burial fill and not deposited as grave goods. During the MPPNB period leaving grave goods with the interred was not a widespread practice (e.g. Simmons *et al.* 1990:70).

From the distribution of the shells it is not thought that marine shells were used as status or rank markers, that is if hoarding or other evidence of accumulation should be understood as wealth keeping (Hamon and Quilliec 2008). In other contemporary settlements such as Ain Ghazal, Beidha and Kfar HaHoresh, hoarding involved utilitarian objects such as flint tools, or figurines and other cult objects such as plastered skulls. Shells are almost entirely absent from such contexts (Goring-Morris 2005:97; Kirkbride 1967:10; Kuijt 2000:151; Rollefson 2000:167).

Conclusion

Marine shells were used at Shkarat Msaied as personal ornaments. Cowrie and *Nerita* shells were among the favoured types. The time and skills required for the production of the shell artefacts proved to be low level, as opposed to the production of stone beads. Their archaeological contexts suggest that the distribution of marine shells to group members was more or less controlled at a collective level. This was possibly done in order to ensure an egalitarian wealth distribution. It has been suggested by Kuijt (2000) that mortuary rituals, which involved among other practices skull removals and skull caching, were used by MPPNB groups as a means of maintaining egalitarian social systems as a response to emerging social differentiation (Kuijt 2000). Should the marine shells be assigned a value at Shkarat Msaied, it was not necessarily due to their distant origin, but rather because of their association with mortuary rituals.

Acknowledgements

Many thanks to Henk Mienis for his help with species identifications. Thanks to David Reese, Ashton J. Spatz and Daniella Bar-Yosef for allowing me to use information from their studies of marine molluscs from sites in the southern Levant as comparative material. Thanks also to the Shkarat Msaied project team, especially Charlott Hoffmann Jensen and Ingolf Thuesen, who gave me access to the marine mollusc shell assemblage. Thanks to Susanne Kerner, Sarah Cox and Kristoffer Damgaard for reviewing this article; the final form of the paper is my responsibility. Thanks to Susanne Kerner for supervising my MA thesis on which this article is based.

[3] It is worth mentioning that a couple of sandstone and bone disc beads were also found at Shkarat Msaied. These beads require the same manufacturing processes and tools as the green stone beads. It is likely that they were produced on site as both materials are abundant at the site.

[4] This study only includes the distribution of ground stone tools.

[5] In the 2010 season a child burial was found in building R.

References

ABU-LABAN, AIYSHA. 2010. Analysis and reconstruction of the use of mollusc shells from the MPPNB site Shkarat Msaied in Southern Jordan. MS thesis, University of Copenhagen.

ANDERSON, DONALD THOMAS, ed. 2001. *Invertebrate Zoology*. Oxford: Oxford University Press.

BANNING, EDWARD. 1998. The Neolithic Period. *Near Eastern Archaeology* 61 (4): 188-237.

BAR-YOSEF MAYER, DANIELLA E. 1999. The role of shells in the reconstruction of socio-economic aspects of Neolithic through Early Bronze Age societies in Sinai. PhD dissertation, Hebrew University of Jerusalem.

BAR-YOSEF, OFER and ANNA BELFER-COHEN. 1989. The Levantine «PPNB» interaction sphere. *In People and culture in change*. I. Hershkovitz, ed. Pp. 59-72. Oxford: Bar International Series 508.

BYRD, BRIAN F. 2005. *Early village life at Beidha, Jordan: Neolithic spatial organization and vernacular architecture. The excavations of Mrs. Diana Kirkbride-Helbæk.* Oxford: Oxford University Press.

CLAASSEN, CHERYL. 1998. *Shells*. Cambridge: Cambridge University Press.

EARLE, TIMOTHY. 1999. Production and exchange in prehistory. In *Companion Encyclopedia of Archaeology*, Vol. 1. G.Barker, ed. Pp. 608-635. London: Routledge.

FRANCIS, PETER. 1982. Experiments with early techniques for making whole shells into beads. *Current Anthropology* 23 (6): 713-714.

———. 1989. The manufacture of beads from shell. *In Proceedings of the 1986 Bead Conference – Selected Papers.* Charles F. Hayes III and Lynn Ceci, eds. Pp. 25-37. Rochester: Rochester Museum and Science Center. Research Records No. 20.

GEBEL, HANS GEORG. 1988. Late Epipalaeolithic – Aceramic Neolithic sites in the Petra-Area. *In The Prehistory of Jordan – The state of research in 1986.* Andrew Garrard and Hans Georg Gebel, eds. Pp. 67-100. Oxford: Bar International Series 396(i).

GORING-MORRIS, NIGEL. 2005. Life, death and the emergence of differential status in the Near Eastern Neolithic: evidence from Kfar HaHoresh, Lower Galilee, Israel. In: *Archaeological Perspectives on the transmission and transformation of culture in the eastern Mediterranean (Levant Supplement Series vol. 2).* Joanne Clarke, ed. Pp. 89-105. Oxford: CBRL & Oxbow Books.

HAMON, CAROLINE, and BENEDICTE QUILLIEC, eds. 2008. *Hoards from the Neolithic to the metal ages, technical and codified practices, session of the XIth Annual Meeting of the European Association of Archaeologists.* Oxford: Bar International Series 1758.

HERMANSEN, BO DAHL, INGOLF THUESEN, CHARLOTT HOFFMANN JENSEN, MORITZ KINZEL, MIKKEL BILLE PETERSEN, MARIE LOUISE JØRKOV, and NIELS LYNNERUP. 2006. Shkarat Msaied: The 2005 season of excavations. A short preliminary report. *Neo-Lithics* 1/06: 3-7.

JENSEN, CHARLOTT HOFFMANN. 2008. Workshops and activity areas in the PPNB period: the excavations at Shkarat Msiaed. In *Proceedings of the 4th International Congress of the Archaeology of the Ancient Near East.* Vol. 2. Hartmut Kühne, Rainer M. Czichon and F.Janoscha Kreppner, eds. Pp 331-343. Wiesbaden: Harrasowitz Verlag.

JENSEN, CHARLOTT HOFFMANN, BO DAHL HERMANSEN, MIKKEL BILLE PETERSEN, MORITZ KINZEL, METTE MARIE HALD, PERNILLE. BANGSGAARD, NIELS LYNNERUP and INGOLF THUESEN. 2005. Preliminary report on the excavations at Shkarat Al-Musay'id, 1999-2004. *Annual of the Department of Antiquities of Jordan* 49: 115-135.

JENSEN, CHARLOTT HOFFMANN. Forthcoming. Three years of excavation at the PPNB site Shaqarat Mazyad, Southern Jordan. *Proceedings of the 3rd International Congress of the Archaeology of the Ancient Near East.*

KENYON, KATHLEEN. 1957. *Digging up Jericho: The Results of the Jericho Excavations,* 1952-1956. New York: Praeger.

KIRKBRIDE, DIANA. 1967. Beidha 1965: An Interim Report. *Palestine Exploration Quarterly* 99: 5:13.

KUIJT, IAN. 2000. Keeping the peace: ritual, skull caching, and community integration in the Levantine Neolithic. In *Life in Neolithic Farming Communities – Social organization, identity and differentiation.* Ian Kuijt, ed. Pp. 137-160. New York: Kluwer Academic/Plenum publishers.

KUIJT, IAN and NIGEL GORING-MORRIS. 2002. Foraging, farming, and social complexity in the Pre-Pottery Neolithic of the Southern Levant: A review and synthesis. *Journal of World Prehistory* 16 (4): 361-440.

REESE, DAVID. Unpublished manuscript. *Neolithic shells from Beidha.*

RENFREW, COLIN. 1977. Alternative models for exchange

and spatial distribution. In *Exchange Systems in Prehistory.* Earle, Timothy K. and Jonathon E. Ericson, eds. Pp. 71-91. New York: Academic Press.

Rollefson, Gary O. 2000. Ritual and social structure at Neolithic 'Ain Ghazal. In *Life in Neolithic Farming Communities – Social organization, identity and differentiation.* Ian Kuijt, ed. Pp. 163-188. New York: Kluwer Academic/Plenum publishers.

Sciama, Lidia D. 1998. Gender in the making, trading and uses of beads. An introductory essay. In *Beads and Beadmakers.* Lidia D. Sciama and Joanne B. Eicher, eds. Pp. 1-47. Oxford: Berg.

Simmons, Alan, Ann Boulton, Carol Roetzel Butler, Zeidan Kafafi and Gary O. Rollefson. 1990. A Plastered Skull from Neolithic 'Ain Ghazal, Jordan. *Journal of Field Archaeology* 17 (1): 107-110.

Spatz, Ashton J. 2008. Ornamental marine mollusc shells from the Pre-Pottery Neolithic B site of Ayn Abu Nukhayla, Southern Jordan, and implications for exchange networks in the Southern Levant. MS Thesis, The University of Tulsa.

Watkins, Trevor. 2008. Supra-Regional Networks in the Neolithic of Southwest Asia. *Journal of World Prehistory* 21: 139-171.

Wright, Katherine I., Pat Critchley and Andrew Garrard. 2008. Stone bead technologies and early craft specialization: Insights from two Neolithic sites in eastern Jordan. *Levant* 40 (2): 131-165.

2 - EVALUATING THE ROLE OF THE MOLLUSCAN SHELL ASSEMBLAGE RECOVERED FROM PADRI, A COASTAL HARAPPAN SETTLEMENT IN GUJARAT, INDIA

Arati DESHPANDE-MUKHERJEE and Vasant SHINDE
Department of Archaeolgy, Deccan College Post Graduate and Research Institute, Pune 411006, Yerwada, India,
adm.muk@gmail.com, vshinde.dc@gmail.com

Abstract : *Excavations at Padri, situated on the Bhavnagar coast in Gujarat, have yielded ceramics, stone beads, copper objects, molluscan shell remains, and mammalian and fish bones. These were recovered from a two fold cultural sequence, namely Early Historic (1st century BCE to 1st century CE) and Harappan (3300 to 2000 BCE). On the basis of its Pre-Harappan/Non Harappan ceramics and radiocarbon dates as old as the second half of the fourth millennium BCE, Padri is considered as one of the earliest Chalcolithic settlements in Gujarat.*
In this paper we present results of the analysis of the shell assemblage recovered from the Harappan and Early Historic levels. The shell assemblage is comprised of isolated molluscan shells, mostly of marine estuarine origin, along with shell objects and shell working debitage. An attempt was made to study the processes by which the various molluscs were introduced into the settlement and the role they played. A reconstruction of shell working during the Harappan and Early Historic period was also attempted. Comparison with other contemporary coastal Harappan sites has helped in elucidating the extent of molluscan exploitation at Padri. Additional insights into the past coastal environment around the site were obtained from the overall shell data.

Keywords: *Gujarat, Harappan, Early historic, Gulf of Khambat, Gulf of Kachchh,* Turbinella pyrum

Introduction

The Gujarat region in western India has a long indented coastline of more than 1500 km that borders the regions of Saurashtra, mainland Gujarat and Kachchh. On its western side it overlooks the Gulf of Kachchh, Arabian Sea and Gulf of Khambat. Hence it is not surprising that in the past marine molluscs were either naturally or due to anthropogenic processes incorporated into many of the archaeological sites in this region. Molluscan shells were first observed and reported in the archaeological context by Robert Bruce Foote (1916) in the early part of the twentieth century. Since then, shells and objects made from shell have been recovered from sites dating from the Mesolithic to the Medieval period. The earliest evidence for the use of marine shells in Gujarat is from the Mesolithic sites of Langhnaj (Sankalia 1965), Tarsang and Kanewal (Sonawane 2002) where *Dentalium* shells are reported.

However it is during the Harappan civilization, dating to the middle of the third millennium BCE, that large-scale exploitation of marine molluscs occurred in the Gujarat region. Even though a large variety of molluscs inhabit the Gujarat coast, only a few species were exploited. These chiefly included the three large marine gastropods *Turbinella pyrum*, *Chicoreus ramosus* and *Pugilina buchephala* as well as several other smaller molluscs. All these shells were mainly used in shell working for the manufacture of a variety of shell objects very similar to those found at Mohenjodaro and Harappa. A highly complex technology was employed by the Harappans in shell cutting and processing into bangles, ladles, inlays, beads and other miscellaneous objects (Kenoyer 1984). Such objects have been recovered from almost all Harappan sites excavated so far, such as Rangpur, Lothal, Nageshwar, Kuntasi, Nagwada, Shikarpur, Dholavira, Zekhada, Bagasra and Bet Dwarka, (fig. 2-1).. An important factor responsible for the large-scale shell working in this region was its proximity to the Gulf of Kachchh, the source area of these large molluscs. Some sites like Nageshwar (Bhan 1992), Kuntasi (Deshpande 1996; 1998), Bagasra (Sonawane *et al.* 2003), Bet Dwarka (Gaur *et al.* 2005) and Nagwada (Bhan and Gowda 2003), which were located close to the gulf, had been involved in shell procurement, manufacture and trade of shell objects to other Harappan settlements. Besides shell working, exploitation of certain marine molluscs like clams and oysters as a food resource is also evident at a few coastal sites (Deshpande-Mukherjee 1998).

Following the decline in the Harappan civilization around the second millennium BCE the use and production of shell objects was not discontinued. In fact the Early Historic period, around 6th century BCE, also witnessed the large-scale use of shell objects made mainly from *T. pyrum*, which were found to continue into the Historic and Medieval periods. Shell objects such as intricately carved shell bangles, beads and other minor objects have been reported from Amreli (Sastri 1936; Rao 1966), Somnath

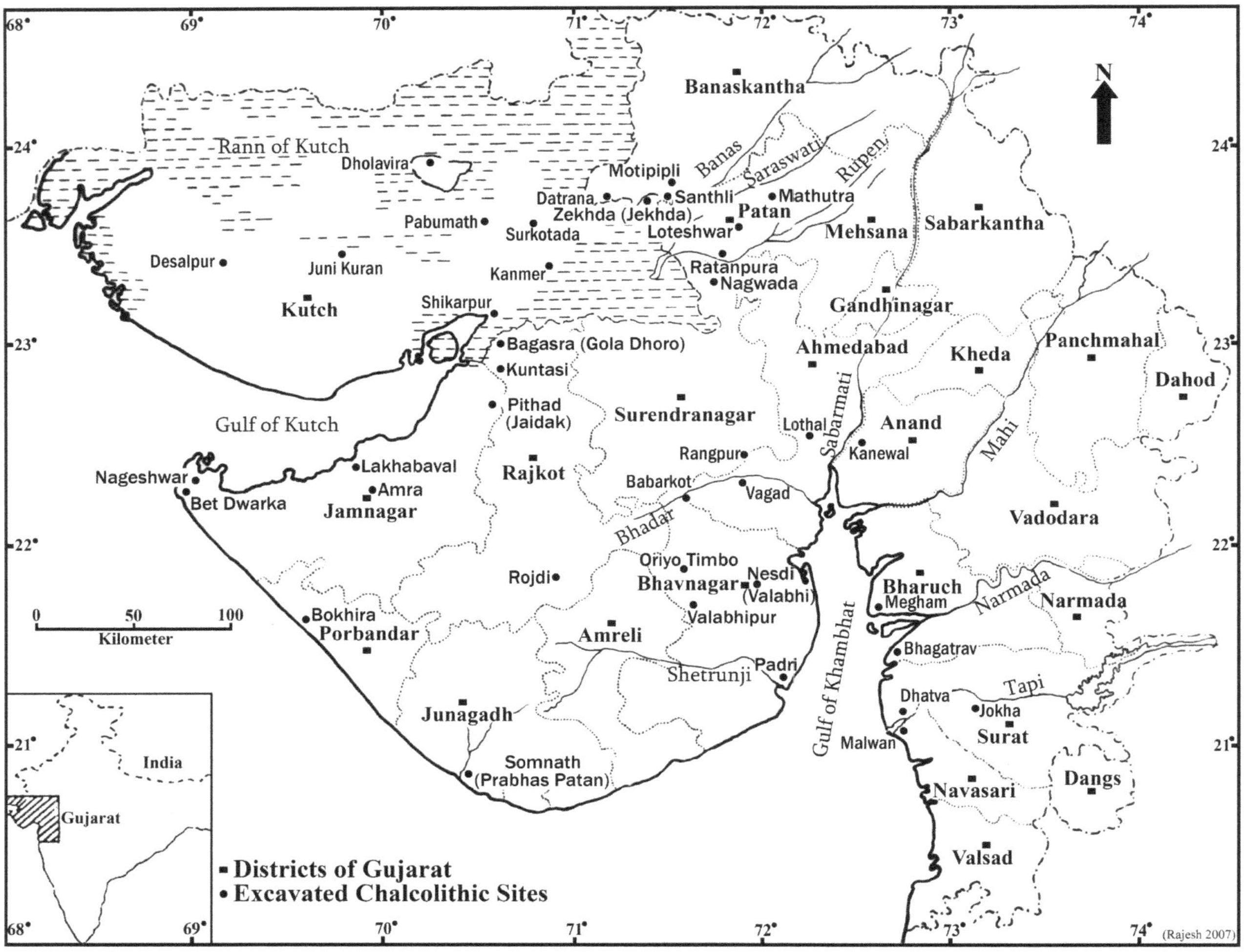

Figure 2-1. Map of Gujarat showing Padri and other excavated Harappan /Chalcolithic sites.

(Nanavati *et al.* 1971), Broach (IAR 1959-60), Nagara (Mehta and Shah 1968), and Hathab (Pramanik 2004; IAR 2002-3).

Today an appraisal of the shell record from Gujarat strongly suggests that this particular region experienced a long tradition of shell use in the past. The archeological record in this region has not only helped preserve this tradition, but has also helped reveal the patterns of shell use through time. A commonly observed feature of some of the Harappan/Chalcolithic sites in Gujarat is their reoccupation in the Historic and Medieval periods after a considerable time gap. At these sites shell remains are found in all cultural periods. The study of these particular remains from such sites can help us reconstruct shell working activity during various cultural periods. One such site is Padri where excavations have revealed a two fold cultural sequence of a Harappan period succeeded by an Early Historic one. At this site shell remains occur throughout its occupation. Hence a detailed study of the shell assemblage recovered from Padri has helped to identify the processes by which shells were introduced, as well as identify their role during both the Harappan and Early Historic periods.

The site and its environs

The ancient site of Padri, locally known as'Kerala-no-Dhoro', is situated on the left bank of the Shetrunji River in the Talaja Taluka, Bhavnagar district of Gujarat (fig. 2-1). The Shetrunji River flows roughly 3 km to the southeast of Padri, rising in the Dundhi hills of the Gir forest with a southeasterly winding course of about 190 km, and flows to Sultanpur where it enters the Gulf of Khambat (Trivedi 1969). The site lies about 2 km from the Gulf of Khambat and 2.5 km to the south of the present village Padri Gohilni (22° 22'N:72° 95' E). It is located in a shallow depression on a slightly elevated sand dune which are commonly found along this coast. The site is protected from the impact of tidal waves by a natural barrier of basalt rock, roughly 6 m high and running parallel to the sea coast. On its northern side is a vast fertile tract where bananas, groundnuts and wheat are being cultivated with the help of artificial irrigation. The site is in a semi-arid region characterized by hot and humid summers with an annual precipitation ranging from 400 to 800 mm (Trivedi 1969). The area around Padri is covered with dry deciduous scrub forest consisting of various species of *Acacia*. The eastern coast bordering the Gulf of Khambat is considered a submergent coast with marshes extending 6-8 km inland which are interlaced with the parallel pattern of dendritic streams draining southeast into the Gulf of Khambat (Ahmed 1972). It is due to this coastal feature that salt making is one of the chief occupations of the modern people of Padri village and the surrounding villages.

In the early 1990's the site was excavated by the second author from the Department of Archaeology, Deccan College, Pune (1990-95). The excavation was spread over an area of 340 m (E-W) by 210 m (N-S) (i.e., 7.14 hectares). The habitation deposit has a thickness of 3.2 m, of which 1.2 m belongs to the Early Historic period. The excavations revealed a two fold cultural sequence, namely Early Historic (1st century BCE to 1st century CE; layers 1-3) and Harappan. Harappan included Late Harappan (layers 4-5), Mature Harappan (2500-2000 BCE; layers 6-7), Transitional (layer 8), and Early Harappan (3300-2600 BCE; layers 9-11). There is a 0.60 m thick black sterile layer between the two cultural periods, suggesting a break in habitation. This was formed after the desertion of the site around 2000 BCE and prior to its reoccupation in the 1st century BCE. A very important finding from Padri is the presence of an Early Harappan occupation at the site, termed the 'Padri culture' (Shinde 1992a; 1992b; Shinde *et al.* 2009). Radiocarbon dates from this occupation layer date to the second half of the fourth millennium BCE, thereby making it one of the earliest Chalcolithic settlements in Gujarat (Ajitprasad 2002).

The Early Historic period is represented in the upper 1.2 m of the deposit (i.e., layers 1-3). It has been dated from the 1st century BCE to 1st century CE on the basis of ceramics as well as a clay tablet bearing a Brahmi inscription along with coins of the Kshatrap King Nahapana (Shinde 1992a).

The excavations have yielded ceramics of Early Harappan, Harappan and Early Historic period. Antiquities include stone beads, copper objects and shell objects. The presence of a large copper fish hook is significant (Shinde 1993). The overall findings from the site along with ethnographic observations have prompted the excavator to identify Padri as an ancient salt manufacturing site (Shinde *et al.* 2009). Considerable numbers of faunal remains consisting of mammalian bones, fish bones and molluscan shells were also recovered from all the cultural levels. The following are the aims and objectives of the shell study that was undertaken.

Aims and objectives

1. Identification of the molluscan taxa and their source areas.
2. Identify the factors responsible for the accumulation of molluscan shell remains at the site.
3. Ascertain the role of molluscs in the Harappan and Early Historic cultural economy at Padri.
4. Establish the nature and extent of shell working activity at the site in both cultural periods.
5. Compare the results from Padri with other shell bearing sites in Gujarat.

The shell Analysis

Around 1000 molluscan shells were recovered from the various trenches excavated. The molluscan shell assemblage includes isolated molluscan shells, shell objects and shell working debitage (Table 2-1). The shells are well preserved and occur as mostly complete shells. The isolated shells were separately packed and recorded according to the trench, layer and depth at which they were excavated. Shell objects were separately recorded under antiquities. Isolated shells and objects were analysed in the archaeozoology laboratory of the Deccan College, Pune.

Cultural period	Isolated shells	bangles	beads	ladle	Perforated shells	*Dentalium* shells	Misc.	*Turbinella pyrum* Shell debitage
Surface	-	16	4	-	0	-	1	1
EARLY HISTORIC (layers 1-3)	199	149	5		16	1	1	8
LATE HARAPPAN (layers 4&5)	122	27	4	1	9	-	-	1
MATURE HARAPPAN (layers 6&7)	136	55	10		6	10	2	-
TRANSITIONAL (layer 8)	52	12	6		-	13	-	-
EARLY HARAPPAN (Layer 9-11)	90	15	1		0	2	0	0
Total	**599**	**274**	**30**	**1**	**31**	**26**	**4**	**10**

Perforated shells: *Oliva* sp., *Nerita, Ellobium aurisjudae, Neritina crepidularia,* Misc: miscellaneous
Table 2-1. Shell assemblage in different cultural periods at Padri.

Shell samples from the following trenches were studied: L1, L2, C3, J2, TTNA2 and LX1. In the laboratory the shells were cleaned under running water and left to dry. Shell identification at the species level was attempted by referring to Apte (1998) and Abbot and Dance (1991). The species of some shells could not be identified due to discolouration, so they are identified only at the generic level. Taphonomic observations such as charring, breakage, perforations and alteration were recorded on all the shells. Estimation of MNI and NISP was carried out. Measurements were taken wherever possible.

Results

The proximity to the sea helped introduce many varieties of molluscan species, both naturally and anthropogenically, throughout the occupation of the site. However their occurrence varied with time. A total of 28 species, of which the majority are marine estuarine in origin, were recorded (Table 2-2). The shells show good preservation but have lost their original colour. Charring is visible in a few shells and shell breakage is observed mostly in bivalve shells. Artificial modification such as abrasion, alteration and perforation is limited, with the latter being common in certain species. The shell assemblage is mainly dominated by small gastropods with very few bivalves. Besides marine taxa, freshwater and land snail shells were also identified. The molluscan shells identified on the basis of their habitat fall into the following categories: marine bivalves (n=4), marine gastropods (n=19), scaphopod (n=1), freshwater bivalve (n=1), freshwater gastropods (n=2) and land snail (n=1) (Table 2-2).

Marine molluscs

Among the marine molluscs 19 gastropod species, mostly comprising small sized shells, were identified. The only large gastropod present is *Turbinella pyrum* which occurs as shell objects and shell working debitage. The most commonly occurring gastropod shells are those of *Oliva* sp., *Nassarius* sp., *Cypraea* sp., *Nerita* sp., and *Neritina crepidularia*. Very few bivalve species were identified. Those present include venerid clams (*Meretrix meretrix* and *Paphia gallus),* arc shells (*Anadara* sp.) and oyster (*Crassostrea* sp.). Shells of *Meretrix meretrix* are the most abundant in all the layers.

From the Harappan period a total of 401 shells were recovered, 355 of which are marine molluscan shells and 46 of which are freshwater and terrestrial. In the Early Harappan phase, as compared to later phases, a few molluscan taxa (n=12) were identified. These belonged to all the three marine, freshwater and terrestrial habitats. Among the marine molluscs, the bivalve species *Meretrix meretrix* is most common while the rest are small gastropod shells of *Oliva* sp., *Nassarius* sp., *Cerithidae* sp. and *Neritina crepidularia.* In this phase the large bivalves *Paphia gallus* and *Anadara* sp. are absent. The latter species is found in the succeeding phases of the Harappan and Early Historic periods.

An overall decrease in the number of shells and the introduction of certain species such as *Anadara* sp., *Telescopium telescopium, Lophiotoma* sp. and *Paphia gallus* was observed in the Transitional phase. The bivalve species *Paphia gallus* was recorded only in this phase.

In the Mature Harappan phase there was an increase in the number of marine species. *Meretrix meretrix* is the most abundant species, while other bivalves like *Anadara* sp. occur as few fragmented shells in the collection. The *Oliva* sp. and *Nassarius* sp. are some of the commonly occurring shells. Rock dwelling molluscs such as *Trochus* sp., *Turbo* sp., *Cypraea* sp., *Nerita* sp. and *Cantharus* sp. were also observed in this phase. In addition, the mangrove dwellers *Telescopium telescopium* and *Ellobium aurisjudae* were also recorded. The latter species also occurs in the Late Harappan phase.

Most of the species recorded in Mature Harappan levels were also found in the Late Harappan phase. In this phase a single shell fragment of *Turbinella pyrum* was recorded. Besides this isolated find, no other shell fragments of this particular shell were recorded in the Harappan period.

In the Early Historic period a total of 207 shells were recorded of which marine were most abundant (n=185). Species recorded in the preceding Harappan levels were also identified in this period. Seventeen species were identified among which *Meretrix meretrix* was the most dominant species. Here too the entire shell assemblage is dominated by marine gastropod shells of *Oliva* sp., *Nassarius* sp., *Turbo* sp., *Nerita* sp., *Astraea* sp., *Moneta moneta, Cypraea* sp., *Neritina crepidularia, Lophiotoma* sp. and *Cerithidae cingulata.* A large oyster shell belonging to *Crassostrea gryphoides* was also found. Shells of the rock dwelling gastropods *Thais* sp., the mangrove species *Terebralia palustris* and limpets were recorded only in this period.

The large marine gastropod *Turbinella pyrum* is represented by shell objects as well as sawn shell fragments. These represent shell working debitage. No complete shells were recovered but a partially broken *T. pyrum* shell was found on the surface.

Scaphopods

A total of 26 shells of *Dentalium* sp., also known as tusk shells, were identified in the assemblage (Table 2-1). These were found mainly in the Early Harappan, Transitional and Mature Harappan phases, and one specimen was found in the Early Historic period. The scarcity of this shell-type in this period is noteworthy.

No:	Species	Type	Habitat	EH	LHP	MHP	T	EHP	total
1	*Turbinella pyrum**	G/M	Intertidal coral reef	8	1	-	-	-	**9**
2	*Meretrix meretrix*	BV/M	Sandy /mudflats	56	44	49	3	46	198
3	*Paphia gallus*	BV/M	Sandy /mudflats				1		**1**
4	*Anadara* sp.	BV/M	Sandy /mudflats		1	2	1		**4**
5	*Telescopium telescopium*	G/M	Mangroves	1		1	1		**3**
6	*Terebralia palustris*	G/M	Mangroves	1					**1**
7	*Lophiotoma* sp.	G/M	Sandy/mudflats	4	1	1	1		**7**
8	*Lamellidens marginalis*	BV/FW	Rivers &lakes	6		1	1	5	**13**
9	*Turbo* sp.	G/M	Rocky intertidal areas	2	1	1			**4**
10	*Trochus* sp.	G/M	Rocky intertidal areas	1	3	1		2	**7**
11	*Thais* sp.	G/M	Rocky intertidal areas	3					**3**
12	*Cantharus* sp.	G/M	Rocky intertidal areas	1	1	1			**3**
13	*Ellobium aurisjudae*	G/M	Mangroves		2	1			**3**
14	*Oliva* sp.	G/M	Sand/mudflats	37	32	39	5	15	**128**
15	*Dentalium* sp.	SC	Sandy beaches			6	13	1	**20**
16	*Cypraea* sp.	G/M	Sand/mud flats	7	10	2		1	**20**
17	*Nassarius* sp.	G/M	Rocky intertidal areas	6	12	21	2	1	**42**
18	*Nerita* sp.	G/M	Rocky intertidal areas	23	2	2			**27**
19	*Neritina crepidularia*	G/M	Estuarine mudflats	28	3	2	2	1	**36**
20	*Cerithidae* indet.	G/M	Estuarine mudflats	4	2	5	2	8	**21**
21	*Melania striatella*	G/Fw	Estuarine mudflats	11	5	1	20	4	**41**
22	*Zootecus insularis*	G/T	Dry arid soils	4	1			5	**10**
23	*Umbonium*	G/M	Sandy beaches	2					**2**
24	*Indoplanorbis exustus*	G/FW	Banks of rivers	1	2			1	**4**
25	limpet	G/M	Rocky intertidal areas	1					**1**
	Total			207	123	136	52	90	**608**

BV-bivalve; G-gastropod; M-marine; FW-freshwater; T-terrestrial; Sc-scaphopod
EH-Early Historic, LHP-Late Harappan, MHP-Mature Harappan, T- Transitional and EHP-Early Harappan
*only shell fragments included in NISP
In addition *Astraea* sp,. *Crassostrea* sp. and *Moneta moneta* were also identified but are not included in Table 2-1 as they were recorded in other trenches not reported here.

Table 2-2. NISP distribution of molluscan taxa from trenches L1, C3, J2, TTNA2, L2 and Lx1.

Freshwater molluscs

Among the freshwater molluscs the bivalve species *Lamellidens* sp. occurs along with two small gastropods, *Indoplanorbis exustus* and *Melania striatella* in both Harappan and Early Historic periods. All three species are associated with a fresh water riverine habitat.

Land snails or terrestrial gastropods

Land snails are represented by a single species, *Zootecus insularis,* which has a thin shell with a cylindrical shape. They inhabit semi-arid regions and were found mainly in the Early Harappan and Early Historic periods.

Identification of source area

All the molluscs found at Padri are common along the modern Gujarat coast. By considering their modern day habitats, the following source areas for the molluscs could be identified.

Rocky intertidal zone - The Gujarat coast is interspersed with sandy beaches and rocky intertidal areas. The Bhavnagar coast around Padri is characterized by Miliolite rocky formations in the intertidal zone. It is from these open coastal areas that a variety of gastropods such as *Nerita* sp., *Turbo* sp., *Astraea* sp., *Lophiotoma* sp., limpets, Cerithidae sp., *Cypraea* sp. and *Nassarius* sp. are found. These molluscs are found attached to hard rocky surfaces.

Estuarine sandy/mud flats at the mouth of the estuary are inhabited by molluscs which are tolerant of salinity changes and attached to a soft substratum. These include bivalves like *Anadara* sp., *Paphia gallus*, *Meretrix meretrix*, *Crassostrea* sp. and gastropods shells of *Oliva* sp.

Mangroves are indicated by the presence of *Telescopium telescopium* and *Terebralia palustris* which are found attached to their roots and tree trunks or buried in the soft sandy muddy substratum at low tide. Currently, mangroves are common along this coast.

Coral reef areas - The large gastropod shells of *T. pyrum* are found living as colonies in shallow coral reef areas in the intertidal zone only in the gulf of Kachchh. Along the Bhavnagar coast such areas are lacking, hence *T. pyrum* shells were not locally available.

Riverine areas - The freshwater bivalve *Lamellidens* sp. and the small coiled gastropod shells of *Indoplanorbis exustus* are found in freshwater rivers, streams and lakes. The nearest probable source was the Shetrunji River.

Ascertaining the role of molluscs at Padri

In order to ascertain the role played by the molluscs in both cultural periods at Padri, an attempt was made to study the process by which the shells were introduced. In this respect the nature of their occurrence, distribution, and modifications were observed. This helped in identifying species that were used by people and those that were not. It was observed that the shell assemblage could be divided into two groups:

1. shells introduced by natural processes.
2. shells introduced by human activity.

Shells introduced by natural processes

At Padri it is observed that many of the shells were introduced due to natural processes such as tidal effect, wave action, inundation and wind. These included gastropod shells of *Nerita* sp., *Cypraea* sp., *Lophiotoma* sp., *Cantharus* sp., *Neritina crepidularia, Nassarius* sp., Cerithidae indet., and *Zootecus insularis*. Due to their small size the shells were easily incorporated into the cultural deposit. Their occurrence in almost all the layers suggests that they had inhabited the coastal landscape throughout the occupation of the site.

Shells introduced by human activity

It appears that a few molluscan species were deliberately collected and brought to the site for three reasons: a. as a dietary resource, b. as finished shell objects, and c. occasionally as a raw material for making objects.

Dietary resource

At many of the coastal sites world over, different varieties of molluscs have been exploited as a food resource. The most commonly occurring shells at Padri are those of the backwater venerid clam *Meretrix meretrix*. Complete and broken shells, a few of them charred, have been recorded from both the Early Historic and Harappan periods (fig. 2-2). These shells represent food refuse after meat from the shell was extracted for dietary purposes. Mostly complete shell valves were found since meat extraction does not always require shell breakage as observed in the case of modern *Meretrix meretrix* shells. These had been collected from the nearby sand mudflats and brought to the site. Similar species are reported from other Harappan sites in Gujarat like Kuntasi, Shikarpur and Bagasra, where they were consumed (Deshpande-Mukherjee 1998). In present times *Meretrix meretrix* is not only a very commonly occurring clam along the Indian west coast but is also

Figure 2-2. *Meretrix meretrix* shell.

collected regularly as food by the local coastal people (Deshpande-Mukherjee 2000).

Another molluscan species that was eaten was the freshwater mussel *Lamellidens* sp. (fig. 2-3). Its exploitation was more in the Early Historic period. Its nearest probable source was the Shetrunji River, located close to the settlement from which the shells were collected. These were also a commonly exploited food resource at many of the Chalcolithic and Neolithic sites in peninsular India. In present times they are also collected as food by the economically depressed communities in parts of Eastern India (Deshpande-Mukherjee 2005a). Besides these two molluscs, marine species such as *Anadara* sp., *Paphia* sp., *Crassostrea* sp. and *Telescopium telescopium* are observed, however their limited numbers makes it difficult to assess their dietary role. Similar species have been found at some of the other coastal Harappan sites in Gujarat where they had served as a food resource (Deshpande-Mukherjee 1996, 1998). In the case of some of the smaller marine gastropods, their dietary role is uncertain due to their small size and lack of edible value.
At Padri it is difficult to estimate the dietary contribution of molluscs to the overall faunal subsistence economy in both cultural periods. Bones of terrestrial mammals like cattle (*Bos indicus/Bubalus bubalis*) and nilgai

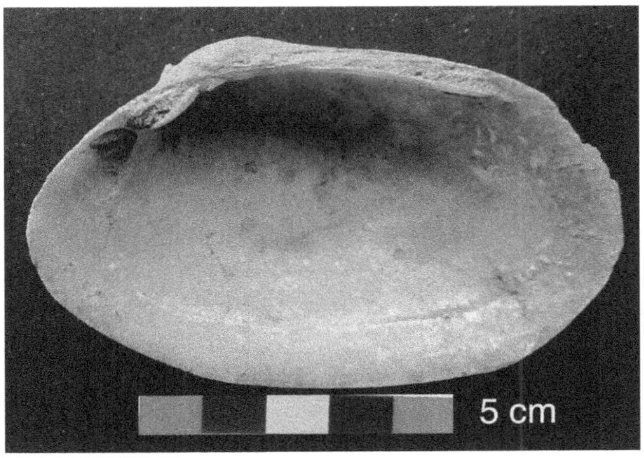

Figure 2-3. Freshwater shell of *Lamellidens* sp.

(*Bos elaphastragocamelus*) have been recovered at Padri along with fish bones, which are found in greater numbers than shells. However these have not been subjected to a detailed study yet. A preliminary study of these animal remains revealed the presence of both domestic (5) and wild terrestrial fauna (10) among which bones of domestic cattle are predominant (Joglekar 1993-94). At most Protohistoric sites in Gujarat, it is observed that mammalian fauna outnumber the remains of molluscs and fish (Thomas and Joglekar 1994; Thomas *et al.* 1997; Thomas 2002). Hence it is very likely that molluscs had probably served as a secondary food resource at most of the coastal sites. The case may have been similar at Padri too during the Harappan and Early Historic period.

Finished shell objects

Shell bangles made from *T. pyrum* are the most commonly occurring shell objects at Padri (Table 2-1). These occur from the Early Harappan phase and become most numerous in the Early Historic period. No complete ones were recovered but a total of 274 bangle fragments are recorded from surface (n=16), Early Historic (n=149) and Harappan (n=109) periods.

Shell bangles from the Harappan levels are plain and range in width from narrow to broad bangles, with the majority in the size range of 5-10 mm (fig. 2-4). Out of a total of 109 bangle fragments a few (n=9) have the 'V' shaped chevron motif characteristic of Harappan shell bangles (fig. 2-5). The bangles are very similar to those reported from other Harappan sites like Lothal, Kuntasi, Nageshwar, Nagwada, etc. A bangle fragment with a plain smooth surface having a maximum width of 28.57 mm was found in the Early Harappan phase (trench LX1, layer 9, depth 2.40 m).

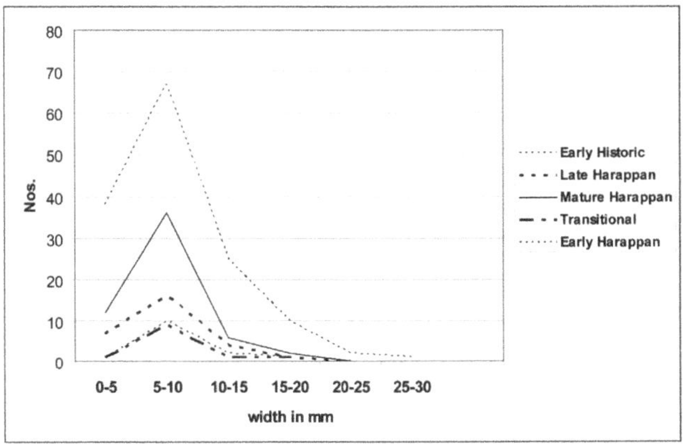

Figure 2-4. Shell bangle widths in different cultural periods at Padri.

The Early Historic period also witnessed the use of shell objects, especially shell bangles. The maximum number of shell bangle fragments were recorded in this period (n=149). Like Harappan ones their widths also ranged from broad to narrow with the majority in the range of 5 to 10 mm (fig. 2-4). Decorated shell bangles are fairly common during the Early Historic, Historic and Medieval period in Gujarat. From a total of 149 fragments, 97 plain, 42 decorated, 8 channeled and 2 grooved ones were recorded. Decorations comprised intricately carved motifs, plain, broad and narrow channeled and three grooves (fig. 2-6 and 7). In this period a broad channeled bangle having a maximum width of 25.65 mm was found (trench A2, layer 2, depth 0.24m) A broad decorated bangle with a maximum width of 23.86 mm was found from trench TTA1, layer 3 at a depth of 1.00 m. Some of decorated fragments share similarities with decorated bangles from other Early Historic sites in Gujarat such as Amreli (Sastri 1936), Somnath (Nanavati *et al.* 1971), Nagara (Mehta and Shah 1968) and Hathab (IAR 2002-3).
Besides bangles a few shell beads were found in the Harappan (n=21) and Early Historic (n=5) periods (Table 2-1). These include small circular, cylindrical, tubular, barrel shaped beads that were found in both periods. Most

of the beads appear to be made from *T. pyrum* shell and were probably not manufactured at Padri. Similar types of beads are reported from the Harappan sites of Rangpur and Lothal.

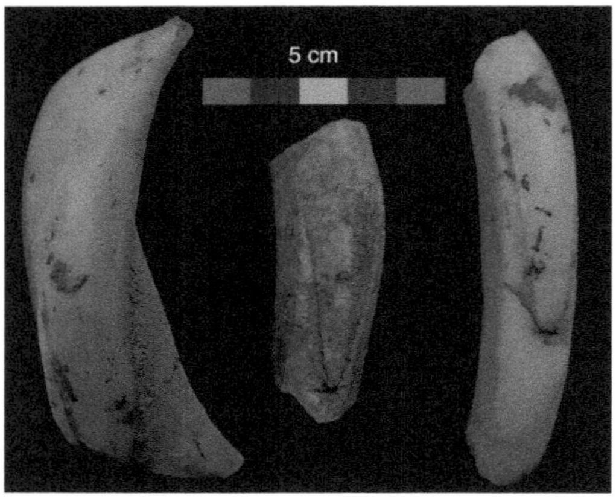

Figure 2-5. Mature Harappan shell bangle fragments.

A single shell ladle was found in the Late Harappan phase from trench N2 NE, layer 4 at a depth of 0.83 cm (fig. 2-8). Its occurrence is significant. Shell ladles were one of the important shell objects manufactured during the Mature Harappan period. These were mostly made from another large marine gastropod shell, *Chicorus ramosus*. At Padri however, the ladle was made using the main whorl of *T. pyrum*. No ladles were observed in the later Early Historic period, an aspect that is noted at other sites in this region

Figure 2-6. Early Historic channeled bangle fragment.

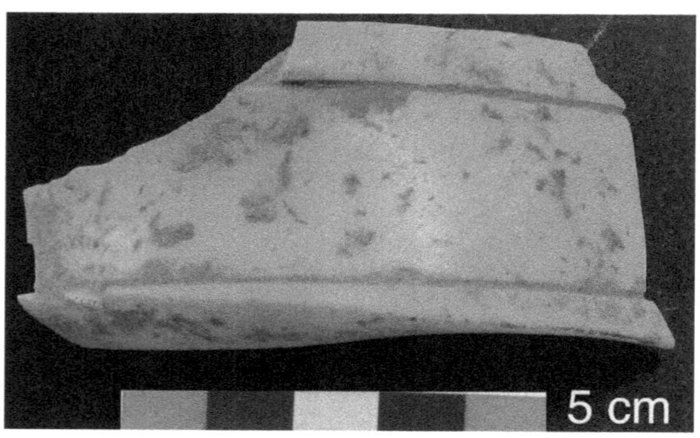

Figure 2-7. Early Historic channeled bangle.

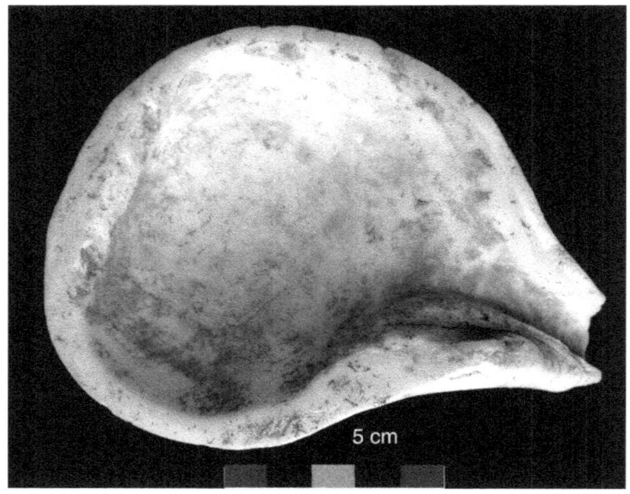

Figure 2-8. Shell ladle.

Besides these major objects a few small miscellaneous objects were also found. A small flat rectangular piece of shell made from *T. pyrum* with a hole in the centre was found in the Mature Harappan period (trench L1, layer 7, depth 1.80m) (fig. 2-9). Two small shell objects, one from Early Historic period (trench N1, layer 2 depth 24cm) and the other from the surface were found. Both objects resemble the one reported at Nagara (Mehta and Shah 1968) and the terracotta spacer bead found at the Early Historic site of Tripuri in Madhya Pradesh (Deo 2000).

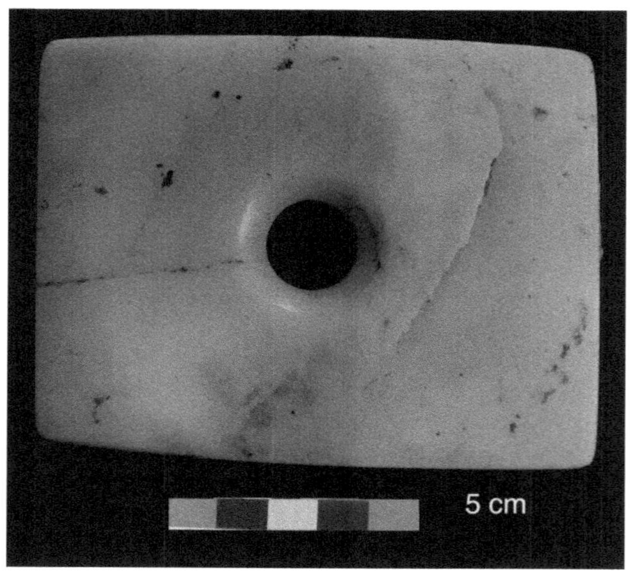

Figure 2-9. Perforated rectangular shell objet.

Shell working

Unlike other Harappan sites such as Lothal, Nageshwar, Kuntasi and Bagasra, no sawn shell fragments representing shell working debitage have been found in the Harappan levels at Padri. Also no specific shell working areas with *T. pyrum* shells in various stages of working have been identified. Hence it is likely that shell objects were procured rather than manufactured locally during the Harappan period. Procurement of shell objects might have been carried out from the above mentioned sites through trade networks. These sites located close to the Gulf of Kacchchh were involved in large-scale shell working due to the easy access to the *T. pyrum* shells.

By contrast, during the Early Historic period the presence of a few *T. pyrum* shell fragments comprising sawn apex portions, columella and part of main whorl with columella could hint at the possibility of shell working at Padri (fig. 2-10, 11 and 12). However, due to the limited number of finds it is difficult to gauge the intensity of this particular activity at the site. During this particular period evidence for shell working is available from some of the nearby

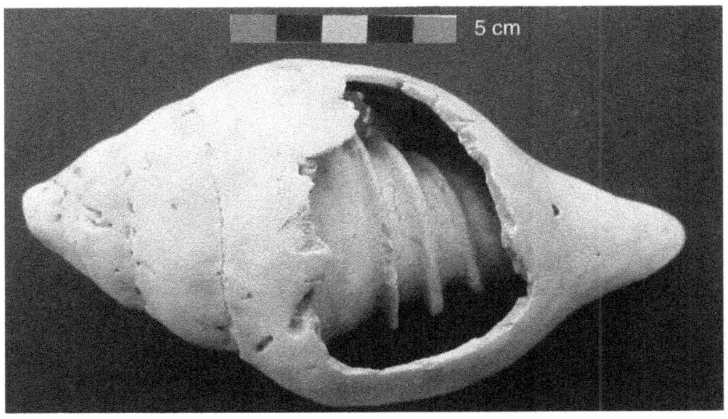

Figure 2-10. *Turbinella pyrum* shell from surface.

Figure 2-11. Swan perforated apex of *Turbinella pyrum*.

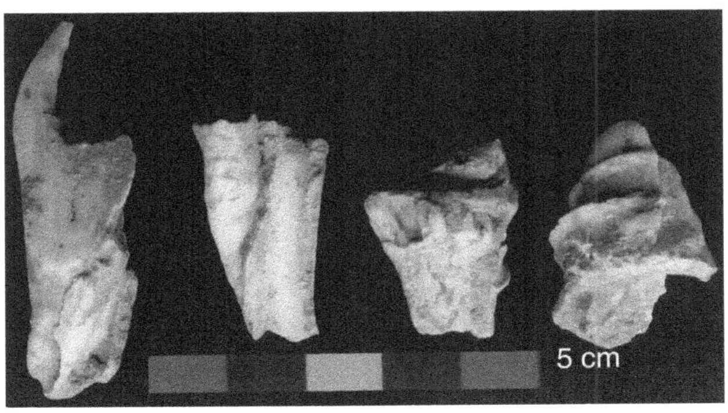

Figure 2-12. Swan columella of *Turbinella pyrum*.

sites such as Amreli (Shastri 1936), Nagara (Mehta and Shah 1968) and Hathab (Pramanik 2004).

Further use of certain shells as beads, pendants or amulets made by using simple techniques is also observed at Padri. A few complete shells with their apex perforated were found in both periods. These comprised 4 marine estuarine gastropods *Oliva* sp, *Ellobium aurisjudae*, *Nerita* sp., and *Neritina crepidularia* and scaphopod shells of *Dentalium* sp. With the exception of *Dentalium* shells, which are tube like and naturally perforated, all other shells were artificially perforated at the apex to be used as beads. These above mentioned species were available locally and were probably collected from the sandy muddy intertidal zone at low tide.

The use of perforated *Oliva* shells is evident in both periods, especially in the Late Harappan phase and Early Historic period (fig. 2-13). Shells of *Ellobium aurisjudae* with their apex perforated are seen only in the Late Harappan period. The exact use of these shells is difficult to determine. However, these may have been of some significance since similarly perforated *Oliva* shells are also reported from some of the Chalcolithic and Early Historic sites in Gujarat and peninsular India (Deshpande-Mukherjee 2005b; 2006-7).

Figure 2-13. Perforated *Oliva* shell.

A perforated shell of *Nerita* sp. was recovered from layer 7 in the Mature Harappan period (fig. 2-14). The surface of its main shell whorl has been abraded and a perforation was made. A similar perforated *Nerita* shell is reported from Chalcolithic levels at Nevasa (Deshpande-Mukherjee 2006-7) and the pre-pottery Neolithic period at Jericho (Biggs 1963). Another similarly perforated shell of *Neritina crepidularia* was found in the Late Harappan period.

In case of *Dentalium* shells, a few were intact while some were cut into small sections. These were found only

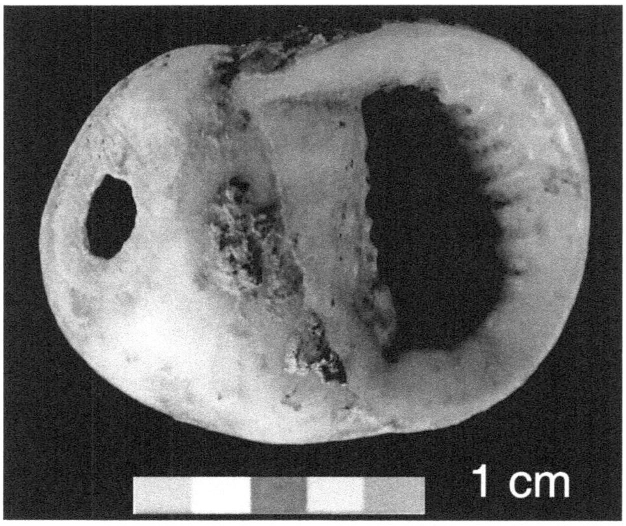

Figure 2-14. Perforated *Nerita* shell.

in the Harappan levels (fig. 2-15). *Dentalium* shells are also reported from other Harappan sites in Gujarat such as Rangpur (Rao 1963), Lothal (Chari 1985), Nageshwar (Hegde *et. al* 1992), Kuntasi (Deshpande 1996), Bagasra (Deshpande-Mukherjee 1999), Jaidak and Khirsara (personal observation). Its use probably had some socio-religious significance attached to it in the Harappan society, an aspect introduced from the preceding Mesolithic cultures in the region. *Dentalium* shells are not observed in the succeeding Early Historic period.

Padri and other shell bearing sites in Gujarat

In Gujarat, around 60 shell bearing sites have been identified. Unfortunately shell assemblages have been studied from only a few sites: Harappan – Lothal (Chari 1985), Rangpur (Rao 1963), Surkotada (Sharma 1990), Malvan (Sharma 1995), Nageshwar (Bhan 1992), Kuntasi (Deshpande 1996), Shikarpur (Deshpande 1995), Bagasra (Deshpande-Mukherjee 1999), Bet Dwarka (Gaur *et al.* 2005) and Nagwada (Bhan and Gowda 2003); Historical – Nagara (Shah 1968) and Kamrej (Deshpande-Mukherjee 2004). A comparison of the molluscan fauna from Padri with some of these sites reveals a few differences. Firstly, the Padri shell assemblage is dominated by small gastropod shells. The use of very few molluscan species and a near absence of large marine gastropods *T. pyrum* and *Chicoreus ramosus* is evident. By contrast, at Harappan sites like Nageshwar, Bet Dwarka, Dholavira, Kuntasi and Bagasra situated close to the Gulf of Kachchh the remains of the large marine gastropods *T. pyrum* and *Chicoreus ramosus* are predominant. Also present at Padri are marine bivalve genera such as *Paphia, Anadara* and *Crassostrea*. This, in a way, reflects the molluscan diversity in the past along the Gulf of Khambat where Padri is located. Today this coast, unlike the Gulf of Kachchh, does not have a large diversity of marine molluscan fauna because the large rivers like the Narmada, Sabarmati, Shetrunji and Tapi regularly deposit their silt into its gulf. This has greatly affected the molluscan diversity in this region.

At Padri the shell objects found in both the periods are similar to those observed at other Harappan and Early Historic sites in Gujarat, however evidence for their manufacture is very limited. The reasons being cited are: a. the nature of the settlement which was not geared for industrial activity or agriculture, b. lack of easy access to the large gastropod shells preferred for shell working, and c. unsuitability of small sized shells for shell working and their lack of food value.

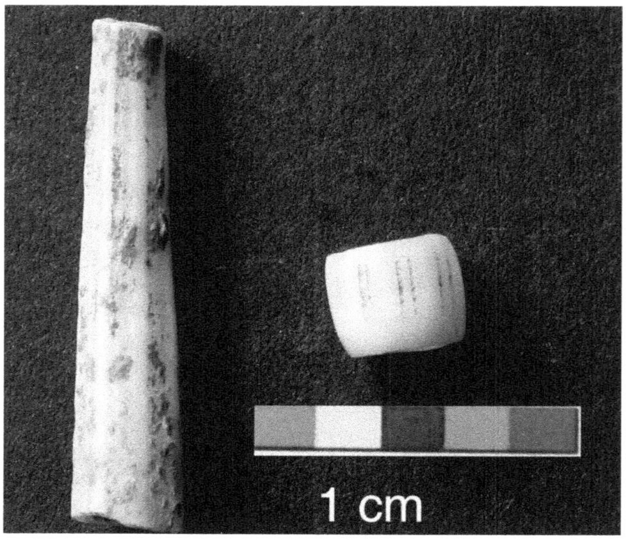

Figure 2-15. *Dentalium* shell and bead from Mature Harappan phase.

Coastal environment around Padri

The diverse range of molluscs found at Padri have been able to provide information on the coastal environment that existed in the past. A majority of the species identified indicate rocky intertidal areas bordering the coast in the past. These areas were inhabited by small marine gastropod molluscs like *Nerita, Turbo* and limpets that are attached to rocks on an open coast. Venerid clams like *Meretrix meretrix* which are found buried in soft substratum such as sand/mud flats at the mouth of estuaries and backwaters indicate estuarine conditions near the mouth of the Shetrunji River. Interestingly, besides the presence of a single shell of *Crassostrea* sp. a near absence of oysters is observed. From its shape and size it was also probably from an estuarine mudflat environment. Molluscs like *Ellobium aurisujdae* and *Neritina crepidularia* are also found in such habitats. Shells of *Telescopium, Terebralia palustris* and *Ellobium aurisjudae* indicate the presence of mangroves in the vicinity of the site. The distribution of shell by cultural period suggests that the sea was closer to the site during the Mature Harappan and Early Historic periods which helps explain the occurrence of wide ranging species in these periods.

The presence of fairly large *Lamellidens* shells indicates

that the Shetrunji River had sufficient fresh water to support these shells. In Gujarat most rivers besides the Narmada and Tapi remain dry all year except during the monsoon period. Further semi dry arid soil conditions around the settlement are indicated by the presence of the land snail *Zootecus insularis*.

Conclusions

The overall picture that emerges from this study is that, although the presence of a variety of molluscan shells is recorded, these played a minor role in the cultural economy at the site. In both Harappan and Early Historic periods the dietary use of marine bivalve *Meretrix meretrix* and freshwater bivalve *Lamellidens* was carried out. Exploitation of a few select species does suggest that molluscs did not constitute a major food resource for the Padri inhabitants as compared to terrestrial mammals and fish. A lack of availability of large bivalves might be one of the reasons for the restricted use of a few species.

Despite Padri being located far from most of the major Harappan sites (the nearest being Lothal), shell objects are similar to other Harappan and Early Historic sites. This has strongly revealed the trade and social contacts that Padri had with these other settlements. It appears that the locally available shells, besides collection as food, were not preferred by the site's inhabitants. Limited shell working activity is attributed to its far location from the procurement source of *T. pyrum* (i.e., Gulf of Kachchh). Further, the nature of the settlement which was probably not geared for industrial activities might have been unable to sustain shell working which involved long distance procurement of shells. Instead demand for *T. pyrum* shell objects prompted the Padri inhabitants to procure them rather than carry out their manufacture.

In the Early Historic period however a certain amount of shell working is observed. By this time a good number of Early Historic settlements had emerged close to Padri such as Hathab, Amreli and Nagara where shell working was carried out. The case with Padri may have been similar, with increased economic activity allowing the small coastal fishing settlement to achieve more prominence.

Acknowledgements.

The kind help of Dr. Probodh Shirwalkar and Ms. Soumi Sengupta during the shell analysis is sincerely acknowledged. Dr. Veena Mushriff-Tripati is also thanked for her help with the figures. We wish to thank the reviewers for their suitable comments and suggestions in the final preparation of the paper.

References Cited

Abbot, R. Tucker and S. Peter Dance. 1991. *ompendium of shells*. London: Charles Letts and Co. Ltd.

Ahmed, E. 1972. *Coastal geomorphology of India.* New Delhi: Orient Longmans.

Ajithprasad, P. 2002. The Pre-Harappan cultures in Gujarat. In *Indian archaeology in retrospect, Vol. II Protohistory, archaeology of the Harappan Civilisation.* S. Settar and Ravi Korisettar eds. Pp. 129-158. New Delhi: Indian Council of Historical Research.

Apte, Deepak. 1998. *The book of Indian shells.* Bombay Natural History Society. Calcutta: Oxford University Press.

Bhan, Kuldeep. K. 1992. Shell industry. In *Excavations at Nageshwar: a Harappan shell working site on the Gulf of Kutch.* Karunakara Thinkgale Manjaya Hegde, Kuldeep K. Bhan, Vishwasrao H. Sonawane, K. Krishnan and Dolly R. Shah, eds. Pp. 125-136. Vadodara: M.S. University Archaeological Series 18.

Bhan, Kuldeep. K. and Dakshayani Gowda 2003. Shell working at Nagwada (North Gujarat) with special reference to shell industries of the Harappan tradition in Gujarat. *Man and Environment* Vol. XXVIII (2): 51-80.

Biggs, H. E. J. 1963. On the mollusca collected during the excavations at Jericho, 1952-1958, and their archaeological significance. *Man* Vol. LXIII: 152-154.

Chari, V. K. 1985. Shell remains. In *Lothal: a Harappan port town 1955-62 Vol II.* Shikaripur Ranganatha Rao, ed. Pp. 614-624. New Delhi: Memoirs of the Archaeological Survey of India No:78.

Deo, Shantaram. Bhalchandra. 2000. Indian beads: a cultural and technological study. Pune: Deccan College.

Deshpande, Arati. 1996. Shell Remains. In *Kuntasi, a Harappan emporium on West Coast.* M. K. Dhavalikar, M. R. Raval and Y. M. Chitalwala, eds. Pp. 331-347. Pune: Deccan College.

Deshpande-Mukherjee, Arati. 1998. Shell fishing and shell craft activities during the Harappan period in Gujarat. *Man and Environment* Vol XXXIII (1): 63-81.

——. 1999. A preliminary study of Marine molluscan shell remains from Bagasra: A Harappan site in Gujarat. *Puratattva* 29: 110-113.

——. 2000. An ethnographic account of contemporary shellfish gathering on the Konkan coast, Maharashtra. *Man and Environment* Vol. XXV(2) :79-92.

——. 2004. Shell remains from Kamrej Excavations-2003. *Journal of Indian Ocean Archaeology* 1: 78-81.

——. 2005a. Significance of freshwater shells in Indian archaeological context. In *River valley Culture of India.* Kalyan. K. Chakravarty and Gyani. l. Badam, eds. Pp. 323-331. New Delhi: Aryan Books International.

——. 2005b. Marine shell utilisation by the Chalcolithic cultures of the Western Deccan region of India. In *Molluscs in their Former Environment.* Daniela Bar Yosef ed. Pp.174-184. London: Oxbow books.

——. 2006-7. Marine molluscs at ancient settlements in the Deccan. *Bulletin of the Deccan College* 66-67: 259-283.

FOOTE, ROBERT. BRUCE. 1916. *The Foote collection of Indian Prehistoric and Protohistoric antiquities.* Chennai: Madras Government Press.

Gaur, ANIRUDDH S., SUNDARESH and KAMALESH H. VORA 2005. *Archaeology of Bet Dwarka Island.* New Delhi: Aryan Books International and National Institute of Oceanography, Dona Paula.

HEGDE, KARUNAKARA THINKGALE MANJAYA, KULDEEP. K. BHAN, VISHWASRAO. H. SONAWANE, K. KRISHNAN and DOLLY. R. SHAH. 1992. *Excavations at Nageshwar: a Harappan shell working site on the Gulf of Kutch.* Vadodara: M.S. University Archaeological Series 18.

Indian Archaeology- A review 1959-60. New Delhi: Archaeological Survey of India.

Indian Archaeology - A review 2002-3. New Delhi: Archaeological Survey of India.

JOGLEKAR, PRAMOD P. 1993-94. A preliminary report on the faunal remains at Padri: a Harappan site in Bhavnagar district, Gujarat. *Puratattva* 24:35-39.

KENOYER, JONATHAN. MARK. 1984. Shell working Industries of the Indus Civilisation: a summary. *Palaeorient* 10 (1): 49-63.

NANAVATI, JAYANDRA, RAMANLAL NAGARJI MEHTA and SOORYAKANTN NARSINH CHOWDHARY. 1971. *Somnath 1956.* Vadodara: Department of Archaeology, Gujarat State and the M.S. University.

PRAMANIK, SHUBRA. 2004. Hathab: an Early Historical port on the Gulf of Khambat. *Journal of Indian Ocean Archaeology* 1:133 -140.

RAJESH, S.V. and AMBIKA PATEL. 2007. A gazetteer of Pre- and Protohistoric sites in Gujarat. *Man and Environment* XXXII: 61-136.

RAO, SHIKARIPUR RANGANATHA. 1963. Excavations at Rangpur and other explorations in Gujarat. *Ancient India* 18-19: 5-207.

——. 1966. Excavations at Amreli. Baroda: *Museum and Picture Gallery Bulletin* Vol 18.

——. 1985. Lothal: a Harappan port town 1955-62, Vol. II. Memoirs of the Archaeological Survey of India No:78. New Delhi: Archaeological Survey of India.

SANKALIA, HASMUKH DHIRAJLAL. 1965. *Excavations at Langhnaj 1944-63.* Pune: Deccan College.

SASTRI, HIRANANDA. 1936. Annual report of the Director of Archaeology, Baroda State 1934-35. Vadodara.

SHAH, DOLLY. R. 1968. Bones and Shells. In *Excavations at Nagara.* R. N. Mehta and D. R. Shah, eds. Pp. 152-163. Vadodara: M.S. University of Baroda.

SHINDE, VASANT. 1992a. Excavations at Padri-1990-91: A preliminary report. *Man and Environment* Vol. XVII (1): 79-86.

——. 1992b. Padri and the Indus Valley civilization. *South Asian Studies* 8: 56-66.

——. 1993. A unique Harappan copper–fish hook from Padri, Gujarat. *Man and Environment* Vol. XVIII:145-147.

SHINDE, VASANT, PABODH SHIRVALKAR and SHARAD. N. RAJAGURU 2009. Padri: a Harappan salt manufacturing centre on West Coast of India. *Journal of Indian Ocean Archaeology* No: 5: 57-84.

SONAWANE, VISHWASRAO H. 2002. Mesolithic culture of Gujarat. In *Mesolithic India.* Vidyadhar. D. Misra and Jayant. N. Pal, eds. Pp.104-153. Department of ancient history, culture and archaeology, Allahabad:University of Allahabad.

SONAWANE, VISHWASRAO. H., P. AJITHPRASAD, KULDEEP. K. BHAN, K. KRISHNAN, S. PRATAPCHANDRAN, ABHIJIT MAJUMDAR, AJITA K. PATEL and JAYA MENON. 2003. Excavations at Bagasra -1996-2003: A preliminary report. *Man and Environment* Vol. XXVIII (2): 21-50.

THOMAS, PAPPYKUZHVEILL K. and PRAMOD. P. JOGLEKAR. 1994. Holocene faunal studies in India. *Man and Environment* XIX (1-2): 179-204.

THOMAS, PAPPYKUZHVEILL. K. PRAMOD. P. JOGLEKAR, YOSHIYUKI. MATSUSHIMA, SEEMA. J. PAWANKAR and

ARATI DESHPANDE. 1997. Subsistence based on animals in Harappan culture of Gujarat, India. *Anthropozoologica* 25-26: 767-776.

THOMAS, PAPPYKUZHVEILL. K. 2002. Investigations into the archaeofauna of Harappan sites in Western India. In *Indian archaeology in retrospect. .Protohistory.* S. Settar and R. Korisettar, eds. Pp.409-420. New Delhi: Indian Council for Historical Research and Manohar Publishers.

TRIVEDI, 1969 *Gujarat State Gazetteer (Bhavnagar District),* Ahmedabad.

3 - THE PROVENANCE AND USE OF FOSSIL SCAPHOPOD SHELLS AT THE LATE NEOLITHIC/ENEOLITHIC SITE VINČA – BELO BRDO, SERBIA

Vesna DIMITRIJEVIĆ
University of Belgrade - Faculty of philosophy, Department of archaeology, Čika Ljubina 18-20, 11000 Belgrade, Serbia, vdimitri@f.bg.ac.rs

Abstract : *Elephant's tusk shells, usually known under the name* Dentalium, *after the best represented scaphopod genus, are frequent finds at the site of Vinča – Belo Brdo. The site is a multi-layered tell, representative of the Late Neolithic/ Eneolithic Vinča culture, that flourished during the second half of the 6th and the first half of the 5th millennia BC in the central Balkans.*

Scaphopod shells are found mostly as tubular shell fragments and used as beads. Bracelets, amulets and beads made of other marine mollusk shells, namely the bivalves Spondylus *and* Glycymeris, *were also common ornamental items in this cultural context. Inhabitants of the settlement at Belo Brdo gained those highly valued items through exchange networks developed in prehistory across the vast area stretching from the Mediterranean and Black Sea coasts to central Europe. Contrary to the shell items obtained by exchange with contemporary prehistoric societies, scaphopod shells were collected in the vicinity of Belo Brdo, at the outcrops of fossiliferous Neogene deposits. This paper presents the distribution of the fossiliferous outcrops near the settlement, and identifies where the fossil scaphopods might have been collected. The provenancing of scaphopod shell is based on the comparison of taphonomic features of palaeontological and archaeological shell.*

Key words: *Scaphopods,* Dentalium, Fissidentalium badense, *Archaeomalacology, Neolithic, Eneolithic, Vinča culture, Central Balkans*

Introduction

Scaphopod shell usage as beads is well known in archaeology and represents one of the topics in archaeomalacology. Beads made from scaphopod shells are common in a number of societies that inhabited coastal regions, but are also found at archaeological sites distant from the coast. At the latter sites, marine shells were acquired by exchange with costal societies or by collecting fossil-shells of organisms that lived in seas that covered these now-inland areas in the geological past (Bar-Yosef Mayer, Gümüş and İslamoğlu 2010).

Such fossil scaphopod shell beads are found in large numbers at the site of Vinča – Belo Brdo, near Belgrade. They are present in all Neolithic and Eneolithic levels of this tell-type site, suggesting continuous exploitation of fossiliferous outcrops for several hundred years.

The archaeological site of Belo Brdo (White Hill) is a multi-layered tell situated in the village of Vinča, on the right bank of the Danube River, 14 km downstream of Belgrade (fig. 3-1). It is the central and representative site for the Late Neolithic/Eneolithic Vinča culture, which flourished in the central Balkans during the second half of the 6th and the first half of the 5th millennium BC (Borić 2009; Chapman 1981; Garašanin 1979). The depth of cultural levels is approximately 9 m, most of which belongs to the Vinča culture. The upper layers are overlaid and occasionally disturbed by later Eneolithic and Bronze Age features and a Late Medieval cemetery.

The excavations at this site first started in 1908 and continued sporadically to the present. In the first half of the 20th century (1908-1934) excavations were performed over an extensive area of approximately 400 m^2, from the top of the tell down to the earliest levels (Vasić 1932, 1936, 1936b, 1936c). In the second phase of research (1978–1986) a 300 m^2 excavation area was opened up adjacent to the previous one. Horizons relating to a Medieval necropolis as well as Bronze Age and Eneolithic features were excavated. Excavations were halted after reaching the upper levels of the Vinča deposits, but were resumed again in 1998 (Tasić 2005; Tasić and Ignjatović 2008).

This paper examines scaphopod beads collected over the course of the excavations at Belo Brdo (1908 to the present) from the whole of the excavation area.

Biology and paleontology of scaphopods

Scaphopods are marine animals; a class within Phylum Mollusca. They are characterized by a conical and slightly curved shell that resembles an elephant's tusk, resulting in

Figure 3-1. The Danube River southeast of Belgrade. The position of the Vinča – Belo Brdo site is circled (Nikolić 2008, p.22-23, photo by Vladimir Miladinović Piki).

their colloquial name, "tusk shells". The shell is open at both ends. The wider end of the shell is pointed downwards and shelters the animal's head, tentacles and foot. The foot, similar in shape and function to the foot of mussels, is the inspiration for the scientific name Scaphopoda (scapho=shovel, poda=leg).

In living animals, the narrow upward (apical) end of the shell protrudes through the bottom sediment into the seawater, or the animal is completely buried and moves through the sediment in search for prey. Scaphopods live from the shoreline to a depth of several thousand meters. They inhabit seawaters at all latitudes, but are most commonly found in tropical and subtropical basins.

Scaphopods are frequently preserved as fossils due to their compact carbonate shell. While the oldest known specimens are found in rocks dating to the Lower Carboniferous (ca. 350 million years ago; Yochelson and Holland 2004), they become more abundant after the Cretaceous (ca. 65 million years ago).

Taxonomic identification is based on shell morphology and soft tissues like the foot and radula in living animals, and solely on shell morphology in archaeological and paleontological specimens. The shell ranges in size from 2 mm to 15 cm, but mostly varies in length between 3 and 6 cm. Important features for species identification are shell shape, dimensions, shell surface sculpture and apical morphology (Barton 1994).

Taxonomic identification of archaeological and paleontological specimens can be problematic due to fragmentation and erosion of surface sculpture and the radical changes in external shell morphology that occur over the course of ontogenetic development.

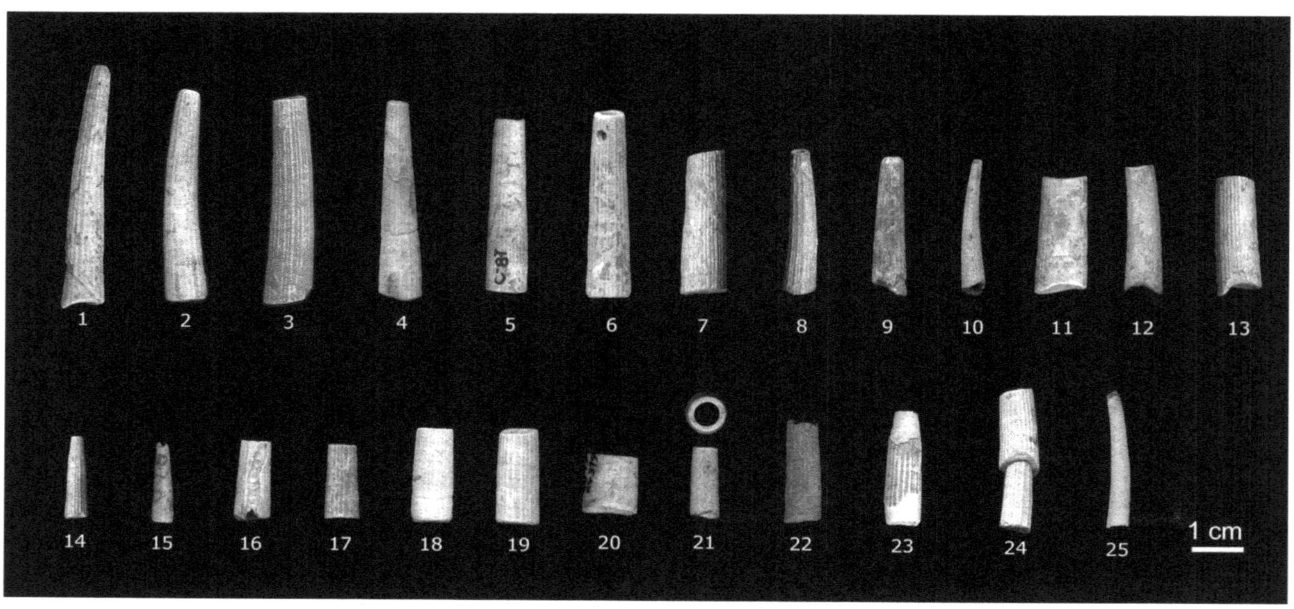

Figure 3-2. Scaphopod beads from the Vinča – Belo Brdo site: 1-24 *Fissidentalium badense*, 25 Scaphopoda indet.

Marine shells at Vinča – Belo Brdo site

Over the course of long-term excavations at Belo Brdo a significant number of decorative objects made of clay, stone, bone and shells have been collected. Among those that could be identified as jewelry the most numerous are bracelets, pendants and beads made from the shells of clams, *Glycymeris* and *Spondylus*, and beads made of scaphopod shells. The total number of scaphopod beads collected is 362. In previous studies, the ornaments made from *Glycymeris* and *Spondylus* were studied (Dimitrijević and Tripković, 2002, 2006). These studies demonstrated that the ornaments were mostly made from the shells of coeval marine organisms, and that the inhabitants of Belo Brdo acquired those through an exchange network with communities inhabiting coastal areas. By contrast, the scaphopod shells used as beads were procured from fossiliferous deposits.

The difference between the shell of a contemporary animal and a fossil shell of an animal that lived millions of years ago is often obvious. Aside from the differences in completeness and preservation of morphological details, modern shells can be identified by their coloration and transparency. When internal structure is compared, modern shells are translucent and elastic, while fossil shells are brittle and opaque. This difference is due to the chemical alterations and mineral replacements that occur during the process of fossilization, and is best recognized on the breakages in broken specimens. However, in archaeological shell it is not always simple to distinguish whether fragmentation, loss of coloration, and erosion of morphological details are due to its archaeological or geological age. In the collection of marine shells from Belo Brdo, transparency, which is best observed when the shell is put in front of the light source, was the most important feature for establishing a difference between the majority of ornaments made from contemporaneous *Glycymeris* and *Spondylus* and amulets made of fossil *Lymnocardium* (Dimitrijević & Tripković 2006). All other features were either deceptive because of the excellent preservation of fossils, or eroded in the process of ornament manufacturing. This distinction is similar in fossil and modern scaphopod shell.

Only one out of the 362 scaphopod beads found at the site of Belo Brdo is not considered to be of fossil origin (fig. 3-2, 25). The modern origin of this shell was inferred from the fact that the shell is thin and translucent. It also shows much finer longitudinal ribs than fossil shells indicating a different taxonomic status.

Fossil shells are identified as the Miocene species *Fissidentalium badense* (Dimitrijević, Tripković and Jovanović, 2010). Conspicuous features are large, massive shells with very strong ribs and fine growth lines. Ribs are tall and slightly rounded. The identification is made on the basis of the morphology of better preserved specimens (fig. 3-2, 1-6). It is supposed that *Fissidentalium badense* is the only species present, as there are no specimens indicating the presence of any other scaphopod species. Some caution is necessary in this respect, as the external ornamentation in some specimens is eroded (fig. 3-2, 12, 15, 20, 22), or the shell surface is chemically altered (fig. 3-2, 23) which prevents specific and even generic identification.

Most of the fossil shells that the Neolithic residents of Belo Brdo used were unaltered from the form they had when found. These are rarely larger parts of shells (fig. 3-2, 1-6) and more often smaller fragments of broken shells. The wider end of the shell, if present, is difficult to identify, while the apical end is recognized for being narrow, tapering (fig. 3-2, 8, 10, 14, 15, 25), and with longitudinal ribs diminishing towards the end (fig. 3-2, 14). Long cylindrical shells are relatively easily broken at the level of growth lines. Fractures are sometimes at right angles to the axis of the shell (fig. 3-2, 17, 18, 19) and sometimes skewed (fig. 3-2, 3, 7). When the break came before fossilization, the edges of broken shell are usually rounded due to friction, either on the sea bottom or on the beach (fig. 3-2, 2-6, 11-20, 24). Friction with clastic particles in the deposit during fossilization may also round broken edges of a shell. Sometimes the growth lines are sinusoidaly curved, and fractures that accompany such growth lines have one (fig. 3-2, 1, 11, 13) or two sinuses, or vaulted outlets (fig. 3-2, 12, 13). Shells were modified rarely, probably when a collected shell had sharp, not rounded, edges, or when it was intentionally broken for getting a desired length. This conclusion is based on traces of scraping or polishing. Traces of modification are sometimes poorly visible to the naked eye, but they are observable under magnification (fig. 3-2, 21).

The length of beads largely depended on the length of fragments of fossil shells. The greatest measured length was 72 mm and the smallest 7.8 mm. Beads in which the length is approximately 3-5 times larger than the diameter prevail. The average length of beads is 24.5 mm.

Scaphopod beads are found at Belo Brdo in all Vinča culture levels, at depths between 0.7 and 9.0 m. They are found scattered in occupation levels where their discard was most likely unintentional. Their sporadic occurrence in the upper levels of the site (i.e., in Late Eneolithic and Bronze Age features) suggests that they may also have been used in later periods.

Marine shells in Balkan prehistory

Before the Neolithic, scaphopod beads are rare in the region of the Balkan peninsula. Their first known occurrence is in the Upper Palaeolithic levels of the Franchthi cave on Peloponnese (Schackleton 1988). They are recorded only in the Epipalaeolithic layer of the rockshelter Cuina Turcului in the Iron Gates (Păunescu 1970). In the Neolithic, Eneolithic and Bronze Ages scaphopod beads are relatively abundant on the Aegean coast (Nikolaidu 2003). They are also very abundant in the Eneolithic and Bronze Ages of the eastern Balkans where they are mostly found in graves like at Durankuluk (Avramova 2002) and Varna (Ivanov and Avramova 2000).

In the central Balkans there are no finds of scaphopod beads before the Vinča culture. When they appear they are made mostly of fossil shells, with the exceptions of the already-mentioned single specimen from Belo Brdo and one found at the site of Anzabegovo in FYR Macedonia. At the later site, the Early Vinča complex (Anza IV) is the uppermost complex covering a unique Early and Middle Neolithic sequence. The bead we believe to be a scaphopod apical end is described as "tubular fragile white shell" (Gimbutas 1976, Fig. 218, 3). It is the only scaphopod bead found at this extensively excavated site.

After the Vinča culture, scaphopod beads are used primarily during the Bronze Age. For example, they are well represented in the inventory of burials from the Mokrin cemetery of the Moriš culture (Girić 1971). Again, fossil shells were used in combination with beads and pendants made from other materials (teeth, bones, kaolinite, stone, bronze, etc.).

There are no dedicated studies devoted to the appearance of scaphopod beads and their origins among the prehistoric cultures of the central Balkans. Rather, such finds have been sporadically illustrated and listed in the inventory of studied sites. Although they seem to be a rare occurrence, their poor representation may instead be a consequence of a failure of researchers to recognize them. A possible reason that these artifacts have been overlooked is that shell beads do not belong to a "major" category of archaeological materials (e.g., ceramics, stone, bone), and thus, "wander" in the inventory of one or the other categories. A common lack of sieving might be another reason for their omission.

How did they wear them?

The excavations at Belo Brdo unearthed a plenitude of well preserved remains of architecture, ceramic figurines and vessels, and stone and bone tools, which showed that the inhabitants of the settlement of Belo Brdo devoted great attention to decorating their houses, objects for daily use, and bodies. Decorative objects made of the shells of marine organisms, bracelets, amulets and beads of *Spondylus* and *Glycymeris*, and beads of scaphopods provide the most comprehensive data about personal ornaments.

Currently, we have only assumed that scaphopod shells were used as beads. Although their shape strongly suggests that they were used as beads, we have no reliable evidence for this. At Belo Brdo they were found mainly scattered around the settlement, not in a context that can elucidate their usage. Therefore, we do not know how the beads were worn and arranged as ornaments. They may have been worn individually or strung in a row, around the neck, arms or legs, or perhaps sewn onto clothing.

At the site Belo Brdo most of the beads were found separately, but this does not mean that they were not carried strung together, as they could be scattered before deposition. At least some of the beads were carried strung in a necklace, as shown by 29 beads found together at the depth of 8.2 m (Vassitz 1930) (fig. 3-3). Strung together, they form a necklace that is approximately 58 cm long. This may suggest that it was worn around the neck, reaching midway between breasts and the waist. A few shell tubes in this necklace are inserted one into the other, with the narrow end of the first bead firmly embedded in the wider end of the second, thus suggesting that they were worn strung. Another group of beads that probably formed a necklace is held in the collection of the National Museum in Belgrade, however no information about its context is available. It includes 8 scaphopod beads. Among those are two sets of two interconnected beads, one small two-pierced clam valve and one small discoid bead. Similarly, two sets of two interconnected scaphopod beads were found at the depths of 4 m and 3.4 m respectively (fig. 3-2, 24).

The only burial at the site of Belo Brdo is a collective burial

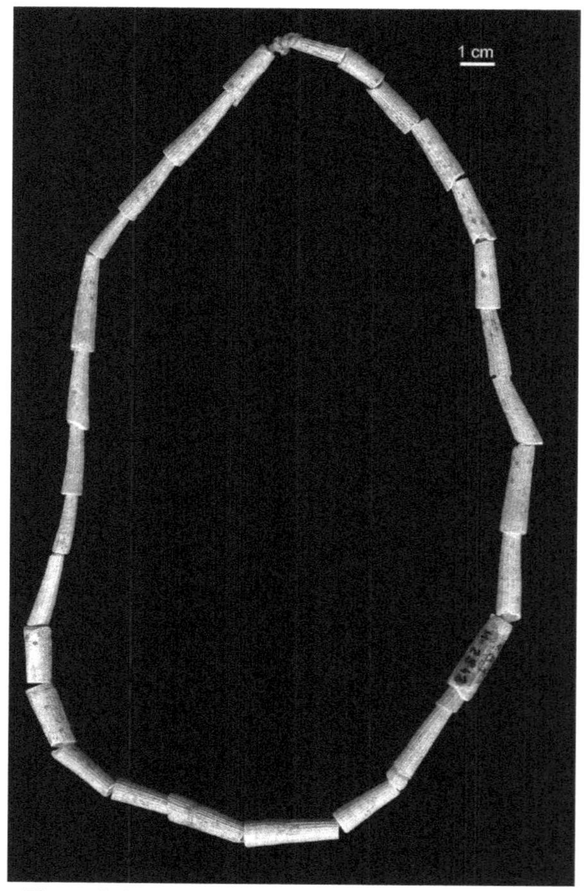

Figure 3-3. Necklace made of 29 scaphopod beads, Vinča – Belo Brdo, depth 8.2 m.

Figure 3-4. Clay figurine with amulet, Vinča – Belo Brdo, depth 2.5 m, figurine height 55.5 cm (Ignjatović 2008: p.234).

Figure 3-5. Clay figurine with necklace and bracelets, Supska – Stublina, height 9.2 cm (D. Garašanin & M. Garašanin, 1979).

at the lowermost level. Burials are also rare at other sites of the Vinča culture (Chapman 1981). Since scaphopod beads have never been associated with the burials of the Vinča culture, the burials provide no indication of the ways that such ornaments might have been worn on the body. The average length of the beads is 24.5 mm. This is much longer than in the Levant where the average length is 5-20 mm at most sites, with very short beads of 1-3 mm being well represented at some sites (Bar-Yosef Mayer 2008). By contrast, very short beads are exceptionally rare at Belo Brdo, possibly because fossil shells have thick walls. Therefore they almost never break in short segments and are not easy to cut in shorter pieces.

Another way of resolving how scaphopod beads were worn is on the basis of ornament depictions on clay figurines. The Vinča culture is famous for a plenitude of anthropomorphic and zoomorphic figurines that are found at every excavated site. At Belo Brdo alone more than a thousand clay figurines were collected in the first phase of research conducted by Vasić (1936b). On those representing women, which are most abundant, clothing is represented by lines and staining. Clothing is tightly fitted on the body and probably made from linen. Among the depicted jewelry are bracelets, necklaces and belts. The most frequent type of necklace is a string around neck with a large round amulet (fig. 3-4). But there is one example of a figurine possibly showing a bead necklace (fig. 3-5). There are no indications of scaphopod beads or other ornament types being sewn onto the clothing or attached to a coiffure, although both the clothing and coiffures were sometimes elaborately depicted with a range of details.

Fossiliferous outcrops in the Vinča surroundings

Fossil scaphopods found at Belo Brdo originate from the Middle Miocene, Badennian sediments, deposited about 15 million years ago. At that time, the Vinča area was covered by a shallow and warm sea of normal salinity. Conditions for life were extremely favorable as indicated by the remains of a variety of diverse plant and animal fossil remains. Vertebrate remains include bony fish and sharks. Invertebrates are very diverse, with over 150 different species identified. The most numerous are snails and clams, then scaphopods, corals, foraminiferans, ostracods, bryozoans and annelids. Remains of plants include cinnamon trees and conifer cones. Fossil associations found at some of the fossiliferous outcrops, such as the Bučvar creek in Veliki Mokri Lug or Višnjica, resemble modern communities inhabiting shallow sandy sea beds, including shells found on beaches of the Adriatic coast today (Stevanović 1977).

Outcrops of Badennian sediments rich in fossils are found in the center of Belgrade and in many of its suburbs (e.g., Višnjica, Rakovica, Jajinci, Leštane, Veliki and Mali Mokri Lug, Železnik, Bele Vode and Slanci; fig. 3-6). Badennian sediments were also discovered south of Belgrade, near

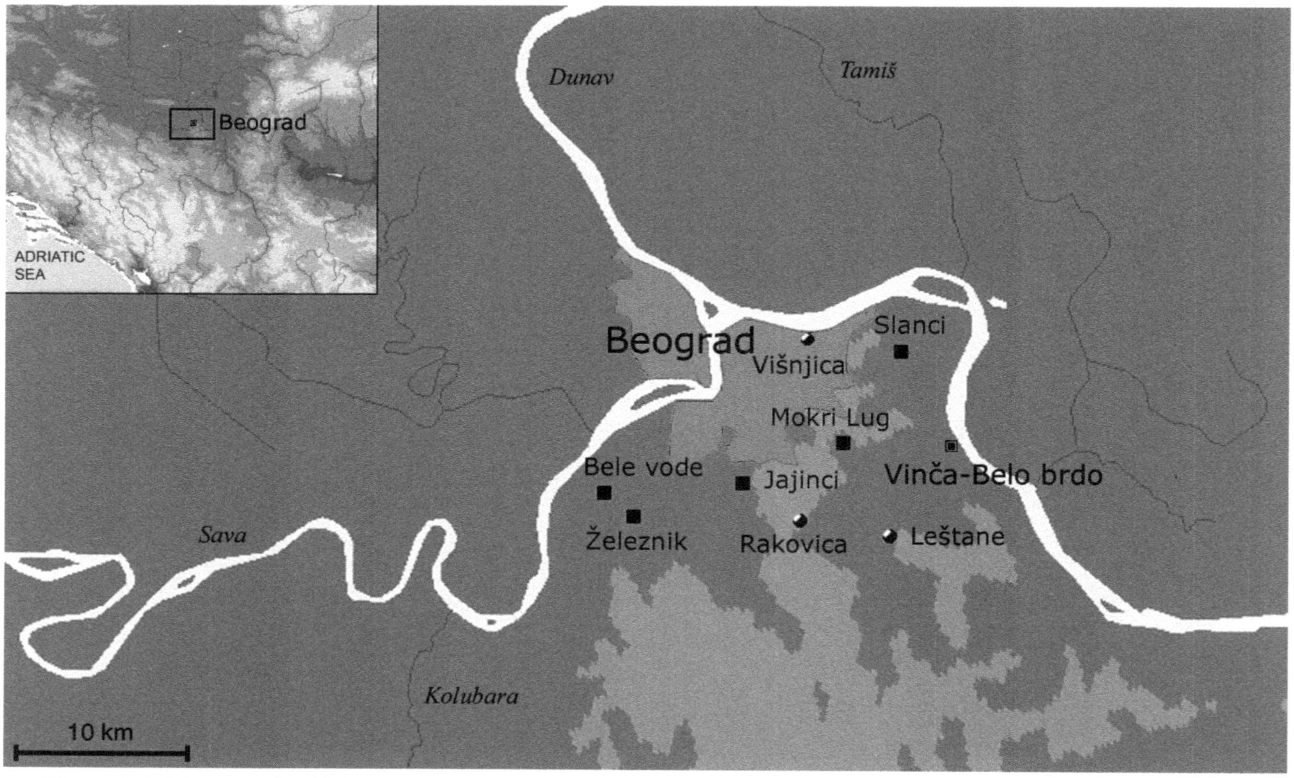

Figure 3-6. Neogene fossiliferous outcrops in the vicinity of the village Vinča (after (Dimitrijević, Tripković, and Jovanović, 2010).

the towns of Loznica, Arandjelovac, Golubac, Zaječar and Negotin (Stevanović 1977). The species of scaphopods identified in clay sediments of Višnjica and sands of Leštane and Rakovica include *Gadilina jani*, *Dentalium mutabile*, *Dentalium entalis*, *Dentalium sexangulum acutangularis* and *Fissidentalium badense* (G. Jovanović & M. Jovanović, 1998).

Taphonomy of palaeontological and archaeological shells

Residents of Belo Brdo obtained larger pieces of jewelry (e.g., bracelets and amulets) made of modern marine clams through exchange with contemporary coastal communities (Dimitrijević and Tripković, 2002, 2006). However, the fossil appearance of scaphopod shells used as beads at Belo Brdo, and the existence of Miocene outcrops in the area of the site, suggests that the residents of Belo Brdo collected scaphopod shells in the vicinity of their settlement, and did not acquire them through exchange.

To confirm this hypothesis, scaphopod fossils in the collection of the Natural History Museum in Belgrade, originating from the aforementioned fossiliferous outcrops and shells from the archaeological site of Belo Brdo were compared (fig. 3-7).
The scaphopod shells found at Belo Brdo, which were identified as *Fissidentalium badense* (Dimitrijević, Tripković, and Jovanović 2010) are fragmented, with rounded breaks mostly at growth lines. Longitudinal ribs that adorn the exterior shell are more or less rounded by erosion. The average length of fragmented segments is generally 3-5 times greater than the diameter. The color of the shells is whitish, light grey or yellowish.

On a number of shells, there are cracks and fractures as a consequence of mechanical damage. These cracks accumulated during the life of the organism and are frequently in the plane of growth lines where the shell is thinner and can easily break (fig. 3-2, 1, 4, 9). Cracking of the shell is rather common in high-cone shaped forms. If the animal survives this type of injury, the shell heals. Traces of healing are usually observed in the form of irregular lines or thickening. Another kind of mechanical damage is fracturing. Fractures are also common in the plane of growth lines, but are created post-mortem due

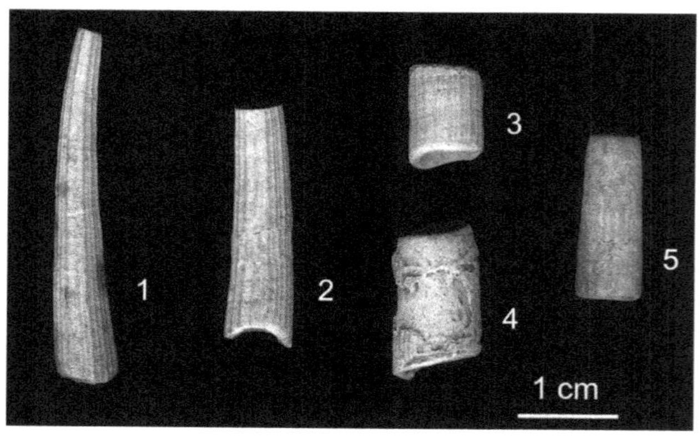

Figure 3-7. Fossils from the Miocene outcrop Višnjica near Belgrade: 1-5 *Fissidentalium badense*, Natural History Museum, Belgrade.

to wave action and other mechanical forces on the empty shells. They differ from cracks by the absence of healing traces, as the animal died prior to, or as a result of, the injury.

In addition to mechanical damage, biotic damages are also observed, indicating that scaphopods were sometimes targeted by predators. A round perforation that tapers toward the inner wall of the shell (fig. 3-2, 6) is an example of an injury created by a moon shell, *Polinices*. Another example of a predator-induced injury is an irregular curving furrow that might have been inflicted by a crab or a fish (fig. 3-2, 16) (Dimitrijević, Tripković, and Jovanović 2010).

All of the damages discussed in the previous paragraphs (e.g., fragmentation, rounded breakages in almost every specimen, and abrasion of longitudinal ridges), suggest that shells were not deposited at the place where the animals died. Instead, they were probably re-mobilized and accumulated on a beach or in shallow water prior to their burial and fossilization.

Scaphopod shells from the fossiliferous Višnjica outcrop show important similarities with archaeological specimens. A sample of 79 shells has been examined for taphonomical features. All shells are fragmented. Their color varies from whitish-grey to yellowish. The average length of the fragments is 16.5 mm. The length of the longest preserved fragment is 37.2 mm (fig. 3-7, 1). The external ornamentation is mostly well preserved, with longitudinal ribs preserved in some specimens (fig. 3-7, 1-3) and eroded in others (fig. 3-7, 4-5). At least 40% of fragments are broken at the level of growth lines. Breakages are flat or sharp in some instances (fig. 3-7, 1-2 narrow end), or rounded in other (fig. 3-7, 3). This indicates that the shells were not deposited where the animals lived, but were instead transported and deposited before fossilization. Approximately 31% of shells show an irregular or sharp breakage that would probably have needed some modification in order to be used as a bead. The remaining shells are suitable for beads as found, especially those with well rounded breakages, which comprise approximately 9.5 % of the total. Some shells are broken at the level of growth lines in such a way that the breakage is sinusoidaly curved, forming one or two rounded outlets (fig. 3-7, 2).

To conclude, the comparison of scaphopod shells from the Belo Brdo archaeological site and the Middle Miocene Višnjica outcrop showed great similarities in the patterns of fragmentation, color, and overall state of preservation. This strongly suggests that the source of archaeological shell was Višnjica, some 10 km northeast of the settlement of Belo Brdo and on the same side of the Danube River.

Conclusion

Taphonomic considerations are of great importance in understanding the origin of shells used as ornaments at archaeological sites. Comparisons of the taphonomical features of archaeological items made from shell and shell from a fossiliferous outcrop can elucidate whether their taphonomical histories are the same. If the same taphonomic conditions are identified, the archaeologist can infer where the specimens were collected in prehistory (i.e., the fossiliferous outcrop).

Taphonomy is generally a well developed field of palaeontological research, with many books and papers written on the topic (e.g., Kidwell 1986; Lyman 1994; Martin 1999; Shipman 1981; Taylor and Wilson 2003). While the basic principles of marine fossil taphonomy are broadly applicable to this kind of study, there are some important differences that should be taken into consideration.

First, taphonomical analysis is usually directed towards the reconstruction of environmental conditions in which the fossil organisms lived. The value of each of the established characters depends on how closely it points to a certain type of environment. In taphonomical analysis of archaeological fossil shells, the goal is different. It is to establish from which fossiliferous outcrop people collected shells. The importance of the established properties of the specimens, thus, depends on how narrowly they define a particular fossiliferous outcrop. For example, the average length of a fragment or the degree of roundness of a breakage might be more important for this study than precise taxonomic identification, which is the prerogative in any palaeontological analysis including taphonomical studies.

Further, taphonomical analyses aimed to research paleontological sites are directed toward identifying associations between organisms. By contrast, in an archaeological shell collection fossils are selected by people in the past and are mostly limited to a few or a single species. Therefore, taphonomical analyses and comparisons between sites must be directed to a particular species and not to associations.

Recently, a study of isotopic composition of *Dentalium* shells used as ornaments at the Upper Palaeolithic site of La Madeleine in Dordogne, France traced the source of the shells to a coastal area, proving that they were not collected at fossiliferous outcrops in the vicinity of the site (Vanhaeren 2004). Taphonomic analysis is another method that can enable one to determine the source of fossils used as ornaments in the past. Researchers, thus, have multiple approaches with which to study fossil shell acquisitions and directions of their eventual exchange in the past. Future research may show that, in addition to the exchange of contemporary marine shell in European prehistory, fossils may have been items of exchange too. The next step in research of the Vinča culture fossil shells, and particularly shells from Belo Brdo is to investigate the possible exchange of fossils with other contemporaneous communities to determine whether the fossils from Belo Brdo were exported farther afield.

Acknowledgements

I would like to thank Nenad Tasić who conducted excavations at the site of Vinča – Belo Brdo since 1998, as well as to numerous colleagues and students who took part in these excavations. My thanks also go to Boban Tripković and Gordana Jovanović with whom I had the pleasure to work with on the topic.

I am also grateful to the Ministry of Science and Technological Development of the Republic of Serbia for funding the project "Transitional processes in the Neolithic of Southeastern Europe", in the scope of which scaphopod shells were analyzed, as well as to the University of Belgrade - Faculty of Philosophy, Department of Archaeology, for providing all necessary conditions for faunal research.

My special thanks go to Dušan Borić and Daniella Bar-Yosef for their valuable comments on earlier versions of this paper.

References

AVRAMOVA, M. 2002. *Der Schmuck aus den Gräbern von Durankulak. Durankulak. Die Prahistorischen Graberfelder von Durankulak.* Band II. Deutsches Archaologisches Institut. Pp. 191–206. Sofia, Berlin.

BAR-YOSEF MAYER, DANIELLA. 2008. Dentalium Shells Used by Hunter-Gatherers and Pastoralists in the Levant. *Archaeofauna*, 17: 103-110.

BAR-YOSEF MAYER, DANIELLA, BUR IN AŞKIM GÜMÜŞ and YEŞIM İSLAMOĞLU. 2010. Fossil Hunting in the Neolithic: Shells from the Taurus Mountains at Çatalhöyük, Turkey. *Geoarchaeology: An International Journal*, 25, 3: 375-392.

BARTON, ANREW JOHN. 1994. *Fishing for Ivory Worms: A review of etnographic and historically recorded Dentalium source locations. Review Literature And Arts Of The Americas.* Pp. 1-176. Master of Arts Thesis, Simon Fraser University.

BORIĆ, DUŠAN. 2009. Absolute dating of Metallurgical innovations in the Vinča culture of the Balkans. In *Metals and Societies, Studies in honour of Barbara S,Ottawa. T. L.* Kienlin T.L. and B.W.Roberts, eds. Universitaetsforschungen zur praehistorischen archaeologie, Band 169: 191-245. Bonn: Dr.Rudolf Habelt GMBH.

CHAPMAN, JOHN. 1981. *The Vinča Culture of South-East Europe. Studies in chronology, economy and society.* British Archaeological Reports International Series 117. Oxford: BAR Publishing.

DIMITRIJEVIĆ, VESNA and BOBAN TRIPKOVIĆ. 2002. New *Spondylus* findings at Vinča-Belo Brdo: 1998-2001 campaigns and regional approach to problem. *Starinar*, 52: 47-62.

——. 2006. *Spondylus* and *Glycymeris* bracelets: trade reflections at Neolithic Vinča-Belo Brdo. *Documenta Praehistorica*, 23: 237-252.

DIMITRIJEVIĆ, VESNA, BOBAN TRIPKOVIĆ and GORDANA JOVANOVIĆ. 2010. Perle od dentalijuma - ljuštura fosilnih morskih mekušaca na nalazištu Vinča - Belo Brdo. *Starinar*, 60: 7-18.

GARAŠANIN, MILUTIN. 1979. Hronologija vinčanske grupe Centralno-balkanska zona. In *Praistrija jugoslavenskih zemalja II – neolitsko doba.* Alojz Benac, ed. Pp. 79–212. Sarajevo: Akademija nauka i umjetnosti BiH – Centar za balkanološka istraživanja.

GARAŠANIN, DRAGA and MILUTIN GARAŠANIN. 1979. *Supska, Stubline - vorgeschichiliche Ansiedlung der Vinca-grup.* Pp. 1-87. Beograd: Narodni muzej.

GIMBUTAS, MARIA. 1976. Ornaments and other small finds. In *Neolithic Macedonia, as reflected by Excavation at Anza, Southeast Yugoslavia.* Maria Gimbutas, ed. Pp. 242-256. Los Angeles: The Institute if Archaeology, The University of California.

GIRIĆ, MIODRAG 1971. *Mokrin: nekropola ranog bronzanog doba.* Dissertationes and monographie XI. Washington, Kikinda, Beograd: Smitsonian Institution, Narodni muzej, Arheološko društvo Jugoslavije.

IGNJATOVIĆ, MILORAD. 2008. Katalog. In *Vinča – praistorijska metropola, istraživanja 1908– 2008.* Dubravka Nikolić, ed. Pp. 203-277. Beograd: Vizartis.

IVANOV, I., and M. AVRAMOVA. 2000. *Varna Necropolis: The Dawn of European Civilization.* Sofia.

JOVANOVIĆ, GORDANA and MIODRAG JOVANOVIĆ. 1998. Tertiary scaphopods in Natural History Museum Belgrade. *Glasnik Prirodnjačkog muzeja*, A, 47-50: 199-204.

KIDWELL, S. M. 1986. Taphonomic feedback in Miocene assemblages: testing the role of dead hardparts in benthic communities. *Palaios*, 1: 239-255.

LYMAN, LEE R. 1994. *Vertebrate Taphonomy.* Cambridge University Press.

MARTIN, RONALD E. 1999. *Taphonomy: A Process Approach.* Pp. 1-524. Cambridge University Press.

NIKOLAIDU M. 2003. Items of Adornment. In *Prehistoric*

Sitagroi: Excavations in Northeast Greece, 1968–1970. Second edition. Collin Renfrew, ed. Pp. 331–360. Los Angeles: University of California, Cotsen Institute of Archaeology.

Nikolić, Dubravka, ed. 2008. *Vinča - praistorijska metropola: istraživanja 1908-2008.* Beograd: Vizartis.

Păunescu, A. 1970. Epipaleoliticul de la Cuina Turkcului-Dubova. *Studii şi Cercetări Istorie Veche şi Archeologice*, 21(1): 3-29.

Shackleton, J.C., 1988. Marine Molluscan Remains from Franchthi Cave. In: Excavations *at Franchthi Cave, Greece*, fasc. 4: 1-152 Indiana University Press, Indiana. Shipman, P. 1981. *Life history of a fossil: An introduction to taphonomy and paleoecology*. Harvard University Press.

Stevanović, Petar. 1977. Miocen okoline Beograda. In *Geologija Srbije, II-3, Stratigrafija, Kenozoik.* Kosta Petković, ed. Pp. 107-145. Beograd: Zavod Regionalnu geologiju i paleontologiju Rudarsko-geološkog fakulteta Univerziteta u Beogradu.

Tasić, Nenad. 2005. Vinča – The Third Glance (Excavations 1998 – 2002). In *Prehistoric Archaeology and Anthropological Theory and Education.* L. Nikolova, J. Fritz and J. Higgins, eds. Pp. 1-8. Salt Lake City: International Institute of Anthropology.

Tasić, Nenad and Milorad Ignjatović. 2008. Od tradicionalne do moderne metodologije. Istraživanja u Vinči 1978–2008. In *Vinča – praistorijska metropola, istraživanja 1908– 2008.* Dubravka Nikolić, ed. Pp. 87-119. Beograd: Vizartis.

Taylor, P. D. and M. A. Wilson. 2003. Palaeoecology and evolution of marine hard substrate communities. *Earth-Science Reviews*, 62: 1-103.

Vanhaeren, M. 2004. Tracing the source of Upper Palaeolithic shell beads by strontium isotope dating. *Journal of Archaeological Science*, 31(10): 1481-1488.

Vasić, Miloje 1930. Excavation on the Neolithic site at Vinča, on the Danube. *Man*, 30: 197-200.

——. 1932. *Preistoriska Vinča I. Industrija cinabarita i kosmetika u Vinči.* Beograd: Državna štamparija Kraljevine Jugoslavije.

——. 1936a. *Preistoriska Vinča II. Oblici grobova. – Mistične oči. – Igra na tabli. – Datovanje Vinče.* Beograd: Državne štamparije Kraljevine Jugoslavije.

——. 1936b. *Preistoriska Vinča III: Plastika.* Beograd: Državna štamparija Kraljevine Jugoslavije.

——. 1936c. *Preistoriska Vinča IV: Keramika.* Beograd: Državna štamparija Kraljevine Jugoslavije.

Yochelson, E. L., and C. H. Holland. 2004. Dentalium saturni GOLDFUSS, 1841 (Eifelian: Mollusca): complex issues from a simple fossil. *Palaeontologische Zeitschrift*, 78 (1) (April): 97-102.

4 - PERFORATED SHELLS FROM AN EARLY MESOLITHIC CEMETERY AT LA VERGNE (CHARENTE-MARITIME, FRANCE): FROM ACQUISITION TO USE AND (SOMETIMES) TO WEAR

Catherine DUPONT, Luc LAPORTE
UMR 6566 CReAAH « Centre de Recherche en Archéologie, Archéosciences, Histoire » CNRS, Université de Rennes 1, Rennes 2, Nantes, Université du Maine et Ministère de la culture, Université de Rennes 1, Campus Beaulieu, bâtiment 24-25 -CS74205, 35042 Rennes Cedex, France,
catherine.dupont@univ-rennes1.fr, luc.laporte@univ-rennes1.fr

Patrice COURTAUD, Henri DUDAY
UMR 5199PACEA – A3P « De la Préhistoire à l'Actuel : Culture, Environnement, Anthropologie –Anthropologie des Populations Passé et Présentes » Université Bordeaux1, Avenue des facultés, 33405 Talence Cedex, France,
p.courtaud@pacea.u-bordeaux1.fr, h.duday@pacea.u-bordeaux1.fr

Yves GRUET
Université de Nantes, 58 rue Stendhal, 44300 Nantes, France,.
achil.lemeur@wanadoo.fr

Abstract : *More than three thousand perforated shells were discovered in association with Early Mesolithic burials (ca. 8500-8000 cal BC) at La Vergne (Charente-Maritime, western France). Despite the status of this excavation (an archaeological rescue operation), the locations of all the artefacts were recorded three dimensionally. This initial information is fundamental to understanding how the shells were associated with the individual human bodies. The determination of species is also important in showing where and how these shells were gathered. The condition of their surfaces allows those marks related to marine erosion, animal and anthropogenic perforations to be distinguished. Identification of use-wear and the degree of wear, in combination with spatial information, permits the reconstruction not only of body ornaments but also of objects associated with them during burial. This paper presents an overview of the different stages in the study of the ornaments from La Vergne, from the excavation to the reconstruction of the original position of the shells.*

Keywords: *Ornament, Shell, Mesolithic, Decorated objects, Burial*

Introduction

The use of perforated marine shells as personal adornment is attested at least as far back as the Middle Palaeolithic period (d'Errico *et al.* 2009). Obviously, perforated marine shells are not always ornaments: some may have been perforated naturally by marine erosion, marine animals or taphonomic processes (Dupont 2006; Dupont *et al.* 2010). In this context, the presence of use-wear seems to be one of the better criteria for identification of a perforated shell as a shell bead. During the Mesolithic period in Europe the majority of such items are complete perforated shells that are found in burials (Newell *et al.* 1990; Taborin 1974a and b). Studies mainly focus on species assemblages and their overall association with the bodies. But what does each assemblage, and the perforated shells themselves, represent? As has already been noted for the Neolithic period (Laporte 1998a), Simpson (1996) emphasises that "the term 'body adornment' encapsulates many elements, including ornaments, clothing, scarification, tattooing and body painting." We will see here that interdisciplinary study of the perforated shells from La Vergne makes a further contribution to the debate. This paper is a synthesis of the methodologies applied to the shell beads from La Vergne, from acquisition to use (Dupont and Gruet submitted; Laporte and Dupont submitted).

The site of "La Grande Pièce" is located at La Vergne near Saint-Jean-d'Angély beside the River Boutonne (fig. 4-1A). Today the site is 45 km from the Atlantic coast. This Mesolithic cemetery was discovered during rescue excavations in advance of road works in 1995, the excavation initially being undertaken to study a Gallo-Roman settlement. The excavation specification was drawn up rapidly, and the time devoted to the excavation of these Mesolithic structures was quite short, considering the rarity and the quality of the burials at La Vergne. The team was composed of anthropologists, archaeozoologists and archaeologists: H. Duday, P. Courtaud, V. Dujardin, H. Martin and K. Robin (Duday *et al.* 1998). The high quality of the work performed under such conditions is quite impressive.

Several graves in this cemetery were studied (Duday and Courtaud 1998, fig. 4-1B). Grave 3 was incomplete owing to destruction during the Gallo-Roman occupation.

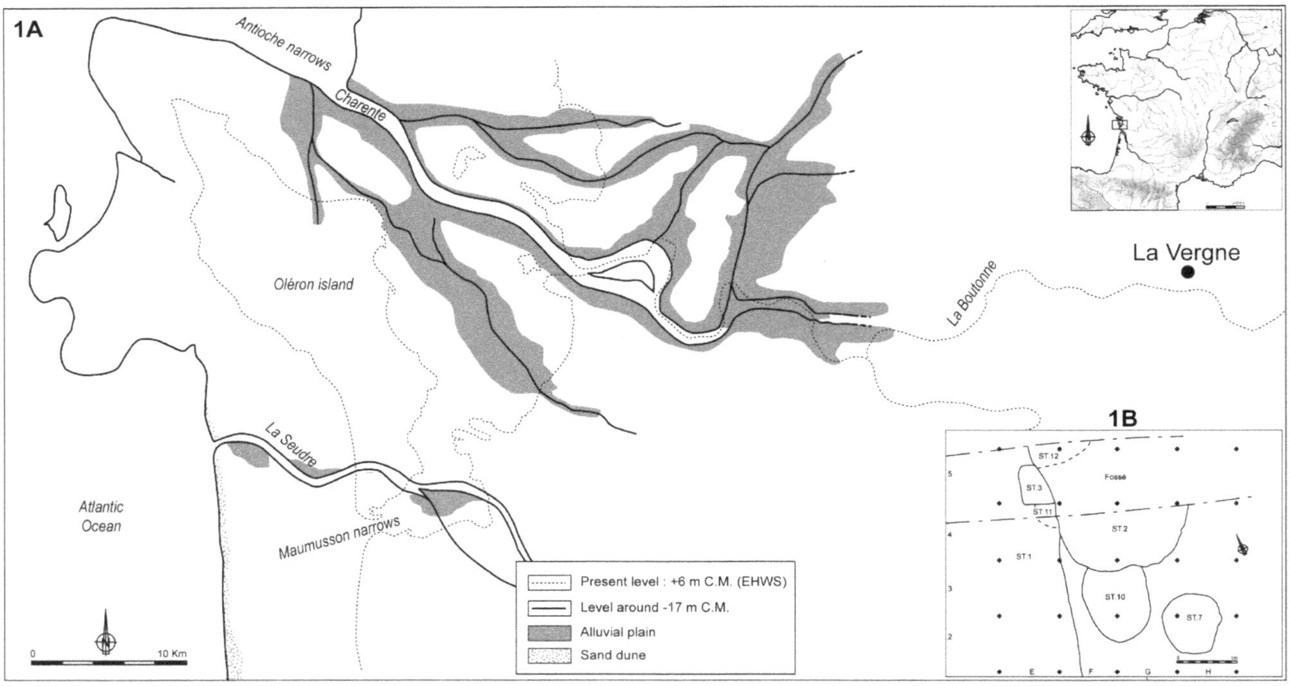

Figure 4-1A: Location of La Vergne site in relation to the former and current coast. 4-1B: Distribution of the graves after André 1986; Duday and Courtaud 1998; CM: isobaths after marine map (CAD L. Quesnel & C. Dupont)

A woman and a foetus, along with a child of 12 to 15 years, comprised the primary burial. A secondary burial deposit contained the remains of an aged individual. Grave 7 contained two primary deposits of adult skeletons. In the same grave, to the south of these, the remains of a very young child of 1 to 2 years old were covered by a secondary cremation. Grave 10 was divided into two parts. The southernmost was occupied by two slaughtered aurochs, whereas two human skeletons lay in the northern part in a primary position (those of a man and a very young child). Grave 11 principally contained the remains of a man truncated by more recent activity. Two other bones belong to a child of 5 to 10 years and to a child that had died at the perinatal phase. All of these human remains are dated to the Early Mesolithic (ca. 8500-8000 cal BC). They were associated with more than three thousand perforated shells and teeth from fox, deer and humans.

The excavation was carried out with awareness of the quantity of the human remains. Small metallic tools and thin sticks of green wood were used to extract the remains (Duday *et al.* 1996). This precaution was crucial in order to limit damage to the original surfaces of the shells. The sediments were removed with brushes and a micro-aspiration system, and were sieved through a 1 mm mesh. All the archaeological remains, shells included, were numbered and drawn on plans at a scale of 1/5. Some were also drawn at a more precise 1/2 scale to record details of the arrangement of beads and bones. From this we can determine, for example, if the valves were articulated. On these plans the general shape of the visible side of the shells was recorded, and the main parts of the shells can be identified from these drawings: anterior and posterior extremities for scaphopods; apex, aperture, ventral and dorsal faces for gastropods; hinge, teeth shape, umbo, and internal and external faces for bivalves. The location of the perforations was also recorded. Additional photographs give even more information at different stages of the excavation. The locations of all the shells were recorded three dimensionally, and the orientation of the main axis of each scaphopod was calculated in degrees with respect to a horizontal reference. The excavators also described the shells and sometimes identified the species. All of these data were crucial: without them it would have been impossible to know precisely how the shells were associated with the bodies or with the various types of funerary deposit. The detailed study of wear marks also provides information on how the shells had been attached.

Acquisition

Species determination

In previous studies the identification of shells used as ornaments has rarely been pursued to the species level This is unfortunate since it fails to provide data about landscape accessibility (which can explain differences between shell assemblages – whether rocky or sandy species are more or less accessible) or about criteria of selection that are no longer apparent (such as the color of shells). Confusion can also be introduced by the existence of a multitude of synonyms. For example, excavators of the Téviec Mesolithic burials in Brittany cited the presence of *Cardium norvegicum* among the shells used as ornaments (Péquart and Péquart 1954). Sometime later, Yves Taborin (1971) referred to the presence there of only ten *Cardium* shells. This later identification provided less information than the original publication since more than ten species of *Cardium* are known from the region and they live in varied environments; some are accessible on the seashore; others live strictly below sea level. We personally verified the initial identification by studying material deposited in the Carnac Museum and old photographs. For researchers working in Europe, Clemam (2010) also is an important tool because it takes a census of all the synonyms of marine shells and indicates the currently valid name. The valid name for *Cardium norvegicum* is now *Laevicardium crassum*. We should also mention a recurrent mistake of identification between *Littorina obtusata* and *Lacuna parva* (Dupont 2006).

At La Vergne, the 3297 shells were identified using reference works in marine biology (Fretter and Graham 1976; 1978; 1981; 1984; Poppe and Goto 1991; 1993) and a reference collection (the Gruet comparative collection). At least fifteen species have been identified: one scaphopod, seven gastropods and seven bivalves (Table 1, fig. 4-2). Identification has not been pursued beyond genus level for *Spisula* as the shells were broken, impeding recognition of the shape of the valve. Use-wear on *Dentalium* has erased the ribs and with them the criteria for identification. The same factor together with the absence of colour explains the difficulty in identifying the limpets. Not all of these species are present in great numbers. Two species dominate the corpus: *Dentalium* and *Nassarius reticulatus*. All are marine species with the exception of *Neritina fluviatilis*. Most of the shells represented (94% of the 1.97 kg of shells) are from species that live on sandy coasts. More than 39% come from subtidal species. Here the determination of shell species suggests three different strategies for the acquisition of the raw material: some shells would have been gathered alive on the beach at low tide; others, the subtidal ones, could also be gathered alive using methods such as diving or dredging; still others would have been gathered when they had been washed up on the beach once the animal was dead.

Biometry and morphometry

All the shells were measured. The aim of this process was to determine whether Mesolithic people had modified the original size of the shell, of *Dentalium* for example (Laporte and Dupont submitted). It also provides information, for each grave, about the degree of selection that had been employed when collecting the individual shells at the coast (Dupont and Gruet submitted).

The measurement of the shells also provides information on variations in shape caused by the environment and consequently on the characteristics of the coast that were exploited by the people searching for the raw material. Thus one of the majority species at La Vergne, the netted dog whelk *Nassarius reticulatus*, has a shape that varies with environmental conditions. A variety of this species, var. *nitidus*, is typical of sheltered inlets (Fretter and Graham 1984). It presents fewer and larger ribs than the netted dog whelk living on exposed shores. Thanks to modern samples from exposed and sheltered positions we can identify the variation in shape of *Nassarius reticulatus* as a function of the degree of exposure (Dupont and Gruet 2000). This variation of shape can be analysed by dividing the number of ribs counted on the last whorl of the spire by the maximum width of the gastropod. Applied to samples from La Vergne it shows that the netted dog whelks used for shell beads seem to have affinities with the *nitidus* variety and to have come from sheltered marine environments. This variety can live in the brackish waters of estuaries and marshes.

Analysis of sea level change indicates that La Vergne was over 55 km from the coast during the Mesolithic (fig. 4-1). Marine influences were closer than today since the estuary

	Grave 3	Grave 7	Grave 10	Grave 11	Structure 1	Structure 2	Structure 4	Structure 12	Others
Nassarius reticulatus	+++	+++	+++	+		+	+	+	+
Phalium saburon			+						
Euspira catena	+				+				
Nassarius incrassatus		+	+						+
Neritina fluviatilis	+								
Ocenebra erinaceus		+							
Patella sp.		+	+		+				
Dentalium sp.	+++	+++	+++		+		+	+	+
Cerastoderma edule		+							
Mytilus sp.					+				
Spisula sp.			+						
Venus verrucosa					+				
Glycymeris glycymeris			+		+	+			
Ostrea edulis			+						
Laevicardium crassum		+	+		+				
Caardidae					+				

Table 4-1. Frequency of shell species in the different structures from La Vergne (+: less than 30 individuals, +++: greater than 140 individuals)

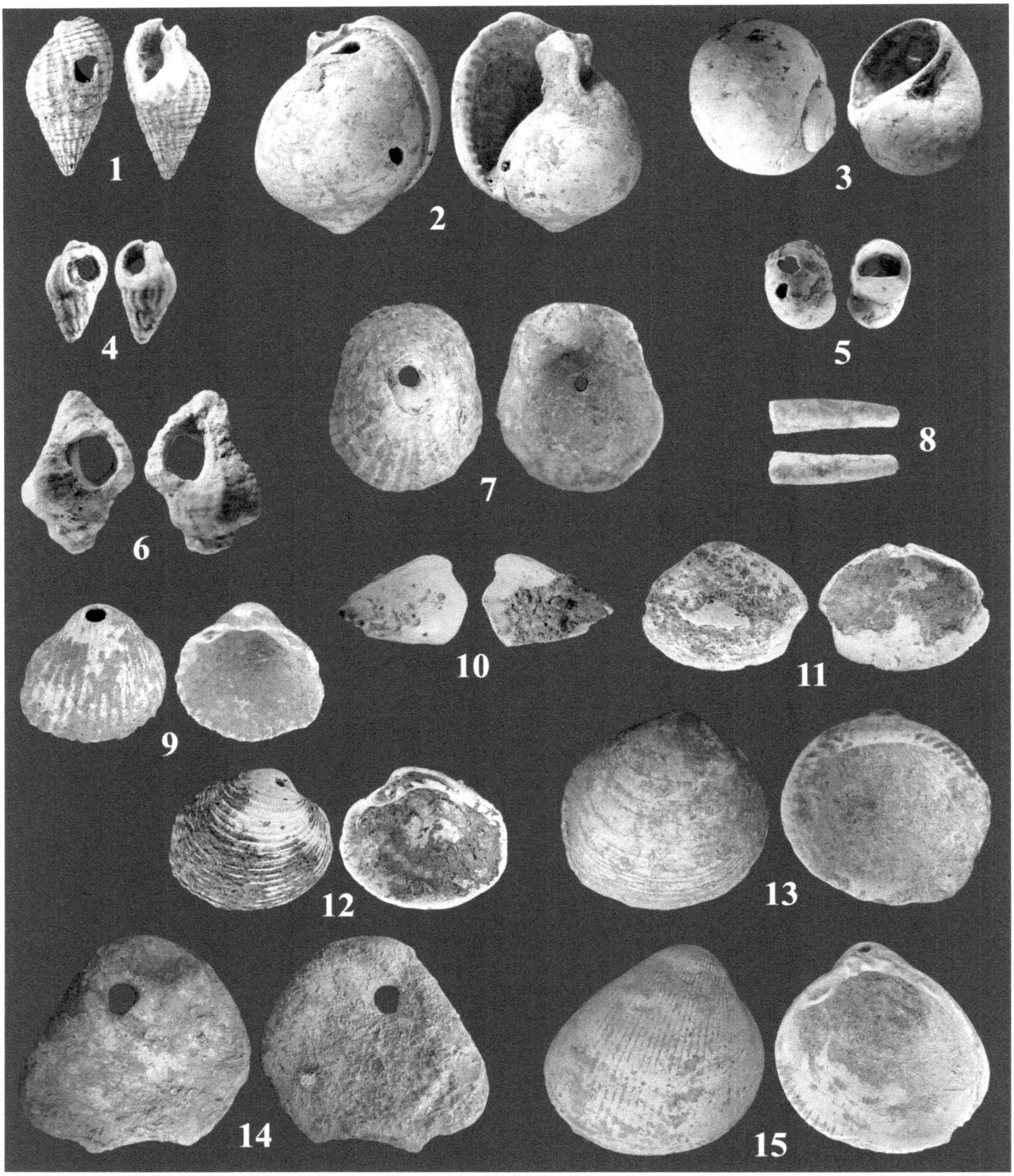

Figure 4-2. Shells identified at La Vergne: 1- *Nassarius reticulatus* (L=22mm), 2- *Phalium saburon* (L=53mm), 3- *Euspira catena* (L=25mm), 4- *Nassarius incrassatus* (L=9mm), 5- *Neritina fluviatilis* (L=8mm), 6- *Ocenebra erinaceus* (L=19mm), 7- *Patella* sp. (L=41mm), 8- *Dentalium* sp. (L=12mm), 9- *Cerastoderma edule* (L=29mm), 10- *Mytilus* sp. (L=20mm), 11- *Spisula* sp. (L=33mm), 12- *Venus verrucosa* (L=36mm), 13- *Glycymeris glycymeris* (L=51mm), 14- *Ostrea edulis* (L=70mm), 15- *Laevicardium crassum* (L=54mm) (CAD C. Dupont).

of the Charente was wider and deeper. Marine water penetrated inland along this estuary. Palaeoenvironmental cores indicate the presence of former shorelines (Gruet and Dupont 2001; Laporte 1998b, Laporte *et al.* 2009). The large estuary of the Charente could have presented sheltered conditions that would explain the dominance of sandy species at La Vergne. This environmental evidence is supported by the presence of *Nassarius* variety *nitidus* biotopes and the presence of *Neritina fluviatilis* further upstream from this estuary. Despite the current distance between the Mesolithic site and the sea, the shell beads are hence not necessarily evidence of exchange between different populations.

Thanatocenosis and taphonomy

The study of the La Vergne assemblage presents another important clue to understanding how these shells were gathered: observation of modifications to their shape and surface. Some of this can be related to the death of the shells and to marine erosion before their acquisition by Mesolithic groups; others are the result of their burial, and due for example to animal activity or soil acidity. To determine whether shells had been modified by marine erosion, different parts of the shells were compared to a reference collection of modern shells gathered dead on a beach (Comparative collection Gruet). Such erosion affects the more prominent areas of the shells: the umbo, the hinge, the edge and the sides of the valves; the apex, the last whorl of the spire, the siphonal canal and the peristome for gastropods. Natural erosion results in an overall polishing of the shell.

In extreme cases it may lead to the breakage of the ventral edge of valve, and of the apex and siphonal canal in gastropods. These breakages are followed by a homogeneous polishing of the ridges. Marine erosion can also cause the removal of teeth from the hinge of a bivalve and from the peristome of gastropods. The reduction in thickness of the shell caused by marine erosion can also result in a perforation localized on the extremity of the umbo for a bivalve and on the last whorl on the dorsal side for a gastropod. These observations allow an anthropogenic perforation to be distinguished from a natural perforation caused by the overall state of the shell surface. At La Vergne we observed different degrees of marine erosion that indicate that some of the shells were already dead when they were gathered on the beach (Dupont and Gruet submitted). There is an emphasis on subtidal species. The hypothesis that they were collected by diving or dredging can be rejected.

Transforming a shell into an adornment

Perforation

All well-preserved specimens of shells at La Vergne are perforated, usually with only a single functional perforation. The position and shape of each perforation was recorded at 80x magnification, and each gastropod shell was also divided into 15 areas for systematic analysis of potential traces of preparation, perforation and regularization of the contour of the perforation. Application of these different criteria to the La Vergne assemblage show that a number of the shells were already perforated when they were gathered (Laporte and Dupont submitted). Some of the original holes were made by shell-boring gastropods, others by marine erosion. Such perforations are preferentially located on the last whorl of spire for turbinate gastropods perpendicular to the peristome, at the apex or at the margin for limpets, and near the umbo for bivalves. Among the major species, shells of *Nassarius reticulatus* reveal various stages of human action (traces of preparation, perforation, and regularization of the contour of the perforation) on only 30 % of the specimens observed. This scarcity is related, as we will see below, to intense use-wear that has erased them. Some surfaces were prepared by abrasion and incision before the piercing of the holes. The more frequent technique is external percussion on the dorsal side of the body whorl of the gastropod. Some perforations have also been regularized with a rotary movement. Some *Dentalium* shells also show marks of preparation. The biometric study includes the length of the shell, anterior, posterior, internal and external diameters. These measurements have been compared to those of modern samples and reveal that most of the *Dentalium* had been trimmed before being used as ornaments. The aim of this action was probably to enlarge the perforation to facilitate the passage of the string and also to straighten the shape.

Use marks

We also systematically observed traces of wear both across the entire surface of the shell and at the perforation (Laporte and Dupont submitted). The majority of these traces correspond to glossy, polished and abraded surfaces. As a first approach we gave priority to binocular inspection of the whole collection. Ninety percent of the *Nassarius reticulatus* shells display traces of use-wear. Following individual observation, we statistically tested the occurrence of the different wear traces observed according to their position on the shell and their intensity. fig. 4-3-B is the result of the statistical test for netted dog whelks from graves 7 and 10. It corresponds to more recurrent localizations and intensities of observed traces. These diagrams allow us to propose alternative possibilities for the use of one or more strings (fig. 4-3-C). Statistical tests have also been applied to some clusters observed in each grave where that seemed significant. In some cases, comparisons were made with wear marks observed on objects selected from ethnographic collections (Laporte and Dupont, submitted). Some *Dentalium* shells also displayed marked lateral abrasion, indicating that they were mobile and must have rubbed against something. Others were embedded in each other when discovered, and had clearly been threaded together. The perforated bivalves occasionally present wear marks suggesting some kind of suspension. The intensity of some of the use-wear demonstrates indisputably that these shells were not only funerary objects. Alone or in groups, they were part of the adornment of more complex objects that had had previous use. Some could have been worn on clothing or as personal ornaments, but not only in the ways we might expect today. Furthermore, we must question whether all of the personal ornaments adorning the corpse of the deceased were representative of clothing worn by the living? At La Vergne, the perforated shells around with the legs of a young child show intense traces of wear.

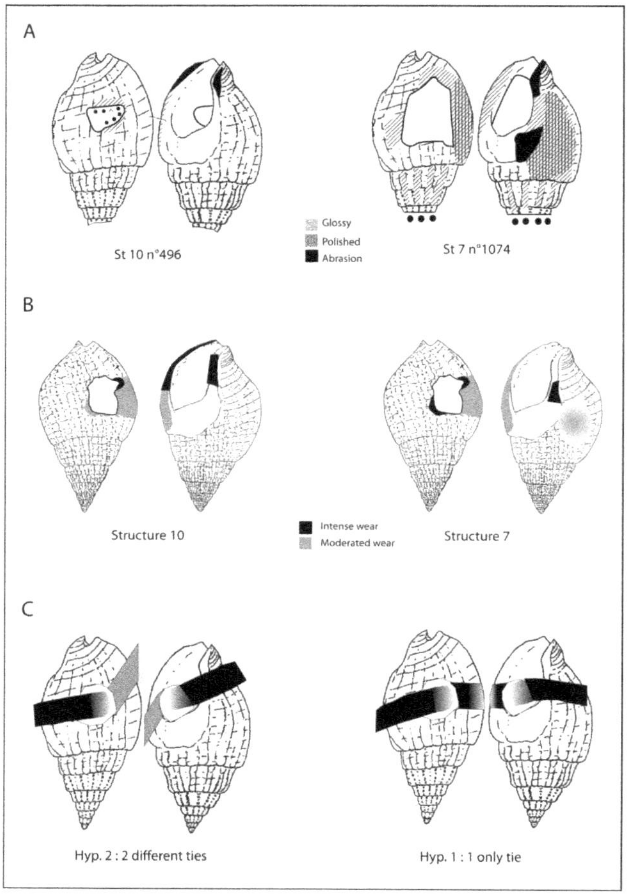

Figure 4-3. From traces to reconstitution: stages of the study; A- Identification and location of marks on each shell; B- Synthesis of the main marks for each structure; C- Reconstruction of the passage of the strings (CAD L. Quesnel & L. Laporte)

Spatial distribution

This study of the material from La Vergne was possible thanks to the detailed nature of the excavation. We synthesized all the observations of species, perforation and use-wear traces as a function of the spatial arrangement of the shells. This must be accommodated within the dimensions of each grave, which in each case was partly occupied by the bodies that had been laid there and the funerary deposits. Some shells seem to be strictly linked to a specific anatomical part such as legs, head or hair, but the pattern is not the same for each skeleton (Laporte and Dupont submitted). Furthermore, many of them may have adorned items used as containers of bodies which were deposited in a contracted position: qualifying these perforated shells as personal ornaments would in that case be valid only during the funerals of the deceased. Others decorated the objects that contained bones, for example in grave 10 where the shells appear be to personal ornaments of a child, but are in fact associated with the sides of a small box of perishable material (fig. 4-4). As a consequence, proximity to human bones is not always sufficient to suggest the function of personal ornament for all of these perforated shells. Spatial analysis of the clusters of large shells deposited before the face of the adult buried in grave 10 revealed that they were part of a discrete deposit; several opposed valves, inserted and fitted together, might have knocked against one another and made a noise.

A limpet is sometime placed at the top, perhaps to accommodate the different strings. Many other perforated shells are spatially disconnected from the bodies but that does not indicate that they were not part of clothes separately deposited. This becomes more tenuous, however, when a cluster of perforated shells is associated with deposits of animal remains, such as those spread along and over the horns of aurochs skulls in this same grave 10. We also observed that the intensity of use wear is not homogeneous for all clusters of perforated shells in the grave. Those *Nassarius* that are closely associated with the bodies show more intense wear than the others, highlighting differences between body ornaments probably buried after a long use by the living person, and other ornamental objects forming funeral deposits. Finally, at La Vergne, adornments that had definitely been used as body ornaments constitute only a minority of the whole shell assemblage. Spatial reconstruction is fundamental to understanding the difference in species composition between the graves.

Discussion

What are the similarities and differences between La Vergne and the two other Mesolithic sites from western France with well-documented perforated shells (Téviec and Hoëdic)? The three sites cumulatively represent considerably more than half of all the perforated shells associated with Mesolithic burials in Western Europe: 3297 at La Vergne, approximately 6987 at Téviec (Dupont 2006: 174), 5068 at Hoëdic (Dupont 2006: 160), and probably even more if we take in account that none of these sites has been totally excavated. In Brittany, a few isolated Mesolithic shell beads from shell middens had perhaps been lost during daily life (Dupont *et al.* 2010). We must first emphasise some limitations to such a comparison. Téviec and Hoëdic are attributed to the Late Mesolithic and La Vergne to the Early Mesolithic; they are hence separated by a space of some four millennia. The Early Neolithic assemblage of 3288 manufactured shell beads from Germignac (Laporte and Gomez 2001, Laporte 2009) is more or less contemporary, on this time-scale, with some of the graves at Hoëdic. Furthermore, the two sites in Brittany are coastal and associated with shell middens. La Vergne is an open air cemetery and the burials were around 45 km from the seashore. Stable isotope analyses

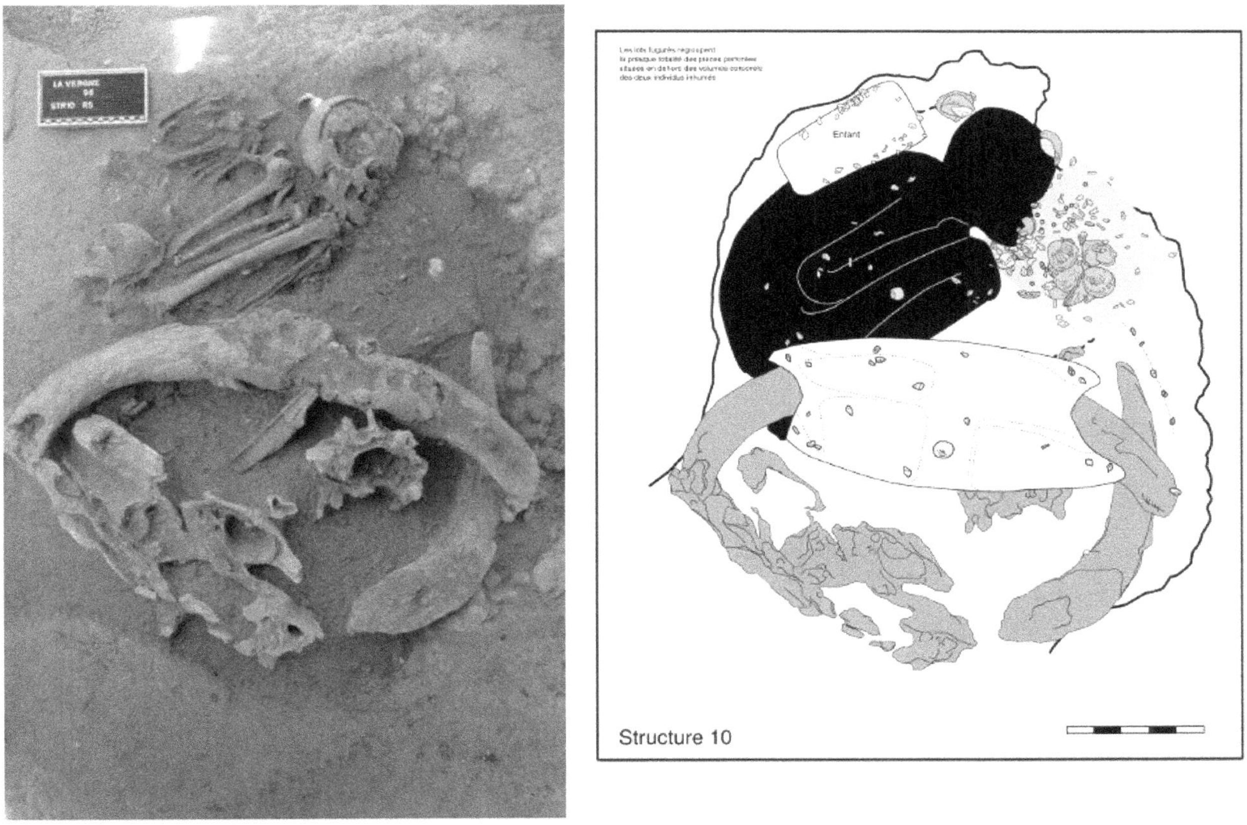

Figure 4-4. Spatial distribution of perforated shells in grave 10 at La Vergne (Charente-Maritime) – Cl. P. Courtaud, CAD L. Laporte, L. Quesnel

carried out on the individuals from La Vergne indicate that this Mesolithic population essentially followed an inland way of life dominated by terrestrial resources (Schulting *et al.* 2008).

A binary composition of the assemblage has already been noticed by Yvette Taborin (1974a and b) at Téviec and Hoëdic, where two small species dominated the corpus (*Trivia monacha* and *Littorina obtusata*). The same is found at La Vergne where *Nassarius reticulatus* and *Dentalium* sp. constitute 97 % of the perforated shells. The abundance of small species can be attributed to their lesser weight and volume. Some large bivalves (*Acanthocardia echinata*, *Laevicardium crassum*, *Pecten maximus*, *Glycymeris glycymeris*) are also represented at the three sites but in smaller numbers. Species that contributed to the diet of coastal Mesolithic populations are rare in these funerary sites, an observation that can be extended to the whole of the Atlantic French façade: shells of species that were eaten were not recycled as ornaments (Dupont 2006). The presence of subtidal species and the marks of marine erosion are also a common feature that we have observed at La Vergne, Téviec and Hoëdic: in all three cases, some of the shells were already perforated when gathered. La Vergne differs from Téviec and Hoëdic in the species composition of the assemblage. Whereas *Nassarius reticulatus* and *Dentalium* sp. have been identified at Téviec and Hoëdic, they are less commonly represented than at La Vergne. This difference can be explained by the accessibility of the relevant environments; the species from Téviec and Hoëdic are mainly accessible on rocky shores; those of La Vergne in sheltered areas such as sandy beaches (Dupont 2006). Whether the general appearance of gastropods like *Littorina obtusata* and *Nassarius reticulatus* would have been seen as so different by those who gathered them for this specific use can also be questioned.

Intra-site analysis shows recurrent variations of assemblages in all cases. At Téviec and Hoëdic, Yvette Taborin described differences both in the proportions of the two main species (*Trivia monacha* and *Littorina obtusata*) and in the way the shell ornaments were supposed to be worn as a function of the sex of the buried individuals (Taborin 1974a: 164-174; 1971: 160). The quantities of perforated shells associated with a skeleton appeared to her to be dependent also on the age of the deceased. At La Vergne, the determination of sex is in progress of revision in light of improved methodologies; what should we say then about sex and age determinations that have not been revised in the same way at Téviec and Hoëdic? At La Vergne, the differences observed in the composition of shell species between the graves relate mainly to large and rare shells (Dupont and Gruet submitted): from the spatial reconstruction we know that they were not attached to the clothing of the deceased, at least when buried (Laporte and Dupont, submitted). On the other hand, we do not have spatial data at our disposal for Téviec and Hoëdic. If associated shell objects were present in the burial, they have not been distinguished from direct body ornaments; nevertheless, examination of photographs taken by the excavators in the 1930s seems to reveal a much more frequent association with the clothing of the dead at burial. Differences of methodology are thus a major obstacle to interpretation and comparison, but not the only one.

At Téviec and Hoëdic, large shells of gastropods or bivalves are sometimes described as lying next to the skull, like those we referred to at La Vergne, but none of the publications describe these perforated shells in any other way than as ornaments and components of necklaces, pectorals, bracelets or anklets, and hair-nets (Péquart *et al.* 1937). This vision of shell beads associated with body ornaments was heavily influenced by the intellectual context of the mid-1950s. It seemed improbable that these "primitive populations" could have sported more complex clothes. Yves Taborin (1971: 162) evoked the possibility of shells "sewn" at the level of the hip. Even so, the dominant interpretation was systematically that of clothing separately deposited in the grave: one has only to visit any museum or ethnographic collection to notice how many other possibilities exist of objects decorated with perforated shells (Laporte and Dupont, submitted). During the Mesolithic period comparable objects at La Vergne, as well possibly as some clothing at Téviec or Hoëdic, contributed to the ostentatious theater associated with the deposition of bodies in the grave. Whether these were the funerals of all the humans buried together in the grave, with animal offerings and valuables, or the funeral of only one of these individuals can also be questioned, as in the case of Mesolithic burials in Denmark (Testart 2004). If so, the elaborate clothing and personal ornaments associated with the main inhumation might have been quite exceptional in comparison to those of his followers, as the whole community contributed to honor just one individual. Were the ornaments worn by all the deceased who were buried in the same grave specifically personal?

New methodologies applied during the excavation of Mesolithic burials and associated objects at La Vergne lead us to modify our vision of these hunter-gatherers. The gathering of the shells intended as ornaments is a well-defined activity in which shells were collected that were already perforated. These ornaments did not only correspond to body decoration. Without exhaustive and detailed excavation the distinction between bracelets or necklaces, beads sewn on to clothing, beads linked to objects, and objects where shells were the key component cannot be proved. Techniques must take into account the voids and objects of perishable material, just as anthropologists do when studying the successive stages of decomposition of human bodies. These different volumes (those of objects and those of bodies) merge into one another, besides, when the object in question is an item of clothing. Though it is difficult to obtain such information from old excavations, we hope that the experience of La Vergne will persuade future excavators to take into account more than just the skeletal remains when undertaking three-dimensional reconstruction.

Acknowledgements

We would like to thank Judith Scarre and Chris Scarre for improving our English language.

Bibliography

ANDRÉ, XAVIER. 1986. *Élaboration et analyse des cartes bathymétriques détaillées du proche plateau vendéo-charentais. Reconstitution des paléorivages de la transgression holocène.*, PhD thesis, Université of Bordeaux I, Bordeaux.

CLEMAM 2010. Check List of European Marine Mollusca. http://www.mnhn.fr/base/malaco.html (accessed April 15, 2010).

D'ERRICO, FRANCESCO, MARIAN VANHAEREN, NICK BARTON, ABDELJALIL BOUZOUGGAR, HENK MIENIS, DANIEL RICHTER, JEAN-JACQUES HUBLIN, SHANNON P. MCPHERRON, and PIERRE LOZOUET. 2009. Additional evidence on the use of personal ornaments in the Middle Paleolithic of North Africa. *Proceedings of the National Academy of Sciences of the USA* August 2009:1-6.

DUDAY, HENRI, and PATRICE COURTAUD. 1998. La nécropole mésolithique de La Vergne (Charente-Maritime). In *Sépultures d'Occident et genèses des mégalithismes (9000-3500 avant notre ère), Séminaire du Collège de France.* Jean Guilaine, ed. Pp. 27-37. Paris: Errance.

DUDAY, HENRI, PATRICE COURTAUD, VÉRONIQUE DUJARDIN, HÉLÈNE MARTIN and KARINE ROBIN. 1996. *La Vergne : La Grande Pièce (Charente-Maritime) (17.405.002 AH-17.465.501 AP) 1995. Complément au D.F.S. de sauvetage urgent (du 10.06.95 au 04.07.95).* URA 376 CNRS, Laboratoire d'Anthropologie de l'Université de Bordeaux I, SRA Poitou-Charente.

DUDAY, HENRI, PATRICE COURTAUD, KARINE ROBIN, VÉRONIQUE DUJARDIN, YVES GRUET, GÉRARD GOURAUD, HÉLÈNE MARTIN, and CRISTINA SAN JUAN-FOUCHER. 1998. La Vergne : La Grande Pièce (Déviation de Saint-Jean d'Angély, Charente-Maritime). *Bulletin de la société Préhistorique Française* 95-3: 423-435.

DUPONT, CATHERINE. 2006. *La malacofaune de sites mésolithiques et néolithiques de la façade atlantique de la France : Contribution à l'économie et à l'identité cultu-*

relle des groupes concernés, British Archaeological Reports Int. Series 1571. Oxford: BAR Publishing.

Dupont, Catherine, and Yves Gruet. 2000. Variations morphologiques de mollusques gastropodes (*Nucella lapillus* et *Hinia reticulata*) : Intérêts pour l'archéologie. *Revue d'Archéométrie* 24: 53-61.

——. *submitted.* Étude de la malacofaune marine. In *Monographie des sépultures de la 'Grande Pièce' (La Vergne ; Charente Maritime).*

Dupont, Catherine, Grégor Marchand, Yolanda Carrion, Nathalie Desse-Berset, Loïc Gaudin, Yves Gruet, Dominique Marguerie, and Christine Oberlin. 2010. Beg-an-Dorchenn : une fenêtre ouverte sur l'exploitation du littoral par les peuples mésolithiques du sixième millénaire dans l'ouest de la France. *Bulletin de la Société Préhistorique Française* 107-2: 227-290.

Fretter, Vera, and Alastair Graham. 1976. The prosobranch molluscs of Britain and Danmark, Part. 1-Pleurotomariacea, Fissurelacea and Patellacea. Department of zoology, University of reading. *The journal of molluscan studies* Supplement 1: 1-38.

——. 1978.-The prosobranch molluscs of Britain and Danmark, Part. 3-Neritacea, Viviparacea, Valvatacea, terrestrial and freshwater Littorinacea and Rissoacea. Department of zoology, University of reading. *The journal of molluscan studies* Supplement 5: 101-152.

——. 1981. The prosobranch molluscs of Britain and Danmark, Part. 6-Cerithiacea, Stombacea, Hipponicacea, Lamellariacea, Cyperaeacea, Naticacea, Tonnacea, Heteropoda. Department of zoology, University of reading. *The journal of molluscan studies* Supplement 9: 285-362.

——. 1984. The prosobranch molluscs of Britain and Danmark, Part. 8-Neogasteropoda. Department of zoology, University of reading. *The journal of molluscan studies* Supplement. 15: 435-556.

Gruet, Yves, and Catherine Dupont. 2001. Au Néolithique dans le Centre-Ouest de la France, la pêche des coquillages reflète-t-elle l'environnement marin ? In *Systèmes fluviaux, estuaires et implantations humaines de la préhistoire aux grandes invasions, Actes des congrès nationaux des sociétés historiques et scientifiques. 124e Nantes 1999.* Jean L'Helgouach and Jacques Briard, eds. Pp. 183-199. Paris: Éditions du Comité des travaux historiques et scientifiques.

Laporte, Luc. 1998a. La parure au Néolithique. In *Les premiers paysans du Golf, le néolithique dans le marais poitevin.* Roger Joussaume, ed. Pp. 45-47. Chauray: Patrimoines et Médias.

——. ed. 1998b. *L'estuaire de la Charente de la Protohistoire au Moyen âge.* dAf 72 (série Grands Travaux), Paris: Éditions de la Maison des Sciences de l'Homme.

——. 2009. La parure néolithique dans le Centre-Ouest de la France - Chapitre : La parure. Approches stylistiques, technologiques et fonctionnelles du mobilier. In *Des premiers paysans aux premiers métallurgistes sur la façade atlantique de la France (3500 -2000 av. J.-C.).* Luc Laporte, ed. Pp. 455-469. Mémoire XXXIII. Chauvigny: Association des publications Chauvinoises,

Laporte, Luc, and José Gomez-de-Soto. 2001. Lamérac et Germignac : perles et anneaux-disques dans le Centre-Ouest de la France. *Revue Archéologique de l'Ouest* 18: 13-26.

Laporte Luc, and Catherine Dupont. *submitted.* Études des perforations et des usures de la parure en coquillages. Répartition spatiale de la parure. In *Monographie des sépultures de la 'Grande Pièce' (La Vergne ; Charente Maritime).*

Laporte, Luc, Christian Caemerlynck, Nicolas Florsch, François Levêque, Didier Néraudeau, Christine Oberlin and Laurent Quesnel. 2009. Occupations préhistoriques et variation des lignes de rivage : l'exemple des marais charentais – Chapitre : Iles ou péninsules ? Evolution des milieux littoraux sur la façade atlantique du Centre-Ouest de la France au cours de l'Holocène. In *Des premiers paysans aux premiers métallurgistes sur la façade atlantique de la France (3500 -2000 av. J.-C.).* Luc Laporte, ed. Pp. 16-28. Mémoire XXXIII . Chauvigny: Association des publications Chauvinoises..

Newell, Raymond R, D. Kielman, Trinette Constandse-Westermann, A. Van Gijn and W.AB. Van Der Sanden. 1990. *An Inquiry into the Ethnic Resolution of Mesolithic Regional Groups.* New York: Brill.

Péquart, Marthe, and Saint-Just Péquart. 1954. *Hoëdic, deuxième station-nécropole du Mésolithique côtier Armoricain.* Anvers: De Sikkel.

Péquart, Marthe, Saint-Just Péquart, Marcellin Boule and Henri Vallois. 1937. *Téviec : station nécropole mésolithique du Morbihan.* Archives de l'Institut de Paléontologie humaine 18, Paris: Masson.

Poppe, Guido T., and Yoshihiro Goto. 1991. *European Seashells : Polyplacophora, Caudofoveata, Solenogastra, Gasteropoda.* Germany: Verlag Christa Hemmen.

——. *European Seashells : Scaphopoda, Bivalvia, Cepha-*

lopoda. Germany: Verlag Christa Hemmen.

SCHULTING, RICK J., STELLA M. BLOCKLEY, HERVÉ BOCHERENS, DOROTHÉE DRUCKER and MIKE RICHARDS. 2008. Stable carbon and nitrogen isotope analysis on human remains from the Early Mesolithic site of La Vergne (Charente-Maritime, France). *Journal of Archaeological Science* 35: 763-772.

SIMPSON, BIDDY. 1996. Self and social identity: an analysis of the Mesolithic body adornment from the Scottish western isles. In *The early prehistory of Scotland.* Tony Pollard and Alex Morrison A., eds. Pp. 237-251. Edinburgh: Edinburgh University Press for the University of Glasgow.

TABORIN, YVETTE. 1971. La parure en coquillage de l'Epipaléolithique au Bronze ancien de France, PhD thesis, Université of Paris.

—. 1974a. La parure en coquillage de l'Epipaléolithique au bronze ancien en France, *Gallia Préhistoire* 17.1: 101-179.

—. 1974b. La parure en coquillage de l'Epipaléolithique au bronze ancien en France, *Gallia Préhistoire* 17.2: 307-417.

TESTART, ALAIN. 2004. *La servitude volontaire 1 : les morts d'accompagnements.* Paris: Errance.

5 - SHELL USE IN WEST MEXICO AND THE SOUTHWESTERN UNITED STATES. AN ARCHAEOLOGICAL COMPARISON

Luis GÓMEZ-GASTÉLUM
Universidad de Guadalajara, Mexico
gastelum@cencar.udg.mx

To Phil C. WEIGAND
In Memoriam

Abstract : *For many years the possibility of the existence of relationships between the Pre-Columbian Mesoamerica and the Southwestern United States has been discussed at length. In this realm, the northern portion of West Mexico has been considered as a corridor for influences from Mesoamerica to the Southwestern United States. Although there are some material elements from one cultural area in the other, like copper bells, macaws, cacao, or turquoise, the challenge for archaeologists is define the ultimate nature of this exchange. In this paper I will discuss some of the assumptions surrounding this proposal, utilizing an alternative material and perspective. I will examine the use of seashells and the associated color symbolism. In this interaction I hope find elements to evaluate the similarities and differences that Pre-Hispanic West Mexico and Southwestern United States societies had in their relationship with shells and their respective cosmovisions. Finally, I discuss the results as a contribution on the problematic nature of Mesoamerican-Southwestern United States prehistoric relationships.*

Key words: *Color, Shells, Interaction.*

Introduction

In archaeological studies analyses of shells are becoming more and more important. They have implications for ancient economies, technologies, social aspects, biology and palaeoenvironments. In terms of shell artefact studies, the major areas of focus are shell object exchange and the technology that was used in manufacture. But additionally the beauty of shells was associated with religious aspects in many places of the world. In this text, I pay attention to two subjects. First, I revise some views about the relationships between Pre-Hispanic Mesoamerica and Southwestern United States. After, I will compare the conceptualization of shells and objects manufactured with them in these two cultural areas. From a revision of the knowledge about the different molluscan genera and species found in Mesoamerican West Mexico and the Southwestern United States, I will offer an interpretation about their use observing the Pre-Columbian worldview in each cultural area. Finally, I will contribute with some ideas for the debate about the interaction between the cultural areas mentioned in the manuscript.

The setting

In this research work I am considering two cultural areas: Mesoamerica and the Southwestern United States. Mesoamerica was defined by Paul Kirchhoff in the 1940s. He determined that the territory between the Sinaloa River, the Panuco River, both in Mexico, and the Nicoyan Gulf in Nicaragua, there was a high level of cultural development, which could be termed a civilization. Its Pre-Columbian chronology was divided in three major periods: Pre-Classic (2500 B. C. to A. D. 200), Classic (A. D. 200 to 650/900) and Post-Classic (A. D. 900/1000 to 1521)[1] (López Austin and López Luján 1996). In these times, some Pre-Columbian societies grew from villages to civilizations, like Zapotec, Teotihuacan, Maya, or Aztec. They constructed great cities like Monte Alban, Teotihuacan, Palenque and Tenochtitlan.

Here, my main interest is in West Mexico (fig. 5-1); a cultural subarea of Mesoamerica. Its territory comprised the Mexican states of Michoacán, Colima, Jalisco and Nayarit. Its chronology was integrated by three periods, similar to the general Mesoamerican ages. During the Early period (1500 B. C. to A. D. 0/600), societies existed at the village level. For the Regional Developments period (A. D. 0/600 to 1100), there were several strongly independent societies that had their own identities. They integrated systems like "interaction spheres". In one case at least, a community reached the civilization level. The third period is the Post-Classic (A. D. 1100 to 1536), in which the Tarascans from Michoacán formed a State that competed with the Aztecs (Gómez-Gastélum 2005).

The Southwestern U. S., after McGuire (1989:40), is a cultural area that includes the U. S. states of Arizona, New Mexico, southeast Utah, southwest Colorado, and trans-Pecos Texas, as well as the Mexican states of Sonora and

[1]Here it is necessary point out that the dates mentioned in the course of the chapter are not radiocarbon dates. They are the accepted chronology of the major horizons in each cultural area.

Figure 5-1. The West Mexico showing the archaeological sites discussed in the text.

Chihuahua (fig. 5-2). U. S. archaeologists recognize four major cultural/spatial units: the Anasazi, the Mogollon, the Patayan, and the Hohokam. Despite each cultural unit having its own chronology, Wheat (1954:577) proposed temporal divisions for the area which I utilize in this work. He divides the Pre-Columbian epoch in four periods: Pioneer (300 B. C. to *ca.* A. D. 500), Colonial (*ca.* A. D. 500 to 900), Sedentary (A. D. 900 to 1100), and Classic (A. D. 1100 until the Spaniards arrived). It is necessary to say that this cultural area has several names. Besides Southwestern U. S., some archaeologists recognize it as a part of the "Gran Chichimeca", others refer it as a portion of the "Noroeste de México", and others still see it as a fraction of "Oasisamerica" (Braniff 1989:72; Braniff 1993).

Figure 5-2. The Southwestern U. S. showing the archaeological sites discussed in the text.

Since the end of the nineteenth century, Southwestern archaeologists have postulated that the Southwest had some kind of relationship with Mesoamerica. Their views on this theme are varied. Some think that the Mesoamericans colonized the Southwest U. S., other consider that the links were weak and did not go beyond some isolated cultural features. In Eric Powell's opinion: Mesoamericans did not colonize the Southwest, and the finds of Mesoamerican remains only suggest some links with the Pre-Hispanic societies in Mexico (Powell 2005).

A common question in this theme is "What, in short, is the effect of Mesoamerica on Southwestern cultures, not on specific large sites?" (Gumerman 1978:24). There have been many attempts to answer this question. George J. Gumerman utilizes a model of long distance exchange, with low and high value variables. For him, the exchange between Southwestern communities and Mesoamerica probably was an example of a high-value exchange sphere with negative reciprocity, and included turquoise from north to south, and copper bells, macaws, and religious ceremonies in the other direction (Gumerman 1978:28). Robert Lister (1978) established a number of traits that were identified as, or presumed to be, Mesoamerican in Chaco Canyon archaeological sites. Between them, he mentioned shell objects like trumpets and beads. Unfortunately, the author did not offer biological identifications to indicate provenance. So, in my opinion, its mere presence is not enough to prove a Mesoamerican origin. For the U. S. archaeologists and some Mexicans scholars, the archaeological evidence for some kind of interaction is very tantalizing, and their studies and models have postulated that the long-distance exchange created a coherent framework for explaining a great many diverse facts. Nevertheless, they suggest that the proposition still requires much more testing before it should be regarded as demonstrated. "It also should become a basis for examining how such connectivity may have become a means for ideological ideas to flow back and forth between the American Southwest and West Mexico" (Wilcox *et al.* 2008:116).

From the universe of material elements, there are two forms of specific interest in this work. They are both of biological origin, and they are exchange objects from Mesoamerica to Southwestern United States: macaws and cacao. Macaws were studied by Lyndon L. Hargrave (1970). In his study, analyzing the birds bones, Hargrave established the presence of two different species of macaws: the Military Macaw (*Ara militaris*), whose nearest occurrence to the United States is the Mexican state of Sonora, and the Scarlet Macaw (*Ara macao*), restricted to the Mexican states of Tamaulipas and Oaxaca, although Phil C. Weigand has reported its presence in Jalisco (Wilcox *et al.* 2008:115). Cacao has been studied by Patricia L. Crown and W. Jeffrey Hurst (2009). They analyzed several fragments of ceramic vessels from Pueblo Bonito in Chaco Canyon, and detected the presence of theobromine, a marker for *Theobroma cacao*. With the biological identification, the authors established that "[r]elative to the Chaco area, the closest cacao cultivation at contact was in Central Mexico [...], including portions of northern Veracruz and Colima" (Crown and Hurst 2009:2111). As you can see, in both cases, the biological identifications obtained were a very important element in corroborating long-distance exchange. Therefore, the scientific name is an indispensable way of securely suggesting the existence of contacts between distant cultural areas. This quality must be applied to shells.

Regarding West Mexico, Meighan (1966:7-10) discussed point by point some cultural similarities between the archaeological site of Amapa, in the Mexican state of Nayarit, and the Hohokam culture. With regards to shells, the author mentioned the existence of genus *Glycymeris* bracelets. Following from this, he established examined the evidence for relationships between the Mesoamerican and southwestern culture areas including cultural patterns of general nature. These were used to argue for diffusion from one culture area to another (Meighan 1999:208).
Wilcox and his colleagues pointed out that the iron-pyrite mirrors with pseudo-cloisonné designs from earlier contexts in the Hohokam area came from West Mexico. In their opinion, these objects were product of a trade relationship between Jalisco and southern Arizona (Wilcox *et al.* 2008:167).
In the same vein, Paul Fish states that West Mexico had considerable interaction with Southwestern United States; mainly with the Hohokam. For Fish: "The Mexican iconography, or symbols, used in pottery and other designs [were] often shared with the Hohokam" (Harrison 2009). These contacts might have taken place in the Civano phase, *ca.* A. D. 1300 (Doyel 1993:44).

Typically, shells have not been used to analyze the process of interaction between Mesoamerica and the Southwestern United States. Shells are considered a luxury item, very frequently being found in burials as offerings or personal effects, associated with high rank individuals. If we accept Gumerman's (1978) proposal, mentioned above, then it is possible to utilize shells in the analysis of long distance exchange. Following from this, here I discuss this proposal using data on shells recovered in both cultural areas. I will deal with the chronology of the Post-Classic period of West Mexico, A. D. 1100 to 1536, which corresponds to the Classic period of the Southwestern United States I choose these times because the United States archaeologists propose that the Mesoamericanization process of the Southwestern United States occurred in this period (Wilcox, Weigand, Wood, and Howard 2008:120).

The model

My research on the Pre-Hispanic use of shells in ancient West Mexico (Gómez-Gastélum 2005) has permitted me to build a model to utilize in analysis. In general terms, it consists of the following steps:

Biological identification

It is necessary to ascertain the correct biological identification of the mollusks, in order to establish further features of the species used, like shape, color and habitat. Identification at the species level is ideal, or at least an identification to genus (Polaco 1991:18-19).

Contextualization

For all recovered shells it is necessary to know the relationships between the find and the archaeological context. An accurate correlation between the shells, the container, and the site, aids in establishing the social action that generated it.

Reconstruct the color

In this step, I follow the analysis of Berlin and Kay (1999, original edition 1969), who established an evolutive scale for basic color terms. The authors were primarily concerned with color perception, and did not deal with the issue of meaning. Building on their foundation, Anna Wierzbicka (1990:100) modified and amplified Berlin and Kay's original scheme, in order to include discussion about color meanings.

Review these elements in the regional worldview

Here, I deal with the symbolism of color, shells, the human body, and other elements. For the model, it is necessary to establish whether the shells were utilized in their original shapes or whether they were modified. If an object was elaborated, we need to ascertain the modified form or nature of representation.

West Mexico

In this part, I present the information about the presence of archaeological shells in sites of West Mexico (fig. 5-1, Table 5-1). I outline the species identified and, for some, I describe their archaeological context. I further assess information from the regional literature in order to present a broad overview of shells used in this cultural subarea. From A. D. 1100 to 1536, there are seven archaeological sites or regions that yielded shells or shells artifacts. From the Apatzingán region of the Mexican state of Michoacán, during the "Chila" phase (A. D. 1100 to the Spaniards arrival), in the site of "El Llano", Kelly found various shell objects and valves from different species. She reported unmodified shells of *Oliva* sp., *Oliva porphyria, Trivia radians* and *Morum tuberculosum;* all shells were used as pendants. Also recovered were shell strings made with winkles (Marginellidae), and atlatl handles elaborated with an unidentified species. Additionally, she found complete valves of *Trivia solandri*, as well as *Turritella* cf. *leucostoma*, *Oliva* sp., *Collumbella* cf. *fuscata*, *Thais* sp., and *Pinctada* sp. (Kelly 1947:114-119). These elements were found as part of the human burials. Unfortunately, we do not know if they were offerings or otherwise-associated paraphernalia (Kelly 1947:214-215).

For the Mexican state of Jalisco there are two regions with Pre-Hispanic shell materials: Tuxcacuesco and Sayula, both in the Jaliscan southern area. In the first, during Tolimán phase (A. D. 1100 to the Spaniards arrival), Kelly observed geometric carved pendants with circular and quadrangular shapes, some found at surface, but other that were part of the human burials. Also, she found *Glycymeris* cf. *gigantea* bracelets, and in some the umbo has been converted in a effigy head that the author qualified as "a far cry from the finely carved shell bracelets of the Arizona Hohokam". She reported the use of *Spondylus* sp. in several modified forms, including shaped pendants and oddments. She also recorded valves of freshwater mollusks of Family Unionidae, and *Unio* sp. With regards to seashells, there were specimens of species within the Cardiidae, Pectinidae, and Veneridae families, *Glycymeris* sp., *Pinctada mazatlanica, Nerita scabricosta, Littorina* sp., and *Turritella* sp. Some of these elements were found as burial paraphernalia in Potrero del Casco I and Paso Real sites (Kelly 1949:128-132).

For Sayula Basin sites, a Mexican-French archaeological team worked in the region between 1990 and 2000 and reported more than 170 archaeological sites. The sites mentioned here relate to the Amacueca phase (A. D. 1100 to Spaniards arrival). The shell materials came from both excavations and collections of surface finds. There are *Spondylus* sp. quadrangular earrings, with central openwork. These were hooked onto copper circles. There were also pectorals manufactured in nacreous species, maybe *Pinctada mazatlanica*. Other objects include necklaces and bracelets made with *Spondylus* beads strings, or with unmodified shells of *Olivella* sp., or *Prunum apicinum*. Also recovered were *Glycymeris gigantea* pendants and bracelets, and *Oliva* sp. strings maybe used as rattles (Gómez-Gastelum, unpublished manuscript, 2000). All of these objects were plentiful at the Fraccionamiento San Juan Atoyac, and Caseta sites, where they were used as burial paraphernalia.

In the Mexican state of Nayarit, the site of Amapa was explored by Meighan and a team of the University of California at Los Angeles (UCLA) during the 1960s. There are two phases relating to the Post-Classic period: Ixcuintla (A. D. 1000 to 1300) and Santiago (A. D. 1300 to 1400 or Spanish arrival). Here, not much shell was reported by the author. As burial paraphernalia, he collected about 17 *Glycymeris* cf. *gigantea* bracelets fragments (Meighan 1976:122). Like burial offerings, he stated that shell beads were infrequent finds but that they were clearly manufactured at the site. The material, found in association with burials, includes unworked shell fragments, disc beads, and square beads with notched edges. The shell species used were *Anadara* cf. *grandis, Spondylus* cf. *calcifer*, and *Pinna rugosa* (Meighan 1976:123). Finally, the author mentioned the use shells of genus *Donax* as food.

Another Nayaritan region of importance is Marismas Nacionales. Here, the Conchera phase (A. D. 900/1000 to Spanish arrival; Scott and Foster 2000:110) is of particular interest. From several sites very many shell objects were reported. These include pendants, pectorals and beads, all in a variety of shapes that were constructed in a massive red or orange unidentified species of shell. These objects were used as paraphernalia in the human burials. As offerings there were also specimens of a small gastropod, *Cerithidea montagnei* (Gill 1974:85-88).

Finally, in the Mexican state of Sinaloa, there are two sites with shell material. Of relevance is the Culiacán archaeological site and particularly the Early Culiacán I (A. D. 1050 to 1300), Middle Culiacán (A. D. 1300 to 1400), and Late Culiacán (A. D. 1400 to Spaniards arrival) phases (Schöndube 1980:128). Here Kelly reported that shell objects were scant. She described bracelets made from *Megapitaria aurantiaca, Chione gnidia*, and *Anadara grandis*. Pendants were constructed from *Agaronia testacea, Oliva spicata*, and *Polinices recluzianus*. Beads were manufactured with Serpulorbis sp. In all cases the shells were preserved in natural form. The only piece with a clear archaeological context was a zoomorphic pendant, with no biological identification, that represents a frog found in an olla burial (Kelly 1945:143-146).

The last archaeological site from West Mexico is Guasave. Here the Guasave phase (A. D. 1000 to 1450) is of interest. According to Carpenter (1996), shell artifacts were popular in burial offerings. There were beads, pendants, bracelets, mosaic plaques, and valves. The beads formed necklaces, anklets, belts, and were sewn into fabric. They took several forms: disc, bi-lobed, tooth-shaped, carved, globular, and tubular. Given their size, biological identification was not possible. Bracelets were made from *Glycymeris gigantea* and *Laevicardium elatum*. The pendants were made in *Ostrea* sp. The unmodified valves were *Laevicardium elatum, Anadara grandis* and *Ostrea* sp. (Carpenter 1996:264-266). Gordon F. Ekholm (1942:111) considered that "In general, the amount of shell work at Guasave is more reminiscent of the Hohokam culture of southern Arizona than of cultures of Central Mexico".

The Southwestern United States

At present, I have synthesized the information on shells from two archaeological traditions from Southwestern United States: Cerro de Trincheras and Hohokam (fig. 5-2, Table 5-1). Cerro de Trincheras is an archaeological site located in the Mexican state of Sonora. McGuire and Villalpando (1998:3), guided by 20 radiocarbon dates, suggest that the site was occupied between A. D. 1300 and 1450. Their excavations yielded a lot of marine shell, mainly manufacturing waste from the production of shell jewelry. However, they also found many fragments of finished items of jewelry and believe that artisans produced this jewelry primarily for local use (McGuire and Villalpando 1998:5). Villalpando stated that there was targeted selection of shell. She suggested the principal use of three bivalves genera and one genus of gastropod. The bivalves were *Glycymeris*, *Laevicardium* and *Trachycardium*, utilized to make beads and pendants. The gastropod was *Conus* sp., used in the production of rattles, rings, pendants, and beads (Vargas 2004:72; Villalpando 2000:539). Additionally, small marine snails such as various *Columbella, Nassarius, Olivella, Turritella,* and *Thais* species were used as beads or pendants (Villalpando 2000:539). Vargas (2004:74) states that, this site "appears to have intensively engaged in shell craft production during the Late Ceramic period, especially in the production of *Conus* ornaments". However, the author indicates that the sites surrounding Cerro de Trincheras were dedicated to the manufacture of *Glycymeris* artifacts including bracelets, pendants, and beads (Vargas 2004:75). Vargas (2004:73) further mentions that the *Conus* rattles "were commonly used as rattlers or shakers, often attached in groups to anklets, bracelets, and as dance regalia decorating clothing".

From Hohokam, I consider evidence from three places: Snaketown, Tonto Basin, and Marana community. Snaketown is an archaeological site located in south central Arizona, south of Phoenix, on the Gila River Indian Community (Center for Desert Archaeology 2010). Occupation spans from A. D. 500 to 1450. Here, Braniff, reanalyzing Gladwin's information, recorded the presence of eight worked species of mollusk. *Glycymeris gigantea* was used for bracelets and rings, *Strombus galeatus* for trumpets, *Aequipecten circularis* to make pendants, *Laevicardium elatum* in pendants and bracelets, *Glycymeris maculata* for bracelets, and *Spondylus princeps, S. calcifer,* and *Trivia solandri* for beads (Braniff 1989:47-56).

The Tonto Basin is a well-watered region in central Arizona. Clark (1998:1) described its position as "defined by the Salt River and Tonto Creek drainages, lies near the boundaries of the Hohokam, Mogollon, and Anazasi areas as they have traditionally been defined". In this region more than 300 human burials were excavated, "The majority of inhumations were associated with four early Classic period compounds (A. D. 1200 to 1325)" (Hall and Clark 1998:6). From these there were two major observations. First, the archaeologists only talk about two molluscan genera: *Glycymeris* and *Conus*. Later, the associations of shell artifacts with burials were considered with regards to burial sex and age. Regarding *Glycymeris*, Hall and Clark (1998:7) pointed out that large pendants and earrings were associated with males. Their forms included flying birds and dog medallion shapes. Thick armlets were worn above the left elbow by males. Personal ornamentation included thin bracelets worn below the left elbow and necklaces made of beads. Pendants seem to be exclusively associated with males; these were rarely found with women and children. Shell toads were the exception to this pattern. Several examples were only associated with young women of childbearing age and infants. With regards to *Conus*, the authors only stated that rattles were also rare and associated with males (Hall and Clark 1998:7).

Marana community is located in the Tucson Basin, adjacent to the Santa Cruz River, northwest of Tucson. Occupation spans from A. D. 1100 to 1400, in the Early Classic period (Fish, Fish, and Madsen 1992). Here, several whole seashells, as well partially carved shells of different species were found in excavations. Shell jewelry included bracelets, beads, earrings, pendants, and rattles (Bayman 1995:2). Also, at least, one specimen of a *Strombus* conch was found. This shell had the apex cut off, and then was used as musical instrument: specifically a trumpet. On the general use of the shell artifacts, Bayman stated: "Among historic Pueblo societies in northern Arizona and northern New Mexico, projectile points and shell jewelry were (and sometimes still are) worn as ornaments on costumes and clothing by participants in ceremonial dances" (Bayman 1995:4). Although Bayman did not identify any of the species used for jewelry, it was likely the same genera used in other Hohokam sites referred in this paper.

In general terms, Bayman (2002:70) assigns four functions to Hohokam jewelry: "1) material symbols of group membership and identity; 2) ritual performance paraphernalia; 3) instruments of power; 4) insignia of office". In association with the first category, he pointed to the shell bracelets and armrings made of *Glycymeris*. He linked the second category to shell beads, pendants, and *Conus* tinklers. He assigned the shell trumpets to the third category and to the fourth, he assigned the *Euvola vogdesi* pendants. About color he said:

Among many native North American societies, certain colors (i. e., white, red, blue, yellow, and black) are symbolically associated with specific deities and/or directions [...] The widespread use of white marine shells in Hohokam society, therefore, was potentially charged with deep religious meaning (Bayman 2002:82).

Finally, Bayman recognizes the use of the following genera and species in the Hohokam world: *Glycymeris* sp., *Conus* sp., *Laevicardium* sp., *Strombus galeatus*, S. *gracilior*, *Melongena patula, Muricanthus* sp., *Olivella incrassata* , and *Euvola vogdesi.*

Elements for the analysis

So far this paper has dealt with the biological identifications and the archaeological contexts provided for various sites under consideration. Now, it is necessary to address the remainding elements of the model. The anthropological study of color has strong roots in American anthropology. In the 1960s, Berlin and Kay (1999) proposed the existence of an evolutionary process in color perception, which was materialized in systems of basic color terms recognized and used by each society at an historical moment. They established seven stages that include from two to eleven basic color terms. Although Berlin and Kay worked with contemporary languages, they indicated that their propositions should apply also in the past. Of particular relevance to this study is stage IV of their model. This encompasses five terms, which include black, white, red, green or blue, and yellow (Berlin and Kay 1999:23). While Berlin and Kay were only concerned with color perception, the issue of meaning was tackled by Wierzbicka (1990). She considered several natural phenomena (like the day, the night, the fire, the sun, things that grow from the ground or the sky) within her discussion. Color in Mesoamerica was specifically analyzed by MacLaury (1986). He concludes that most of the languages of state-level societies in ancient Mesoamerica named five color categories (MacLaury 1986:61).

In Pre-Hispanic Mesoamerica color has deep symbolical meaning. Indeed, color is present in the cosmos origin myths. , Its importance is also evident through the actions of Mesoamerican gods where colors were linked with directions, personages, or objects. In Pre-Hispanic times, five colors were held to be sacred: yellow, red, white, blue, and black (Garibay 1985). Other sacred aspects of color were analyzed by Hosler (1994). She refers to the importance of golden and silvery metallic colors, associated with the Sun and the Moon as deities in Post-Classic Mesoamerica.

Further north, the indigenous peoples of the Southwestern United States recognize color symbolism too. Berlin and Kay (1999:73-74) indicate that the native languages existing in this zone have systems of the IV stage, with five basic color terms. Riley (1963) suggested that symbolic meanings were associated with colors at least as early as the 15th century. For these peoples, the sacred colors are yellow, white, red, black, and blue.

Other important characteristics, closely related to color, are the directions of the world or world quarters. In Mesoamerican beliefs, the universe was divided into four quarters and the center. Although these coincide with the cardinal directions, they are not orientation instruments, but instead they are fields where four cosmic forces – four gods that generate the cosmos equilibrium – engage in combat for universal domination (León-Portilla 2001:111-112). At a human level, several natural phenomena were associated with world quarters. It was believed that the rain or the wind could come from good or bad directions. Each world quarter was also associated with a color. Several Mesoamerican societies had their own order in the direction/color association, but all coincide in assigning red to the East quarter. Indeed, it is the place where the sun is born.

Groups of the Pre-Columbian Southwestern United States had beliefs about the world quarters too. According to McGee, Jr. (1985:280), the typical Pre-Columbian construction named "kiva": "represents the seven directions: north, west, south, east, nadir, zenith, and center". The contemporary Pueblo Indians of the Southwest had myths and games that are materializations of the world quarters beliefs. The Emergence myth or the Patol game, are examples of these (McGee, Jr. 1985). This cultural area associates the universe directions with color too. Like in Mesoamerica, each community had their own order, but they generally all associate white with East (Riley 1963:59-60).

Another important aspect is symbolism related to the human body. In 16th century Central Mexico, the Nahua people believed that human life was maintained by three animistic substances or 'souls', named *tonalli*, *teyolía*, and *ihíyotl* (López Austin 1996:217). These existed in the human body, in specific organs or places. The *tonalli* was situated in head, the *teyolía* in heart, and the *ihíyotl* in liver. These three substances were also found in the zones of body articulation and the calves. The behavior of these animistic substances, and the way in which they were cared for by the person, was a precondition of human life. In the Southwestern United States, I can find no equivalen information concerning this phenomenon. But, Fewkes (1896:161-162, 166) offered some data that suggest the existence of similar beliefs in this cultural area. Nevertheless, at this moment, I cannot say if Pre-Columbian views about human body in both areas were similar like approaches to color or the world quarters conception.

Discussion but no conclusion

For comparison between West Mexico and Southwestern U. S., it is necessary that both cultural regions present the conditions for the model referred paragraphs above. Here I will consider each step one by one.

Biological identification: In both areas we have specific identifications of the mollusks used by the Pre-Columbian societies. When we compare them it can be seen that only four species and four genera are shared by both cultural areas: *Glycymeris gigantea, Spondylus calcifer, Trivia solandri, Laevicardium elatum, Glycymeris* sp., *Olivella* sp., *Thais* sp., and *Turritella* sp., Table 5-1). However, this does not necessarily mean the existence of long-distance exchange. Keen (1971:55) observes that *Glycymeris gigantea* has a geographical distribution "confined to the Gulf of California area, from Bahía Magdalena, Baja California, to Acapulco, Mexico". Skoglund (2001:S14) extends this territory to Peña Negra, Piura, Peru. *Spondylus calcifer* is present from the Gulf of California to Ecuador and Peru (Keen 1971:96-98; Skoglund 2001:S32-S33). Keen (1971:487) states that *Trivia solandri* extends from "Southern California throughout the Gulf of California and south to Peru", but in Skoglund's opinion (2002:S85) its distribution goes from "upper Golfo de California to Mazatlán, Sinaloa, Mexico". *Laevicardium elatum* has a distribution from Southern California as far as Panama (Keen 1971:160), but Skoglund (2001:S53) states that there are "No established populations north of Laguna Ojo de Liebre, Baja California Sur, Mexico". In other hand, genus *Glycymeris*, in general, extends from the Gulf of California to Peru (Keen 1971:55-57; Skoglund 2001:S14-S15). Genus *Olivella* has the same distribution (Keen 1971:626-632; Skoglund 2002:S149-S150). Genus *Turritella*, following Keen (1971:391-394) has a wide distribution, from the Gulf of California to Peru. Skoglund (2002:S48-S49) agrees with this distribution. Finally, genus *Thais* extends from the Gulf of California to Chile (Keen 1971:549-550; Skoglund 2002:S118-S119).

In both areas we have information on the archaeological contexts in which the shells were found. They were used as paraphernalia and offerings in human burials.

From their scientific name, we can extrapolate that in West Mexico there are shells that are grouped in colors: red, white, yellow, darker, and mottled. In the United States Southwest, the genera and species are colored reddish, mottled, and white.

For West Mexico, the Mesoamerican worldview details elements that link the symbolism of color, world quarters, and the human body. When we connect these elements, we can propose an interpretation about the symbolism of shells and shell artifacts. In general terms, this symbolism could be linked with the representation of the universe. The world quarters were identified with colors, and they were also represented by elements like corn, clouds, arrows, and others. When archaeological shells are considered by color, it is possible to perceive the same patterns. However, at this moment, I do not have the same baseline information for the Southwestern United States. I am aware of the color symbolism, and the world quarters, but not about the human body symbolism.

In a comparative revision, I can make the following observations:

a) Due to the geographical distribution of the mollusks found in both cultural areas, I cannot confirm the existence of a shell long-distance exchange between West Mexico and Southwestern United States. In fact, several studies (Brand 1938; Braniff 1989; Pastrana and Villalpando 2011) have demonstrated that the main source of seashells for United States Southwest was the Gulf of California, while West Mexico utilized, among others, the Colima coast for the same purpose (Gómez-Gastélum 2004).

b) West Mexico utilized a wide variety of mollusks, while the Southwestern United States used a more restricted set of species. In archaeological contexts of West Mexico, I find the use of five families, 12 genera, and 20 species of mollusks. While in the Southwestern United States, I count only 11 genera, and 12 species (Table 5-1).

	West Mexico	U. S. Southwest
Genera	*Oliva* sp. *Thais* sp.* *Pinctada* sp. *Unio* sp. *Glycymeris* sp.* *Littorina* sp. *Turritella* sp.* *Spondylus* sp. *Olivella* sp.* *Donax* sp. *Serpulorbis* sp. *Ostrea* sp.	*Glycymeris* sp.* *Laevicardium* sp. *Trachycardium* sp. *Conus* sp. *Columbella* sp. *Olivella* sp.* *Turritella* sp.* *Thais* sp.* *Strombus* sp. *Muricantus* sp.
Species	*Oliva porphyria* *Trivia radians* *Morum tuberculosum* *Trivia solandri** *Columbella fuscata* *Glycymeris gigantea** *Pinctada mazatlanica* *Nerita scabricosta* *Prunum apicinum* *Anadara grandis* *Spondylus calcifer** *Pinna rugosa* *Cerithidea montagnei* *Megapitaria aurantiaca* *Chione gnidia* *Agaronia testacea* *Oliva spicata* *Polinices recluzianus* *Laevicardium elatum**	*Glycymeris gigantea** *Strombus galeatus* *Aequipecten circularis* *Laevicardium elatum** *Glycymeris maculata* *Spondylus princeps* *Spondylus calcifer** *Trivia solandri** *Euvola vogdesi* *Strombus gracilior* *Melongena patula* *Olivella incrassata*

Table 5-1. Shells present in archaeological contexts from West Mexico and United States Southwest. Asterisks mark when the genera or species are present in both cultural areas.

c) In West Mexico the use of shells included paraphernalia and human burial offerings, approximately with the same importance. In Southwestern United States it seems that shell use outside of mortuary contexts was of greater importance, mainly in dance regalia to decorate clothing.

d) Similarly, West Mexico and the Southwestern United States had its own color and world quarters symbolism, that assigns a color to the different directions. In West Mexico red is associated with the direction where the sun is born, while in the Southwestern United States is white. According to different studies, red shells and white shells were very important in each area (Bayman 2002; Gómez-Gastélum 2005). For contemporary times, Bahti (1999:Chart B), points out that the Navajo tribe associates white shells with the East, and Abalone shells with the West.

e) Both in West Mexico and in the Southwestern United States only one particular representation stands out. I agree with Ekholm (1942:111) when he says: "[d] isc beads, pendants of various kinds, and partially worked shells occur practically through Middle America and the Southwest, and consequently they are items of little diagnostic value". However, the frog representation exists in both cultural areas, and in these territories batrachians were culturally important, where they were associated with the aquatic world and fertility. It is thus unsurprising that were represented in shell objects, and offered as funereal gifts. But this feature does not mean a direct ideological contact between West Mexico and Southwestern United States Due to the existence of local toads and frog species in each cultural area, the symbolism associated with these animals could have appeared independently from direct observation of their life cycle.

After consideration of the various lines of evidence, I do not see evidence for long-distance exchange between West Mexico and the Southwestern United States. The sites within each cultural area obtained shell from local sources, and, except for general coincidences, I see no parallels in the shell objects used. Indeed, I find a more restrictive approach to the use of shell in the Southwestern United States when compared to West Mexico. With regards to color and world quarters beliefs, there are similarities between both areas, but the variability in the color-direction pairings is great within both West Mexico and the Southwestern U. S. In addition to this, following Riley (1963) the antiquity of color-world quarters beliefs is earlier in Mesoamerica than in the Southwestern United States.

So how far into the symbol structure or cosmology of the United States Southwest did Mesoamerican concepts penetrate, and were shells and their color associations part of this penetration? I consider this to be an open question. The existence of contacts and long-distance exchange between Mesoamerica and Southwestern U. S. cannot be questioned, as evidenced by the verifiable presence of Mesoamerican macaws and cacao in the Southwest, or Southwestern turquoise in Mesoamerica. These facts prove an economic relationship existed. However if we turn to the archaeological contexts, the original symbolism does not seem shared. In my opinion, each cultural area integrated the long-distance exchange objects to conform to their own cosmovision. If the various societies involved in this long-distance exchange relationship knew the original symbolic contexts of the objects, and they wanted retain these, it seems they modified to the usages to adapt them to their worldview. The shell studies undertaken in Southwestern United States share these characteristics: they are predominantly dealt with in economic terms and they have not generally considered issues of symbolism in this field of study. For these reasons, there are currently few comparable studies of shell use between West Mexico and the United States Southwest. These cannot confirm a relationship between both cultural areas, as was suggested by classical southwestern archaeologists. As far as the field of archaeological mollusks is concerned, I cannot say that Pre-Hispanic West Mexican perceptions about shells and their colors were part of a Mesoamerican ideological penetration into the Southwestern United States.

To close, the issue of shell symbolism in the Southwestern United States remains an open question. Refocused study will permit a better understanding of the Pre-Columbian Southwestern United States societies and will permit discussion on a more solid basis when assessing the possibility of Mesoamerican influence in the Southwestern United States.

Acknowledgments

I want to thank to Cristina Ramírez-Munguía her support in the bibliography localization. To Valentina Marín-Muñoz for the first English version revision of this text. To Peter Jiménez-Betts for his revision of English language. From two anonymous peer-reviewers, I thank their comments and suggestions. Of course, misinterpretation and errors are my responsibility.

References

BAHTI, MARK. 1999. Spirit in the stone. *A handbook of Southwest Indian animal carvings and beliefs.* Tucson, AZ: Rio Nuevo.

BAYMAN, JAMES M. 1995. The trade and manufacture of shell and obsidian in Classic Hohokam Society. *Archaeology in Tucson* 9(1):1-5.

——. 2002. Hohokam craft economies and the materialization of power. *Journal of Archaeological Method and*

Theory 9(1):69-95.

BERLIN, BRENT, and PAUL KAY. 1999 (1969). *Basic color terms. Their universality and evolution.* Stanford, CA: CSLI Publications.

BRAND, DONALD D. 1938. Aboriginal trade routes for sea shells in the Southwest. *Yearbook of the Association of Pacific Geographers* 4:3-10.

BRANIFF, BEATRIZ. 1989. *Arqueomoluscos de Sonora, Noroeste y Occidente de Mesoamérica.* Mexico City, Mexico: Instituto Nacional de Antropología e Historia.

——. 1993. The Mesoamerican Northern frontier and the Gran Chichimeca. In *Culture and contact. Charles C. Di Peso's Gran Chichimeca.* Anne I. Woosley and John C. Ravesloot, eds. Pp. 65-82. Dragoon, AZ: Amerind Foundation.

CARPENTER, JOHN PHILLIP. 1996. El Ombligo en la labor. Differentiation, interaction and integration in prehispanic Sinaloa, Mexico. PhD dissertation, The University of Arizona, Tucson.

Center for Desert Archaeology (CDA). 2010. Snaketown Artistic Heritage Project. http://www.cdarc.org/what-we-do/past-projects/snaketown-artistic-heritage-project/ (accessed August 8, 2010).

CLARK, JEFFERY J. 1998. The Tonto Basin revisited: The Tonto Creek Archaeological Project. *Archaeology in Tucson* 12(4):1-3.

CROWN, PATRICIA L., and W. JEFFREY HURST. 2009. Evidence of cacao use in the Prehispanic American Southwest. *Proceedings of the National Academy of Sciences of the United States of America* 106(7): 2110-2113.

DOYEL, DAVID E. 1993. Interpreting prehistoric cultural diversity in the Arizona desert. In *Culture and contact. Charles C. Di Peso's Gran Chichimeca.* Anne I. Woosley and John C. Ravesloot, eds. Pp. 39-64. Dragoon, AZ: Amerind Foundation.

EKHOLM, GORDON F. 1942. *Excavations at Guasave, Sinaloa, Mexico.* Anthropological Papers, vol. 38, part 2. New York City: The American Museum of Natural History.

FEWKES, J. WALTER. 1896. The prehistoric culture of Tusayan. *American Anthropologist* 9(5): 151-173.

FISH, SUZANNE K., PAUL R. FISH, and JOHN H. MADSEN. 1992. An introduction to time, place, and research. In *The Marana Community in the Hohokam world.* Suzanne K. Fish, Paul R. Fish, and John H. Madsen, eds. Anthropological Papers 56. Tucson: The University of Arizona Press. http://www.uapress.arizona.edu/onlinebks/Fish/chapter1.htm (accessed August 8, 2010).

GARIBAY, ÁNGEL MA., ed. 1985. *Teogonía e historia de los mexicanos. Tres opúsculos del siglo XVI.* Mexico City, Mexico: Editorial Porrúa, S. A.

GILL, GEORGE W. 1974. Toltec-Period burial customs within the Marismas Nacionales of Western Mexico. In *The archaeology of West Mexico.* Betty Bell, ed. Pp. 83-105. Ajijic, Mexico: Sociedad de Estudios Avanzados del Occidente de México, A. C.

GÓMEZ-GASTÉLUM, LUIS. 2004. Las conchas marinas en el Occidente de México durante el Clásico y el Postclásico temprano. In *Bienes estratégicos del Occidente de México. Producción e intercambio.* Eduardo Williams, ed. Pp. 229-260. Zamora, Mexico: El Colegio de Michoacán.

——. 2005. Conchas y caracoles en el antiguo occidente de México. Un ensayo de antropología simbólica. Tesis de doctorado en antropología. Escuela Nacional de Antropología e Historia, México.

GUMERMAN, GEORGE J. 1978. Regional variation in the Southwest and the question of Mesoamerican relationships. In *Across the Chichimec Sea. Papers in honor of J. Charles Kelley.* Carroll L. Riley and Basil C. Hedricks, eds. Pp. 22-33. Carbondale: Southern Illinois University Press.

HALL, SUSAN D., and JEFFERY J. CLARK. 1998. Life and death along Tonto Creek. *Archaeology in Tucson* 12(4):6-7.

HARGRAVE, LYNDON L. 1970. *Mexican macaws. Comparative osteology and survey of remains from the Southwest.* Anthropological Papers 20. Tucson: The University of Arizona Press.

HARRISON, JEFF. 2009. Archaeologists mark 75-year anniversary of seminal Hohokam excavations. *UANews,* July 12. http://uanews.org/node/26098.

HOSLER, DOROTHY. 1994. *The sounds and colors of power. The sacred metallurgical technology of Ancient West Mexico.* Cambridge, MA: The MIT Press.

KEEN, A. MYRA. 1971. *Sea shells of tropical West America. Marine mollusks from Baja California to Peru.* Stanford, CA: Stanford University Press.

KELLY, ISABEL. 1945. *Excavations at Culiacán, Sinaloa.* Ibero-Americana 25. Berkeley: University of California Press.

——. 1947. *Excavations at Apatzingan, Michoacan.* Publications in anthropology 7. New York: Viking Fund.

——. 1949. *The archaeology of the Autlán-Tuxcacuesco area of Jalisco. II: The Tuxcacuesco-Zapotitlán zone.* Ibero-Americana 27. Berkeley: University of California Press.

LISTER, ROBERT H. 1978. Mesoamerican influence at Chaco Canyon, New Mexico. In *Across the Chichimec Sea. Papers in honor of J. Charles Kelley.* Carroll L. Riley and Basil C. Hedricks, eds. Pp. 233-241. Carbondale: Southern Illinois University Press.

LEÓN-PORTILLA, MIGUEL. 2001. *La filosofía náhuatl, estudiada en sus fuentes.* Mexico City, Mexico: Universidad Nacional Autónoma de México, 9th ed.

LÓPEZ AUSTIN, ALFREDO. 1996. *Cuerpo humano e ideología. Las concepciones de los antiguos nahuas.* Mexico City, Mexico: Universidad Nacional Autónoma de México.

LÓPEZ AUSTIN, ALFREDO and LEONARDO LÓPEZ LUJÁN. 1996. *El pasado indígena.* Mexico City: El Colegio de México, Fideicomiso Historia de las Américas, & Fondo de Cultura Económica.

MACLAURY, ROBERT ETHAN.1986. Color in Mesoamerica, vol. I: A theory of composite categorization. PhD dissertation, University of California, Berkeley.

MCGEE, Jr., HAROLD FRANKLIN. 1985. Playing at knowing: *Patol* and Pueblo world view. In *Contributions to the archaeology and ethnohistory of Greater Mesoamerica.* William J. Folan, ed. Pp. 263-292. Carbondale: Southern Illinois University Press.

MCGUIRE, RANDALL H. 1989. The greater Southwest as a periphery of Mesoamerica. In *Centre and periphery. Comparative studies in archaeology.* T. C. Champion, ed. Pp. 40-66. London: Routledge.

MCGUIRE, RANDALL H. and ELISA VILLALPANDO. 1998. Cerro de Trincheras: A PreHispanic terraced town in Sonora, Mexico. *Archaeology in Tucson* 12(1):1-5.

MEIGHAN, CLEMENT W. 1966. Influencias entre culturas de Nayarit, México y de Arizona, E. U. A. *Eco* 24:7-10.

——. 1976. *The archaeology of Amapa, Nayarit.* Monumenta Archaeologica, no. 2. Los Angeles: The University of California.

——. 1999. The Mexican west coast and the Hohokam region. In *The Casas Grandes World.* Curtis F. Schaafsma and Carroll L. Riley, Eds. Pp. 206-212. Salt Lake City: The University of Utah Press.

PASTRANA, MAYELA, and ELISA VILLALPANDO. 2011. Espirales en el desierto. La manufactura de ornamentos en conchas del Golfo de California en el Noroeste de México. In *Moluscos arqueológicos de América.* Adrián Velázquez-Castro, Emiliano Ricardo Melgar-Tísoc, and Luis Gómez-Gastélum, eds. Estudios del Hombre, no. 29. Guadalajara, Mexico: Universidad de Guadalajara. Forthcoming.

POLACO, ÓSCAR J. 1991. La fauna en el Templo Mayor, una aproximación metodológica. In *La fauna en el Templo Mayor.* Óscar J. Polaco, Ed. Pp. 15-32. Mexico City, Mexico: Instituto Nacional de Antropología e Historia.

POWELL, ERIC A. 2005. The turquoise trail. *Archaeology* 58(1): 24-29. *Academic Search Complete.* EBSCO*host* (accessed April 21, 2011).

RILEY, CARROLL L. 1963. Color-direction symbolism. An example of Mexican-south-Western contacts. *América Indígena* 23(1):49-60.

SCHÖNDUBE B., OTTO. 1980. Época prehispánica. In *Historia de Jalisco. I. Desde los tiempos prehistóricos hasta fines del siglo XVI.* José Ma. Murià, ed. Pp. 111-130. Guadalajara, Mexico: Unidad Editorial del Gobierno de Jalisco.

SCOTT, STUART D. and MICHAEL S. FOSTER. 2000. The prehistory of Mexico's Northwest coast. A view from the Marismas Nacionales of Sinaloa and Nayarit. In *Greater Mesoamerica. The archaeology of West and Northwest Mexico.* Michael S. Foster and Shirley Gorenstein, eds. Pp. 107-135. Salt Lake City: The University of Utah Press.

SKOGLUND, CAROL. 2001. Panamic province molluscan literature. Additions and changes from 1971 through 2000. *The Festivus 32* (suppl.).

——. 2002. Panamic province molluscan literature. Additions and changes from 1971 through 2001. *The Festivus* 33 (suppl.).

VARGAS, VICTORIA D. 2004. Shell ornaments, power, and the rise of the Cerro de Trincheras. Patterns through time at Trincheras sites in the Magdalena River Valley, Sonora. In *Surveying the archaeology of Northwest Mexico.* Gillian E. Newell and Emiliano Gallaga, eds. Pp. 65-76. Salt Lake City: The University of Utah Press.

VILLALPANDO, MARÍA ELISA. 2000. Conchas y caracoles. Relaciones entre nómadas y sedentarios en el noroeste de México. In *Nómadas y sedentarios en el Norte de México. Homenaje a Beatriz Braniff.* Marie-Areti Hers, José Luis Mirafuentes, María de los Dolores Soto, and Miguel Vallebueno, eds. Pp. 525-546. Mexico City, Mexico: Universidad Nacional Autónoma de México.

Wheat, Joe Ben. 1954. Southwestern cultural interrelationships and the question of area co-tradition. *American Anthropologist* 56(4):576-586.

Wierzbicka, Anna. 1990. The meaning of color terms: semantics, culture, and cognition. *Cognitive Linguistics* 1(1):99-150.

Wilcox, David R., Phil C. Weigand, J. Scott Wood, and Jerry B. Howard. 2008. Ancient cultural interplay of the American Southwest in the Mexican Northwest. *Journal of the Southwest* 50(2):103-206.

6 - OCCURRENCES OF EXOGENOUS FRESHWATER MUSSEL SHELLS (BIVALVIA: UNIONOIDA) DURING THE PRECOLUMBIAN CERAMIC AGE OF THE LESSER ANTILLES

Nathalie SERRAND
INRAP DOM TOM, Route de Dolé, 97113 Gourbeyre, Guadeloupe, UMR 7209 CNRS-MNHN Paris,
nathalie.serrand@inrap.fr

Kevin S. CUMMINGS
Illinois Natural History Survey, University of Illinois, 607 E. Peabody Dr., Champaign, IL 61820 USA,
ksc@inhs.uiuc.edu

Abstract : *Various exchange systems involving shells are known to have existed in prehistoric as well as sub-recent societies, including island communities. In the Lesser Antilles, the pre-Columbian societies of the Ceramic age (5th BC - AD 15th centuries) developed comparable short- and long-distance exchange systems among islands and with the South American mainland. These systems are mainly inferred from the presence at island sites of raw and modified materials (stone, ceramic, animals) known to occur or be produced only in some of the Lesser Antilles islands or South America. With regards to shells, the identification of a few worked specimens of freshwater mussels shells (Unionoida) in a growing number of contexts of the Lesser Antilles has drawn our attention. The identified species,* Prisodon syrmatophorus, Anodontites infossus, Lamproscapha ensiformis, *are not known to occur on any of the Lesser Antillean islands but are found on the South American mainland. These data suggest that exogenous shells of a continental origin circulated through interaction networks as elements of economic or symbolic value. It adds to the evolving picture of the socio-economic dynamics of the Lesser Antillean Ceramic age groups, and more largely, of the various interlocked interaction spheres that were involved in broader pan-Caribbean network systems.*

Keywords : *Pre-Columbian, Lesser Antilles, Exogenous Unionoids, Ornements, Exchanges*

Introduction

Complex cultural processes took place during the pre-Columbian occupation of the Lesser Antilles islands (fig. 6-1A, B) which, on current reckonings, lasted 7500 years. In particular, during the Ceramic age, the occurrence at island sites of raw and modified materials (stone, ceramic, animals) known to occur or be produced only in some of the Lesser Antilles islands or South America point to the existence of elaborate systems of exchange and cultural interactions between the Lesser Antillean groups and with the South American mainland.

Chrono-cultural context

Following a long period of settlement by preceramic groups of mobile fishers and gatherers beginning around the seventh millennium B.C. (Boomert 2000, Keegan 1994), the Ceramic age starts, from about 500 B.C, with the arrival, in the Lesser Antilles (fig. 6-1A, B), of horticultural, pottery-making peoples. Known in the mainland (Venezuela, Lower Orinoco; Rouse and Cruxent 1963) as the Saladoid series, these peoples subsequently migrated from the Venezuelan Orinoco Basin (Roosevelt 1980:193-196) into the Guianas and to the northern coasts of Venezuela, developing here a Cedrosan Saladoid subseries (Oliver 1989; Rouse 1986, 1992). Typical ceramics of this subseries are subsequently found in Trinidad, Tobago, and most of the Lesser Antilles as far north as the eastern tip of Hispaniola (Bonnissent 2010; Keegan 1994, 2000; Rouse 1986, 1992:77-90; Siegel 1991; Veloz Maggiolo 1991; fig. 6-1). During the early Ceramic age, an overall cultural unity appears in the ceramic assemblages (Rouse 1992) along with similarities in the prevailing settlement patterns and in the village economies, the latter combining horticulture, hunting, fishing and food collecting (de France, Keegan, and Newsom 1996; Petersen 1997; Wing and Wing 1995). Nevertheless, more or less strong cultural and stylistic influences can be seen, especially after ca. A.D. 300 (Barrancoid influences in the southern Lesser Antilles), which suggest interactions with various groups of the mainland (Boomert 2000:217-251). After ca. A.D. 600/800 and up to the contact period (1492), a regional diversification in ceramic assemblages and styles, a substantial growth in the number of known sites, along with colonization of the entire Lesser and Greater Antilles and the Bahamas initiate the late Ceramic age (Ostionoid and Troumassoid series). It appears as a time of divergent local developments, population movements, and an increase in social complexity (traditionally associated with the Taíno societies of the Greater Antilles), none of which are yet fully understood (Keegan 2000).

As substantial research on the topic of mobility, interaction and exchange, conducted over the last decades, has shown, multiple lines of evidence indicate that dynamic

Figure 6-1. Location of the Lesser Antilles and Lesser Antillean islands and sites where archaeological Unionoida fragments have been found.

interactions (i.e. movements of populations, ideas, goods, etc.) were operating during the early and late Ceramic ages among various areas of the Antilles and continental America (Curet and Hauser 2011; Hofman, Bright, Ramos 2010; Hofman and van Duijvenbode 2011; Rodriguez 1991; Watters 1997).

In the Lesser Antilles, they are particularly evidenced by the presence of exogenous materials, resources or specific technological products in the islands sites: semi-precious lithics (carnelian, amethyst, turquoise) with a continental origin (Boomert 1987; Harlow *et al.* 2006); local stone materials with specific restricted origins (i.e. flint; Cody 1993; Knippenberg 2007, 2011; Rodriguez Ramos 2011); and various other ceramic, bone and guanine artifacts implying the circulation of raw materials and finished products among the archipelago and with the southern American continent (see Hofman, Bright, Ramos 2010; Hofman and van Duijvenbode 2011). In addition, enduring and recursively revitalized south-American

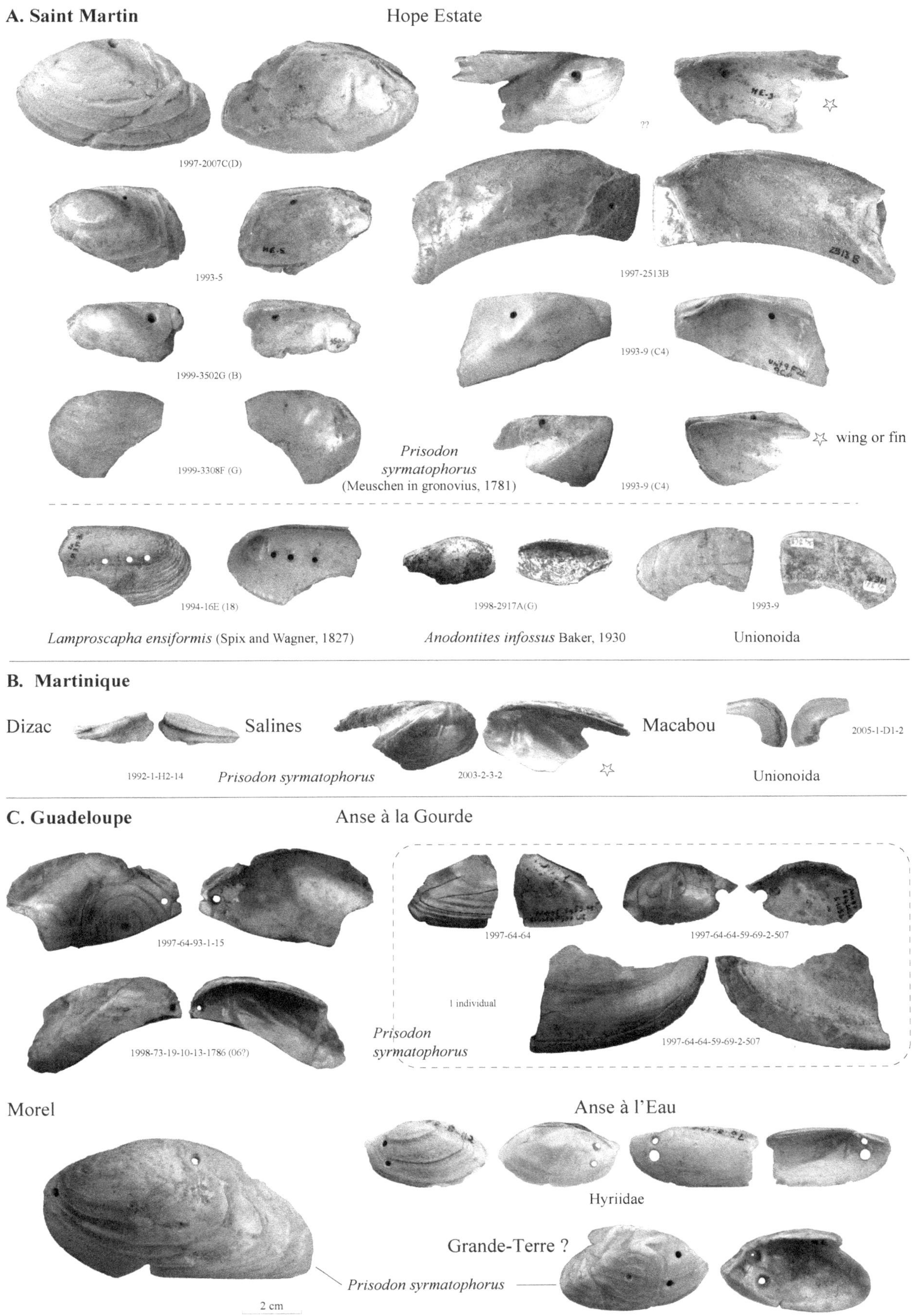

Figure 6-2. Specimens of the archaeological Unionoida fragments ; all pictures © N. Serrand.

A *Prisodon syrmatophorus* (Gmelin, 1791) Family Hyriidae, Subfamily Hyriinae

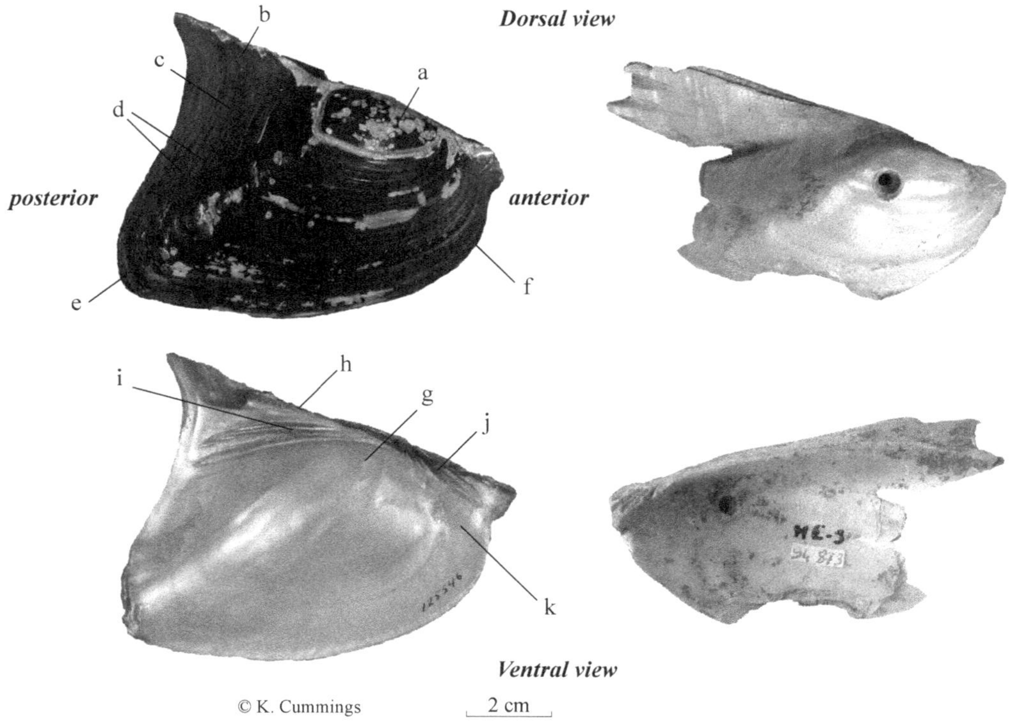

a. umbo: prominent raised area along dorsal edge, usually eroded - b. dorso posterior ridge - c. dorso posterior slope - d. lamellar ridges; e. posterio-ventral end - f. anterior margin - g. beak cavity: recessed area on the inside of the valve, under the beak - h. hinge: area of connection between the two valves, includes connective tissue and hinge teeth - i. lateral teeth: thin elongate teeth located along the hinge - j. pseudocardinal teeth: short stout teeth located below or just anterior to the beak - k. anterior adductor scar.

B *Anodontites infossus* Baker, 1930
Family Mycetopodidae
Subfamily Anodontitinae

C *Lamproscapha ensiformis* (Spix and Wagner, 1827), Family Mycetopodidae
Subfamily Anodontitinae

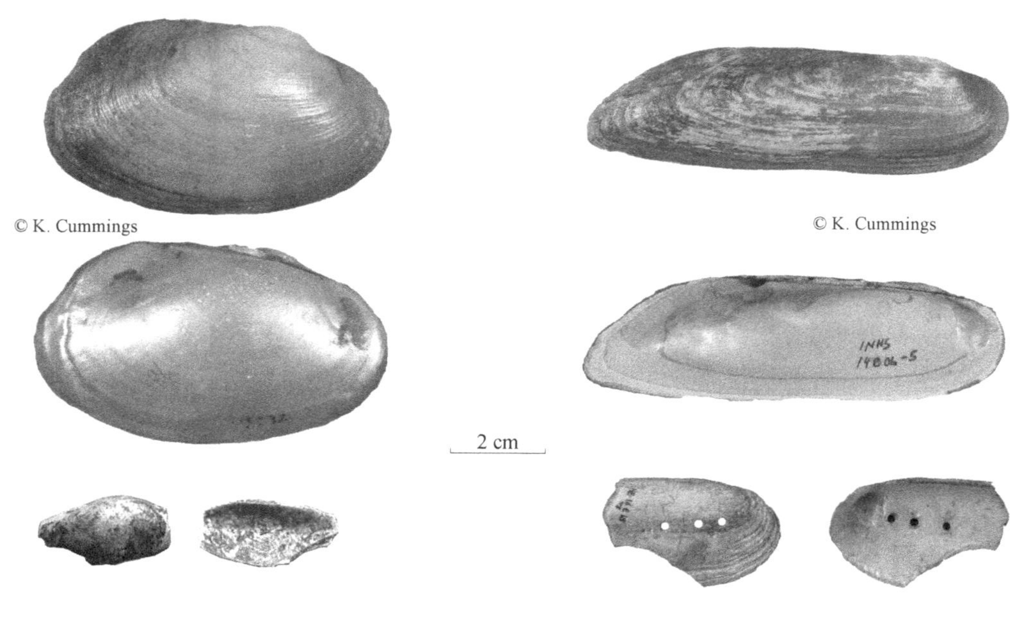

Figure 6-3. Characteristics of the present day and archaeological identified Unionoida.

legacies are occasionally introduced into the subsistence systems (translocated South American mammals and semi-domesticated plants; Newsom and Wing 2004) and iconographic themes of the island Ceramic groups.

Additional data now evidence the occurrence of shell materials of exogenous origin in Ceramic age sites of the Lesser Antilles.

Exogenous shells in a few early and late Ceramic age sites of the Lesser Antilles

It all started with the discovery of a few specimens at four Ceramic sites (Serrand 2001). Eleven specimens were found at the early Saladoid site of Hope Estate (fig. 6-1C and fig. 6-2A), occupied between the 6th century BC and the 8th century AD in St Martin (Bonnissent 2010; Henocq and Petit 1999). Three additional specimens were recorded from three different sites in Martinique (fig. 6-1C and fig. 6-2B), occupied between the 5th and 15th centuries AD: the modified / late Saladoid Dizac site (Vidal 1999), the Troumassoid site of Les Salines (Bérard 2002), and the Troumassoid / Suazoid site of Macabou (Grouard, Serrand, and Bérard 2006).

At these sites, numerous remains attest to the exploitation of local marine molluscs, for food and artefact manufacture (Serrand 2007). But these 14 fragments of nacreous bivalve shells, all seemingly worked, stood out and quickly appeared as being rare representatives of freshwater mussels of the Order Unionoida.

Although altered or fragmentary, most specimens exhibit human-made modifications. Most of them have conical or biconical perforations (fig. 6-2). One of the Hope Estate specimens has three aligned perforations associated with a pigmented line, while another unperforated fragment is incised (fig. 6-2A). Finally, three specimens show a similar shape with their posterior end likely cut into a wing or fin, sometimes accentuated by incisions (fig. 6-2A).

Further identification of these freshwater mussel fragments was a difficult task given unionoids in the Neotropics form a diverse group of species for which distributions and characteristics are poorly known (Johnson 1981; Haas 1969a, 1969b; Olsson and Wurtz 1951; Ortmann 1921; Parodiz and Bonetto 1963). Complicating matters, most studies of South American and Caribbean freshwater bivalves were conducted nearly a century ago (Baker 1930; Ortmann 1921; von Martens 1900), and the geographical distributions of South American freshwater mussels are poorly known due to a lack of adequate collections.

Recent research by the second author has furthered the knowledge of northern South American unionoid systematics and ecology (Cummings 2003; Graf 2000; Graf and Cummings 2006, 2007). By conducting field work throughout Venezuela during the past twenty years, the second author and his colleagues added over 65 collection sites and thousands of additional specimens to the collection, clarifying some distributional and taxonomic problems. Twenty-three species of Etherioidea have thus been reported from Venezuela including 9 species in the family Hyriidae (2 *Castalia*, 2 *Diplodon*, 2 *Rhipidodonta*, 2 *Prisodon*, and 1 *Triplodon*); and 14 in the family Mycetopodidae (10 *Anodontites*, 1 *Lamproscapha*, 2 *Mycetopoda* and 1 *Tamsiella*).

The second author identified all archaeological shells and all were found to be members of the superfamily Etherioidea. Further examination revealed that 10 of the shells (8 from Hope Estate, 1 from Dizac, 1 from Macabou) exhibited common characteristics (fig. 6-2A, B and fig. 6-3A): full, low umbos; a strong, angular dorso-posterior ridge; lamellar ridges stretching behind the umbo to the posterior end; a long and straight hinge; and well-formed and compressed pseudocardinal teeth. These 10 pieces were identified as members of the Family Hyriidae, Subfamily Hyriinae and species *Prisodon syrmatophorus* (Meuschen in gronovius, 1781) (fig. 6-3A). Two other specimens from the Hope Estate site belonged to the family Mycetopodidae (fig. 6-2A, fig. 6-3B and C): one is flattened, lacks a pronounced posterior ridge, has a toothless hinge, and most closely resembles *Anodontites infossus* Baker, 1930. The other has an elongate shape that is only found in one species, *Lamproscapha ensiformis* (Spix and Wagner, 1827). Two further fragments (Hope Estate, Macabou) lack diagnostic characteristics and can only be attributed to the order.

Where did these exogenous shells come from?

Unionoid bivalves are widely distributed in the Neotropics but absent from the Lesser Antilles (Haas 1969a; Maze 1883:5). In the Greater Antilles, they have been found only in Cuba, in the western part of the island (Pinar del Rio and La Habana provinces ; Arango 1865; Johnson 1981:270-280; Morelet 1849, 1851; von Martens 1900), where two endemic species are known, *Nephronaias scamnata* (Morelet 1849) and *Nephronaias gundlachi* (Dunker 1858) (Arango 1865:144; Johnson 1981:270-275, 275-278; Lea 1859:77; Pointier, Yong, and Gutiérrez 2005; von Martens 1900:507-508). However, the generic placement and phylogenetic relationships of the Cuban unionoids are uncertain and are in need of further study. Direct comparisons of the two species known in Cuba (specimens made available by Dr. J.-P. Pointier and by the Illinois Natural History Survey, INHS Nos 21712 - 21714), with the archaeological pieces showed no similarities. No unionoids have been reported for the Dominican Republic, Puerto Rico (Crosse 1891:69-211, 1892:5-71, van der Schalie 1948), or Jamaica (Vendryes 1899:590-607). The only other Caribbean island with mussels is Trinidad, where two species in the family Mycetopodidae are present with one supposed endemic *Anodontites leotaudi* (Guppy, 1864) and *Mycetopoda siliquosa* (Spix and Wagner, 1827) also found in Venezuela.

Examination by K. Cummings thus revealed that the archaeological families, genera and species are only found in South America (Johnson 1981:280-287). Of the two archaeological species in the family Mycetopodidae, *Anodontites infossus*, in particular, was described from a tributary of the Rio Aroa (fig. 6-4). Specimens of *Prisodon syrmatophorus* (Meuschen in Gronovius, 1781) are known from the Orinoco River system (fig. 6-4) (Academy of Natural Sciences, Philadelphia ANSP no. 125546; MNHN uncat.; INHS 16980, Bolivar State, University of Colorado 36453 and 36454, both Anzoategui State) and *Anodontites infossus* Baker, 1930 was described from a tributary of the Rio Aroa in the state of Yaracuy (Baker 1930; INHS no. 17032; INHS 17033; ANSP 147732; University of Michigan, UMMZ 48007,112619, and 112620; Museum of Comparative Zoology, Harvard University MCZ 63010; Carnegie Museum, CM 61.97).

Therefore, based on comparisons with the freshwater mussels found in northern South America, the archaeological specimens very likely had a continental origin and may have come directly from the Venezuelan drainages mentioned above. In any case, they did not originate from any island in the Lesser Antilles, the Greater Antilles, or Cuba.

A large scale phenomenon in the Lesser Antilles?

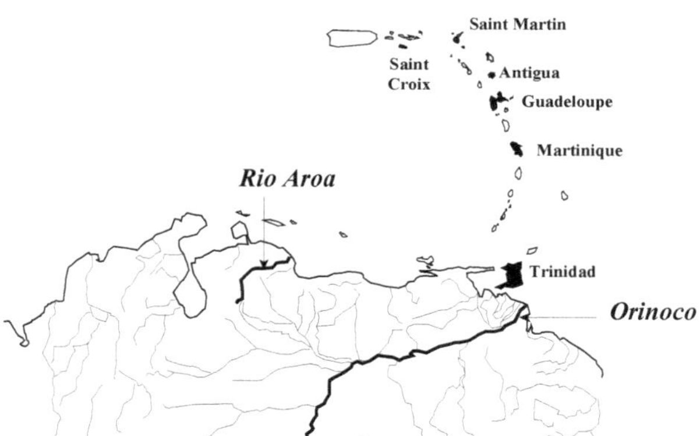

Figure 6-4. Location of the possible natural source areas for the archaeological Unionoida in northern South America.

This discovery shed new light on the former sparse unpublished or unanalyzed indications of such specimens occurring in a few other Ceramic sites on the islands of St. Croix (U.S. Virgin Islands), Antigua, Guadeloupe, and Trinidad (fig. 6-1).

Indeed, in 1979, archaeologists Garry S. Vescelius and Linda S. Robinson communicated about close to 200 nacreous shell artefacts from St Croix, U.S. Virgin Islands, from the Richmond and St Georges middle Saladoid sites, dated to between AD 350-800 (they are part of the Folmer Andersen's Collection, property of the National Park Service). Their communication (Vescelius and Robinson, unpublished manuscript 1979) at the International Congress for the Study of the Pre-Columbian Cultures of the Lesser Antilles, in St Kitts, was never published. We only have a manuscript without illustration describing specimens carefully worked with perforations, geometric designs, and animal and human representations. They were, at the time, considered as "unlikely to have been of local origin" and identified by David H. Stansbery, of the Ohio State University, as South American freshwater bivalves of the genus *Prisodon* Schumacher, 1817 (Unionoida: Hyriidae) known from the Amazon and Orinoco basins (Vescelius and Robinson 1979, unpublished manuscript).

Vescelius and Robinson also pointed out two other unrecognized unionoid remains: a carved pendant from the early Saladoid Indian Creek site (Rouse and Faber Morse 1999), in Antigua, appearing in a publication by Olsen in 1974 (1974:46, Fig. 71); and a perforated artefact from Guadeloupe, appearing in a publication by Clerc in 1974 (1974:129-130, Fig. D-7), coming from an unidentified site assigned to a chronological span of A.D. 300 to 800 (Clerc 1974:132). Both stand out by virtue of their size and morphology, and are likely unionoids.

Although Vescelius and Robinson's unpublished observations clearly drew attention to the presence of exogenous shells at Antillean Ceramic sites, the implications for exchanges with South America or the Greater Antilles remained unexplored – and mainly unknown, even to us.

The recent data from St Martin and Martinique thus shed light on this earlier information. Additional evidence was then sought. Indeed, in Trinidad, two ornaments mentioned in the literature from the early Saladoid site of Palo Seco (Boomert 2000:408, Fig.65; Bullbrook 1953:51, Fig. 12A) are identified to the genus *Anodon* by Bullbrook (1953:51) and another unidentified specimen is known from the post-Saladoid Golden Grove site, in Tobago (Boomert 2000:410, 412, Fig. 65-13). Several local scholars also reported to us the existence of 4 additional worked and unworked *Prisodon* elements from 4 sites in Trinidad: an unworked valve of *Prisodon syrmatophorus* from the Saladoid Erin site (Boomert and Faber Morse, pers. comm. 2005); a possible ornament from the late Saladoid / Arauquinoid site of St Catherine's and two unworked specimens from the late Saladoid (Barrancoid) / Arauquinoid sites of Guayaguayare and Icacos tentatively identified as *Prisodon syrmatophorus* specimens (Harris, pers. comm. 2004).

We found additional reports in the literature of unrecognized unionoids from Guadeloupe. Two were published as mother of pearl ornaments in Y. Lammers-

Keijsers's work (2007). They originate from the sites of Morel (Lammers-Keijsers 2007: 116, Fig. 5.2), an early Saladoid settlement dated to 300 BC - AD 700 (Hofman, Hoogland, and Delpuech 1999), and of Anse à la Gourde (Lammers-Keijsers 2007: 67, Fig. 4.4), a late Saladoid / post Saladoid site which was occupied between AD cal 450 and 1350 (Hofman, Hoogland, and Delpuech 2001). While examining the specimens in the collection of the Service régional de l'Archéologie, in Guadeloupe, we found an additional one from the site of Anse à la Gourde (Hofman et al.'s 1998 excavation). They are all *Prisodon* ornaments (fig. 6-2C). Three additional specimens, collected and stored by E. Clerc in 1978 in the collections of the Musée Edgar Clerc, in Guadeloupe, were recovered and examined as well (fig. 6-2C): an additional *Prisodon* specimen from Morel, and two from the Saladoid site of Anse à l'Eau, in Guadeloupe (Clerc 1968) made of species of unionoids not yet identified. In these collections, the piece from the unidentified Guadeloupean site mentioned by Clerc (1974, see above) was recovered and confirmed as being a *Prisodon* specimen. Finally, a small piece of a *Prisodon* specimen was recently dug out in the Troumassoid site of Pointe du Canonnier, in St Martin (Serrand 2011).

One of many lines of evidence for elaborate networks of social interactions. Discussion

Therefore, the present record, although limited, shows that South American unionoids are present at sites in the Virgin Islands, Lesser Antilles, and Trinidad dating from early to late Ceramic age contexts. These exogenous shells were clearly circulating as elements of value among Lesser Antillean and South American Ceramic groups.

So far, only theoretical scenarios can be postulated about their circulation:
(i) The shells, either as finished products or raw materials, were transferred from South America during the initial dispersal in the islands as elements of the symbolic repertoire. However, the recovered shells are not confined to early contexts. Nevertheless, it is possible that their use persisted beyond the initial transfer and afterwards and that they functioned for many generations (possible heirlooms *sensu* Fitzpatrick *et al.* 2009?). It is also possible that constant supply was sustained through ulterior exchanges with the continent, which would join scenario (ii).

(ii) The shells may have been acquired as differentially available raw materials or finished objects after the initial colonization. This may have been accomplished directly by insular groups travelling to the continent to exchange insular products; or through distant commercial exchanges involving contacts with intermediate groups and probably equal economic or symbolic repayments, which have yet to be identified.
At any rate, the raw or modified shells were valued for various characteristics: their thick iridescent layer, the skills needed for their acquisition and work, their exogenous status, etc. Extracted from South America, they may also have represented a symbolic bond with the mainland / homeland. The finished artefacts themselves were probably meaningful elements as the similarities in the ornamental modifications suggest; they were possibly integrated into ceremonies or symbolic transfers.

Nevertheless, there are still insufficient data to fully define the temporal range of these shells, their specific significance or the routes and partnerships involved. A better understanding requires careful examination of the Caribbean shell assemblages to differentiate fragments of unionoids from local Antillean oysters; precise association with stratified and dated cultural contexts for a better definition of the shells' temporal occurrence; and identification of debitage remains at sites to determine possible continental or island production centers. Combining these data will help explore possible links with other exchanged items, and identify possible down-the-line trade(s).As for the meaning of this circulation of unionoids, ethnohistoric exchange systems involving shells such as the Melanesian *kula* of the South Pacific (Leach and Leach 1983; Malinowski 1922), the Papua New Guinean *moka* (Strathern 1975), or the north-eastern North American *wampum* exchanges (Iroquois confederacy; Herman 1956; Smith 1983) suggest that the Antillean unionoids of Ceramic times are the material traces of what were once elaborate networks of alliances and social interactions. These networks certainly did not only find their origin in the differential occurrence of natural resources and the possible resulting craft specialization of groups (Boomert 2000:422; Lathrap 1973) but aimed at maintaining relationships between communities through the circulation of materials, ideas, persons, and values (Boomert 2000:422).

This circulation of unionoid shells is one of many lines of archaeological evidence of regional and inter-regional interactions between the Antillean archipelago and continental America during pre-Columbian times (see Hofman, Bright, Ramos 2010; Hofman and van Duijvenbode 2011). Ethnohistorical records confirm, on the other hand, the extension and permanence of such interactions among Amerindian communities of the continent and adjacent areas, well into the contact period: trade of pearls from Cubagua and Margarita into the interior and along the coast as far as Trinidad and the Lower Orinoco valley (Sotomayor in Breton 1921); trade of *quiripá* shell beads strings in the Venezuelan llanos and Orinoco, Colombia, as far as the South Caribbean Trinidad, coast of the Guianas, Boomert 2000:433-434); exchange of gold, ceramics, salt, canoes, and 'jade' in the sixteenth and seventeenth centuries along the Amazon tributaries (Fritz 1922; Myers 1979) and in the South Caribbean (Boomert 2000:425). For more recent times, abundant ethnographic evidence documents exchanges from one group to the other, serving the needs for social interaction, in northern South America (Humboldt and Bonpland 1852; Lathrap 1973;

Meggers 1974; Palmatary 1965; Rice 1928), the Amazon and Orinoco drainage basins or the Guiana Highlands (Im Thurn 1883:270-271; Roth 1924).

Little is still known about the interactions patterns, routes and partnerships involved in the circulation of south-American unionoids in the pre-Columbian Lesser Antilles. Nevertheless, further research into this additional archaeological evidence of Ceramic age cultural interactions will contribute to our reconstructing of the various local and extended interaction spheres and their interlocking within broader pan-Caribbean network systems.

Acknowledgments

Warm thanks go to the following colleagues for making available data from the sites: D. Bonnissent, C. Stouvenot, C. Henocq, Association Archéologique de Hope Estate (Hope Estate); B. Bérard, N. Vidal (Dizac, Salines) ; A. Boomert, P. Harris, B. Bafer-Morse (Trinidad / Tobago information); C. Hofman, A. Delpuech (Anse à la Gourde, Morel), and S. Guimaraes for letting us look at and photograph the specimens in the collections of the Musée départemental E. Clerc, Guadeloupe. We are grateful to Dr. P. Lozouet, Laboratoire de Biologie des Invertébrés Marins et Malacologie, for giving us access to the comparative collections of the M.N.H.N. in Paris; to Dr. J.P. Pointier, Laboratoire de Biologie Marine and Malacologie (E.P.H.E.,UMR 5555 CNRS), for making available the Cuban specimens he collected; and to G. Rosenberg, Academy of Natural Sciences, Philadelphia, Dr. Robert Guralnick, University of Colorado, Diarmaid O'Foighil, University of Michigan, Adam Baldinger and Richard Johnson, Museum of Comparative Zoology, Harvard University, and Tim Pearce, Carnegie Museum, for the loan of specimens and access to collections under their care.

Bibliography

ARANGO, R. 1865. Catálogo de los moluscos terrestres y fluviatiles de la isla de Cuba. *Reportorio Fisico-natural de la Isla de Cuba*: 81-112, 123-149

BAKER, H. B. 1930. The Mollusks collected by the University of Michigan-Williamson Expedition in Venezuela VI. *Occasional Papers of the Museum of Zoology* 210.

BÉRARD, B. 2002. Le néolithique de la Martinique dans son contexte antillais. In *Bilan scientifique de la région Martinique 2000,* Ministère de la culture et de la communication, Sous-direction de l'Archéologie, SRA Martinique. Eds. Pp.13-15, Martinique.

BONNISSENT, D. 2010. *Archéologie précolombienne de l'île de Saint-Martin, Petites Antilles. Des campements des nomades des mers aux villages des agriculteurs-potiers (3300 BC-1600 AD).* Sarrebruck : Editions Universitaires Européennes.

BOOMERT, A. 1987. Gifts of the Amazons: «green stone» pendants and beads as items of ceremonial exchange in Amazonia and the Caribbean. *Antropologica* 67:33-54.

——.2000. *Trinidad, Tobago and the Lower Orinoco interaction sphere. An archaeological / ethnohistorical study. Alkmaar*, The Netherlands: Cairi Publications.

BRETON, A. C. 1921. The Aruac Indians of Venezuela. *Man* 21:9-12.

BULLBROOK, J. A. 1953. *On the excavation of a shell mound at Palo Seco, Trinidad,* B.W.I. Yale University Publications in Anthropology 50

CLERC, E. 1968. Sites précolombiens de la côte Nord-est de la Grand-Terre de Guadeloupe. *Proceedings of the 2nd International Congress for the Study of the Pre-Columbian Cultures of the Lesser Antilles.* Pp. 47-60. Barbados: Barbados Museum.

——.1974. Le travail du coquillage dans les sites précolombiens de la Grande-Terre de Guadeloupe. *Proceedings of the 5th International Congress for the Study of the Precolumbian Cultures of the Lesser Antilles.* Pp. 127-131. Antigua: The Antigua Archaeological Society.

CODY, A. 1993. Distribution of exotic stone artifacts through the Lesser Antilles: their implications for prehistoric interaction and exchange. *Proceedings of the 14th International Congress for Caribbean Archaeology.* A. Cummins and P. King eds. Pp. 204-226. Barbados: Barbados Museum and Historical Society.

CROSSE, H. 1891. Faune malacologique terrestre et fluviatile de l'île de Saint-Domingue. *Journal de Conchyliologie* 39:67-211.

——.1892. Faune malacologique terrestre et fluviatile de l'île de Porto Rico. Journal de Conchyliologie 40:5-71.

CUMMINGS, K.S. 2003. *The freshwater mussels (Bivalvia: Etherioidea) of Venezuela. Lecture presented at the Carnegie Museum of Natural History*, 22 October 2003.

CURET, L. A. and M. W. HAUSER, eds. 2011. *Islands at the crossroads. Migration, seafaring, and interaction in the Caribbean.* Tuscaloosa: University of Alabama press, Caribbean Archaeology and Ethnohistory

Series.

de France, S. D., W. F. Keegan, and L. A. Newsom. 1996. The Archaeobotanical, bone isotope, and zooarchaeological records from Caribbean sites in comparative perspective. *In Case Studies in Environmental Archaeology.* E. J. Reitz, L.A. Newsom, and S. Scudder, eds. Pp. 289-304. New York: Plenum Press.

Dunker, W. 1858. Einige neue species der naiaden. *Malakozoologische Blätter* 5:225-229.

Fitzpatrick, S. M., Q. Kaye, J. Feathers, J. A. Pavia, and K. M. Marsaglia, 2009. Evidence for inter-island transport of heirlooms: luminescence dating and petrographic analysis of ceramic inhaling bowls from Carriacou, West Indies. *Journal of Archaeological Science* 36(3):596-606

Fritz, S. 1922. *Journal of the travels and labors of Father Samuel Fritz in the river of the Amazons between 1686 and 1723.* The Hakluyt Society, 2nd Series, 51.

Graf, D.L. 2000. The Etherioidea revisited: a phylogenetic analysis of hyriid relationships (Mollusca: Bivalvia: Paleoheterodonta: Unionoida). *Occasional Papers of the University of Michigan Museum of Zoology* 729:1-21.

Graf, D.L. and K.S. Cummings. 2006. Palaeoheterodont Diversity (Mollusca: Trigonioida + Unionoida): what we know and what we wish we knew about freshwater mussel evolution. *Zoological Journal of the Linnean Society* 148: 343-394.

——.2007. Review of the systematics and global diversity of freshwater mussel species (Bivalvia: Unionoida). *Journal of Molluscan Studies* 73: 291-314.

Grouard S., N. Serrand, and B. Bérard. 2006. *Le site précolombien de Macabou, sud-est de la Martinique, Rapport de fouilles programmées pour l'année 2005, demande pour l'année 2006.* Martinique : Ministère de la Culture, Direction du Patrimoine, Sous-Dir. de l'Archéologie de Martinique

Haas, F. 1969a. Superfamily Unionacea Fleming, 1828. In *Treatise on Invertebrate Paleontology*, Part N, Vol.1, Mollusca 6, Bivalvia. R. C. Moore, ed.. Pp.411-470. Boulder: The Geological Society of America, Inc., the University of Kansas.

——.1969b. Superfamilia Unionacea. In *Das Tierreich, Eine Zusammenstellung und Kennzeichnung der Rezenten Tierformen*, H. Wermuth ed.. Lieferung 88, Seite I-X. Pp. 1-663, Berlin: Walter de Gruyter & Co.

Harlow, G. E., A. R. Murphy, D. J. Hozjan, C. N. de Mille, and A. A. Levinson, 2006. Pre-Columbian jadeite axes from Antigua, West Indies: description and possible sources. *The Canadian Mineralogist* 44(2):305-321.

Henocq, C. and F. Petit. 1999. Présentation de six sites archéologiques de Saint-Martin et de leur environnement. In: *Proceedings of the 16th International Congress for Caribbean Archaeology.* Conseil Régional de la Guadeloupe, eds. Pp. 300-315, Basse-Terre, Guadeloupe : Mission Archéologique et du Patrimoine.

Herman, M. W. 1956. Wampum as money in Northeastern North America. *Ethnohistory* 3:21-33.

Hofman, C. L., A. J. Bright, and R. Rodriguez Ramos, 2010. Crossing the Caribbean sea: towards a holistic view of pre-colonial mobility and exchange. *Journal of Caribbean Archaeology*, Special Publication 3: 1-18.

Hofman, C.L., M.L.P. Hoogland, and A. Delpuech. 1999. New perspectives on a Huecan Saladoid assemblage on Guadeloupe: the case of Morel I. *In Archaeological Investigations on St. Martin (Lesser Antilles). The Sites of Norman Estate, Anse des Pères and Hope Estate with a contribution to the 'La Hueca problem'.* C.L. Hofman and M. L.P. Hoogland, eds. Pp. 303-310. Leiden : Archaeological Studies Leiden University.

——.2001. Spatial organisation at a Troumassoid settlement, the case of Anse à la Gourde, Guadeloupe. In *Proceedings of the 19th International Congress for Caribbean Archaeology.* L. Alofs and R. A.C.F. Dijkhoff, eds. Pp. 124-130. Aruba: Publication of the Museo Arqueologico Aruba 9.

Hofman, C. L. and A. van Duijvenbode, eds. 2011. *Communities in Contact. Essays in archaeology, ethnohistory and ethnography of the Amerindian circum-Caribbean.* Leiden: Sidestone Press.

Humboldt, A. von, and A. Bonpland. 1852. *Personal narrative of travels to the equinoctial regions of America during the years 1799-1804.* London: Henry G. Bohn.

Im Thurn, E. F. 1883. *Among the Indians of Guiana: being sketches chiefly anthropologic from the interior of British Guiana.* London: Kegan Paul.

Johnson, R. I. 1981. Recent and fossil Unionacea and Mutelacea (freshwater bivalves) of the Caribbean Islands. *Occasional Papers on Mollusks* 4(60): 269-288. Cambridge: Department of Mollusks, Museum of Comparative Zoology, Harvard University.

KEEGAN, W. F. 1994. West Indian Archaeology. 1. Overview and foragers. *Journal of Archaeological Research* 2(3):255-284.

——.2000. West Indian Archaeology. 3. Ceramic Age. *Journal of Archaeological Research* 8(2):135-167.

KNIPPENBERG, S. 2007. *Stone artefact production and exchange among the northern Lesser Antilles.* Leiden: Amsterdam University Press, Archaeological Studies Leiden University Press.

——.2011. Much to choose from. The use and distribution of siliceous stone in the Lesser Antilles. In: *Communities in Contact. Essays in archaeology, ethnohistory and ethnography of the Amerindian circum-Caribbean.* C. L. Hofman & A. van Duijvenbode, eds. Pp. 171-185. Leiden : Sidestone Press.

LAMMERS-KEIJSERS, Y., 2007. *Tracing traces from present to past. A functional analysis of pre-Columbian shell and stone artefacts from Anse à la Gourde and Morel, Guadeloupe, FWI.* The Netherlands: Amsterdam University Press, Leiden University Press Academic.

LATHRAP, D. W. 1973. The antiquity and importance of long-distance trade relationships in the moist tropics of pre-Columbian South America. *World Archaeology* 5:170-186.

LEA, I. 1859. Observations on the Genus *Unio*. Descriptions of new genera and species in the families Naiades, *Melaniana and Colimacea.* 7(1):1-43. Philadelphia: J. Jay Jr. & Co.

LEACH, J. W., and E. LEACH, ed. 1983. T*he Kula: new perspectives on Massim exchange.* Cambridge: Cambridge University Press.

MALINOWSKI, B. 1922. *Argonauts of the Western Pacific.* London: London Routledge.

MAZE, H. 1883. Mollusques terrestres et fluviatiles de la Guadeloupe et de ses dépendances. *Journal de Conchyliologie* 31:5-54.

MEGGERS, B. J. 1974. Environment and culture in Amazonia. In *Man in the Amazon.* C. Wagley, ed. Pp. 91-110. Gainesville: University of Florida Press.

MORELET, A. 1849. *Testacea Novissima Insulae Cubanae et Americae centralis.* Pars I. Paris : Librairie de l'Académie Nationale de Médecine. J.B. Baillière.

——.1851. *Testacea Novissima Insulae Cubanae et Americae centralis.* Pars II. Paris: Librairie de l'Académie Nationale de Médecine. J.B. Baillière.

MYERS T. P. 1979. Aboriginal trade networks in Amazonia. In *Networks of the Past: Regional Interaction in Archaeology, Proceedings of the 12th Annual Conference.* P.D. Francis, F.J. Kense and P.G. Duke, eds. Pp. 19-30. Calgary: Archaeological Association of the University of Calgary.

NEWSOM, L.A. and E. S. WING. 2004. *On land and sea: native American uses of biological resources in the West Indies.* Tuscaloosa: University of Alabama Press.

OLIVER, J. R. 1989. *The archaeological, linguistic and ethnohistorical evidence for the expansion of Arawakan into Northwestern Venezuela and Northeastern Columbia.* Ph. D dissertation, Urbana Campaign: University of llinois.

OLSEN, F. 1974. *Indian Creek: an Arawak site on Antigua, West Indies.* Norman University of Oklahoma Press.

OLSSON, A. A., and C. B. WURTZ. 1951. New Colombian naiades, with observations on other species. *Notulae Naturae* 239:1-9.

ORTMANN, A. E. 1921. South American naiades. A contribution to the knowledge of the freshwater mussels of South America. Memoirs of the Carnegie Museum VIII(3), *Publications of the Carnegie Museum* 109. W.J. Holland, ed. Pp. 451-684. Pittsburgh: Carnegie Institute.

PALMATARY, H. C. 1965. *The river of the Amazons, its discovery and early explorations: 1500-1743.* New York: Carlton Press.

PARODIZ, J. J., and A. A. BONETTO. 1963. Taxonomy and zoogeographic relationships of the South American naiades (Pelecypoda: Unionacea and Mutelacea). *Malacologia* 1(2):179-213.

PETERSEN, J. B. 1997. Taino, island Carib, and prehistoric amerindian economics in the West Indies: tropical forest adaptations to island environments. In *The Indigenous People of the Caribbean.* S. M. Wilson, ed. Pp. 118-130. Gainesville: University Press of Florida.

POINTIER, J.-P., M. YONG, and A. GUTIÉRREZ. 2005. Guide to the freshwater molluscs of Cuba. Hackenheim: Conchbooks.

RICE, A. H. 1928. The Rio Branco, Yuraricuera and Parima: surveyed by the expedition to the Brazilian Guyana from August 1924 to June 1925. *The Geographical Journal* 71(2):113-143.

RODRIGUEZ LÓPEZ, M. 1991. Early trade networks in the Caribbean. *In Proceedings of the 14th International Congress for Caribbean Archaeology.* A. Cummins and P. King, eds.Pp. 306-314. Barbados: Barbados Museum and Historical Society.

Rodríguez Ramos, R. 2011. The circulation of jadeitite across the Caribbeanscape. In *Communities in Contact. Essays in archaeology, ethnohistory and ethnography of the Amerindian circum-Caribbean*. C. L. Hofman & A. van Duijvenbode, eds. Pp. 117-135. Leiden: Sidestone Press.

Roosevelt, A. C. 1980. *Parmana: prehistoric maize and manioc subsistence along the Amazon and Orinoco.* New York, London: Academic Press, Studies in Archaeology

Roth, W. E. 1924. An introductory study of the arts, crafts and customs of the Guiana Indians. *38th Annual Report of the Bureau of American Ethnology (1916-1917)*. Pp. 25-754. Washington D.C.: Smithsonian Institution.

Rouse, I. 1986. *Migrations in prehistory: inferring population movements from cultural remains. New Haven: Yale University Press.*

——.1992. *The Taïnos: rise and decline of the people who greeted Columbus*. New Haven: Yale University Press.

Rouse, I., and J. Cruxent. 1963. Venezuelan archaeology. Caribbean Series 6. New Haven: Yale University Press.

Rouse, I., and B. Faber Morse. 1999. Excavations at the Indian Creek site, Antigua, West Indies. Yale University *Publications in Anthropology* 82.

Serrand, N. 2001. Occurrence of exogenous freshwater Bivalves (Unionoida) in the Lesser Antilles during the first millenium A.D.: example from the Hope Estate Saladoid site (St Martin, French Lesser Antilles). In P*roceedings of the 18th International Congress for Caribbean Archaeology.* Conseil régional de la Guadeloupe. eds. Pp. 136-152, St George, Grenada: St George's University Campus, Mission Archéologique.

——.2007. L'économie des sociétés précolombiennes des Petites Antilles. Contribution des données sur l'exploitation des invertébrés marins et terrestres. In *Archéologie des départements français d'Amérique,* S. Rostain, and N. Vidal eds. Pp. 78-90. Les Nouvelles de l'Archéologie 108-109.

——.2011. *Pointe-du-Cannonier, COM, Saint Martin, Les-Terres-Basses.* Rapport final d'opération. Diagnostic archéologique, Inrap GSO, SRA Guadeloupe.

Siegel, P. E. 1991. Migration research in Saladoid archaeology: a review. *The Florida Anthropologist* 44(1):79-91.

Smith, T. J. 1983. Wampum as primitive valuables. In *Research in Economic Anthropology* 5, G. Dalton ed. Pp. 225-246. Greenwich: Jai Press.

Strathern, A. 1975. The rope of Moka. Big-men and ceremonial exchange in Mount Hagen, New-Guinea. *Cambridge Studies in Social Anthropology* 4. Cambridge: Cambridge University Press.

van der Schalie, H. 1948. The land and freshwater mollusks of Puerto Rico. *Miscellaneous Publications of the Museum of Zoology* 70: 9-129. Ann Arbor: University of Michigan Press.

Veloz Maggiolo, M. 1991. *Panorama historico del Caribe precolombino.* Santo Domingo: Banco Central de la Republica Dominicana.

Vendryes, H. 1899. Systematic catalogue of the land and freshwater shells of Jamaica. J*ournal of the Institute of Jamaica* 2: 590-607. Kingston, Jamaica: Institute of Jamaica.

Vescelius, G. S., and L. S. Robinson. 1979. Exotic items in archaeological collections from St. Croix: prehistoric imports and their implications. *Paper presented at the 8th International Congress for the Study of the Precolumbian Cultures of the Lesser Antilles*. St. Kitts.

Vidal, N. 1999. Le site précolombien de la Plage Dizac au Diamant, Martinique. In *Proceedings of the 16th International Congress for Caribbean Archaeology, Conseil Régional de Guadeloupe, eds. Pp. 7-16. Basse-Terre, Guadeloupe*: Mission Archéologique et du Patrimoine.

von Martens, E. 1900. *Biologia Centrali-Americana. Land and Freshwater Mollusca.* London : Pub. for the editors by R. H. Porter.

Watters, D. R. 1997. Maritime trade in the prehistoric Eastern Caribbean. In *The Indigenous People of the Caribbean*. S. M. Wilson, ed. Pp. 88-99. Gainesville: University Press of Florida.

Wing E. S., and S. R. Wing. 1995. Prehistoric Ceramic Age adaptation to varying diversity of animal resources along the West Indian archipelago. J*ournal of Ethnobiology* 15(1):119-148.

7 - DEAD FROM THE SEA: WORN SHELLS IN AEGEAN PREHISTORY

Tatiana THEODOROPOULOU
Wiener Laboratory (The American School of Classical Studies at Athens, Greece), Equipe de Protohistoire Egéenne (ArScAn-UMR7041, France),
tatheod@hotmail.com

Abstract : *The exploitation of shells as raw material is a common feature of Aegean prehistory. Shell items include ornaments, tools, containers, toys, or musical instruments. Many of these shell objects are made from water- or beach-worn specimens, both bivalves and gastropods. However, non-modified worn shells also appear often in the Greek archaeological record. This paper explores the paths of acquisition of this raw material through its distribution in the coastal and inland sites of the northern circum-Aegean region, relating collection of dead shells to other possible sea-shore activities and trade routes. The choice to procure this raw material cannot be seen independently from the final products. Yet in the case of sea-worn shells, their cultural function is not always straightforward, as Aegean examples range from shells without any modification to very elaborate objects. A multi-dimensional approach tries to explain the preference for this raw material for specific uses and shapes, bringing together technological and symbolic parameters. Both are considered vital in the appropriation of a raw material into the cultural sphere.*

Keywords : *Beach-worn shells, Natural, Cultural, Aegean prehistory, Neolithic, Bronze Age.*

Introduction

In various periods and cultures across the world the shells of molluscs have been exploited for raw materials either directly as a live animal or as empty shells along the shore, or secondarily, after the consumption of the flesh of the mollusc. This paper discusses material that falls under the first category. The collection of marine shells for ornamental or non-utilitarian purposes has been recognized in modern human sites as early as the Middle Stone Age (Bouzouggar *et al.* 2007; Jerardino and Marean 2010) and has been extensively discussed in various chronological and cultural contexts. On the other hand, although the cognitive, technological, practical and symbolic implications of the use of such shells form an integral part of molluscan analyses, reports are usually confined to a simple description of their state at collection (for discussion, Çakirlar 2009). Additionally, few are the studies dedicated specifically to worn, often unmodified shells, and their peculiar quality of belonging to a site's material culture while still retaining a "raw" quality from their natural environment of origin.

This paper attempts an evaluation of the transformation of naturally modified shells into cultural objects. The geographical area of study is the prehistoric Aegean with an emphasis on its northern part. Beach-worn specimens often feature among shell assemblages found in Neolithic and Bronze Age sites in Northern Greece. Although worn valves and shell fragments without any secondary modification are usual finds, objects made from worn shells are also present, thus implying a presumably deliberate exploitation of this raw material in Aegean prehistory. Until now emphasis was put on transformed finds, while unmodified worn specimens were only briefly mentioned as part of the shell assemblages, resulting in an under-appreciation of their significance within Greek cultural deposits. In the following paragraphs, an overview of natural and culturally-generated shapes and possible uses of worn shells is presented. It will be argued that, despite the humble character of the repertoire, the study of these items may offer insights into ways of procurement, prehistoric craftsmanship and functions. On a broader level, the goal of this paper is to set a number of questions regarding the encoded meanings behind the exploitation of this raw material.

The life and death of shells

Worn shells are the result of various natural processes that may cause alteration of the shell surface and structure. Although most of these processes are seemingly irrelevant to cultural ones studied by archaeologists, the presence of worn shells in archaeological contexts calls for a need to identify and evaluate these specimens among other archaeological shells. The discrimination between fresh and worn specimens, and naturally versus culturally modified shells within an archaeological assemblage, is of crucial importance, for, as it will be argued, in addition to enriching our knowledge on the use of raw materials in the past, it may provide an insight into the world of human cognition and aesthetics. For this reason, it is important to define the various steps in the natural history of worn shells with respect to their cultural life. These could be summarised in the following: a) transformations that occur in natural environments (naturally worn natural finds), irrelevant to, but potentially crucial for the potential cultural life of shells; b) naturally transformed shells, accidentally introduced into a cultural environment, (naturally worn natural finds), as opposed to c) naturally worn shells intentionally introduced into a cultural site, unaltered or modified by humans (c1-2: naturally worn unmodified and modified cultural finds); finally, d) shells that undergo transformations after their cultural period of use, thus not in direct connection to cultural processes (naturally worn archaeological finds). The following paragraph further describes these steps.

Shelled invertebrates are subject to a number of natural processes before and after their death (step a) (Claassen 1998; Vermeij 1993; Zuschin *et al.* 2003), described by Çakirlar (2009) as follows: a) predation by various marine organisms (including carnivorous molluscs, arthropods, sponges, octopods, worms, etc.) that cause distinctive alterations (perforations, fragmentation, encrustation) on the shell of living molluscs, b) abrasion and dissolution in marine sediments, which further modifies shell morphology during the life of molluscs, and c) shell alteration of dead shells deposited as debris in coastal or deeper marine beds that undergo mechanical pressures due to wave and sand action, causing abrasion, fragmentation, loss of sculpture, bioerosion, etc. Most of these shells remain a part of the natural depositional system, while some of these naturally modified shells may end up in a culturally formed assemblage. This is the part of the life of shells of interest to the archaeomalacologist. Worn shells may be the intentional result of human collection (step c), potentially not unrelated to the natural transformations described. To the degree that these specimens are intentional introductions in a site, they are considered part of a cultural context. They may remain unaltered or additional human modifications may further transform the shape of already naturally worn specimens. However, some remains may be the result of animal activity by predators or scavengers known to transport shell carcasses onto land (step b) (Erlandson and Moss 2001). Other scenarios include intense wave action close to a human settlement that may have washed shells with evidence of sea weathering. Human transport of marine sediment (e. g. sand, algae) containing small dead shells is also possible.These cases, when identified, should rather be categorized under natural finds. The final degradation of a shell may occur after burial in a culturally formed deposit (step d) (Muckle 1985; Claassen 1998; Çakirlar 2009). Shell sculpture can become worn, abraded or facetted in disturbed deposits and plough soil. This will again be a natural alteration, not to be connected to the cultural life of shells.

Although the various processes described are generally known to specialists, identifying the various steps in the transformation of shell is not easy, as distinctive criteria may overlap. To begin with, the basic distinction between fresh and worn specimens is generally possible on a macroscopic level. Some of the aforementioned natural events during and after the life of a shell are recognizable to the archaeomalacologist due to the presence of natural holes, worn or smoothed surfaces, abraded edges, erosion, encrustations, particularly on the thinner zones of the shell, namely the umbo, apex and outer zones (Claassen 1998; Light 2005). Other approaches include a comparative morphological study of archaeological specimens and modern dead shell assemblages from the assumed location of collection in order to observe similar alterations on fresh and archaeological shells (Light 2003, 2005). However, it is rightly stressed that "the distinction between 'fresh' or live-collected shells and 'worn' or beach-rolled empty shells is somewhat arbitrary, since one condition grades into the other and some 'fresh' shells can be picked up empty from the beach whereas others attain a 'worn' appearance during life" (Ridout-Sharpe 1998). Moreover, it is sometimes extremely difficult to distinguish these attributes, especially on worked specimens. Natural alterations can sometimes resemble anthropogenic modifications either prior or after utilisation of the shell. Adding to the complexity, naturally worn recent shells can sometimes be mistaken for fossil shells. Fossil shell-bearing geological deposits are present in several areas, including Greece, and might have been exploited in the past. Microscopic examination and various chemical analyses are some of the methods to discriminate recent worn from fossil specimens (for a review of these methods, Douka 2011).

Despite any limitations in the identification of worn shells from archaeological assemblages, it cannot be stressed enough that all the aforementioned variations are to be seen with respect to their connection to a cultural context. Beyond the strict distinction of worn versus fresh or fossil shells, their presence in an archaeological site itself raises an important issue: the terms of choice, acquisition and "acculturation" of these naturally modified shells by human groups. The following sections address these issues in the light of evidence from prehistoric Northern Aegean.

Harvesting dead shells

The presence of a raw material within a site of human activity generally involves a deliberate chain of events, from the choice of a specific material and the ability to locate its source, to its transport to a place of transformation and production, and ultimately to its final use. This is very much the case of marine organisms that cannot be seen *per se* as naturally belonging to a terrestrial environment. There is nevertheless a potential for ambiguity inherent in the coastal environment of beach-worn shells, which is often perceived as an interface zone between two major ecosystems of land and sea, and easily accessible to coastal communities. Although accidental transport by animals, with beach sediments, algae, or harvesting of shell beds is possible (Erlandson and Moss 2001), deliberate collection of washed-up shells along the beach is not uncommon across the periods and cultures of the world. Therefore, the presence of naturally modified shells in archaeological contexts may be a deliberate behaviour and it thus cannot be seen independently from the environment of acquisition of these materials, the route from a natural to a cultural environment, to be finally incorporated as social objects within a given human community. Let us examine this crucial episode in the life of worn shells, the process of transition from natural finds to cultural ones.

One might think that in Aegean prehistory the original source of collection of either fresh or worn shells would always lie close to the settlements, as coastal archaeological sites are a common feature in this region. Coastal prehistoric sites that have yielded worm marine specimens are indeed numerous in Northern Greece (fig. 7-1, Table 7-1).[1] It will probably remain unknown whether beach-picked shells were the object of fortuitous collection of odd or attractive-looking shells during shellfishing or simple strolls on the beach by men, women or children, valued and regarded as worth taking back to the village, or the mindful acquisition of a needed raw material. Although the numbers of worn specimens in coastal sites are never too high to suggest the latter (Table 7-1), modification occurring on several specimens suggests that some choices and decisions were made, either before collection or right after a fortuitous acquisition that proved adequate for transformation into an object. There are also cases of unmodified worn shells that might have been brought accidentally to the site, but seem to have found some kind of functional utility, as will be argued.

Notably, marine shells, both fresh and beach-collected are also present in inland sites, indicating that these might have travelled a distance of more than 100 km (e.g. 80-120 km from the Aegean to the lakeside settlement at Dispilio), although a mean distance for transport to inland prehistoric sites is estimated to around 25 km (e.g. Sitagroi). For a number of sites, namely those situated in the western part of Northern Greece, it has been suggested that shells were obtained instead from the Adriatic Sea (for a review, Ifantidis 2006), thus the covered distance would have been greater. It is interesting that sites at distances greater than 50-60 km from the sea exhibit some degree of contact with the marine element, especially from the Middle Neolithic to the beginning of the Bronze Age period. In this respect, transport of worn shells to inland sites has to be evaluated differently from coastal occurrences. In these sites, many shell specimens, either fresh or beach-worn, are worked down to ornaments or other objects (Table 7-1). It would be interesting to know whether worn and fresh shells arrived at the site at the same time. If this would hold true, the deliberate choice of worn specimens or objects made

Figure 7-1. Map of Northern Greece with prehistoric sites mentioned in the text.

from worn shells could be more straight forward. Acquisition of shells by inland groups would be achieved by either organising harvesting expeditions to the coasts (potentially combined with other activities) or by bringing the desired

[1] (Andreou *et al.* 1996). The current list is by no means an exhaustive record of finds, but rather a panorama of examples from the chosen area of study. Specimens judged to have been fortuitous introductions with marine sediments or due to mass-harvesting of shell beds (several examples in Veropoulidou 2011) are not considered in this paper.

Site	Type[1]	Period[2]	Total of shell assemblage[3]	Worn shells[4]	Natural modification[5]	Human modification[5]	Use[5]	Reference
Aggelochori	Inland	LBA	540	most of 41 *Cerastoderma glaucum*, 1 *Hexaplex trunculus*	worn surface, holes	perforation	accidental, pendants	Veropoulidou 2010
Agios Athanasios	(Coastal)	EBA	1776	15 (no specification of species)			accidental introductions	Veropoulidou 2011
Archontiko	(Coastal)	EBA	10 307	3 (no specification of species)			accidental introductions	Veropoulidou 2011
Assiros	Inland	LBA	442	*Arca noae, Cerithium vulgatum, Unio sp.*			no further specification	Halstead and Jones 1980
Augi	Inland	MN to LN	24	*1 Cerastoderma glaucum, 1 Hexaplex trunculus, 13 Spondylus gaederopus*	worn surface, holes	grounding, abrasion, perforation	annulets, "buttons", pendants, other?	Theodoropoulou 2011, and study in progress
Aliakmon valley (Pontokomi, Xirolimni, Megalo Nisi Galanis)	Inland	LN	72	2 *Spondylus gaederopus*, 4-5 *Cerastoderma glaucum*, 2 *Acanthocardia* sp.	worn surface, holes	grounding, abrasion, perforation	beads, annulets, "buttons", pendants, other?	Theodoropoulou 2007
Agios Mamas	Coastal	MBA	863	*Arca noae, Buccinulum corneum,Cerastoderma glaucum, Cerithium vulgatum, Charonia tritonis. Conus mediterraneus, Glycymeris* sp., *Spondylus gaederopus, Ostrea edulis*	worn surface, holes	grounding, polishing	pendants, tools (spoons, spools, lamps), utensils, other?	Becker 1996; Krolla and Becker 2008
Dikili Tash	Inland	LN to EBA	2501	*Cerastoderma edule, Glycymeris* sp., *Spondylus gaederopus*	(worn surface, holes)	grounding, abrasion, perforation	beads, annulets, tools (bat, cup)	Karali-Yannacopoulou 1992
Dimitra	Inland	MN to LN	305	*Cerastoderma edule*, 3 *Conus ventricosus*, 46 *Dentalium vulgare*, 2 *Euthria cornea, Glycymeris* sp., 1 *Nassa* sp., *several Spondylus gaederopus*	(worn surface, holes)	grounding, abrasion, perforation	(annulets), beads, pendants, tools (polishers)	Karali 1997
Dispilio	Inland	MN to LN	313	55 *Cerastoderma glaucum*, 4 *Dentalium dentale*, 3 *Glycymeris glycymeris*, 1 *Glycymeris pilosa*, 1 *Luria lurida*, 162 *Spondylus gaederopus*	(worn surface, holes)	abrasion, polishing, perforation	beads, annulets, "buttons", pendants	Veropoulidou and Ifantidis 2004
Kastanas	(Coastal)	MBA to LBA	1790	*(Cerastoderma glaucum, Luria lurida, Tricolia speciosa, Cyclope neritea, Cerithium vulgatum, Littorina neretoides.* Note: the above marine species are present but it is not specified if they are worn)			no further specification	Becker 1986
Koukonisi	Island	EBA to LBA	3603	*Aequipecten opercularis, Arca noae, Cerastoderma glaucum, Cerithium vulgatum, Glycymeris glycymeris, Mactra glauca, Ostrea edulis, Patella caerulea, Spondylus gaederopus, Thais haemastoma*	worn surface, holes	no	beads, pendants, other?	Theodoropoulou 2007

Kastri and Larnaki	Island	EBA	2912	no specification				Karali 1999a-c
Kryoneri	Inland	LN to EBA	163	1 *Acanthocardia tuberculata*, 1 *Arca noae*, 3 *Cerithium vulgatum*, 1 *Columbella rustica*, 3 *Hexaplex trunculus*, 7 *Luria lurida*, 1 *Ocinebrina* sp., 2 *Patella caerulea*, 5 *Spondylus gaederopus*	worn surface, holes	grounding, abrasion, polishing, perforation	annulets, pendants, tools (polishers), other?	Theodoropoulou 2007
Limenaria	Island	(MN) LN to EBA	14 195	1 *Cerastoderma glaucum*, 10 *Cerithium vulgatum*, 1 *Cyclope neritea*, 4 *Glycymeris glycymeris*, 17 *Hexaplex trunculus*, 1 *Thais haemastoma*, 1 *Ostrea edulis*, 39 *Spondylus gaederopus*, 1 *indet.*	Worn surface, holes	grounding, abrasion, perforation	pendants, tools (polishers), other?	Theodoropoulou 2007; Theodoropoulou 2011
Makriyalos	Coastal	LN	795 585	956 (no specification of number/species but numbers of worked shells of which several worn: 1074 *Cerastoderma glaucum*, 856 *Spondylus gaederopus*-34% worn, 118 *Cyclope neritea*, 173 *Glycymeris* sp. Also, *Astraea rugosa*, *Cerithium vulgatum*, *Columbella rustica*, *Hexaplex trunculus*,)	worn surface, holes	grounding, abrasion, polishing, perforation	annulets, beads, buttons, earrings, other	Pappa and Veropoulidou 2011;
Veropoulidou 2011								
Mesimeriani	Inland	EBA to LBA	1454	no specification				Karali 2002b
Mikro Vouni	Island		81 268	1 *Arca noae*, 1 *Buccinulum corneum*, 1 *Cerastoderma glaucum*, 13 *Cerithium vulgatum*, 3 *Chama gryphoides*, 2 *Conus mediterraneus*, 1 *Cymatium corrugatum*, 20 *Glycymeris bimaculata*, several *Hexaplex trunculus*, 1 *Luria lurida*, a few *Ostrea edulis*, 100+ *Spondylus gaederopus* (whole/fragments), *(Thais haemastoma)*	worn surface, holes	grounding, abrasion, perforation	pendants, tools (spoon), utensils, other?	Theodoropoulou 2007
Nea Nikomedeia	(Coastal)	AN to MN	yes	(1 *Ostrea edulis*)				Clench 1962
Olynthos	Coastal		yes	no specification				Mylonas 1929
Paliambela	Inland	EN to LN	20 803	1 *Glycymeris* sp., 1 *Spondylus gaederopus*, 1 not specified			ornaments	Veropoulidou 2011
Paradeisos	Inland	LN	377	2 *Arca noae*, 24 *Cerastoderma glaucum* and *Acanthocardia* sp., 2 *Columbella/Pisania*, 1 *Conus mediterraneus*, 1 *Dentalium* sp., 1 *Ensis/Solen*, 54 *Glycymeris* sp., 2 *Murex brandaris*, 2 *Mytilus galloprovincialis*, 2 *Pecten jacobeus*, 1 *Spondylus gaederopus*, 1 *Thais haemastoma*, 2 *Venus* sp.	worn surface, holes	grounding, perforation	pendants, tools?	Reese 1987
Pentapolis	Inland	EBA	12	no specification				Karali 1981
Poliochni	Island	MBA to LBA	20 211	no specification				Bernabò-Brea 1964; Sorrentino 1997

Promachon-Topol-nitsa	Inland	LN	144	1 *Hexaplex trunculus*, 2 *Ostrea edulis*, 3 *Spondylus gaederopus*, 1 *Thais haemastoma*	worn surface, holes	grounding, abrasion, polishing, perforation	annulets, pendants, rings, tools, utensils, other?	Theodoropoulou 2007
Servia	Inland			no specification				Mould et al. 2000
Sitagroi	Inland	MN to EBA	1787	(4 *Hexaplex trunnculus*, a few *Glycymeris* sp., *Spondylus gaederopus*)	worn surface, holes?	grounding, abrasion, polishing, perforation	beads, pendants	Miller 2003; Nikolaidou 2003; Shackleton 2003
Skala Sotiros	Island	EBA	35 413	no specification				Karali 1999
Stauroupoli	Inland	MN to LN	21 889	no specification				Karali 2002, 2004
Toumba Thessaloniki	Coastal	LBA	341 006	602 (2 *Acanthocardia tuberculata*, 2 *Arca noae*, 5 *Cerastoderma glaucum*, 25 *Cerithium vulgatum*, 2 *Chama* sp., 115 *Hexaplex trunculus*, 7 *Murex brandaris*, 3 *Spondylus gaederopus*, 16 *Venus verrucosa*)	worn surface, holes, encrustations	grounding, perforation		
(2,5% of total ornaments bear natural holes)	beads, pendants, tools, other?	Vero-poulidou 2002, 2011						
Yannitsa	Coastal	AN to LN	7010	85 *Cerastoderma glaucum*	worn surface, holes	no	beads, pendants	Theodoropoulou 2007
Vassilika	Inland	MN	492	no specification				

[1]at the time of occupation; [2]timespan of studied material (AN: Ancient Neolithic; MN: Middle Neolithic; LN: Late Neolithic; EBA: Early Bronze Age; MBA: Middle Bronze Age; LBA: Late Bronze Age); [3]when lists/counts are available; [4] when information available; [5] when specified.

Table 7-1. Presence of worn shells in prehistoric sites in Northern Greece.

raw material or finished items to the settlements via exchange networks, connecting coastal sites with inland villages. The transport of marine curios collected by inland villagers visiting the shores is also considered, as unmodified worn shells found inlands did not always receive a human transformation.

The question of harvesting and circulation of worn shells in prehistoric Northern Aegean cannot be definitively answered by available data, as information on the presence of worn shells in many studied assemblages is lacking. Despite these limitations, thus far recorded worn specimens do not betray any distinctive pattern of presence according to the distance of the sites from the sea, cultural sub-period or geographical region and the ways of procurement and transport of these shells are not straightforward in either coastal or inland sites. It remains unknown whether these raw materials were acquired by the inhabitants themselves or arrived to the coastal or inland villages through exchange networks. In any case, they should not necessarily be regarded as coincidental or simultaneous with the transport of fresh shells.

Thus far available studies on long-distance trade of specific species, namely *Spondylus gaederopus*, transformed into annulets found across the Balkans up to Central Europe during the Neolithic (Müller 1997; Theodoropoulou 2011) suggest that fresh and worn shells should be analysed as different categories. Yet, the question of procurement and transport of unmodified worn spiny oysters versus transformed objects of the same raw material has not been effectively addressed either by malacologists or archaeologists (a more detailed study is offered for Sitagroi, Miller 2003).

Despite any discrepancies in the available record of worn shells, their omnipresence, from coastal to inland sites, and diversity in modification, ranging from unmodified to fully transformed objects, should be seen as more than isolated, meaningless events. It thus becomes intriguing to explore the repertoire of forms and steps in the transformation of dead shells.

From mass to shapes

The archaeomalacological record from prehistoric North-

FAMILY/GENUS/SPECIES	OCCURENCE	HUMAN MODIFICATION	TYPE OF SITE
Bivalves:			
Cerastoderma glaucum (cockle)	+++	yes/no	Coastal/Inland
Spondylus gaederopus (spiny oyster)	+++	yes/no	Coastal/Inland
Acanthocardia sp. (rough cockle)	++	yes/(no)	Coastal/Inland
Glycymeris sp. (dog cockle/bittersweet)	++	yes/(no)	Coastal/Inland
Arca noae (Noah's ark)	++	yes/no	Coastal/Inland
Ostrea edulis (oyster)	+	yes/no	Coastal/Inland
Venus sp. (venus clam/carpet shell)	+	yes/no	Coastal
Mactra sp. (trough shell)	+	yes	Coastal
Mytilus galloprovincialis (mussel)	+	yes	Coastal/Inland
Chama gryphoides (jewel box)	+	yes/no	Coastal
Pectinidae (scallop)	+	yes	Coastal/Inland
Ensis sp./*Solen* sp. (razor shell)	+	no	Coastal
Scaphopods:			
Dentalium sp. (tusk shell)	+(+)	(yes)/no	Coastal/Inland
Gastropods:			
Muricidae (murex/purple-dye shell)	+++	no	Coastal/Inland
Conus mediterraneus (cone shell)	++	yes/no	Coastal/Inland
Columbella rustica (dove shell)	++	yes/no	Coastal/Inland
Cypraeidae (cowrie)	++	yes/no	Coastal/Inland
Cerithium vulgatum (horn shell)	++	no	Coastal
Patellidae (limpet)	+	(yes)/no	Coastal
Nassariidae (dog whelk)	+	yes/no	Coastal
Littorinidae (periwinkle)	+	yes	Coastal
Buccinulum corneum (whelk)	+	yes/no	Coastal
Cymatiidae (triton/trumpet shell)	+	no	Coastal
Other	+	yes/no	Coastal

Table 7-2. Occurence of worn shells according to type of settlement (+ < +++).

ern Greece offers quite a diverse range of beach-collected shells, both unmodified and transformed. Although the data presented in this paper are not an exhaustive listing of worn shells from Northern Greek prehistoric sites, they point to specific species collected post-mortem, as well as to potentially repetitive types of transformation. In the following, different types of natural and cultural worn finds will be presented, and their potential uses will be investigated.

The most frequently encountered worn shells in the prehistoric assemblages from Northern Greece are species of the Cardiidae, Muricidae and Spondylidae familieas (Table 7-2). A number of other species are quite often found in the assemblages: *Cerithium vulgatum, Glycymeris* sp., *Arca noae, Luria lurida, Conus mediterraneus* and *Columbella rustica.* Occasional finds include various bivalves *(Ostrea edulis, Venus verrucosa, Mactra* sp., *Chama gryphoides, Aequipecten opercularis, Mytilus galloprovincialis, Pecten jabobeus, Ensis* sp./*Solen* sp.) and gastropods (*Patella* sp., *Cyclope neritea, Nassa* sp., *Littorina neretoides, Tricolia speciosa, Buccinulum corneum, Cymatium* sp., *Charonia tritonis)*, as well as the scaphopod *Dentalium* sp.

The listed shells exhibit various types of wear and they range from simple naturally worn specimens to elaborate artificial shapes given to the original worn material. Becker (1996), with respect to the Kastanas shell assemblage, distinguishes four sub-types within the worn shell assemblage: a) shells without any distinctive natural shape or hole, b) shells with a naturally rounded, regular shape, possibly used without any or with minimal modification, c) shells of various shapes with natural holes, and d) shells with peculiar natural shapes that have undergone human working. The shell species that falls under all the aforementioned descriptive types is *Spondylus gaederopus*. With respect to the first type sensu Becker (step b or c1, as defined in this article), the spiny oyster can be found without any typical wear alteration but its naturally worn surface. Several worn valves or fragments of this species are encountered in northern Greek contexts. Other beach-worn shells, such as whole or fragmented murex, cowries, tusk shells, whelks and tritons, bearing no sign of natural or human modification, are also found in several sites (fig. 7-2a). Both *Spondylus* and other shells are often additionally naturally rounded or broken (type b *sensu* Becker, step b or c1 in this article). Because of the difficulty in attributing any functional or ornamental use to these specimens, they are not always recorded in older reports, or are regarded as accidental gatherings, especially when they found together with fresh specimens of the same species (Veropoulidou 2011). An additional difficulty lies, as previously seen, in distinguishing natural fragments of worn shells from intentionally broken pieces that have undergone post-depositional alterations (step d; Muckle 1985).

Things are clearer when it comes to naturally modified, unworked specimens (types b-c *sensu* Becker, step b or c1 in this article). Several examples from Northern Greece are included in this group (fig. 7-2b). Holes on the valves of cockles and dog cockles and less often of other bivalves (arks, trough shells, carpet shells, spiny oysters), are the most common natural transformation of this group. The hole is usually situated on the umbo, although occasionally it is at the centre of the valve. Gastropod species are also often naturally-perforated, including smaller dog whelks, periwinkles, pheasant shells and limpets, and larger murex, horn shells, cowries, whelks and tritons. Holes on gastropods either occur on the dorsal side of the body whorl and the aperture, or occur when the apex/spire is naturally worn or missing. Although this type of find may be quite common in northern Greek assemblages, it is not always recorded or there seems to be confusion with anthropogenic modifications (Vermeij 1993; Claassen 1998; Morton *et al.* 2007; Spanier 1986).

Human perforation or other types of intervention are indeed common among worn shells (type d *sensu* Becker, step c2 in this article) (fig. 7-2c). *Spondylus* valves or other shells often undergo a minimal modification: abrasion, flattening, polishing, perforation or just a widening of an already existing natural hole. On the other hand, several shell specimens, usually bivalves (such as the spiny oyster, occasionally the dog cockle and scallop) are ground down to produce a different shape to the natural one. The spiny oyster offers one of the most diversified records of produced, often elaborate shapes, such as annulets,[2] beads and pendants (for examples see Ifantidis and Nikolaidou 2011). In these cases, the difficulty often lies in the recognition of the original worn material, as heavy grinding and polishing results in the obscuring of the shell's natural wear. This difficulty may be responsible for the scant attention paid to the distinction between fresh and worn specimens used for the manufacture of *Spondylus* annulets in the Neolithic (Theodoropoulou 2011).

The noting of the presence of the aforementioned beach-found shells in a cultural deposit and description of their state (as, for example, by Becker's sub-types for Kastanas), however useful, does not offer a definite answer as to the natural (step b) or cultural conditions (step c) of introduction within a site, especially when it comes to shells unmodified by humans. This issue has been partly remedied by modern research which attempts to explain the uses of these objects within a cultural context. Human cognition attributes specific uses to every shaped artefact. As stated above, the original shape of the shell may be preserved and slightly adapted to the desired use, or the shell may be given a new shape fulfilling a specific need. Archaeologists tend to recognize more easily familiar cultural shapes. Worn shells have been used in prehistoric Northern Greece for the manufacture of a range of objects (fig. 7-2c). Among them, ornaments seem to be more easily identified and are often recorded, such as the beads and pendants made from worn spiny oyster and murex or *Spondylus* and *Glycymeris* annulets, as highly-regarded objects of Neolithic technology and trade net-

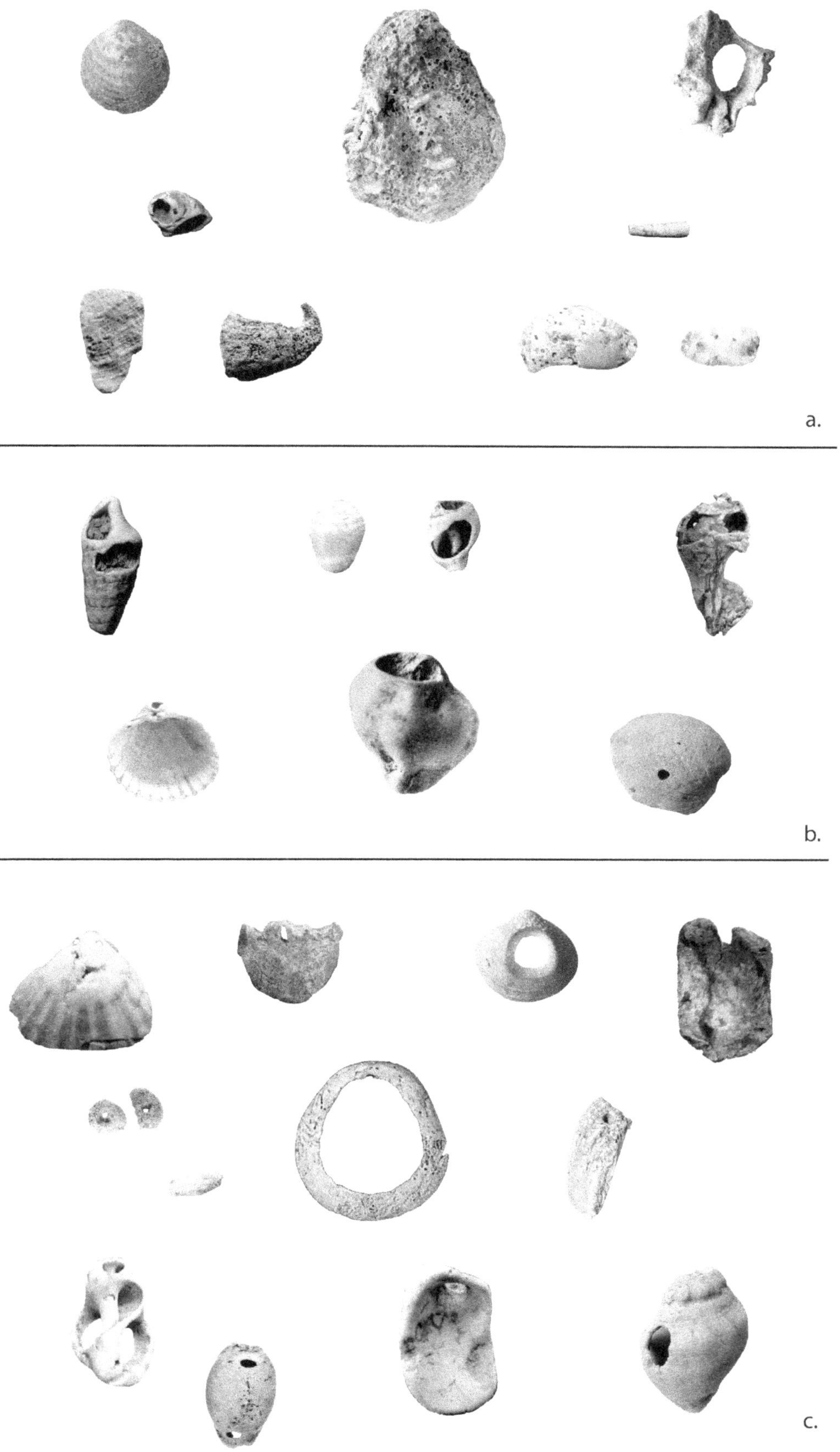

Figure 7-2. Different types of worn shells from prehistoric sites in Northern Greece (Augi, Kryoneri, Limenaria, Mikro Vouni, Promachon-Topolnitsa, Yannitsa: captions T. Theodoropoulou; Dikili Tash: Karali 2002; Dimitra: Karali 1997; Servia: Mould *et al.* 2000; Sitagroi: Miller 2003).

works. On the other hand, rarer shapes are not always securely attributed a specific use, such as the "spoon" from Mikro Vouni or the "fishing weight" from Limenaria (Theodoropoulou 2007, 2011). Available data seem to suggest that non-ornamental shapes are less numerous in this region than ornaments, thus implying some standardisation in the choice of worn shells to cover specific uses. However, different types of wear and familiar shapes of worn shells may have had different uses in space and time.

If deciphering the uses of artefacts of transformed worn shell in prehistoric Northern Greece can be intriguing and complex, this is even more the case when it comes to interpreting unshaped shells. Of course, "unshaped" refers only to the lack of human modification and not to the absence of shape as such. It may be argued that, even if these objects had not been conceived to serve a use, it may be their very shape that generated the idea for a cultural function. The latter is well exemplified by a range of naturally modified shells, both bivalves and gastropods found in northern Greek assemblages, which may have served as ready-to-use ornaments thanks to their natural shape, type of fracture or perforation (fig. 7-2b). On the other hand, the use of completely unmodified worn shells is usually more difficult to interpret (fig. 7-2a). Even in this case, their natural shape might be indicative of their function. Beach-found *Dentalium* offer natural beads without any further modification. Large worn *Spondylus* valves or concave gastropods have been interpreted as spoons, spools, scrapers, polishers, or other utensils and recipients (Becker 1996; Karali 1999b; Theodoropoulou 2007). In some cases, minimal human modification has rendered their shape more 'usable'. On the other hand,

TYPES OF WORN SHELL	COASTAL/ ISLAND	INLAND
Beads/pendants	+	+
Annulets	+	+
Spools/plates	+	+
Scrapers/burnishers	+	+
Tools/recipients?	+	+
Unmodified	+	+

Table 7-3. Types of worn shells according to type of settlement.

unmodified shells might have fulfilled non-utilitarian functions far from specific shapes, as it will be argued in the following section.

Summing up, descriptions of shapes and interpretation of related uses only partly explains the presence of either unmodified or transformed worn shells evidenced in most coastal and inland sites in Northern Greece (Table 7-3). Proximity to the source or access to a regional exchange network, desired shapes and fulfilled uses may be good reasons to suggest the intentional incorporation into a site of some of these specimens, but not of others. The choice to incorporate post-mortem collected shells as a whole in the material culture of a human group thus needs to be addressed.

Giving life to dead matter

Deciphering the meanings behind the choice and transformation of a raw material is not easy. Why would prehistoric groups collect worn shells or broken fragments of shells on the beach? Would the choice, transport, transformation and use of these items be considered as part of a deliberate, shared, and transmitted non-utilitarian behaviour, as suggested elsewhere (Jerardino and Marean 2010)?

According to Taborin (2004:67) "nature has offered humans shells as natural jewels, an available source, constantly renewed, diversified and coloured, providing species of multiple sizes and shapes". Qualities of a raw material are indeed a tangible and meaningful attribute leading to its selection. Shell shape and structure provided ancient populations a valuable material for the manufacture of ornaments and tools. Among infinite shell shapes, water or beach-worn shells not only often retain their original shape, surface, colours and natural properties (hard, dense, light), thus being an excellent surface to work on, but, to make Taborin's argument even more explicit, beach-collected shells can indeed be naturally finished items in themselves, used for adornment or as resistant tools. In fact, shells could be selected on the beach for their shape, whole or fragmented, or alteration (perforation, modified shell sculpture, described in step a in this article as, "naturally worn natural finds"), and be used as utilitarian or personal items without any further modification (step c1, "naturally worn unmodified cultural finds") (fig. 7-2b). In this case, it would be interesting to know whether it was the range of natural shell wear that inspired prehistoric people in the choice of specific uses and transformation into various objects, or the other way round. As Çakirlar (2009) notes: "it might be rightfully argued that the first cowry beads were beach-collected versions of the naturally-modified type, constituting the inspiration for subsequent cowry bead production."

Additional working of shells in a cultural context may of course lead to partial or more important transformation of the beach-worn raw material (step c2, "naturally worn modified cultural finds"). Such examples are not absent from the Northern Aegean prehistoric record (fig. 7-2c). The choice of the material as well as the selection of one or more methods of working and the degree of transformation is generally dictated by cognitive parameters, such as the technical skill of the craftsperson, the technological level of a given human community, and the desired object. On the other hand, experimental studies suggest that beach-worn specimens are more friable and breakable than fresh ones (Miller 2003), and if so underlining the choice to use the specific material despite any technological drawbacks. This special choice can be suggested in northern Greek sites, where broken and weathered specimens are found together with beautiful fresh shells (fig. 7-2a). Although it might be argued that "people were satisfied with picking up "second best choice" after whole fresh specimens" (Jerardino and Marean 2010), it can equally be suggested that worn shells were actually their first choice, driven by utilitarian or non-utilitarian motivations.

Non-utilitarian driven choices, such as the perception of the material or the natural shape, the encoded message of the raw material and shape through a complex network of aesthetics and symbolisms, are indeed a decisive factor in human cultural expression (Claassen 1998; Nikolaidou 2003). Sensory perception alone is a determining component, as it is imbued with cultural significance. The sight itself may be linked with different trains of associations, and particular sensory elements, such as colour, texture and shape, may acquire a special value in different contexts (Classen 1997). For instance, although heavily worn shells are not always appreciated by the modern eye, their beauty might have been more evident to past people. Although it cannot be claimed that all unmodified worn shells from archaeological sites are cultural introductions, some of them might have been chosen to be incorporated into a site for what they were, rather for what they were meant to become. The original nature of these items needs to be considered.

Beyond natural properties and visual attributes, it is almost impossible to get an insight into the complex world of representation and symbolism of the matter that lie at the base of aesthetic appreciations (Gell 1998:66). It should be remembered that beach-found shells belong to both the sphere of land and the sea. Dead shells collected on the beach, at the junction of two different environments, might have carried an archetypical symbolic meaning, linking two worlds, the remote sea and the 'familiar' land. In the bold assumption that people in the past had acquired the knowledge that fresh and worn shells were different forms of the same thing, worn shells may have then wrapped a seemingly remote and perfect material, such as seashells, with a more "tangible" texture, an appreciated wear older than human lifetime can perceive. Viewed individually as objects beached or "offered" by the sea, it might be that the symbolic attributes that surrounded this raw material were different than the ones given to freshly-collected shells (Theodoropoulou 2007). Paraphrasing Eliade (1961:141), the use of worn marine items may present "a case of degradation of the original metaphysical meaning, of its eviction by a secondary, exclusively magical meaning".

Conclusions

Worn shells are a much overlooked category of archaeological find. Their presence within an archaeological assemblage is either as assumed to be a natural intrusion or arbitrarily considered as a product of the material culture. Further analysis is usually provided only when such shells are given specific shapes through human modification. Although worn shells might sometimes undergo additional modification, important to our knowledge of ancient technology and material culture, archaeologists need to consider natural modifications as well, as rightly pointed out by Çakirlar (2009).

Careful analysis of naturally worn shells may help reconstruct the history of beach-found items and the integration of this natural material into a specific cultural assemblage, such as prehistoric sites in Northern Greece. Identifying the terms of presence within a cultural deposit can allow us to reconstruct the choices related to the exploitation of natural resources and raw materials. The presence of worn specimens in coastal and inland sites in this region is can in several cases be acknowledged as the result of deliberate transport, although fortuitous presence is not excluded, especially in coastal settlements. Explaining the choice to exploit worn rather than fresh shells may help unveil specific attributes ascribed to what seems to the modern eye as a undervalued marine material. Description of the state and degree of intervention on these natural items, unmodified or transformed into artefacts, may help us comprehend the intentions behind these choices. Beach-worn shells found in prehistoric northern Greek sites range from heavily worn unworked pieces to fully transformed personal or functional objects. Although descriptive typologies and suggested uses are helpful, what needs to be clarified in Aegean archaeology is the missing link between the various steps of transformation and related meaning. The latter is far from straightforward, as the higher degree of transformation of this raw material does not necessarily equal a more 'useful', 'cultural' function. Rather, it is argued that different degrees of transformation may imply diverse perceptions of the worn, and this is what needs to be codified. In other words, the simple question of why these people chose to keep unaltered or modify worn shells needs to be addressed.

Much more than inventorying shapes and deciphering possible uses, understanding the role of both raw and worked worn shells may allow for insights into the dialectics of matter and culture. For, beyond shapes and uses, every item integrated into the cultural sphere becomes part of a system of beliefs and symbolism. If artificial shell shapes offer a more direct insight into technological, practical and social processes, unmodified worn shells found in cultural contexts may shed their own light on non-utilitarian, sensory behaviour in the past.

Acknowledgements

This paper brings together old and new data from excavation publications, independent studies (articles, MA and PhD dissertations), as well as personal observations. I am grateful to all excavators who have entrusted me with the study of shell assemblages from Northern Greece included in my PhD dissertation (2007). For other data from Northern Aegean, any errors and omissions of published sites or projects in progress are my sole responsibility. The idea for a closer insight into the fascinating world of worn shells came after reading Canan Çakirlar's article providing a "fresh" look on the "worn", and I should thank her for triggering the thoughts presented in this paper. Finally, I should thank the two anonymous reviewers for their insightful comments on the final version of the text.

References

Andreou, Stelios, Michael Fotiadis and Kostas Kotsakis. 1996. Review of Aegean Prehistory V: The Neolithic and Bronze Age of Northern Greece. *American Journal of Archaeology* 100:537–597.

Becker, Cornelia. 1986. *Kastanas. Ausgrabungen in einem Siedlungshügel der Bronze- und Eisenzeit Makedoniens 1975-1979: Die Tierknochenfunde.* Prähistorische Archäologie in Südosteuropa 5. Berlin: Spiess.

——. 1996. Nourriture, cuillères, ornements... les témoignages d'une exploitation variée des mollusques marins à Ayios Mamas (Chalcidique, Grèce). *Anthropozoologica* 24:3–17.

Bernabò-Brea, Luigi. 1964. *Poliochni I: Città preistorica nell'isola di Lemnos.* Monografie della Scuola Archeologica di Atene e delle Missioni Italiane in Oriente II. Roma: Scuola Archeologica di Atene.

Bouzouggar, Abdeljalil, Nick Barton, Marian Vanhaeren, Francesco d'Errico, Simon Collcutt, T Higham, Edward Hodge, Simon Parfitt, Edward Rhodes, Jean-Luc Schwenninger, Chris Stringer, Elaine Turner, Steven Ward, Abdelkrim Moutmir, and Abdelhamid Stambouli. 2007. 82,000-year-old shell beads from North Africa and implications for the origins of modern human behaviour. *PNAS* 104(24):9964–9969.

Çakirlar, Canan. 2009. Problems in determining the chain of production in shell "objects". Observations on shell assemblages from coastal sites in the Eastern Mediterranean. In *Knochen pflastern ihren Weg. Festschrift für M. and H.-P. Uerpmann.* R. de Beauclair, Roland, S. C. Münzel and H. Napierala, eds. Special issue, *BioArchaeologica* 5:45–49.

Claassen, Cheryl. 1998. *Shells.* Cambridge Manuals in Archaeology. Cambridge: Cambridge University Press.

Classen, Constance. 1997. Foundations for an anthropology of the senses. *International Social Science Journal* 49:401–412.

Clench, William J. 1962. Molluscs. Appendix to R.J. Rodden (and contributors). Excavations at the early neolithic site of Nea Nikomedeia, Greek Macedonia (1961 season). *Proceedings of the Prehistoric Society* 28:274.

Douka, Katerina. 2011. The contribution of archaeometry to the study of prehistoric marine shells. In *Spondylus in Prehistory: New data and approaches – Contributions to the archaeology of shell technologies.* F. Ifantidis and M. Nikolaidou, eds. Pp. 171–180. British Archaeological Reports-International Series 2216. Oxford: BAR Publishing.

Eliade, Mircea. 1961. *Images and symbols: Studies in religious symbolism.* Princeton: Princeton University Press.

Erlandson, John M. and Madonna L. Moss. 2001. Shellfish feeders, carrion eaters, and the archaeology of aquatic adaptations. *American Antiquity* 66(3):413–432.

Gell, Alfred. 1998. Art and agency. *An anthropological theory.* Oxford: Clarendon.

Halstead, Paul and Glynis Jones. 1980. Appendix: Bioarchaeological remains from Assiros Toumba. In K. A. Wardle, Excavations at Assiros, 1975-1979. A settlement site in Central Macedonia and its significance for the prehistory of South-east Europe. *BSA* 75:265–267.

Ifantidis, Fotis. 2006. Ta Kosmimata tou Neolithikou Oikismou tou Dispiliou Kastorias: Paragogi kai Chrisi mias 'Aisthitikis Ergaleiothikis'. MA thesis, Aristotle University of Thessaloniki, Thessaloniki.

Ifantidis, Fotis. 2011. Cosmos in Fragments: Spondylus and Glycymeris Adornment at Neolithic Dispilio, Greece. In *Spondylus in Prehistory: New data and approaches – Contributions to the archaeology of shell technologies.* F. Ifantidis and M. Nikolaidou, eds. Pp. 123-137. British Archaeological Reports International Series 2216. Oxford: BAR Publishing.

Ifantidis, Fotis and Mariana Nikolaidou. 2011. *Spondylus in Prehistory: New data and approaches – Contributions to the archaeology of shell technologies.* British Archaeological Reports International Series 2216. Oxford: BAR Publishing.

Jerardino, Antonieta and Curtis W. Marean. 2010. Shellfish gathering, marine paleoecology and modern human behaviour: perspectives from cave PP13B, Pinnacle Point, South Africa. *Journal of Human Evolution* 59(3-4):412–424.

Karali, Lilian. 1997. Dimitra: Matériel malacologique. In *Neolithiki Makedonia.* D. V. Grammenos, ed. Pp. 200–211. Dimosieumata tou Archaeologikou Deltiou 56. Athens: Ministry of Culture, Tameio Archaeologikon

Poron kai Apallotrioseon.

——. 1999a. Meleti tou malakologikou ulikou apo ta nekrotafeia Kastri kai Larnaki. Parartima III. In *Proistoriki Thasos I. Ta Nekrotafeia tou Proistorikou Oikismou Kastri kai tis Thesis Larnaki*. Ch. Koukouli-Chrysanthaki, ed. Pp. 756–759. Dimosieumata tou Archaeologikou Deltiou 45. Athens: Ministry of Culture, Tameio Archaeologikon Poron kai Apallotrioseon.

——. 1999b. *Shells in Aegean Prehistory.* British Archaeological Reports International Series 761. Oxford, BAR Publishing.

——. 1999c. To malakologiko yliko tis Thasou: Ta ostrea os proti yli. In THASOS: *Matières premières et technologie de la préhistoire à nos jours, Actes du colloque international 26-29/9/1995 Thasos, Limenaria.* Ch. Koukouli-Chrysanthaki, ed. Pp. 389–400. Athens: Editions de l'Ecole Française d'Athènes.

——. 2002a. Anaskafi Stauroupolis: Malakologiko yliko. In *Sostikes Anaskafes sto Neolithiko Oikismo Stauroupolis Thessalonikis*. D. V. Grammenos and S. Kotsos, eds. Pp. 745–775. Dimosieumata tou Archaiologikou Institoutou Boreias Elladas 2. Thessaloniki: Archaiologiko Institouto Boreias Elladas.

——. 2002b. Mesimeriani Toumba: malakologiko yliko. In *Anaskafi ston proistoriko oikismo Mesimeriani Toumba Trilofou N. Thessalonikis, Anaskafikes periodoi 1992, 1994–1996. 2000. 2001.* D. V. Grammenos and S. Kotsos, eds. Pp. 359–437. Dimosieumata tou Archaiologikou Institoutou Boreias Elladas 1. Thessaloniki : Archaiologiko Institouto Boreias Elladas.

——. 2004. Anaskafi Stauroupolis: Malakologiko yliko. In *Sostikes Anaskafes sto Neolithiko Oikismo Stauroupolis Thessalonikis:* Meros II (1998–2003). D. V. Grammenos and S. Kotsos, eds. Pp. 527–603. Dimosieumata tou Archaiologikou Institoutou Boreias Elladas 6. Thessaloniki: Archaiologiko Institouto Boreias Elladas.

KARALI-YANNAKOPOULOU, LILIAN. 1981. Malakologiko yliko. In *Anaskafi se oikismo tis Epohis Halkou (Proimis) sitn Pentapoli tou nomou Serron.* D. V. Grammenos, ed. Pp. 115–118. Archaiologiki Efimerida. Athina: I en Athinais Archaiologiki Etaireia.

——. 1992. Les mollusques au site de Dikili Tash, village préhistorique de Macédoine Orientale; La parure. In *Dikili Tash, Village Préhistorique de Macédoine Orientale, I: Fouilles de Jean Deshayes 1961–1975.* R. Treuil, ed. Pp. 112 [outillage], 153–157 [alimentation], 159–164 [parure]. Bulletin de Correspondance Hellénique Supplément 24. Athènes: École Française d'Athènes.

KROLL, HELMUT and CORNELIA BECKER. 2008. *Das prahistorische Olynth: ausgrabungen in der Toumba Agios Mamas 1994–1996. Ernährung and Rohstoffnutzung im Wandel.* Prähistorische Archäologie in Südosteuropa 22. Rahden.

LIGHT, JAN. 2003. Dog Cockle Shells as Occasional Finds in Romano-British Shell Middens from Newquay, North Cornwall, UK. *Environmental Archaeology* 8(1):51–59.

——. 2005. Marine mussel shells. Wear is the evidence. In *Archaeomalacology: Molluscs in Former Environments of Human Behaviour. Proceedings of the 9th ICAZ Conference, Durham* 2002. D. E. Bar-Yosef Mayer, ed. Pp. 56–62. Oxford: Oxbow.

MILLER, MICHELLE A. 2003. Technical aspects of ornament production at Sitagroi. In *Prehistoric Sitagroi: Excavations in Northeast Greece, 1968–1970.* Volume 2: The Final Report. E. Elster and C. Renfrew, eds. Pp. 369–382. Monumenta Archaeologica 20. Los Angeles: Cotsen Institute of Archaeology, University of California.

MORTON, BRIAN, MELITA PEHARDA and ELIZABETH M. HARPER. 2007. Drilling and chipping patterns of bivalve prey predation by *Hexaplex trunculus* (Mollusca: Gastropoda: Muricidae). *Journal of Marine Biological Association UK* 87:933–940.

MOULD, CRESSIDA A., CATHERINE RIDLEY and KEN A. WARDLE. 2000. The shell small finds. In *Servia I: Anglo-Hellenic Rescue Excavations 1971–73.* C. Ridley, K. A. Wardle and C. A. Mould, eds. Pp. 276–287. BSA Supplementary Volume 32. London: British School at Athens.

MUCKLE, ROBERT JAMES. 1985. *Archaeological considerations of bivalve shell taphonomy.* MA thesis, Simon Fraser University.

MÜLLER, JOHANNES. 1997. Neolithische und chalkolithische Spondylus-Artefakte: Anmerkungen zu Verbreitung, Tauschgebiet und sozialer Funktion. In *ΧΡΟΝΟΣ: Festschrift für Bernard Hänsel.* C. Becker, M.-L. Dunkelmann, C. Metzner-Nebelsick, H. Peter-Röcher, M. Roeder and B. Teržan, eds. Pp. 91–106. Internationale Archäologie, Studia Honoraria 1. Espelkamp: Verlag Marie Leidorf.

MYLONAS, GEORGIOS. 1929. *Excavations at Olynthus, I: The Neolithic Settlement at Olynthos.* University Studies in Archaeology 6. Baltimore, Maryland: The John Hopkins University Press.

NIKOLAIDOU, MARIANNA. 2003. Items of adornment. In *Prehistoric Sitagroi: Excavations in Northeast Greece,*

1968–1970. Volume 2: *The Final Report*. E. Elster and C. Renfrew, eds. Pp. 331–360. Monumenta Archaeologica 20. Los Angeles: Cotsen Institute of Archaeology, University of California.

Pappa, Maria and Rena Veropoulidou. 2011. The Neolithic settlement at Makriyalos, Northern Greece: evidence from the *Spondylus gaederopus* artefacts. In *Spondylus in Prehistory: New data and approaches – Contributions to the archaeology of shell technologies*. F. Ifantidis and M. Nikolaidou, eds. Pp. 105–121. British Archaeological Reports International Series 2216. Oxford: BAR Publishing.

Pappa, Maria, Stratos Nanoglou and Aglaia Nitsou. 2000. Anaskafi neolithikou oikismou Thermis. *To Archaiologiko Ergo sti Makedonia kai Thraki* 14:179–186.

Reese, David S. 1987. Marine and fresh-water molluscs. In *Paradeisos: A Late Neolithic Settlement in Aegean Thrace.* P. Hellstrom, ed. Pp. 119–133. Memoir 7. Stockholm: Medelhavsmuseet.

Ridout-Sharpe, Jane S. 1998. The Mollusca. In *Lemba Archaeological Project (Cyprus) II.1B Parts 1 and 2). Excavations at Kissonerga Mosphilia, 1979–1992.* E. Peltenburg, ed. Pp. 338–351. Department of Archaeology Occasional Publication 19. Edinburgh: University of Edinburgh.

Shackleton, Nicholas J. 2003. Preliminary report on the molluscan remains at Sitagroi. In *Prehistoric Sitagroi: Excavations in Northeast Greece, 1968–1970.* Volume 2: The Final Report. E. Elster and C. Renfrew, eds. Pp. 361–368. Monumenta Archaeologica 20. Los Angeles: Cotsen Institute of Archaeology, University of California.

Sorrentino, Claudio. 1997. Poliochni: Il Materiale Faunistico. In *Poliochni e l'antica età del Bronzo nell' Egeo settentrionale (convegno internzionale, Atene 22-25 Aprile, 1996).* C. G. Doumas and V. La Rosa, eds. Pp. 157–167. Atene: Scuola Archeologica Italiana di Atene.

Spanier, Ehud. 1986. Cannibalism in muricid snails as a possible explanation for archaeological findings. *Journal of Archaeological Science* 135:463–468.

Taborin, Yvette. 2004. *Langage sans parole: La parure aux temps préhistoriques.* Paris: La Maison des Roches.

Theodoropoulou, Tatiana. 2007. *L'exploitation des ressources aquatiques en Égée septentrionale aux périodes pré- et protohistoriques* (2 volumes). PhD thesis, Université Paris I Pathéon-Sorbonne, Paris.

——. 2011. *Spondylus gaederopus* in Aegean Prehistory: Deciphering shapes from Northern Greece. In *Spondylus in Prehistory: New data and approaches – Contributions to the archaeology of shell technologies.* F. Ifantidis and M. Nikolaidou, eds. Pp. 93–104. British Archaeological Reports International Series 2216. Oxford: BAR Publishing.

Vermeij, Geerat J. 1993. A natural *history of shells.* Princeton: Princeton University Press.

Veropoulidou, Rena 2002. *Ostrea apo to Ktirio A tis Toumbas Thessalonikis.* MA thesis, Aristotle University of Thessaloniki, Thessaloniki.

Veropoulidou, Rena 2011. *Ostrea apo tous oikismous tou Thermaikou kolpou. Anasynthetontas tin katanalosi ton malakion sti Neolithiki kai tin Epochi tou Chalkou.* PhD dissertation, Aristotle University of Thessaloniki, Thessaloniki.

Veropoulidou, Rena and Fotis Ifantidis. 2004. *Shell Assemblage Analysis of the Neolithic Lakeside Settlement of Dispilio, Kastoria.* INSTAP Grant Report. Thessaloniki.

Zuschin, Martin, Michael Stachowitsch, and Robert J. Jr. Stanton. 2003. Patterns and processes of shell fragmentation in modern and ancient marine environments. Earth-Science Reviews 63:33–82.

8 - TEMPORAL CHANGES IN SHELL BEAD TECHNOLOGIES BASED ON LEVANTINE EXAMPLES

Daniella E. BAR-YOSEF MAYER
Steinhardt Museum of Natural History, Tel Aviv University, Tel Aviv 69778, Israel,
baryosef@post.tau.ac.il

Abstract: *The main techniques used for producing shell beads were established in several studies, based both on observation and experimentation. The techniques include hammering, gouging, incising, grinding, drilling, the use of naturally perforated shells, or any combination of these techniques. A temporal study shows the evolution of these techniques and the connection between the way shells were perceived, and the way they were exploited through time. The earliest shell beads, dating to the Middle Palaeolithic were naturally perforated ones. Next, small gastropod shells were gouged or scratched, probably with a flint tool. During the Upper Palaeolithic we encounter the first evidence for incisions with flint blades, but this is uncommon probably because it rapidly wastes the blades. During the Late Natufian culture of the Levant (13-11.5ka cal BP) shells disk beads were produced for the first time. This required a combination of methods including cutting of blanks, drilling the hole, and grinding the perimeter, or filing it in a grooved stone. The Neolithic periods see an intensification of drilling and grinding and the grinding could have been practiced on the same ground stone tools that were used for grinding of cereals and other foods. The emergence of metal tools allow slightly more accurate results, and the production of other artefacts made of shell, however, the basic technologies for making shell artefacts do not change.*

Keywords: *Shell bead, Technology, Levant*

Introduction

Shell beads used as personal ornaments are considered to indicate human cognitive abilities in that they are among the oldest means of decoration used by humans and had symbolic uses (e.g., Bar-Yosef Mayer *et al.* 2009; d'Errico *et al.* 2009). It is not surprising that shells were the first raw materials chosen to serve as a personal ornament due to their size, shape, hardness and color. Shells are small, light and durable, especially so with the earliest species collected in the Middle Palaeolithic: small gastropods (snails) up to 2 cm in height, and bivalves up to about 5 cm. They do not require special skills for collecting from the sea shore, and their size is suitable for decorating the human body. The specific shapes of shells (rounded or elongate) and their overall smoothness make them attractive to serve as decorations. Mollusc shells are made of calcium carbonate in the form of aragonite with a hardness of 3-4 on the Mohs scale, a relatively soft material which can easily be worked into beads.

Specific shell species were used during certain periods and can be viewed as cultural markers (Bar-Yosef 1989; Bar-Yosef Mayer 2005; Taborin 1993; Suárez Diez 2002) and the technology for producing shell beads was studied by many, but never along a chronological axis. Previous studies only rarely correlate between the decision making process of shell species choice with that of the production technique used to manipulate the shell (Szabo 2008). Not only the choice of species changes over time, but also the technology for producing shell artefacts. This will be demonstrated here, and is based mainly on case studies from the Levant. The data point to the transition from hunting and gathering to farming (the "Neolithic Revolution") as a period that not only affected the socio-economic fabric, but also changed shell exploitation and shell bead production. Further research will enable us to determine if the development of shell bead technologies is universal and is applicable in other regions as well.

Techniques for producing shell artefacts

Techniques for producing shell beads have been studied by many in the past. Most notably Peter Francis Jr. (1982, 1989) conducted experiments in which he manufactured shell beads using old techniques. Additional experiments were carried out by d'Errico *et al.* (1993), Yerkes (1993), and Kozuch (2003). Others, (Lucas and Harris 1962:44; Taborin 1993), made observations on the end products, especially the shapes of the holes and the microwear on the shells, to infer the various techniques that were used to produce the perforated shells. Yet wear signs and various depositional and post-depositional processes in any archaeological site sometimes erase the traces of the manufacturing processes.

Humans either exploited naturally perforated shells, or employed any of five main techniques described below. Natural abrasion is the result of an empty shell being washed ashore and its friction against the sand, cobbles, and sea waves and it often results in an irregular hole (fig. 8-1). Another type of natural hole is that produced by carnivorous gastropods. These holes are perfectly round but too small and placed in awkward positions on the shells, so that they were not used for suspension (fig. 8-2).

Figure 8-1. Naturally perforated bead: *Glycymeris insubrica* from Qafzeh Cave, Middle Palaeolithic.

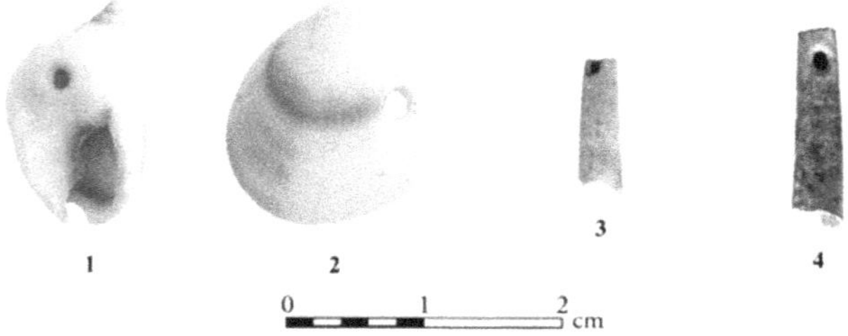

Figure 8-2. Shells naturally perforated by carnivorous gastropods. 1. Recent *Nassarius gibbosulus* from the Mediterranean coast. 2. Recent *Glycymeris insubrica* from the Mediterranean coast. 3, 4. *Antalis* shells from Hayonim Cave.

Artificial methods include:

a. Hammering: This method requires the use of a hammer stone. The hole will be irregular, but sometimes it is possible to use a chisel, which will result in a finer hole. Using a pointed limestone fragment minimized the shattering of shells while hammering a hole (Benghiat *et al.* 2009). It is difficult to differentiate between intentional hammering and a naturally broken shell (fig. 8-3).

Figure 8-3. Hammered holes: 1. *Clanculus pharaonius* from Ujrat el Mehed, of PPNB age. 2. *Columbella rustica* from Hayonim cave, Natufian. It is not possible to distinguish hammering from a natural hole in this species.

b. Gouging: A pointed tool is placed on the shell and pressed with a twist (Francis 1982). This could also be used for enlarging natural holes in shells, and sometimes results in scratches near the hole, probably as a result of missed motions (e.g., Kuhn *et al.* 2009, d'Errico *et al.* 2009).

c. Sawing: This creates a groove in the shell, its width being that of the tool used, and it is also employed for cutting large parts of a shell. The disadvantage of this method is that it requires frequent renewing of the blades used for sawing (Francis 1982). Its advantage is that it produces very accurate results. In Melanesia, this method was applied by using a wooden tool, a bone tool, or rope, to which silicate sand was adhered (fig. 8-4; Semenov 1970).

d. Grinding: The convex side of a shell is ground on a hard surface until a hole is made. The grinding creates a smooth surface around the hole. This method is relatively quick and efficient. The working tools may be made of limestone, sandstone or basalt; the latter two are more abrasive. The advantage of this method is that one does not have to change working tools frequently (fig. 8-5).

e. Drilling is a more refined method, which requires the use of a drill - a thin pointed object. The hole is made in a rotating motion, with or without the aid of other utensils such as a handle, a pump drill or a bow drill (Yerkes 1993). This method is also very wasteful of stone tools. One can often see the rotating striation marks inside a hole if it was drilled (fig. 8-6). Sometimes holes were drilled from two ends, in which case the shape of the hole would be bi-conical (Gorelick and Gwinnett 1989).

Some methods are particularly efficient when combined with other methods: For instance, striking the "dorsum"

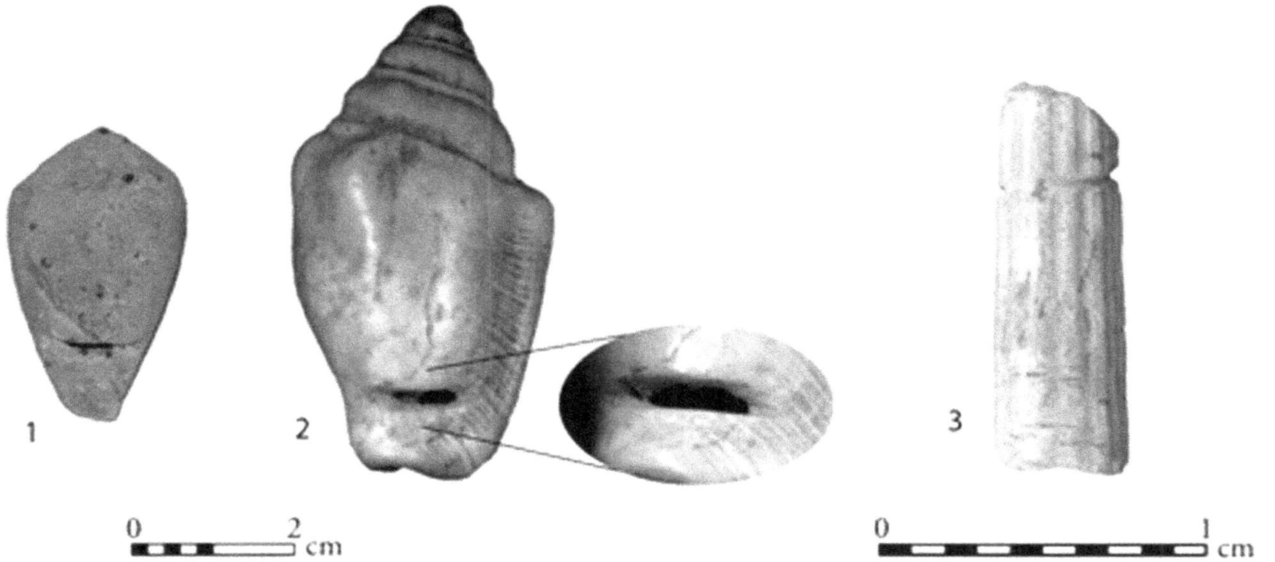

Figure 8-4. Incised holes: 1. *Conus mediterraneus*, from Hayonim cave, Natufian. 2. *Strombus mutabilis* from Ujrat el Mehed, Sinai, of PPNB age. 3. Fossil *Dentalium* shell from Çatalhöyük, Turkey, with incisions, probably unfinished. Photo credit: Çatalhöyük Research Project, Photographed by Jason Quinlan.

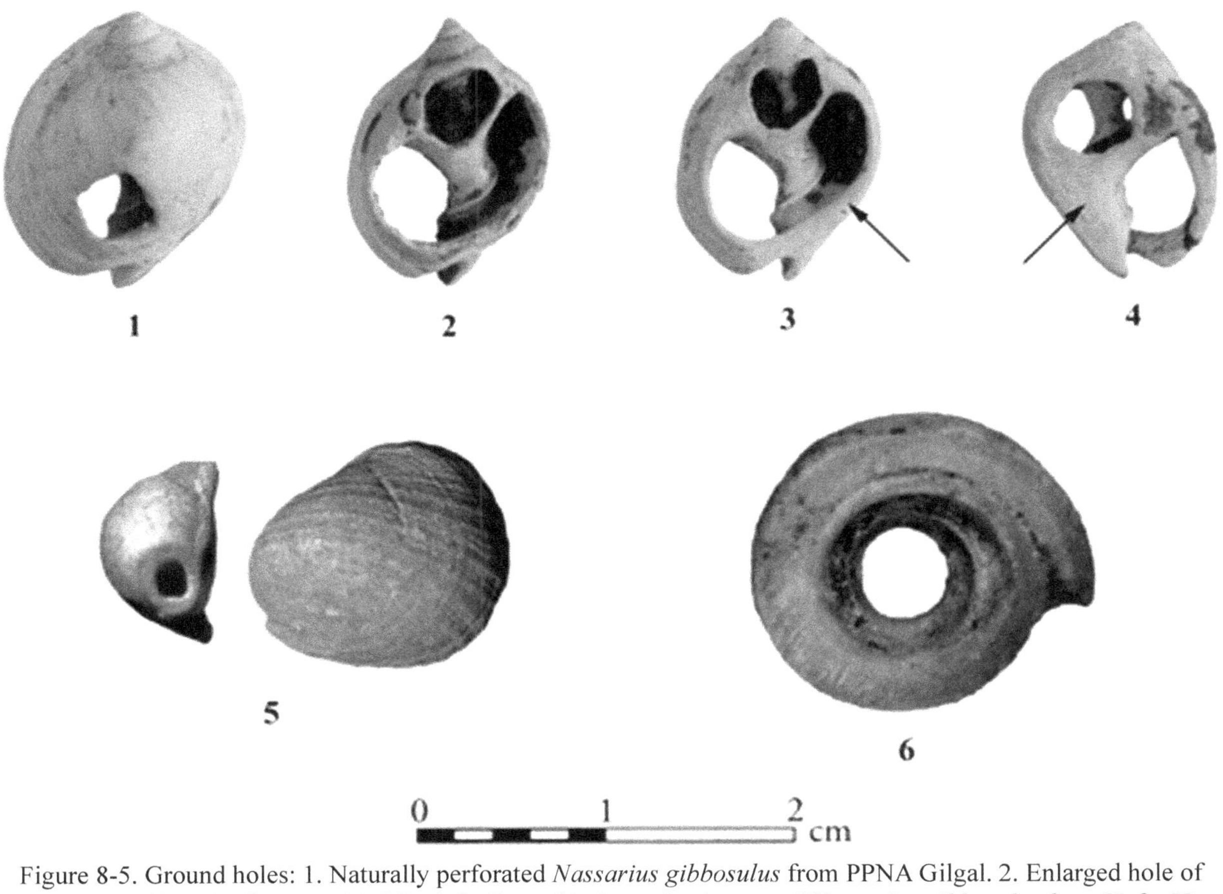

Figure 8-5. Ground holes: 1. Naturally perforated *Nassarius gibbosulus* from PPNA Gilgal. 2. Enlarged hole of *Nassarius gibbosulus* from PPNA Gilgal. 3. Ground hole on the dorsum of *Nassarius gibbosulus* from Kefar Ha-Horesh, PPNB. 4. Other side of *Nassarius gibbosulus* from Kefar Ha-Horesh, PPNB, with a ground hole. 5. *Nerita sanguinolenta* from PPNB Ujrat el Mehed, Sinai. 6. Ground *Conus* disc bead from PPNB Ujrat el mehed, Sinai.

of a cowrie, and then grinding it to create an even hole (Francis 1982). The combination of several techniques is known from ethno-archaeological studies (Kenoyer 1983). Combination of methods makes it impossible to determine the technique used, as only the last step will be visible.

The use of the term bead has been widely debated (Dubin

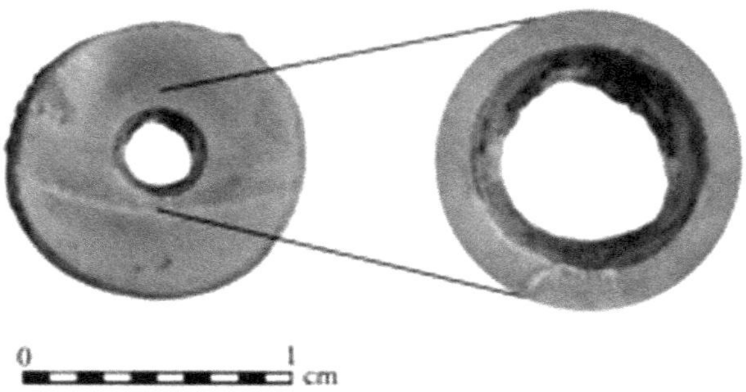

Figure 8-6. Drilled *Unio* disc bead from Eynan, final Natufian.

1987) and it is broadly agreed that "A bead is any object that can be strung to adorn or decorate a person, thing, or place" (Bead Society of Greater Washington; http://bsgw.org/whatisabead.html). Before reviewing the archaeological evidence for shell bead technology I will clarify some basic bead terms following Kenoyer (1991).

1. A simple bead is a shell with a hole in it suitable for stringing.
2. A complex bead is a bead usually made of a part of a shell (often the columella or the body whorl) of a gastropod, or the middle of a valve in the case of bivalves, thus the original shape of the shell is absent. It includes various bead shapes such as cylindrical, globular, etc.
3. A disc bead is a specific category of complex bead. These are round, usually small (ca. 1-2 cm in diameter), thin beads in which the height is less than a third of the diameter (Beck 1928). They are enumerated in many ethnographic records (Woodhouse 1997). The manufacturing technique has been described in detail by numerous authors (e.g., Francis 1989; Yerkes 1993). It is also used for making disc beads of other materials, including ostrich egg shell and various minerals. Shells are cut into rough blank pieces, those blanks are drilled, the they are then strung together and rubbed on a grooved stone (fig. 8-7).
4. Pendant is yet another variant of a complex bead. Whereas beads are usually basically round, pendants are most often elongated (or trapezoidal, oval, etc.). They are usually perforated at one end, and may have more than one perforation (fig. 8-8).
5. Bracelet is a circlet made of numerous components (usually beads) and is worn on the arm or ankle.
6. Bangle is a circlet made of a continuous homogenous material and is worn on the arm or ankle. (fig. 8-9).

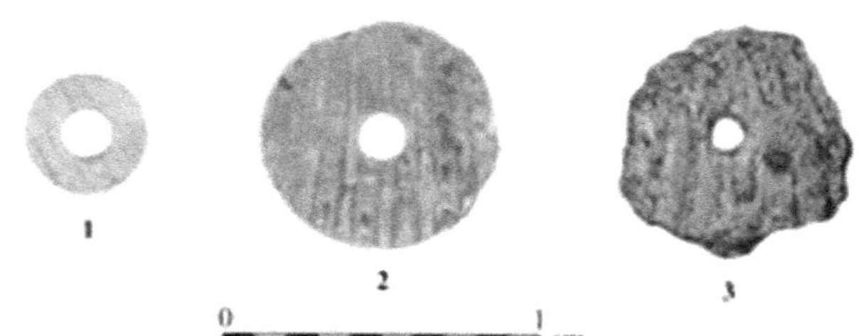

Figure 8-7. Disc beads made of shells from the Cardiidae family, from Eynan, final Natufian. 1,2. Finished beads. 3. Bead before filing.

Evidence for temporal changes

Evidence for shell working presented here begins with the earliest known assemblages. The examples presented here are mostly from the Levant, where a great deal of research has been conducted, and where I am most familiar with the data. I believe that the principles guiding technological innovations in this region have broader implications for the evolution of the various techniques in other parts of the world. Technological determination is based mostly on low magnification microscope observations.

Middle Palaeolithic

The earliest occurrence of shells in a prehistoric site in the Levant, are a few marine shells from the Middle Palaeolithic of Skhul Cave at Mt. Carmel (110ka BP uncalibrated). These include *Nassarius gibbosulus*, *Cardium* sp., and *Pecten jacobaeus* (Garrod and Bate 1937:224; Vanhaeren *et al.* 2006). Another assemblage is that of ten *Glycymeris insubrica* valves found in Qafzeh Cave, TL-dated to ca. 92ka BP (Bar-Yosef Mayer *et al.* 2009). These shells were naturally perforated as a result of abrasion on the beach. Similar shells, especially *Nassarius*, were discovered also in North Africa (d'Errico *et al.* 2009), South Africa (d'Errico *et al.* 2005), and there unperforated *Glycymeris* were also discovered (Jerardino and Marean 2010). One shell from Taforalt bears scratch marks of possible human manipulation (d'Errico *et al.* 2009).

Upper and Epi-Palaeolithic

The use of naturally perforated shells above all continues during the Upper Palaeolithic (45-20 ka BP

Figure 8-8. Pendant made of mother of pearl (probably *Pinctada margaritifera*) from Wadi Tbeik, PPNB.

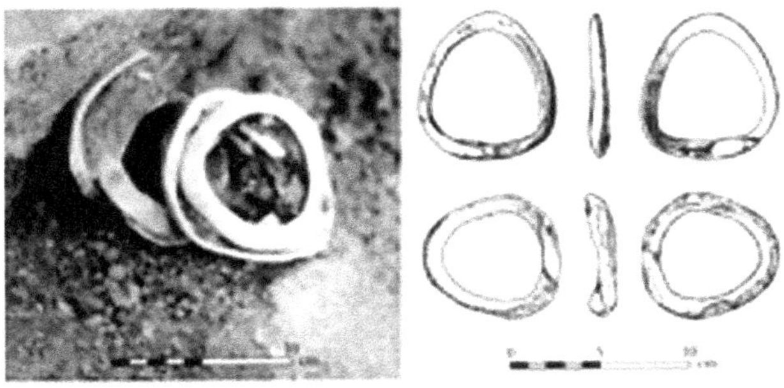

Figure 8-9. Bangles made of *Lambis truncata* from the nawamis burial site of El Abar, 4th millennium B.C.E., as found *in situ* and drawing of similar bangles. Photo credit: Israel Hershkovitz.

uncalibrated) and Epi-Palaeolithic (23-14.5ka cal BP) with occasional gouging evidenced by scratches near the holes. The repertoire of species collected increases and in the Levant includes *N. gibbosulus*, *Columbella rustica*, *Mitrella scripta*, *Antalis* sp., *Glycymeris insubrica* and *Cerastoderma glaucum* while in other parts of the Mediterranean other species are sometimes preferred (Colonese *et al.* 2011). The earliest evidence for perforation by sawing or incising is in the Ahmarian of Üçağızlı cave in Turkey (Kuhn *et al.* 2009:101). Sawing is more commonly evidenced in *Antalis* shells (also known as *Dentalium*, scaphopods, or tusk shells) that are usually about 5-20 mm long, as they are found on the beach. At Ohalo II (*ca.* 23ka cal BP) slicing of *Antalis* shells into short rings of 1-3 mm appears for the first time (Nadel *et al.* 2002). *Antalis* can be cut by snapping it by hand (e.g., Balme and Morse 2006), but achieving **thin slices** requires the use of flint blades. In a few instances partially cut scaphopods have been discovered, with typical v-shaped incisions being visible. (fig. 8-2c).

The Natufian

The Natufian culture dominates the end of the Epi-Palaeolithic period (14.5-11.6 cal BP) in the Levant, and is considered to be the precursor to the "Neolithic Revolution" (O. Bar-Yosef 1998). Scaphopod shells are abundant and at the Final Natufian of Eynan most shells are 1-3 mm long, similar to the older Ohalo II, and here too, they were sliced or sawed (fig. 8-10).

Natufian assemblages exhibit modest numbers of gastropods and bivalves as in earlier Epi-Palaeolithic assemblages, continuing the tradition of gouged or naturally perforated holes. Sliced v-shaped holes are rare (fig. 8-2:1). In the Late Natufian of Eynan we encounter for the first time *N. gibbosulus* with the dorsum removed and the columella exposed (Mienis 1987). This could be an enlargement of a natural perforation by gouging.

One innovation at the Final Natufian of Eynan is the production of disc beads, hence shell is now used as "raw material" (fig. 8-7). These were made of *Cerastoderma glaucum* from the Mediterranean, and in one case from freshwater *Unio* sp. (fig. 8-4; Valla *et al.* 2007: 315-319), and at Huzuk Mussa a disc bead workshop was discovered (Groman-Yaroslavski *et al.* 2013). These emerge at the same time as the first stone beads in the Levant (Bar-Yosef Mayer and Porat 2008).

U-shaped "shaft straighteners" from Natufian sites may have been used to polish the perimeter of the stone and

Figure 8-10. 1-6. Thinly sliced *Antalis* shells from Eynan, Final Natufian.

shell disc beads. The u-shaped groove is about 1 cm in diameter, which corresponds to the diameter of the beads, and similar artefacts were discovered in later Neolithic sites of Azraq, Jordan, where it has been demonstrated that they are found in the context of stone bead manufacturing (Wright *et al.* 2008). Thus, three types of procedures were involved in the production of shell disc beads: Cutting preforms, drilling using a pump drill or bow drill, and polishing or filing. Similar observations were made at Franchthi cave in Greece (Miller 1996).

Neolithic

The Neolithic sees a fundamental change in the economic base, the domestication of cereals, then of sheep and goats, that in turn influences all other aspects of life (e.g., Bouquet-Appel and Bar-Yosef 2008). New strategies characterize the manufacture and use of shells and other decorations. Ground stone tools previously used mostly for grinding foodstuffs were now also used for grinding shells (fig. 8-3:1-4).

Shell assemblages of Pre-Pottery Neolithic B (PPNB) and PPNC sites (ca. 10.6-7.5 cal BP) are referred to here as one unit. These periods are typified by an increase in shell quantities, and an increase in exploitation of shells from the Red Sea (Spatz *et al.* 2014). Hunter-gatherers in the deserts of Jordan, the Negev and the Sinai (Bar-Yosef Mayer 1997) relied on shells as a means of exchange. The symbolic and spiritual value of shells was expressed through their use for decorating plastered skulls and statues.

In the PPNB all the technology that could be used for producing shell beads was already known and used. One innovation is the use of mother of pearl to produce pendants and other decorations (e.g., Bar-Yosef Mayer 1997). For these elements to last, careful filing of the edges was necessary. This is a refinement of cutting and incising that was performed on a relatively new material (an isolated case was seen in earlier Eynan).

During this period we encounter significantly larger numbers of perforations made by incising, especially in the desert sites. A combination of hammering, then grinding was used to produce cowrie beads by breaking their dorsa, then grinding them down (similar to the treatment of *Nassarius*). Disc beads during this period were produced from the spires of small *Conus* shells: The grinding of the apex creates a hole, and the grinding of the body turns them into a thin (short) bead, and therefore it is not necessary to file or perforate them (fig. 8-3:6).

A new way of using pendants is seen in the working of cassid lips (Reese 1989) (fig. 8-11). These are the outer lips of *Phalium granulatum* that are naturally abraded and can be found on the beach. The ends of the lips were grooved to facilitate tying them on a string without drilling a hole. A few items are known from Kefar Ha-Horesh and Nahal Zehora II (Bar-Yosef Mayer 2007). A similar phenomenon of incising shells was encountered among several cowries from Tel Aswad in Syria, also dating to the PPNB (Alarashi, 2010).

Along with the technological innovations, one should also consider some shells that were not worked but may have served an ornamental or symbolic function. A point in case is the recent discovery of *Melanopsis*, a freshwater snail, placed in the eye socket of a plastered skull from Yiftah'el (Khalaily *et al.* 2008).

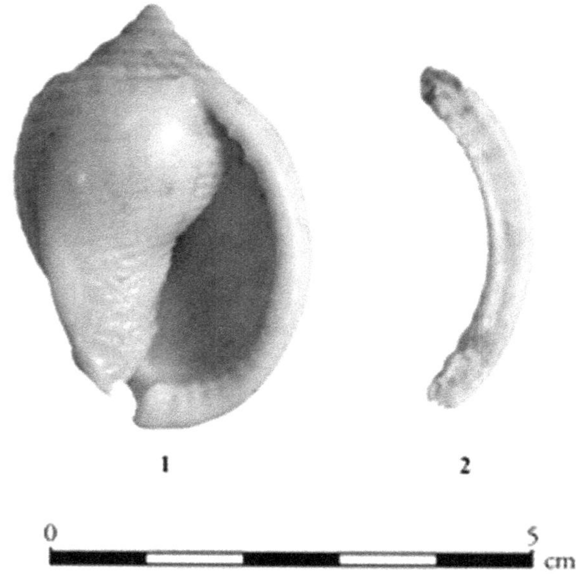

Figure 8-11. A complete recent *Phalium granulatum* and a "cassid lip", the naturally abraded lip of that shell, from Tel Dor, Late Bronze Age.

Chalcolithic and later periods

The use of shells as simple beads continues during the Chalcolithic along with various new artefact types made of shell (Bar-Yosef Mayer 2002). Pendants made of mother of pearl are of special interest. Trapezoidal pendants, sometimes decorated with patterns of incisions are made of both *Chambarida rubens*, a bivalve from the Nile river, and of *Pinctada margaritifera*, a bivalve from the Red Sea. The activities necessary for their production were the sawing and filing of the contour, and drilling of the holes, all of which were well-established technologies by this time. In other areas shell bangles were produced (of *Lambis* in Sinai, of *Spondylus* in the Balkans and of *Turbinella* in the Indus valley; Kenoyer 1983), all using available technologies.

Decoration of shell by incision, first observed in the PPNB, intensifies with a number of oval pendants discovered in the *nawamis* burial grounds in southern Sinai decorated with semi-drilled holes (fig. 8-12). This type of activity on shell is encountered later, for example, on Middle Bronze Age "shell lamps" from Ur (Danti and Zettler 1998:Fig. 117). Towards the end of the Iron Age in the Levant elaborate engravings on *Tridacna* shells and *Lambis* discs were widely distributed (e.g. Brandl 2001). In the Roman period tiny semi-drilled punctures adorn *Pinctada* shells (e.g., Michaelides 1995). Both simple shell beads as well as more elaborate shell artefacts continue throughout the archaeological record (Bar-Yosef Mayer 2007).

Discussion and Conclusions

Techniques for turning shells into shell beads are limited to a few methods and combinations thereof. The Levant is one of the areas with the longest record of the use of shells as artefacts, and therefore the examples presented above allow us to follow the steps in development of manufacturing processes of shell beads. During most of the Palaeolithic shells were mostly collected as "ready-to-use" elements from the beach. Occasionally they were perforated by hammering, gouging, and later in the Upper Palaeolithic, also by sawing. The manipulation of shells for use as beads during this period may be seen as indication for the successful spread of modern humans, in that these populations are now fully modern, both anatomically and culturally (e.g. Conard 2008). While some authors believe that Neanderthals also produced shells beads (Zilhao *et al.* 2010) this topic is controversial and beyond the scope of this paper.

The intensification of shell collection towards the end of the Epi-Palaeolithic and Neolithic periods brought on more elaborate methods, especially grinding and drilling

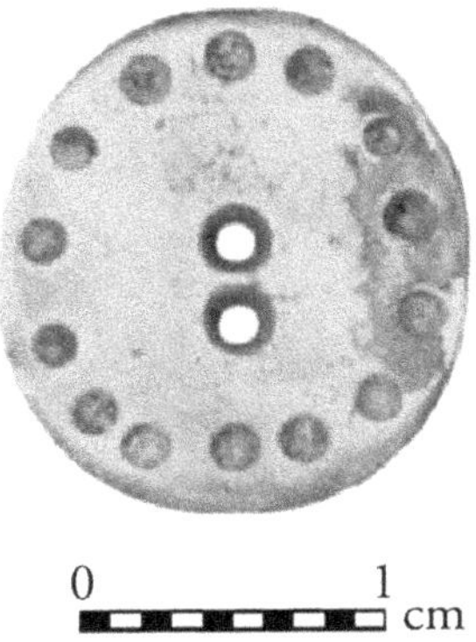

Figure 8-12. A decorated shell pendant from the Nawamis in southern Sinai, fourth millennium BCE.

that produce more accurate results, in turn providing more pleasing end-products, and the first use of shell as raw material for disc beads. The technologies for simple shell bead production formed the basis for all other types of shell artefacts that followed. Later when metal tools came into use, they allowed for more elaborate shell products.

It is often the case that there is a delay between the onset of the innovation and its adaptation for other purposes. The application of grinding stones that first appear in the Upper Palaeolithic (Wright 1991) for production of shell beads begins in the Neolithic period. Once they are used more intensively for processing foodstuffs, they are also used for producing smooth holes in shell beads. This is also the case for flint blades that were used for incising/sawing of shell for the first time in the Upper Palaeolithic following a long tradition of blade use.

The choice of technology for designing shell beads depends to some extent on the shell's structure (Szabó 2008) but could also be influenced by how shell beads were perceived, and their significance to the society. Whether their value is symbolic or economic, shell exploitation increases at the end of the Epi-Palaeolithic and the Neolithic periods. This is attested by an increase in absolute numbers of shells in archaeological sites, greater diversity of species (Bar-Yosef Mayer 2005), as well as the introduction of more elaborate technologies. The major socio-economic change, namely the transition to farming, was also responsible for changes in shell exploitation. The transition to farming allowed for the rise of craft production, as observed by many (e.g. Wright and Garrard 2003) and it is within this framework that the technology of shell bead production sees the emergence of disc beads and other shell artefacts that continue through the archaeological record.

To conclude, humans during prehistoric periods exhausted the basic technologies appropriate for mollusc shell bead production by the Neolithic. This is probably because shells were one of the earliest materials to serve as personal ornaments. All other innovations in later periods were elaborations of the basic techniques of hammering, gouging, incising, grinding and drilling. Because of the symbolic value attributed to shells and their relative availability, they serve in various adornment capacities to this day.

Acknowledgements

This paper stems from a workshop on bead production carried out at the British Museum and organized by Emma Twigger and Holly Miller. I thank two anonymous reviewers for their comments on a previous draft.

References

Alarashi, H. 2010. Shell beads in the Pre-Pottery Neolithic B in Central Levant: Cypraeidae of Tell Aswad (Damascus, Syria), Alvarez Fernandez E and Carvajal Contreras D.R., eds, *Not only food: Marine, Terrestrial and Freshwater molluscs in Archaeological sites*. Munibe Supplemento Gehigaria XX. Pp.88-98.

Balme, J. and Morse, K. 2006. Shell beads and social behaviour in Pleistocene Australia. *Antiquity* 80:799-811.

Bar-Yosef, D. E. 1989. Late Palaeolithic and Neolithic Marine shells in the Southern Levant as Cultural Markers, H.C.F. III, ed, *Proceedings of the 1986 Shell Bead Conference*. Rochester, New York: Rochester Museum and Science Center. Pp 169-74.

Bar-Yosef, O. 1998. The Natufian Culture in the Levant, Threshold to the Origins of Agriculture. *Evolutionary Anthropology* 6:159-77.

Bar-Yosef Mayer, D. E. 1997. Neolithic Shell Bead Production in Sinai. *Journal of Archaeological Science* 24:97-111.

——. 2002. Egyptian-Canaanite Interaction during the Fourth and Third Millennia B.C.: The Shell Connection, E.C.M. van den Brink and T.E. Levy, eds, *Egypt and the Levant: Interrelations from the 4th through the Early 3rd Millennium B.C.E.* London: Continuum. Pp 129-35.

——. 2005. The Exploitation of Shells as Beads in the Palaeolithic and Neolithic of the Levant. *Paléorient* 31:176-85.

——. 2007. Archaeomalacological research in Israel: The current state of research. *Israel Journal of Earth Sciences* 56:191-206.

Bar-Yosef Mayer, D. E. and Porat, N. 2008. Green Stone Beads at the Dawn of Agriculture. *Proceedings of the National Academy of Sciences of the U.S.A.* 105:8548-51.

BAR-YOSEF MAYER, D. E., VANDERMEERSCH, B. and BAR-YOSEF, O. 2009. Shells and Ochre in Middle Paleolithic Qafzeh Cave, Israel: Indications for Modern Behavior. *Journal of Human Evolution* 56:307-14.

BARGE-MAHIEU, H., BELLIER C., CAMPS-FABRER, H., CATTELAIN, P., MONS, L., PROVENZANO, N. and TABORIN, Y. 1991. Objets de parure. Camps-Fabrer, H., ed., *Fiches Typologiques de l'Industrie Osseuse Préhistorique*. Cahier IV. Union Internationale des Sciences Préhistoriques et Protohistoriques. Commission de Nomenclature sur l'Industrie de l'Os Préhistorique. Aix-en-Provence : Publications de l'Université de Provence.

BECK, H. C. 1928. Classification and Nomenclature of Beads and Pendants. *Archaeologia* 2nd series Vol. 1:1-76.

BENGHIAT, S., KOMŠO, D. and MIRACLE, P. T. 2009. An experimetal analysis of perforated shells from the site of Šebrn Abri, Istria, Croatia, S. McCartan, R. Schulting, G. Warren & P. Woodman, eds, *Mesolithic Horizons, Seventh international conference on the Mesolithic in Europe*, Belfast *2005*. Oxford: Oxbow books. Pp 730-6.

BOUQUET-APPEL, P. J. and BAR-YOSEF, O. 2008. Prehistoric demography in the time of globalization, P.J. Bouquet-Appel and O. Bar-Yosef, eds, *The Neolithic Demographic Transition and its Consequences*. Dordrecht: Springer. Pp 1-10.

BRANDL, B. 2001. Two Engraved Tridacna Shells from Tel Miqne-Ekron. *Bulletin of the American Schools of Oriental Research* 323:49-62.

COLONESE, A. C., MANNINO, M. A., BAR-YOSEF MAYER, D. E., FA, D., FINLAYSON, J. C., LUBELL, D. and STINER, M. C. 2011. Marine mollusc exploitation in Mediterranean prehistory: an overview. *Quaternary International* 239: 86-103.

CONARD, N. 2008. A critical view of the evidence for a Southern Afrcian origin of behavioural modernity. *South African Archaeological Society Goodwin Series* 10:175-9.

DANTI, K. and ZETTLER, R. L. 1998. Shell Vessels and Containers, R.L. Zettler and L. Horne, eds, *Treasures from the Royal Tombs of Ur*. Philadelphia: University of Pennsylvania Museum of Arhchaeology and Anthropology. Pp 143-6.

D'ERRICO, F., JARDÓN-GINER, P. and SOLER-MAYOR, B. 1993. Critères à base expérimentale pour l'étude des perforations naturelles et intentionnelles sur coquillages, P.C. Anderson, S. Beyries, M. Otte and H. Plisson, eds, *Traces et fonction: les gestes retrouvés*. Liège: ERAUL. Pp 243-54.

D'ERRICO F., HENSHILWOOD, C. S., VANHAEREN, M. and VAN NIEKERK, K. 2005. *Nassarius kraussianus* shell beads from Blombos Cave: evidence for symbolic behaviour in the Middle Stone Age. *Journal of Human Evolution* 48:3-24.

D'ERRICO F., VANHAEREN, M., BARTON, N., BOUZOUGHAR, A., MIENIS, H. K., RICHTER, D., HUBLIN, J. J., MCPHERRON, S. P. and LOZOUETH, P. 2009. Additional evidence on the use of personal ornaments in the Middle Paleolithic of North Africa. *Proceedings of the National Academy of Science of the U.S.A.* 109:16051-6.

DUBIN, L.S. 1987. The History of Beads from 30,000 B. C. to the Present. New York: Harry N. Abrams.

FRANCIS, P. 1982. Experiments with Early Techniques for making whole shells into Beads. *Current Anthropology* 23:713-4.

——. 1989. The Manufacture of Beads from Shell, C.F.I. Hayes, ed, *Proceedings of the 1986 Shell Bead Conference*. Rochester: Rochester Museum and Science Center. Pp 25-35.

GARROD, D.A.E. and BATE, D.M.A. 1937. *The Stone Age of Mount Carmel*. Oxford: Clarendon Press.

GORELICK, L. and GWINNETT, A. J. 1989. Collars in the Holes of Near Eastern Cylinder Seals. *Archeomaterials* 3:39-46.

GROMAN-YAROSLAVSKI, I., ROSENBERG D. and NADEL, D. 2013. A functional investigation of perforators from the Late Natufian/Pre-Pottery Neolithic A site of Huzuk Musa – a preliminary report. In: Stone Tools in Transition: From Hunter-Gatherers to Farming Societies in the Near East / Ferran Borrell, Juan José

Ibáñez, Miquel Molist (eds.) — Bellaterra (Barcelona) : Universitat Autònoma de Barcelona. Servei de Publicacions, pp. 165-176.

JERARDINO, A., and MAREAN, C. W., 2010. Shellfish gathering, marine paleoecology and modern human behavior: perspectives from cave PP13B, Pinnacle Point, South Africa. *Journal of Human Evolution*, 59:412-424.

KENOYER, J. M. 1991. Ornament Styles of the Indus Valley Tradition: Evidence from Recent Excavations at Harappa, Pakistan. *Paléorient* 17:79-98.

——. 1983. *Shell Working Industries of the Indus Civilization: An Archaeological and Ethnographic*

Perspective. University of California, Berkeley.

Khalaily, H., Milevski, I., Getzov, N., Hershkovitz, I., Barzilai, O., Yarosevich, A., Shlomi, V., Najjar, A., Zidan, O., Smithline, H. and Liran, R. 2008. Recent Excavations at the Neolithic Site of Yiftahel (Khalet Khalladyiah), Lower Galilee. *Neo-Lithics* 2/08:3-11.

Kozuch, L. 2003. Use of fire in shell bead manufacture at Cahokia. *Bulletin of the Florida Museum of Natural History* 44:81-90.

Kuhn, S. L., Stiner, M. C., Güleç, E., Özer, I., Yilmaz, H., Baykara, I., Acikkol, A., Goldberg, P., Martinez Molina, K., Ünay, E. and Suata-Alpaslan, F. 2009. The Early Upper Paleolithic Occupations at Üçagızlı Cave (Hatay, Turkey). *Journal of Human Evolution* 56:87-113.

Lucas, A. and Harris, J. R. 1962. *Ancient Egyptian Materials and Technologies*. London: Arnold Publishers.

Michaelides, D. 1995. Cyprus and the Persian Gulf in the Hellenistic and Roman Periods: The Case of *Pinctada Margaritifera*, V. Karageroghis and D. Michaelides, eds, *Proceedings of the International Symposium, Cyprus and the Sea*. Nicosia: University of Cyprus. Pp 211-26.

Mienis, H. K. 1987. Molluscs from the Excavation of Mallaha (Eynan), J. Bouchud, *ed, La Faune du Gisement Natoufien de Mallaha (Eynan) Israel*. Pp 157-78.

Miller, M. 1996. The Manufacture of Cockle Shell Beads at Early Neolithic Franchthi Cave, Greece: A Case of Craft Specialization? *Journal of Mediterranean Archaeology* 9:7-37.

Nadel, D., Tsatskin, A., Bar-Yosef Mayer, D. E., Belmaker, M., Boaretto, E., Kislev, M. E., Hershkovitz, I., Rabinovich, R., Simmons, T., Weiss, E., Zohar, I., Asfur, O., Emmer, G., Ghraieb, T., Grinberg, U., Halabi, H., Weissbrod, L. and Zaidner, Y. 2002. The Ohalo II 1999 - 2000 Seasons of Excavation: A Preliminary Report. *Mitekufat Haeven, Journal of the Israel Prehistoric Society* 32:17-48.

Reese, D. S. 1989. On Cassid lips and helmet shells. *Bulletin of the American Schools of Oriental Research* 275:33-9.

Semenov, S. A. 1970. *Prehistoric Technology*. London: Bath, Adams & Dart.

Spatz, A., Bar-Yosef Mayer, D. E., Nowell, A. and Henry, D. O. 2014. Ornaments of shell and stone: Social and Economic insight, In *The Sands of Time: The Desert Neolithic Settlement at Ayn Abū Nukhayla*, edited by D.O. Henry and Joseph E. Beaver. Berlin: Ex Oriente. pp. 245-258.

Suárez Diez, L. 2002. *Tipología de los objetos prehispanicos de concha*. Mexico City: Conaculata, INAH.

Szabo, K. 2008. Shell as a Raw Material: Mechanical Properties and Working Techniques in the Tropical Indo-West Pacific. *Archaeofauna* 17:125-38.

Taborin, Y. 1993. *La parure en coquillage au Paléolithique*. Paris: CNRS Editions.

Valla, F. R., Khalaily, H., Valladas, H., Kaltnecker, E., Bocoquentin, F., Cabellos, T., Bar-Yosef Mayer, D. E., Le Dosseur, G., Regev, L., Chu, V., Weiner, S., Boaretto, E., Samuelian, N., Valentin, B., Delerue S., Poupeau, G., Bridault, A., Rabinovich, R., Simmons, T., Zohar, I., Ashkenazi, S., Delgado Huertas, A., Spiro, B., Mienis, H. K., Rosen, A. M., Porat, N. and Belfer-Cohen, A. 2007. Les Fouilles de Ain Mallha (Eynan) de 2003 à 2005 : quatrième rapport préliminaire. *Journal of the Israel Prehistoirc Society* 37:135-383.

Vanhaeren, M., D'Errico, F., Stringer, C., James, S. L., Todd, J. A. and Mienis, H. K. 2006. Middle Paleolithic Shell Beads in Israel and Algeria. *Science* 312:1785-8.

Woodhouse, H. C. 1997. Ostrich eggshell beads in southern Africa. *Rock Art Research* 14:41-3.

Wright, K. 1991. The Origins and Development of Ground Stone Assemblages in Late Pleistocene Southwest Asia. *Paléorient* 17:19-45.

Wright, K. and Garrard, A. 2003. Social Identities and the Expansion of Stone Bead-Making in Neolithic Western Asia: New Evidence from Jordan. *Antiquity* 77:267-84.

Wright, K. I., Critchley, P. and Garrard, A. 2008. Stone Bead Technologies and Early Craft Specialization: Insights from Two Neolithic Sites in Eastern Jordan. *Levant* 40:131-65.

Yerkes, R. W. 1993. Methods of Manufacturing shell beads at prehistoric Mississippian Sites in Southeastern North America, P.C. Anderson, S. Beyries, M. Otte and H. Plisson, eds, *Traces et fonction: les gestes retrouvés*. Liège: ERAUL. Pp 235-42.

Zilhão, J., Angelucci, D. E., Badal-Garcia, E., D'Errico, F., Daniel, F., Dayet, L., Douka, K., Higham, T. F. G., Martinez-Sanchez, M. J., Montes-Bernandez, R., Murcia-Mascaros, S., Perez-Sirvent, C., Roldan-Garcia, C., Vanhaeren, M., Villaverde, V., Wood,

R. AND ZAPATA, J. 2010. Symbolic use of marine shells and mineral pigments by Iberian Neandertals. *Proceedings of the National Academy of Science of the U.S.A.* 107:1023-8.

Internet resources:

Bead Society of Greater Washington; http://bsgw.org/whatisabead.html Accessed August 31, 2011.

9 - SHELL TOOLS IN AN EARLY NEOLITHIC COASTAL SITE IN THE CANTABRIAN REGION (NORTHERN SPAIN): AN EXPERIMENTAL PROGRAM FOR USE-WEAR ANALYSIS AT SANTIMAMIÑE CAVE

David CUENCA-SOLANA
CNRS, UMR 6566 CReAAH. Centre de Recherche en Archéologie Archéosciences Histoire. Université de Rennes 1. Avenue du général Leclerc Campus de Beaulieu bâtiment 24-25 Université de Rennes 1 CS74205 - 35042 Rennes Cedex
david.cuencasolana@univ-rennes1.fr

Igor GUTIÉRREZ-ZUGASTI
Instituto Internacional de Investigaciones Prehistóricas de Cantabria, Ed. Interfacultativo Universidad de Cantabria, Avda. de los Castros s/n, 39005 Santander (Cantabria - Spain).
igorgutierrez.zug@gmail.com

Ignacio CLEMENTE
Departamento de Arqueología y Antropología. IMFCSIC. C/Egipcíaques, 15. E-08001 (Barcelona – Spain),
ignacio@imf.csic

Abstract : *One of the most common debates in the Mesolithic and the early Neolithic of the Cantabrian region (Northern Spain) has focused on the scarcity of lithic and osseous technologies in contexts with large accumulations of shells. To date, several explanations have been proposed, such as differences in the use of the space, a greater use of perishable materials like wood or changes in subsistence strategies. However, seven shells used as tools have been identified in the early Neolithic at Santimamiñe cave (Basque Country, Spain), which is the first evidence of this kind in the region. We therefore propose the hypothesis that shells were used as tools by these human groups in productive activities. To confirm or reject the results obtained through the application of functional analysis on these seven tools, we have developed an experimental program with different mollusc shells that have been tested on wood, dry/fresh animal skin and non-woody plants. From a traceological perspective, we have checked the variables implied in these activities and they have provided enough information to compare the experimental results with the archaeological shells. Finally, the results from the experimental program have allowed us to confirm the utilization of these shell tools in varied activities related with processing wood, plants and skin.*

Key words *: Shell tools, Archaeomalacology, Experimental program, Functional analysis, Neolithic*

Introduction

The use of shells as implements has been studied in different geographical areas of the world for several decades (Charpentier, Méry and Phillips 2004; Choi and Driwantoro 2007; Cristiani *et al.* 2005; Gruet 1993; Jones and Keegan 2001; Lammers-Keijsers 2008; Mansur and Clemente 2009; Smith and Allen 1999; Szabó, Brumm and Bellwood 2007; Szabó 2008; Toth and Woods 1989; Vigié 1987, 1995; Vigié and Courtin 1986, 1987 among others). However, the approach used by archaeologists to study these kinds of tools has generally involved describing the use made of the different artefacts by human groups or developing typological classifications based mainly on their shape and appearance and attempting to infer their use in practice (Dacal Moure 1978; Dacal Moure and Rivero de la Calle 1984; Marquardt and Payne 1992; Prous 1992, among others). This type of approach has been shown to be insufficient as a way to know for which activity and in which way a tool has been used. Out of this insufficiency arose the need of applying use-wear analysis or traceology (Semenov 1964).

In Cantabrian Spain, several likely shell tools were identified during the archaeomalacological study of the deposit at Santimamiñe cave (Bizkaia, Spain) (López-Quintana and Guenaga 2007) (fig. 9-1), after observing alterations such as rounding in some samples. In total, nine shell fragments were studied, from the Neolithic levels Lsm and Slm. These levels were termed according to their sedimentary characteristics, following the principles of Analytic Stratigraphy (Sáenz de Buruaga *et al.* 1998) and cover a chronologic interval ranging from 4380 to 3660 cal BC (dates calibrated to 2σ using CalPal Hulu 2007) (Weninger and Jöris 2008; Weninger, Jöris, and Danzaglocke 2008). From the Lsm level came two hinge fragments of right valves, a fragment of an impression of the adductor muscle of a right valve and an edge fragment, all belonging to the species *Ostrea edulis* (Linné, 1758). An edge of *Patella* sp. and another of *Ruditapes decussatus* (Linné 1758) came from the same level. From the Slm level, a right valve hinge and a fragment of valve with an impression of the adductor muscle of *Ostrea edulis*, and a *Mytilus galloprovincialis* (Lamark 1819) fragment were selected. Seven of these were interpreted

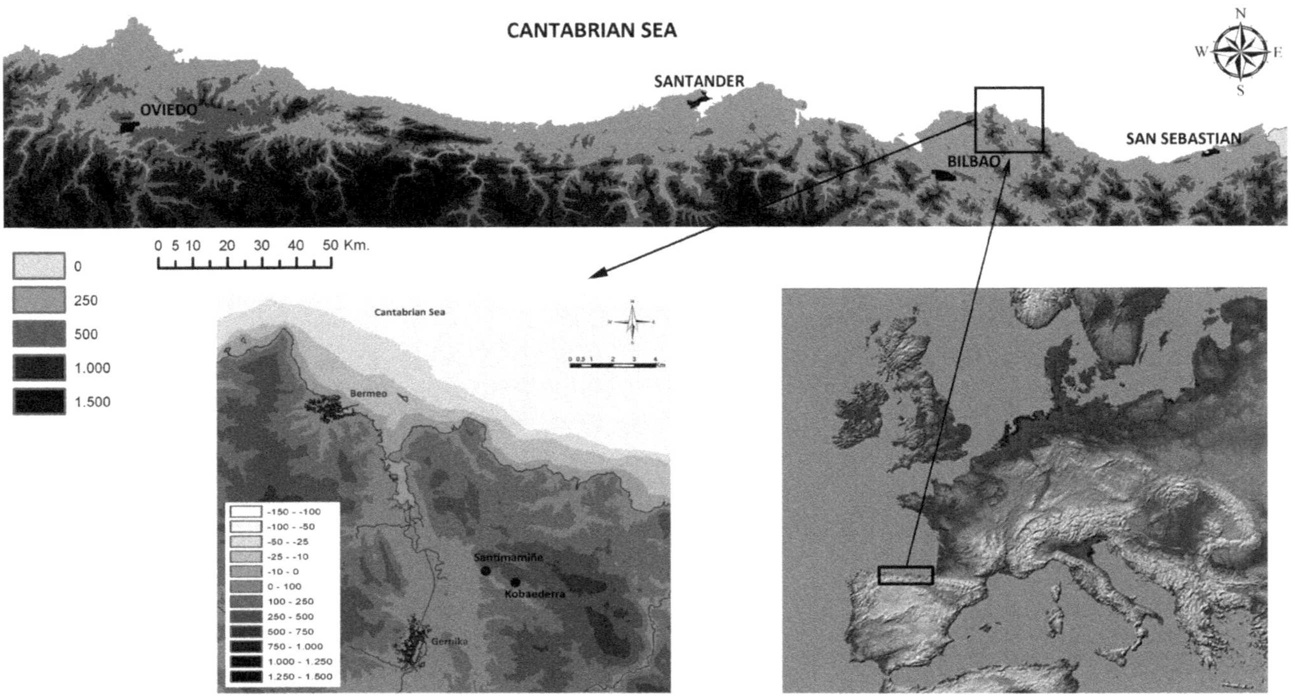

Figure 9-1. Location of Santimamiñe cave (Bizkaia, Northern Spain).

by use-wear analysis as tools used in transversal scraping actions, in some cases alternating longitudinal action to process various soft-medium hardness substances of animal origin and also plant matter (Gutiérrez Zugasti *et al.* 2011) (fig. 9-2). Three *Ostrea edulis* fragments displayed marks on one of the natural edges and a further one on the hinge which was noticeably rounded. These were characterized by closed polish, a matt finish and greasy aspect. In addition, the fragments of *Patella* sp. and *Mytilus galloprovincialis* showed the marks of transversal movement on one and two edges respectively, produced by working some kind of abrasive animal matter. Finally, a *Ruditapes decussatus* fragment with one strongly-rounded edge, exhibited wear with a compact polish together with striations with a dark bottom, oblique and perpendicular to the edge. This fragment was identified as a tool used to process plant matter with a high silica content (Gutiérrez-Zugasti *et al.* 2011). Evidence of use-wear was not found on the two remaining fragments.

In this paper, we present the experimental program designed to identify if these shells were used by hunter gatherers occupying the cave during the start of the Neolithic and the activities they carried out with these tools.

Materials and methods

In order to prove/disprove the hypothesis of the technological use of these remains, an experimental program was developed in parallel to obtain a description of the use-wear marks generated by such implements and the processes that create them. Several variables, both modifiable and unmodifiable (Clemente 1997), were taken into account. The control of these variables meant the experiment was analytical (González-Urquijo and Ibáñez-Estévez 1994). In the first place, the shell implements were observed macroscopically at between 5 and 72.5X with a Leica MZ16A binocular magnifying-glass, in order to analyse and photograph macro-traces, mainly rounding and scarring. Later, the objects were observed and photographed at between 100 and 200X with a Leica DM2500M microscope to analyse the micro-wear. The surfaces of the micro-polish and the bottom of the striations were recorded with a Leica duplicator at up to 400X. The shells used in the experiment were cleaned with water. Before the experimental program began, the most general taphonomic traits and alterations were recorded for each species, by observing and photographing specimens that had been gathered in the same circumstances but not used. The objective was to be able to differentiate natural alterations found on the shells from those caused by their use as tools. So, the biological activity of molluscs and its deposition in archaeological layers can produce alterations on the shell surface. However, these alterations are clearly different from alterations produced by the use of shells as tools. Natural alterations are chaotic and without any order, they usually also appear in marginal areas of the shell and they do not produce well developed polish.

In order to carry out the experiments, a number of plant and animal substances were selected, based on the results obtained by use-wear analysis applied to other kinds of materials, such as flint. Thus, based on the directionality of the marks seen on the tools being studied, the following tasks were performed: bidirectional transversal scraping action with shells of *Ostrea edulis*, *Mytilus*

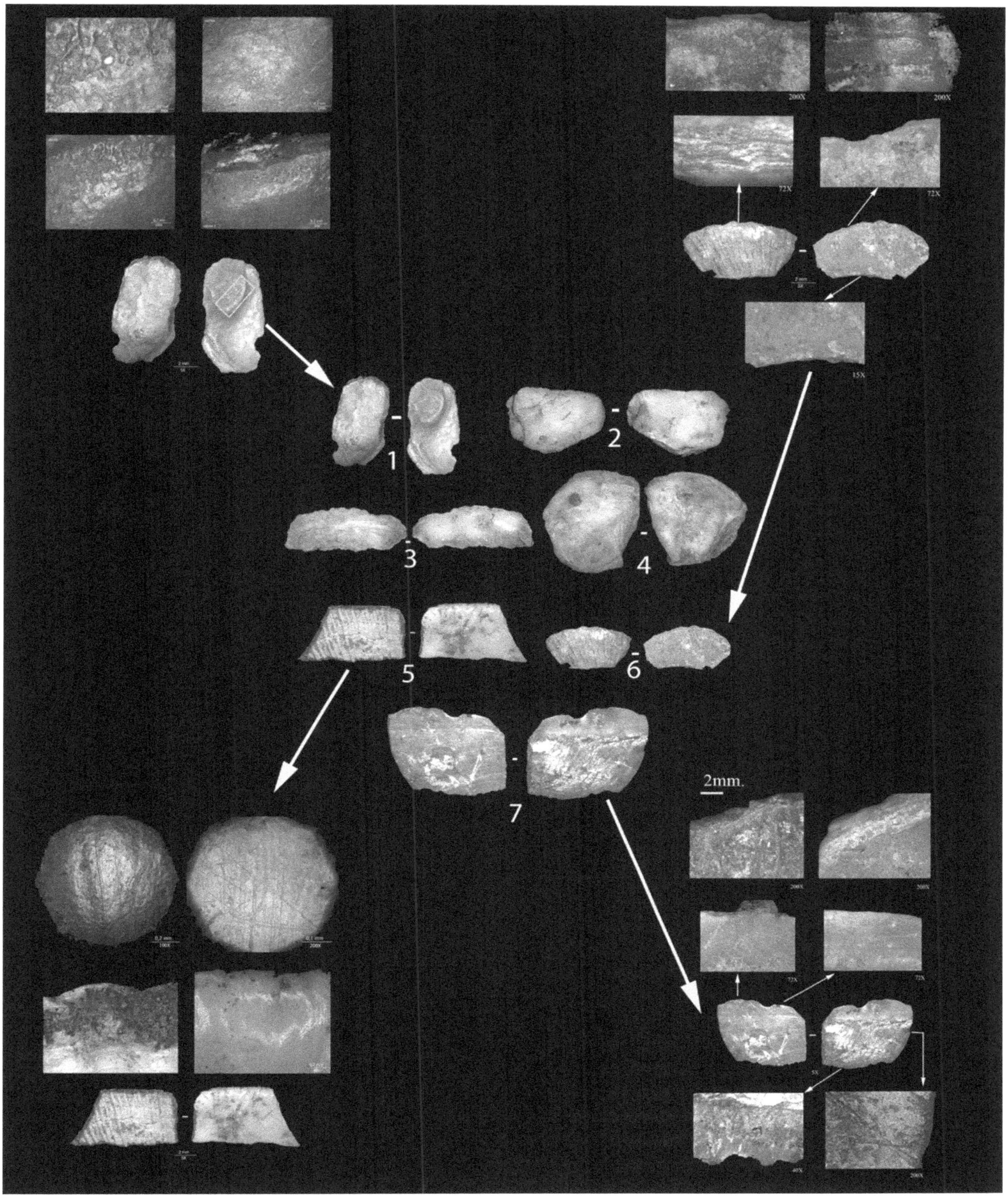

Figure 9-2. Shell tools from the Neolithic levels at Santimamiñe (Slm-Lsm). 1) Hinge of *Ostrea edulis*. 2) Impression of the adductor muscle of *Ostrea edulis*. 3) Edge of *Ostrea edulis*. 4) Impression of the adductor muscle of *Ostrea edulis*. 5) Edge of *Ruditapes decussatus*. 6) Edge of *Patella* sp. 7) Edge of *Mytilus galloprovincialis*.

galloprovincialis, and *Patella* sp. to clean fatty matter and remains of meat, and to thin down fresh skin of *Ovis aries* (Linné 1758) and dry hide of *Capreolus capreolus* (Linné 1758). Bidirectional transversal action was used with an *Ostrea edulis* hinge fragment to thin down and soften strips of hide of *Cervus elaphus* (Linné 1758). A unidirectional transversal scraping action with the natural edge of *Ruditapes decussatus* was employed to extract plant fibres from *Juncus* sp. (reed). The same species was used with a unidirectional transversal action to whittle wands of *Corylus avellana* (hazel) (Linné 1753). Actions with longitudinal movements were used in the experiments to cut the plant fibres from *Juncus* sp. using the edge obtained by fracturing a *Ruditapes decussatus* shell by percussion. In all cases the application of force has involved working with the implement on each substance for three lengths of time of 5, 10 and 15 minutes. The contact angles of the implement with the processed matter were right angles (about 90°), in both the transversal and the longitudinal actions. Only in the experiment of working hide with the *Ostrea edulis* hinge fragment has a flat angle (about 180°) been used, as the hinge surface is flat (Cuenca 2009; Cuenca, Clemente, and Gutiérrez-Zugasti 2010).

Results

***Wood* (Corylus avellana)**

As movement was unidirectional scraping, the polish is located on the internal side of the *Ruditapes decussatus* shell (fig. 9-3.1). It is glossy, flat and compact, with smooth microtopography (fig. 9-3.3). The edge became extremely rounded after working with the implement for fifteen minutes. Regarding the microscopic traces, the implements used to process wood exhibit thin, long striations with a dark bottom,

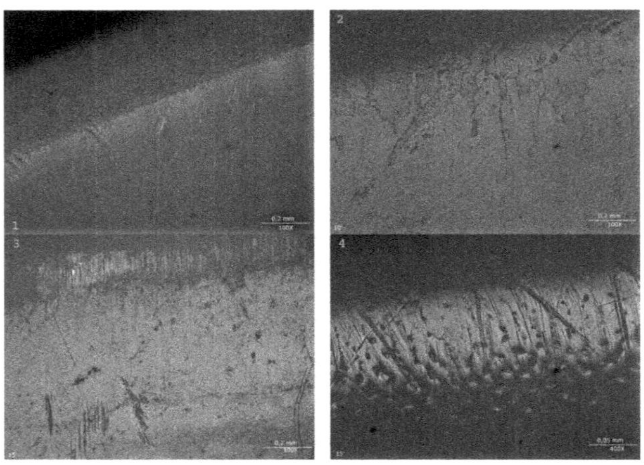

Figure 9-3. Use-wear in *Ruditapes decussatus* experimental tools employed for wood-working. 1) Polish in the area of the edge on the internal face after 10 minutes' work, at 100X. 2) "Chipping" produced on the superficial layer of the internal face, at 100X. 3) Polish on the internal face after 15 minutes' work, at 100X. 4) Use-wear marks in the area of the edge after 15 minutes, at 400X.

perpendicular to the edge, together with finer striations, which almost look like scratches, oblique to the edge (fig. 9-3.4). The most noticeable trait seen after processing wood is the loss of the superficial layer on the internal side of the shell and the areas that have been most in contact with the wood have a "chipped" appearance (fig. 9-3.2 / Description of use-wear in Table 9-1).

***Non wood plant* (Juncus *sp.*)**

The polish is located mainly on the internal edge of the *Ruditapes decussatus*, and also in the area of the cutting edge and to a lesser extent on the external side as the rough layer has been lost.

The polish is closed, with smooth microtopography, in the area of the edge (fig. 9-4.1). Towards the interior of the shell, the polish opens up, and becomes semi-closed. The polish is closed and glossy with a rough microtopography on the cutting edge of the shell (fig. 9-4.2). On the external side, the polish is less glossy; almost matt, closed and with rough microtopography. The rounding of the edge is more noticeable on the internal side of the shell, but is not very marked. These tools exhibit small crescent or semi-circular shaped scars along the edge which tend to line up progressively (fig. 9-4.3).

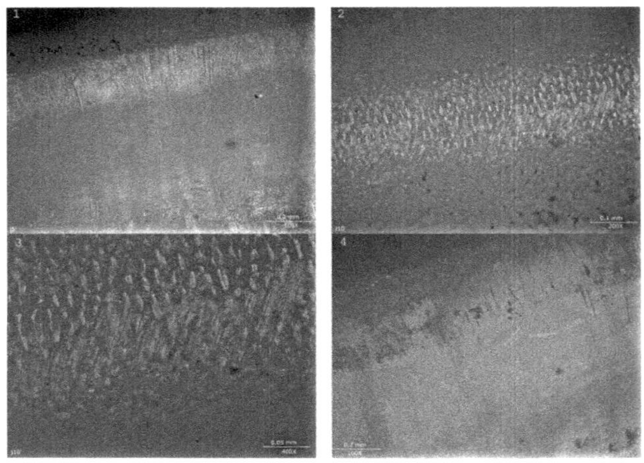

Figure 9-4. Use-wear in *Ruditapes decussatus* experimental tools employed for processing non-woody plants. 1) Polish in the area of the internal side of the edge after 5 minutes' work, at 100X. 2) Alteration in the area of internal side of the edge after 10 minutes' work, at 200X. 3) Details of the anterior area, at 400X. 4) Use-wear marks in the area of the internal side of the edge after 15 minutes' work, at 100X.

On the rounded parts of the edge, compound short striations are formed with a glossy bottom. Towards the interior of the shell, the striations become longer and finer, with a dark bottom, distributed longitudinally (fig. 9-4.4). In the instruments used in longitudinal cutting-actions they are also oblique. On the edge, the striations are less numerous, but also fine and long, with a dark bottom, aligned longitudinally to the edge (description of use-wear in Table 9-2).

***Fresh hide* (Ovis aries)**

The polish began to develop on the internal side after ten minutes' working, located marginally on the edge of the shell (fig. 9-5.1). After fifteen minutes, the polish spread to the external side and on the rounded cutting edge. It

LOCATION OF THE POLISH	CHARACTERISTICS OF THE POLISH	MACRO-WEAR TRACES	MICRO-WEAR TRACES
Internal Face	-Flat and compact polish with glossy appearance -Smooth microtopography	-High level of rounding after 15 minutes	-Fine, long striations with a dark bottom, perpendicular to the edge -Finer striations with oblique orientation -Loss of the superficial layer on the internal side of the shell in the areas of greatest contact with the processed matter

Table 9-1. Description of use-wear in experimental tools used for wood-working.

is closed, glossy and with a very greasy appearance. The microtopography is slightly rough, and tends to be smoother on higher parts of the shell's topography. In the same way, on the *Mytilus* shells, the polish becomes compact on the higher parts of the topography of the internal side (fig. 9-5.3 and 5.4).

The polish developed more quickly on the internal face of *Mytilus galloprovincialis* shells than on *Ostrea edulis* and *Patella* sp. shells. This is probably related to the flat morphology of *Mytilus galloprovincialis* that generates les mechanical resistance, allowing for a greater degree of contact between the shell surface and the worked material than in other more concave and thicker shells, as *Ostrea edulis* or *Patella* sp. So, the flatter shape of the mussels produces a faster development of the use-wear in the tools (Cuenca 2009; Cuenca, Clemente, and Gutiérrez-Zugasti, 2010).

After fifteen minutes, the polish could be seen on the edge of the external side. This is closed, glossy, with a greasy appearance and rough microtopography. The edge became highly rounded after working with the tool for fifteen minutes.

On the *Ostrea edulis* shells, the rough layers of the external face were fractured and became smoothed. Some isolated semi-circular scarring was seen on the rough layer of the external side of the shells, both of *Ostrea edulis* and *Patella* sp. After fifteen minutes, striations developed on the internal face of *Ostrea edulis*. These were deeper and wider, perpendicular to the edge, while other oblique striations were more superficial and finer. Both types were irregular and had a dark bottom. On the *Mytilus galloprovincialis* and *Patella* sp. tools, irregular micro-holes with a dark bottom were documented on the internal side of the shell. However, this kind of microscopic

LOCATION OF THE POLISH	CHARACTERISTICS OF THE POLISH	MACRO-WEAR TRACES	MICRO-WEAR TRACES
-Internal face -On the edge -To a lesser extent, on the external face when the periostracum is lost	-Closed and smooth microtopography in the area of the edge. -Towards the interior of the shell the polish is semi-closed - Closed and glossy on the edge, with rough microtopography. -On the external face the wear is less glossy, almost matt with a closed polish and rough microtopography	-Low/medium level of rounding located mainly on the internal face of the shell edge.	-Scarring in a half moon or semi-circular shape, which lines up progressively -Short compound striations with a glossy bottom in the area of the edge -On the internal face they are longer and finer with a dark bottom -They are less numerous on the edge, and are fine and very long -On the external face they have a dark bottom, are fine and long and longitudinal to the edge, except in the experimental objects used for cutting, where they are also oblique.

Table 9-2. Description of use-wear in experimental tools used for plant processing.

LOCATION OF THE POLISH	CHARACTERISTICS OF THE POLISH	MACRO-WEAR TRACES	MICRO-WEAR TRACES
-On the internal face, after working for 10 minutes located marginally on the shell edge. -After 15 minutes, the wear develops on the outer face and on the edge -The wear develops more quickly in the case of *Mytilus galloprovincialis* than for *Ostrea edulis* and *Patella* sp.	-Glossy and closed polish with a very greasy appearance. -The microtopography is slightly rough, tending to become smoother in the higher parts of the shell's surface. -On the *Mytilus* shells, more compact polish in the highest parts of the surface of the internal face. -After 15 minutes, wear is seen to develop on the edge of the external face. It has a glossy and closed polish with a greasy appearance and rough microtopography.	-A high level of rounding of the edge is reached after working with the implement for 15 minutes	-The rough layers on the external face of the *Ostrea edulis* shells are fractured. -The surface tends to be smoothed with continuous circles that make up the external face. -Occasional scarring with a semi-circular shape occurs on the edge of the rough layer of the external face of both *Ostrea edulis* and *Patella* sp. shells. -After 15 minutes deeper and wider striations develop on the internal face, perpendicular to the edge, together with more superficial and finer oblique striations. Both kinds are irregular and have a dark bottom -On the *Mytilus galloprovincialis*, *Patella* sp. and *Ostrea edulis* implements, irregular micro-holes with a dark bottom are seen on the internal face of the shell.

Table 9-3. Description of use-wear in experimental tools used for fresh hide processing.

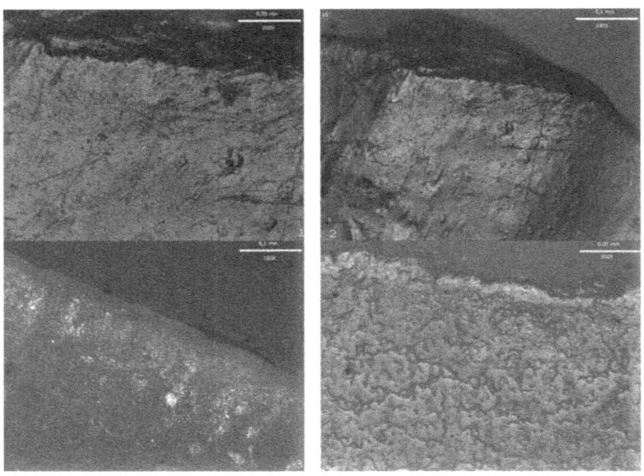

Figure 9-5. Use-wear in *Patella* sp. (top) and *Mytilus galloprovincialis* (bottom) experimental tools employed for fresh hide processing. 1) Polish in the area of the edge of the internal face of *Patella* sp. after 15 minutes' work, at 200X. 2) Edge of the internal face of *Patella* sp. with matt polish and striations with a dark bottom, after 15 minutes' work, at 100X. 3) Polish starting to develop on the internal face of *Mytilus galloprovincialis* after 5 minutes' work, at 100X. 4) Use-wear marks in the area of a *Mytilus galloprovincialis* edge after 10 minutes' work, at 200X.

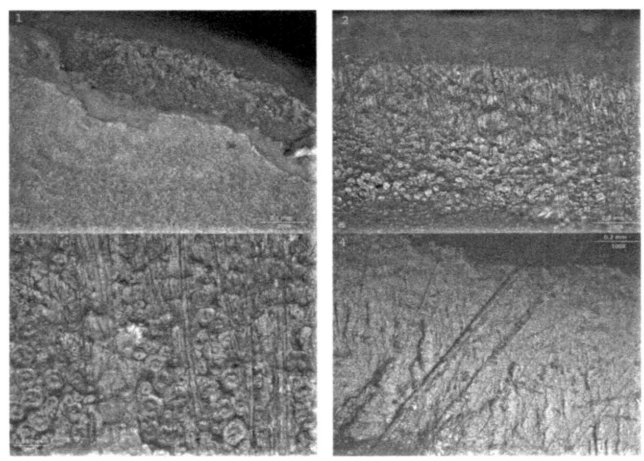

Figure 9-6. Use-wear in *Ostrea edulis* experimental tools employed for dry hide processing. 1) Fracture in the superficial face in the area of the *Ostrea edulis* edge and the unequal development of polish after 10 minutes' work, at 200X. 2) Use-wear marks on the external face of *Ostrea edulis* after 10 minutes' work, at 100X. 3) Detail of the anterior part, with smoothing of the external face of the shell and numerous striations with a dark bottom, at 300X. 4) Details of the striations with a dark bottom, in the area of the *Ostrea edulis* edge, after 15 minutes' work, at 100X.

alteration did not occur in the case of the *Ostrea edulis* tools (Description of use-wear in Table 9-3).

***Dry hide* (Capreolus capreolus)**

The loss of the superficial layer can be seen in some parts of the internal side of the *Ostrea edulis* tools, causing different alterations (fig. 9-6.2). In this way, on the preserved superficial layer the polish is matt and closed with rough microtopography, whereas in places where the superficial layer has been lost, the polish is less well-developed and semi-closed, with rough microtopography. In the case of the *Mytilus* shells, the polish is found mainly on the higher parts of the undulating topography, whereas on *Patella* sp. shells it becomes compact in the higher parts of the microtopography. The polish is closed, with a greasy appearance. In the same way, the cutting edge becomes highly rounded at fifteen minutes' work.

On both the *Ostrea edulis* and *Patella* sp. shells, the external layer of the shell is fractured, which forms two cutting edges, one above the other (fig. 9-6.1). However, marks caused by hand pressure on the tool are only seen on the *Ostrea edulis* tools, where they become visible in the area of the umbo on the external face and the area of the muscle on the internal face, after fifteen minutes' work. On both the *Mytilus* and the *Patella* sp. tools, semicircular or crescent shaped semi-closed scarring is seen in isolation along the edge, tending to line up progressively. However, in the *Mytilus* shells, the cutting edge exhibits abrupt scarring which occasionally produces a dentate appearance.
On the internal side, wide, short and deep striations with a dark bottom can been seen in the area of the edge, perpendicular to the cutting edge (fig. 9-6.3 and 4). On the external side, the striations are very abundant, very fine and long, with a dark bottom.

Circular micro-holes have been documented on the *Mytilus* and *Patella* sp. tools. This kind of alterations are usual in lithic tools used to work dry hide (Clemente 1997; González-Urquijo and Ibáñez-Estévez 1994; Van Gijn 1989), as this is a soft but very abrasive material. The abrasive effects are compounded when elements as salt, ash or ochre are used to work the hide (Vaughan 1985) (description of use-wear in Table 9-4).

LOCATION OF THE POLISH	CHARACTERISTICS OF THE POLISH	MACRO-WEAR TRACES	MICRO-WEAR TRACES
-Loss of the superficial layer in some parts on the internal face of ***Ostrea edulis***, resulting in different alterations and development of the wear. -On the ***Mytilus*** shells, the wear is concentrated mainly on the elevations of their undulating surface. -On the ***Patella*** sp. specimens, the wear is located in the area of the edge on the internal face.	-On the preserved superficial layer the wear is matt, with a closed polish and rough microtopography. -On the areas where the superficial layer has been lost the wear is less developed, the polish is semi-closed and the microtopography is rough. -On the ***Patella*** sp. shells, compact polish in the highest parts of their surface. - Closed polish and greasy appearance on the ***Patella*** sp. shells.	-The rounding of the edge reaches a high level after 15 minutes -After 15 minutes working time, the ***Ostrea edulis*** implements exhibit marks caused by hand pressure on the implement, visible in the area of the hinge on the external face and in the area of the adductor muscle on the internal face	-On the ***Ostrea edulis*** and ***Patella*** sp. shells, the outer layer of the shell is fractured, forming two superimposed edges. - Semi-circular flaking in ***Mytilus*** and ***Patella*** sp. located occasionally on the edge and tending to line up progressively. -On the ***Mytilus*** shells, the edge displays abrupt scarring -On the internal face, the striations have a dark bottom, are short, wide and deep, perpendicular to the edge. -On the external face, the striations are more numerous, very fine and long, with a dark bottom. -On the ***Mytilus*** and ***Patella*** sp. implements, round micro-holes have been recorded.

Table 9-4. Description of use-wear in experimental tools used for dry hide processing.

Discussion and conclusion

The results of the experimental program have confirmed that the shell fragments found in the Neolithic level at Santimamiñe Cave (Bizkaia, Spain) were used as tools. These examples represent the first evidence of shells being used for various productive tasks that have been identified in northern Spain (Gutiérrez-Zugasti *et al.* 2011).

The *Ruditapes decussatus* fragment was probably used to process a non-woody plant. Despite the similarity between the marks in the archaeological and experimental material, they were more developed in the archaeological specimen. This was therefore probably used for a harder non-woody plant than *Juncus* sp., or it was used for a longer length of time than in the experiment. The *Ostrea edulis* hinge displayed very similar diagnostic marks, in this case perhaps somewhat more developed in the experimental specimens. The greater or lesser dryness of the animal hide might be an important factor as the abrasiveness of this substance can cause the development of deeper marks on the shell in a shorter space of time. The experiment has shown that the other fragments, interpreted as implements used to scrape animal matter, were more likely used for dry hide than fresh skins, or that some abrasive was used to tan the hides (such as ochre) which causes more developed marks and more striated surfaces (Cuenca 2009; Cuenca, Clemente and Gutiérrez-Zugasti 2010).

The documentation of shell tools at archaeological sites in this part of Europe, where they had never been identified previously, highlights once more the need to reconsider the view that molluscs are almost exclusively food waste or ornaments (Cuenca, Gutiérrez-Zugasti, and Clemente 2011). Only through the application of functional analysis methodology is it possible to define the use made of any kind of tool. This type of analysis is aimed at inferring the tasks the tools were used for through the study of the use-wear marks generated on their active surfaces. The analysis should therefore be based on experimentation as a way of producing a sufficiently large field of inference. In addition, the use of observation techniques at both high and low magnification is important, as each case corresponds to a different purpose and these are necessarily compatible for a complete study. Finally, it is vitally important that the material should be conserved correctly, and it is therefore essential that it is treated appropriately during the archaeological work itself (particularly in the excavation and washing of the material) and during its storage, in order to guarantee the satisfactory conservation of the malacological remains (Cuenca 2010).

The importance of introducing shell tools into the program of studies is founded on the belief that only through an understanding of all the means of production can we assess the level of technological development reached by different past societies more objectively.

Acknowledgements

The authors would like to thank to the University of Cantabria for financial support, Alejandro García-Moreno for his help and K. Szabó and C. Dupont for their comments. Igor Gutiérrez Zugasti a researcher from the Juan de la Cierva programme, funded by the Spanish Government. David Cuenca-Solana is funded by a Fondation Fyssen.

References

CHARPENTIER, VINCENT, SOPHIE MÉRY and CARL PHILLIPS. 2004. Des coquillages... outillages des Ichtyophages? Mise en évidence d'industries sur Veneridae du Néolithique à l'Âge du Fer (Ymen, Oman, E.A.U.), *Arabian Archaeology and epigraphy* 15: 1–10.

CHOI, KILDO and DUBEL DRIWANTORO. 2007. Shell tool use by early members of *Homo erectus* in Sangiran, central Java, Indonesia: cut mark evidence, *Journal of Archaeological Science* 34: 48-58.

CLEMENTE, IGNACIO. 1997. *Los instrumentos líticos de Túnel VII: una aproximación etnoarqueológica.* Treballs d'Etnoarqueologia 2, CSIC-UAB. Barcelona.

CRISTIANI, EMANUELA, CRISTINA LEMORINI, FABIO MARTINI and LUCIA SARTI. 2005. Scrappers of *Callista chione* from Grotta del Cavallo (Middle Paleolithic cave in Apulia): evaluating use-wear potential. In HEIDI LUIK, ALICE.CHAYKE, COLLEEN BATEY and LEMBI LOUGAS (Eds.): *From hooves to horns, from mollusc, to mammoth. Manufacture and use of bone artefacts from prehistoric times to the present. Proceedings of the 4th meeting of the ICAZ Worked bone Research Group at Tallin, 26th- 31st of August 2003.* Tallin Book Printers. Tallin: 319-324.

CUENCA, DAVID. 2009. *Las "tecnologías invisibles" en los grupos de cazadores recolectores del litoral durante los inicios del Holoceno (9.500-5.000 uncal BP) en la región Cantábrica. Utilización de las conchas de molusco en la realización de actividades productivas.* Unpublished MA Thesis. Santander. Universidad de Cantabria.

—. 2010. Los efectos del trabajo arqueológico en conchas de *Patella* sp. y *Mytilus galloprovincialis* y su incidencia en el análisis funcional. *Férvedes. Revista del Museo de Prehistoria e Arqueoloxia de Vilalba (Lugo),* 6:43-51.

CUENCA, DAVID, IGNACIO CLEMENTE and IGOR GUTIÉRREZ ZUGASTI. 2010. Utilización de instrumentos de concha durante el Mesolítico y Neolítico inicial en contextos litorales de la región cantábrica: programa experimental para el análisis de huellas de uso en materiales malacológicos. *Trabajos de Prehistoria* 67, 1: 211-225.

Cuenca, David; Igor Gutiérrez-Zugasti, and Ignacio Clemente. 2011 The use of molluscs as tools by coastal human groups: contribution of ethnographical studies to research on Mesolithic and early Neolithic contexts in Northern Spain. *Journal of Anthropological Research* 67, 1: 77-102.

Dacal-Moure, Ramón. 1978. *Artefactos de concha en las comunidades aborígenes cubanas.* Universidad de la Habana. Cuba.

Dacal-Moure, Ramón and Manuel Rivero de le Calle. 1984. *Arqueología aborigen de Cuba.* Editorial Gente Nueva.Cuba.

Gijn, Annelou, van. 1989. *The wear and tear of Flint.* Analecta Praehistorica Leidensia 22. University of Leiden.

González-Urquijo, Jesús E. and Juan J. Ibáñez-Estévez. 1994. *Metodología de análisis funcional de instrumentos tallados en sílex.* Universidad de Deusto. Cuadernos de Arqueología 14. Bilbao.

Gruet, Yves. 1993. Les coquillages marins: objets archéologiques à ne pas négliger. Quelques exemples d'exploitation et d'utilisation dans l'Ouest de la France. *Revue Archéologique de l'Ouest* 10:157-161.

Gutiérrez-Zugasti, Igor, David Cuenca, Ignacio Clemente, César González-Sainz and Juan C. López-Quintana. 2011. Instrumentos de trabajo y elementos de adorno en conchas de molusco de la cueva de Santimamiñe (Kortezubi, Bizkaia). In Juan C. López Quintana (Ed.): La cueva de Santimamiñe: revisión y actualización (2004-2006). *Kobie* serie anejos. Diputación Foral de Bizkaia: 155-170.

Jones, Sharyn and William F. Keegan. 2001. Expedient Shell Tools from the Northern West Indies, *Latin American Antiquity* 12, (3): 274-290.

Lammers-Keijsers, Yvonne. 2008. *Tracing Traces from Present to past. A funtional analysis of pre-Columbian shell and stone artefacts from Anse á la Gourde and Morel, Guadaloupe, FWI.* University Press. Leiden, The Netherlands.

López-Quintana, Juan C. and Amagoia Guenaga. 2007. Avance a la secuencia estratigráfica de la cueva de Santimamiñe (Kortezubi), tras la revisión de su depósito arqueológico en las campañas de 2004 a 2006. *KREI* 9: 73-103.

Mansur, María E. and Ignacio Clemente. 2009. ¿Tecnologías invisibles? Confección, uso y conservación de instrumentos de valva en Tierra del Fuego. In Fernando Oliva, Nélida de Grandís and Jorge Rodríguez, (Eds.): *Arqueología Argentina en los inicios de un nuevo siglo.* Publicaciones del XIV Congreso Nacional de Arqueología Argentina. Universidad Nacional de Rosario, Facultad de Humanidades y Artes, escuela de Arqueología. Rosario: Laborde Libros Editor,Tomo II: 359-368.

Marquardt, William and Claudine Payne. 1992. *Culture and environment in the domain of the Calusa.* Institute of Archaeology and Paleoenvironmental Studies. University of Florida.

Prous, André 1992. Os moluscos e a arqueologia brasileira. *Arquivos do Museu de História Natural* (1986/1990) 11: 241-298.

Sáenz de Buruaga, Andoni, Mikel Aguirre, Carmen Grima, Juan C. López Quintana, Aitor Ormazabal and Blanca Pastor. 1998. Método y práctica de la Estratigrafía Analítica. *KREI* 3: 7-41.

Semenov, Serghei Aristarkhavich, 1964. *Prehistoric Technology.* London: Cory Adams and Mackay.

Smith, Anita and Jim Allen 1999. Pleistocene shell technologies: Evidence from Island Melanesia. In J. Hall and I. McNiven (eds). *Australian Coastal Archaeology.* ANH Publications, Department of Archaeology and Natural History, RSPAS, The Australian National University.

Szabó, Katherine. 2008. Shell as a Raw Material: Mechanical Properties and Working Techniques in the Tropical Indo-West Pacific. *Archaeofauna* 17: 125-138.

Szabó, Katherine, Adam Brumm and Peter Bellwood 2007. Shell artefact production at 32.000-28.000 BP in Island Southeast Asia. Thinking across media?, *Current Anthropology* 48 (5): 701-723.

Toth, Nicholas and Michael Woods. 1989. Molluscan shell knives and experimental cut-marks on bones. *Journal of Field Archaeology* 16: 250-255.

Vigié, Béatrice 1995. Du déchet alimentaire à l'objet coquillier: le statut des coquillages en milieu archéologique. In Camps G. (ed.) : *L'Homme préhistorique et la mer.* 120e Congrés National des Sociétés savantes. (Aix-en-Provence, 1995): 351-354. Paris.

—. 1987. Essai d'étude méthodologique d'outils sur coquillages de la grotte de Camprafaud (Ferrières-Poussarou, Hérault). *L'Anthropologie* 91 (1): 253-272.

Vigié, Béatrice and Jean Courtin. 1986. Les outils sur coquilles marines dans le Néolithique du Midi de la

France. *Mesogée* 46: 51-61.

—. 1987. Le problème des coquillages à bord dentelé dans la préhistoire du Midi de la France. *Mesogée* 47: 93-98.

WENINGER, BERNHARD and OLAF JÖRIS 2008. C_{14} age calibration curve for the last 60 ka: the Greenland-Hulu U/Th timescale and its impact on understanding the Middle to Upper Palaeolithic in Western Eurasia. *Journal of Human Evol*ution 55: 772-781.

WENINGER, BERNHARD, OLAF JÖRIS and UWE DANZAGLOCKE 2008. CalPal-2007, *Cologne Radiocarbon Calibration & Paleoclimate Research Package.* http:/www.calpal.de 26/11/08.

10 - SHELL MIDDENS AND THE USE OF MOLLUSCS IN THE LATE MIDDLE HOLOCENE IN THE RIO DE LA PLATA: AN ETHNOARCHAEOLOGICAL CONTRIBUTION

Laura BEOVIDE
Department of Archeology, National Museum of Anthropology, Uruguay. Avda. de las Instrucciones 948, Montevideo, CP 12900,
beovide@mec.gub.uy

Abstract : *Shell middens have been identified in archaeological sites in the Uruguayan coast of the Río de la Plata, with ages between ca. 2900 and 2200 cal. years BP. The middens are composed mainly of the shells of the bivalve* Erodona mactroides. *This species is currently exploited by artisanal fishers to supply restaurants, as well as by communities who live within 150 km eastwards of the archaeological sites. The artisanal activity produces important shell middens of this species in ecosystems that may be comparable to those proposed for the time of archaeological sites occupation in the late Middle Holocene. The article investigates the ethnoarchaeological study of these fishing communities in relation to the material products and area involved in the use, consumption and discard of the shells of* Erodona mactroides. *Taphonomic observations on the formation of the modern shell middens provide new analytical elements for the interpretation of archaeological middens. It also allows discussion of aspects of the archaeomalacological record that may be associated with strictly non-economic dimensions of past human life.*

Keywords : *Middens, Ethnoarchaeology, Río de la Plata Archaeology.*

Introduction

Although the presence of archaeological shell middens is well documented for the Atlantic coast of Uruguay, Brazil and Argentina (e.g. Acosta, Loponte, and Tchilingurian 2010; Bracco 2000; Gaspar 2000; Gaspar and De Blasis 2006; Lopez, Villarmarzo, and Brum 2009; Orquera and Piana 2000), identification of archaeological middens on the coast of the Río de la Plata river is more recent (Beovide 2008, 2010a).

As a result of the systematic archaeological investigation during the last decade (Beovide 2009, 2010b), pre-Hispanic shell middens have been dated to between ca. 2,900 – 2,200 cal. years BP (fig. 10-1; Table 10-1). They are mostly located in the lower reaches of one of the major tributaries of the Río de la Plata in Uruguay: the Santa Lucía river.

The earliest records of human occupation on the lower basin of Santa Lucía river are dated ca. 5000 cal. years BP (Beovide 2009). They are hunter-fisher-gatherer societies who have incorporated into their economy the raising of different cultigens and pottery production. The human

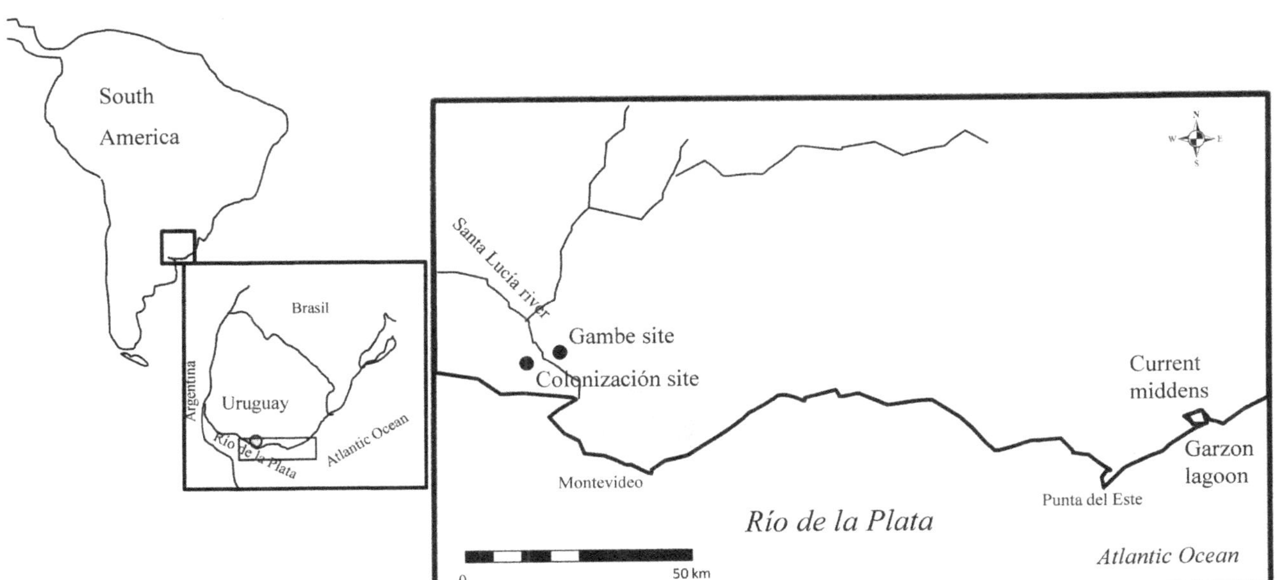

Figure 10-1. Location of the archaeological and current shell middens.

Site	Exc.	Level	^{14}C (years BP)	Calibrated dates: (68%) 1σ	Material/ species	Lab./ Nr.
Colonización	III	7	2,744± 50 (AMS)	2,797-2,907 cal BP 902 ±55 cal BC	Shell/ *Anomalocardia brasiliana*	AA/ 86685
Colonización	S4	7	2,716± 36 (AMS)	2,785-2,851cal BP 968 ±33 cal BC	Shell/*Erodona mactroides*	AA/ 86686
Gambe	III	7	2,710± 60	2,779-2879 cal BP 879±50cal BC	Shell/*Erodona mactroides*	URU / 0309
Colonización	IV	10	2,620± 50	2,718-2,783cal BP 801 ±32 cal BC	Shell/*Erodona mactroides*	URU/ 0551
Colonización	IV	15	2,600±50	2,598-2,762 cal BP 730±82 cal BC	Shell/*Erodona mactroides*	URU/ 0552
Gambe	S8	7	2,409± 32 (AMS)	2,382-2,598 cal BP 540 ±108 cal BC	Shell/*Erodona mactroides*	AA/ 86684
Gambe	III	5	2,365± 32 (AMS)	2,354-2,446 cal BP 450 ±46 cal BC	Shell/*Erodona mactroides*	AA/ 86683
Colonización	I	5	2,310± 60	2,211-2,416 cal BP 364 ±102 cal BC	Shell/*Erodona mactroides*	URU/ 0310

Table 10-1: Radiocarbon and calibrated dates for Gambe and Colonización sites and additional data (based on Beovide 2010a). Calibrated with CalPal 2007.

settlements were located in the myxohaline estuarine ecotones formed after the Climatic Optimum.

The presence of Pre-Hispanic shell middens in the area is related to changes in the economies of the human groups and in the natural coastal ecosystems. By ca. 2900 years BP, when the climate was more arid than today, and after a marine ingression on the Río de la Plata, a number of paleolagoons, which are now fossils and not active, appeared in the landscape (Beovide 2009, 2010b). Gambé and Colonización archaeological shell middens are located on the shores of these lagoons (fig. 10-1).

The formation of these shell middens is attributable to pottery-making societies who practiced small-scale horticulture incorporating *Zea mays* within their set of cultigens (Beovide 2010b). Besides horticultural crops, they exploited different resources such as minerals, plants and wildlife with greater intensity within areas between 10 and 20 km from the sites.

The wide flood plains allowed the exploitation of a species of myxohaline mollusc: *Erodona mactroides* which is the predominant species of anthropic shell middens (fig. 10-2). *Erodona mactroides* (Daudin in Bosc, 1801) is a fluvio-marine bivalve which is 3 cm long and lives in the coastal sandy substrates and is indicative of the temperate climates of the Argentinean province (Martínez *et al.* 2006).

Figure 10-2. *Erodona mactroides*. Scale in mm (size 33mm).

In the latter periods of the occupation of the lower basin of the Santa Lucía river, after ca 1600 BP and at the time of the Spanish - American Indian Contact, the local indigenous groups lived in wetlands with lower salinity levels compared to previous periods. A more humid climate together with changes in coastal geomorphology which reduced flood plains were factors that reduced and transformed the areas where *Erodona mactroides* was collected. These factors have contributed to the disintensification in the exploitation of this resource in later times of the occupation of the area (Beovide 2010b).

As mentioned above, changes in temperature as well as a decrease in salinity levels were among the main causes of the decline in abundance of *Erodona mactroides* banks in the middle stretch of the Río de la Plata.

Therefore the exploitation of *Erodona mactroides* is currently carried out 150 km eastwards of the archaeological sites. This bivalve is collected by traditional fishing communities living at the mouth of the Río de la Plata into the Atlantic Ocean, mainly in Garzón lagoon. These populations do not have any ethnic connection with the pre- Hispanic groups who formed the shell middens as these latter were ethnically exterminated almost two hundred years ago (Cabrera and Barreto 2006).

This artisanal exploitation has taken place in that area since the second half of the twentieth century and is to the result of adaptation of the fishing communities to

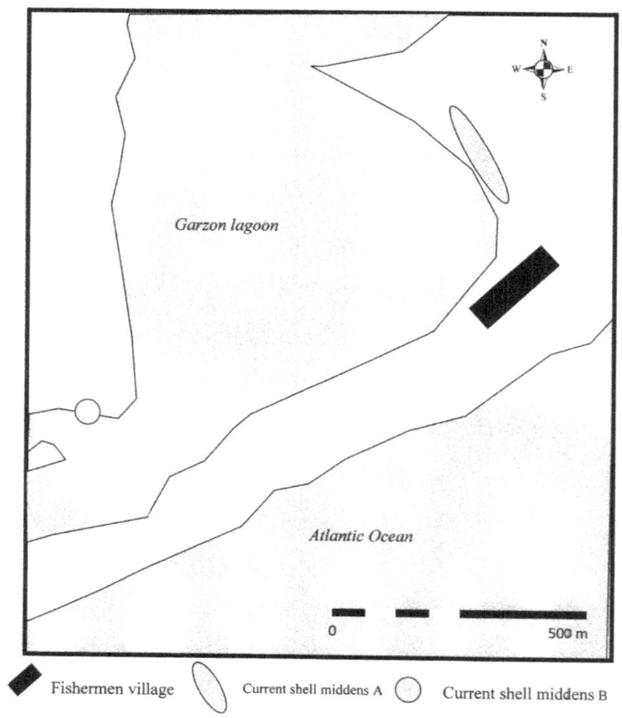

Figure 10-3. Location (schematically shown): current shell middens A, B and fishing village.

local natural resources. This adaptation includes the manufacture of handmade tools that allow them to extract and process shellfish more effectively. The shell middens formed by the current fishermen are an important source of actualistic information which helps us to evaluate the formation processes of the archaeological shell middens as well as the action of taphonomic processes, in this case of *Erodona mactroides*. The fishermen of the Garzón lagoon exploit *Erodona mactroide*s in more saline conditions if we compare them to the prevalent environment in the paleolagoons of the marine ingression into the Río de la Plata in the late Mid–Holocene. The current fishermen deposit the shellfish on the sandy coastal dunes bordering the *Garzón* lagoon (fig. 10-3 and fig. 10-4) with ecosystemic and taphonomic situations which are comparable with the archaeological context seen in the Santa Lucía river area.

This paper presents a research design focused on these modern shell middens using an ethnoarchaeological approach whose main goal is to solve problems related to the formation processes of the archaeological shell middens. In recent years the ethnoarchaeological approach has seen much in the way of theoretical developments in Latin American archaeology (Estevez and Vila 1996; Politis 2002, 2004; Prado 2003; Williams 2005) which have contributed to the discussion of problems of the formation of archaeological sites together with other actualistic studies (e.g. Piana and Orquera 2010; Politis and Jaimes 2005).

In this paper ethnographic analogy is considered as a heuristic (Gandara 1990) which allows the generation of a hypothesis to be tested regarding the processes and factors (anthropogenic/ natural) involved in the formation of rubbish dumps, which can then be applied to the case of archaeological shell middens. This article also considers

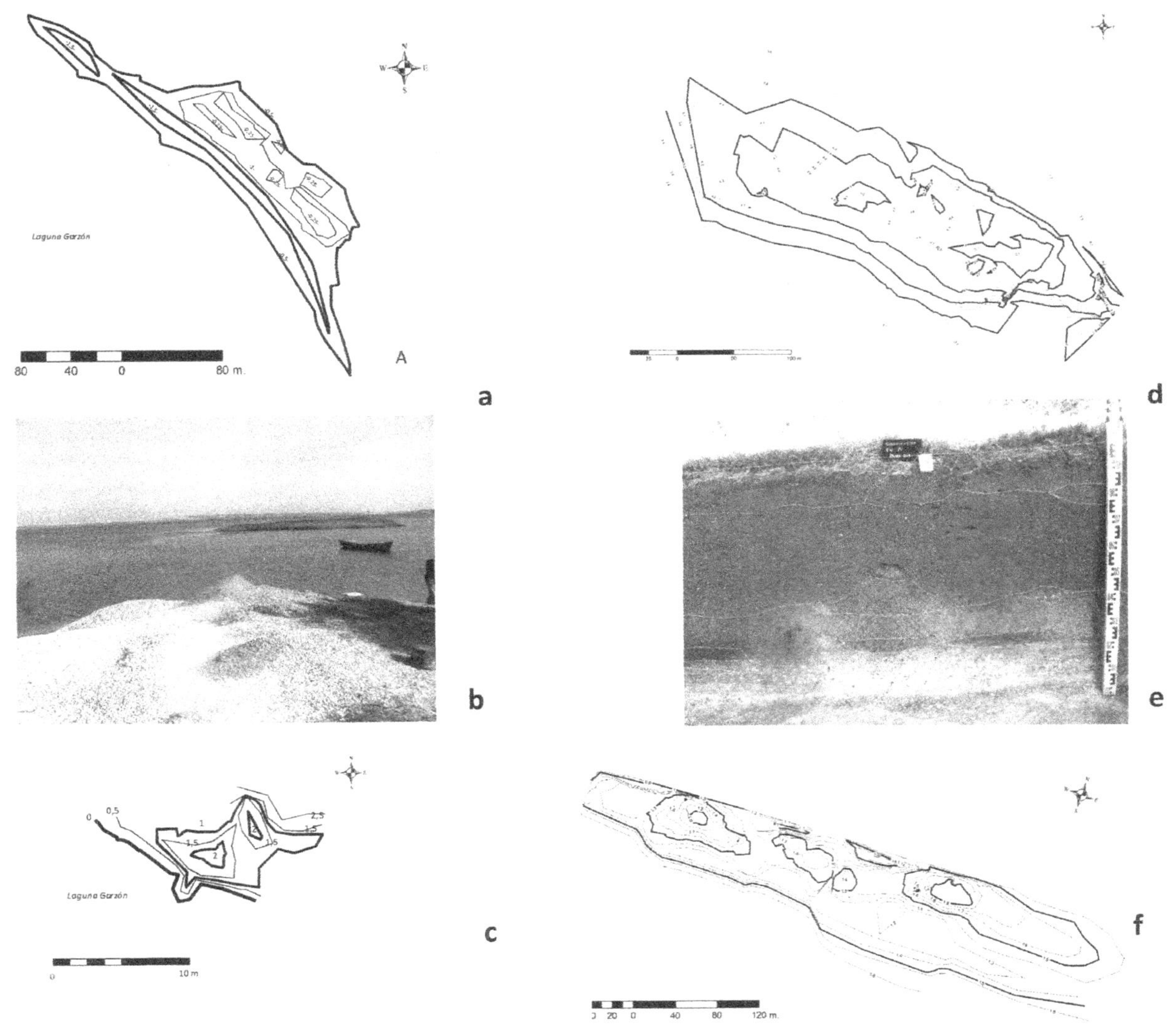

Figure 10-4. Topography of the current (a: shell midden A, c: shell midden B) and archaeological (d: *Gambe* site, f: *Colonización* site) shell middens. Photo: current (b) and archaeological (e: excavation IV profile, *Colonización* site) middens.

the idea that the current material record is the product of the techno-economic, social and ideational order of the actors that produced it (e.g. Gould 1978; Politis and Jaimes 2005). Therefore a series of hypotheses derived from the record of that current "order" of shell midden formation arise from the ethnoarchaeological approach. These hypotheses can be tested on the archaeological record.

The study of modern shell middens from Garzón lagoon includes the record of the material products and of the area involved in the gathering, as well as the use and disposal of *Erodona mactroides* shells. Some of the material products of these activities are compared with different samples of prehistoric shell middens from the middle stretch of the Río de la Plata. This has led to the generation of new elements of analysis and subsequent interpretation of the formation processes in the archaeological sites.

Traditional exploitation of *Erodona mactroides*.

According to the information collected in 2008 when this study was carried out, artisanal exploitation of shellfish in *Garzón* lagoon is performed throughout the year by a community of ten people. The number of inhabitants has been constant for about twenty years (Jorcín 1996).
Garzón is a shallow brackish lagoon that connects to the sea when sand bars that separates it from the ocean sporadically open. In figure 10-3 the areas used for the exploitation of *Erodona mactroides* are schematically shown.

The fishermens' houses total five and they are located on the shores of the lagoon on a coastal dune occupying an area of 1,702 m^2. In the housing area, catches are processed and we identified a number of scales of disposal. The exploitation of the molluscs is carried out between 200 and 600 m from the houses. The shells are discarded in the same place that they are exploited. The deposits cover an area of 4,900 m^2 (shell midden A) and 105.212 m^2 (shell midden B) and they are between 1 and 1.5 m above sea level (see topography in fig. 10-4).

Shellfish are pulled-up from the bottom of the lagoon by a rake that lifts the sediment into a bucket with a mesh of 3 cm. This allows size-selection of the shells which are collected. Once on the shore, the shells are sieved using a mesh of 1 cm in order to remove little pebbles and get the shells. The molluscs are washed in a sink and then they are boiled in pans placed in an oven to protect them from the wind. Once opened, the meat is removed and put into bags. The shells are discarded in adjacent heaps (fig. 10-5).

During our fieldwork, we observed that from a box of 20 kg of *Erodona mactroides* with shell sizes between 2.5 and 3 cm; fishermen obtained 2 kg of meat. Therefore a group of four or five people working four days get 200 kg of *Erodona mactroides* meat.

Studies conducted by Jorcín (1996) and extended by Fabiano and Santana (2006) about artisanal fishery productivity, mentioned that the concentration of molluscs in the banks of Garzón lagoon is 100 individuals per m^2 (average densities of 390 m^2 and maximum densities of 785 m^2 in autumn). Both studies mention that between 1990 and 1994, ten people worked collecting molluscs with sizes between about 1.8 and 2.6 cm in length.

Methods

Firstly we compared the areas covered by the current middens with those covered by the archaeological middens. We also compared the stratigraphic arrangements of the shell assemblages in both kinds of middens (based on Henderson, Anderson, and Mc. Gimsey 2002; and Kidwell, Fürsich, and Aigne 1986).

All the mollusc samples taken weighed 2.5 kg and had a volume of ca. 30x30x30 cm. We took two samples from rubbish dumps at Gambé site (M1 and M2) and two others from the same site but from a mound excavation, level 7 (M3 and M4).

All these samples are from the same occupation level dating to ca. 2700 ^{14}C years BP (radiocarbon dates in Table 10-1). These archaeological samples are compared with four samples from the current exploitation of *Erodona mactroides*. One of these samples was drawn from shells with thermal alteration taken from the current shell midden B (M5) and the other three came from the heaps of molluscs which were boiled and then discarded undergoing pedogenesis processes in heaps in both shell middens B (M6) and A (M7 and M8).

We constructed two data collection matrices. One of them regarding the composition of each sample in grams: *Erodona mactroides* concentrations, sediments and lithiclasts. The other matrix took into account: the Minimum Number of Individuals MNI: articulated valves plus the right and left valves of bivalves were counted separately taking into account the maximum number, the Number of Remains (NR: each unit assigned to that species in the sample) and the sizes of the *Erodona mactroides* shells. In the same matrix some other taphonomic aspects were also recorded (Bender and Guimarães 2006; Martinez *et al.* 2006, Zuschin, Stachowitsch and Stanton 2003) such as: disarticulation, shell weathering level due to biological and physical-chemical factors (weathering stages from 1 to 6 ranging from *Erodona mactroides* fresh shell to its disintegration at touch Beovide 2010a), types of fractures (Beovide 2010a) and the presence of holes.

We explored the data carrying out univariate and multivariate analysis by using the computer program PAST 2.04 (Hammer 2010).

Figure 10-5. Exploitation of the molluscs. a) fishing village, b) rake that combs the sediment and molluscs, c) oven, sink and sieve, d) current shells middens. The spatial arrangement of the valves: e) archaeological (Colonizacion site) and f) current shell middens.

Results

The space occupied by the current shell middens is 4,900 m^3 (midden A) and 105.212 m^3 (midden B) while for the archaeological middens it is 32,131.04 m^3 (Colonización) and 2,785.89 m^3 (Gambé). Therefore the archaeological shell middens are of a larger volume than the current ones.

In figure 10-5 (e and f) we outline the stratigraphy of the modern shell middens compared to the stratigraphy of the archaeological shell middens. The spatial arrangement of the valves in both types of deposits is chaotic, has no preferential orientation and shell packages are relatively dense.

Taking the information from Table 10-2 into account, we compared the composition of the sets of samples taken from current middens with that of the archaeological middens. The difference in composition (log transformed-data 1+x) between the samples was not significant (Kruskal-Wallis test $p = 0.36$, $H = 7.602$ $Hc = 7.955$).

For each of the samples taken from both the archaeological and the current shell middens, we separated the sets of shells of *Erodona mactroides*. With each of these sets we assessed different morphometric features, composition (MNI, NR) and taphonomic attributes of the shells (Table 10-3 in figure 10-6). If we compare the information obtained, there are no significant differences between

Samples weight (g)

	M1	M2	M3	M4	M5	M6	M7	M8
shell	1,200	330.8	1,002.4	596	2,200	2,460	2,420	2,462
sediment	1,248	1,828.2	1,431	1789	200	40	80	38
clast	248	340	57.6	103.92	50	0	0	0
clast <0,5 cm	243	298	57.6	0,72	0	0	0	0
clast 0,5-2 cm	3	0	0	102	0	0	0	0
clast >2 cm	2	42	0	1,2	50	0	0	0

Table 10-2. Composition of the samples. M1-M4 (archeological samples), M5-M8 (current midden samples).

current and archaeological *Erodona mactroides* shells (Kruskal-Wallis test $p = 0.92$, $H = 2.521$ $Hc = 2.731$).

By exploring data in the Table 10-3 (log transformed-data (1+x) the results of Cluster Analysis show a group (1) which is formed by samples M6, 7 and M8 which all belong to the current shell middens. Group 2 comprises all the archaeological samples plus sample M5, this latter is a current shell midden sample but it was taken from the oven, so the shells were affected by heat (fig. 10-6).

	M1	M2	M3	M4	M5	M6	M7	M8
NMI	775	138	480	283	2,156	897	1,587	2,218
NR	1,105	241	786	423	3,031	1,785	3,075	3,960
VD	330	132	461	140	1,970	3,999	879	1,742
VI	775	103	306	280	3,200	830	1587	2,218
T1	0	62	138	143	1,530	0	3	0
T2	0	179	332	163	1,550	0	0	0
T3	1,105	0	317	243	2,090	4910	1,587	2,218
E1	0	0	0	0	0	0	1,225	1,504
E2	0	0	0	0	10	1,785	1,849	2,456
E3	1,105	241	786	423	3,021	0	0	0
E4	0	0	0	0	0	0	0	0
E5	0	0	0	0	0	0	0	0
E6	0	0	0	0	0	0	1	0
SF	709	188	715	3	994	1,585	2,865	2,288
F1	8	7	3	40	0	0	75	0
F2	44	11	6	30	0	0	67	16
F3	48	5	4	36	1,834	16	24	278
F4	64	0	4	22	0	0	0	0
F5	12	0	5	27	35	0	8	0
F6	48	0	19	43	0	80	16	0
F7	16	11	2	18	70	48	0	200
F8	20	4	4	18	0	0	0	56
F9	40	0	5	33	0	16	0	90
F10	84	5	14	42	14	8	20	168
F11	0	0	0	5	63	0	0	864
F12	0	0	0	0	21	32	0	0
F13	0	0	0	0	0	0	0	0
F14	12	10	5	6	0	0	0	0
Orf.	9	102	10	2	0	0	0	0
OVD	2	58	4	1	0	0	0	0
OVI	7	32	6	1	0	0	0	0
IND	0	12	0	0	0	0	0	0

Table 10-3. Minimum number of individuals, number of remains and taphonomic attributes analyzed (*Erodona mactroides*). MNI (minimum number of individuals), NR (number of remains), VD (right valve), VI (left valve), E 1-6 (weathering stages), F1-14 (types of fractures), Orf. (number of holes per sample), OVD (holes in right valve), OVI (holes in left valves), IND (holes in shell). M1-M4 (archeological samples), M5-M8 (current middens samples).

The results of the Principal Components analysis are shown in figure 10-7. The variables of the first principal component (it accounts for the 68% of the covariance) which show a positive correlation in the distinction of the groups are: the NR, the weathering level (E3), the presence of holes and the presence of holes located on the right shells. The holes are only identified in the archaeological samples. However, they do not have the same characteristics as the natural holes identified for *Erodona mactroides* in the region (Farinati, Spagnulo, and Aliotta 2006).

Most of the holes distinguished in the samples are located on the right valve 53% (65). This trend is verified by other studies (Beovide 2010a) when the amount of archaeological samples are increased. However, there could be some variations restricted to certain sectors of the sites (Beovide 2010a). In these sectors the shells which have holes have been studied in depth (based on Velázquez 2007) and they show signs of anthropogenic origin which probably have to do with the manufacture of ornaments (Beovide 2010a). Shell beads made with *Erodona mactroides* can be found in private and public archaeological collections (Beovide 2010a). Beads made from other mollusc species are also frequent in the region (e.g. Bonomo 2007).

Discussion and conclusion

The present work explores the mollusc assemblages as part of the discard activities of the current and past exploitation of *Erodona mactroides*, as well as the taphonomic processes connected to these molluscs. The areas occupied by the archaeological shell middens are bigger than those of the current shell middens. This could be related to the fact that this resource could have been exploited for a longer period of time in the past and /or a higher density of population was devoted to the exploitation of *Erodona mactroides* in the past than nowadays. Although the kind of tools used by the modern fishermen in the extraction of the meat are not likely to be the same as in the past, current shell middens display the same taphonomic attributes as the prehistoric shell

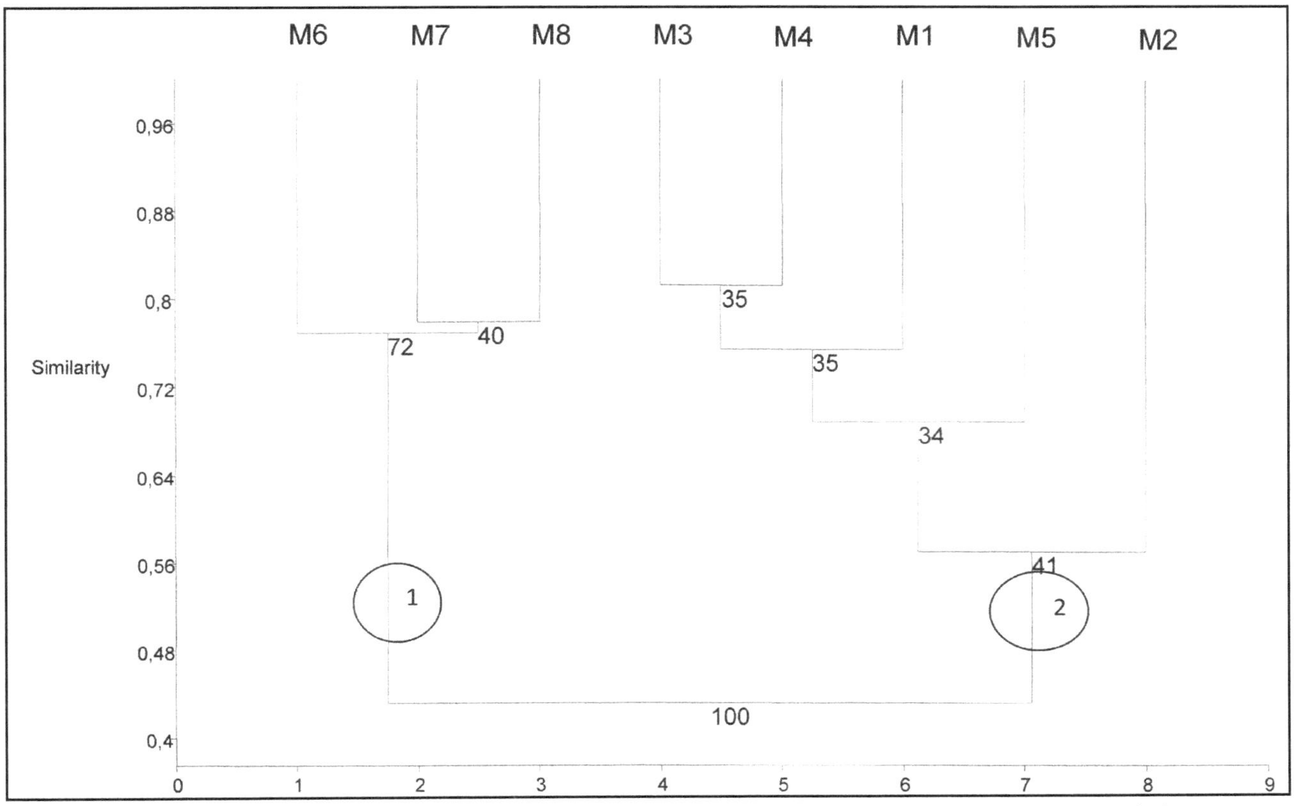

Figure 10-6. Cluster Analysis. M1-M4 (archeological samples), M5-M8 (current middens samples).

PC	Eigenvalue	% variance
1	2,93524	68,887
2	0,713566	16,747
3	0,523395	12,283
4	0,0476175	1,1175
5	0,0384755	0,90298
6	0,00267399	0,062755

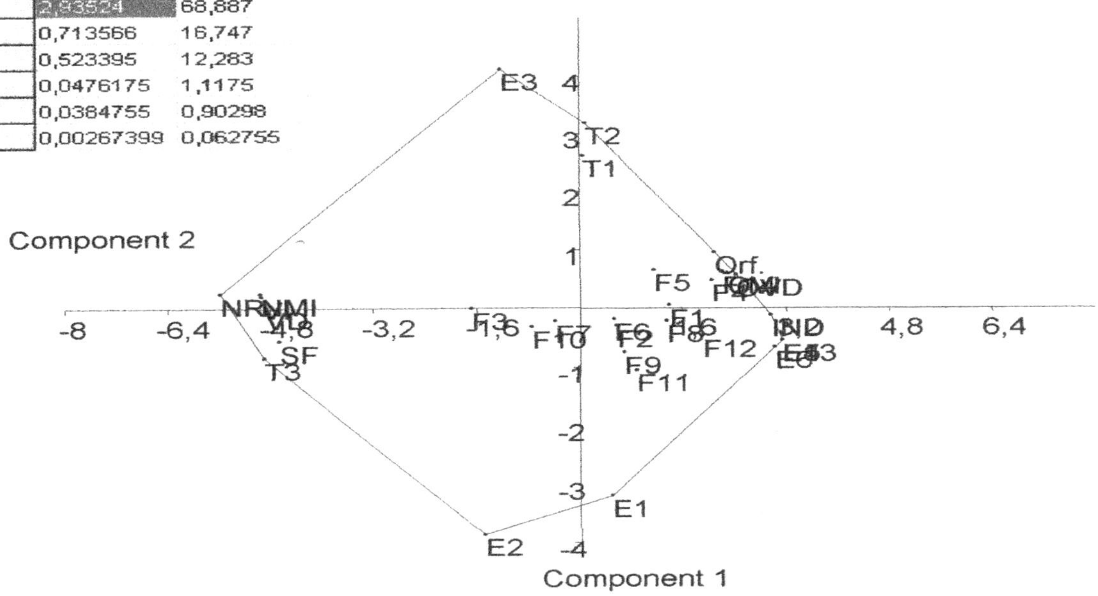

Figure 10-7. Principal Components analysis.

middens, and this has to do with the disposal processes which generated both deposits of shells.

The comparative study of the shell arrangements in the stratigraphic profile of the current and the archaeological shell middens shows that in both cases the shells have a chaotic arrangement where there is no preferential orientation and most of the shells are disarticulated. All these features are generally observed in shell middens of anthropic origin (Henderson, Anderson, and McGimsey 2002).

Moreover, the results obtained are in concordance with the hypothesis initially suggested about the similarity between the environmental contexts of both current and archaeological middens. As mentioned above, if we compare the composition of the samples weighing 2.5 kg (molluscs, sediments and lithiclasts) the differences between the current and the archaeological shell middens are not statistically significant. This suggests common taphonomic and paleoenvironmental processes that led to the incorporation of sediment, lithiclasts and molluscs in similar ratios. When exploring the current and the archaeological samples of *Erodona mactroides* in relation to the MNI, NR, sizes, weathering levels, types of fractures and presence of holes, the difference between the two sets is not significant ($p > 0.05$). Additionally, the results of the multivariate analysis show that the archaeological samples form part of the same group as the burnt samples of modern shell middens. This suggests that the shells were subjected to thermal alteration in the past as is also implied by other lines of evidence in the study of the archaeological contexts (presence of charcoal, burnt bones and lithic tools).

Despite these similarities, there is a difference regarding the use that shells had in the past, where these were also used in the manufacture of ornaments (as suggested by the presence of holes which are not natural). The shells with holes appear to relate to the products of the discard of the manufacturing processes. When comparing archaeological with modern shells (which are the product of subsistence activities only), we suggest that a broader dimension of the use of this resource in the past where shells could have played other roles, possibly linked to the suprastructural aspects of culture.

While many of the conclusions in this study will be tested with more samples and the development of new lines of evidence, the approach here presented allows us to incorporate actualistic data so as to understand past processes of shell midden formation.

Acknowledgements

To the symposium organization of ICAZ for the invitation to publish. To Agencia Nacional de Investigación e Innovación (Grant 161 and FCE-2007-186). Especially to Lic. Elena Vallvé for the English version revision and to Dr. Walter Norbis and Dr. Sergio Martínez.

References

ACOSTA, ALEJANDRO, DANIEL LOPONTE, and PAUL TCHILINGURIAN. 2010. Análisis comparativo sobre la estructura y los procesos de formación de los depósitos arqueológoicos en el humedal del río Paraná inferior (delta del Paraná). In *Arqueología de cazadores-recolectores en la Cuenca del Plata.* G. Cocco and M.R. Feuillet Terzaghi, eds. Pp. 191-208. Centro de Estudios Prehispánicos, Santa Fe. Argentina.

BENDER, CARLA., and MARCELLO GUIMARÃES. 2006. Taphonomic signatures of the recent freshwater mollusks, Touro Passo Stream. RS Brazil. *Revista Brasilera de Paleontología* 9: 243-260.

BEOVIDE, LAURA. 2008. Conductas marisqueadoras y procesos tafonómicos: explorando el registro arqueomalacológico Platense. *Proceedings of the 1er. Congreso Nacional de Zooarqueología Argentina.* Pp. 7-8. Mendoza, Argentina.

—. 2009. Transformaciones productivas y dinámica costera: más allá del concepto de cazadores-recolectores prehispánicos. In *XXIII Simposio de Investigaciones Arqueológicas en Guatemala.* Juan Pedro Laporte, Bárbara Arroyo and Héctor Mejía, eds. 1: 223-236, Guatemala.

—. 2010a. Arqueozoología de los Depósitos Conchilífe-ros de la Cuenca Inferior del Río Santa Lucía. PhD dissertation, University of UDELAR-PEDECIBA, Uruguay.

—. 2010b. La presencia humana en el curso medio del Río de la Plata (Uruguay) durante el Holoceno Medio – reciente: una perspectiva de la continuidad y el cambio. In XVII *Congreso Nacional de Arqueología Argentina, Arqueología Argentina en el Bicentenario de la Revolución de Mayo.* Roberto Bárcena and Horacio Chiavazza, eds. 1: 333-338, Mendoza, Argentina.

BONOMO, MARIANO. 2007. El uso de los moluscos marinos por los cazadores-recolectores pampeanos. *Revista de Antropología Chilena Chungara.* 39 (1):87-102.

BOSC, LOUIS. 1801. *Atlas des mollusques : composé de 51 planches, représentant la plupart des mollusques nus et des coquilles décrits dans le Manuel d'histoire naturelle,* pl 43, fig. Pp 317. Paris, Roret Libraire.

BRACCO, ROBERTO. 2000. Aproximación al Registro Arqueológico del Sitio La Esmeralda desde su dimensión temporal, costa Atlántica del Uruguay. In *Anales de Arqueología y Etnología* 54-55: 13-29. Argentina.

CABRERA, LEONEL, and ISABEL BARRETO. 2006. El ocaso del mundo indígena y las formas de integración a la sociedad urbana montevideana. *Revista TEFROS* 4 (2).

ESTÉVEZ, JORDI, and ASSUMPCIÓ VILA. 1996. Etnoarqueología: el nombre de la cosa. In *Encuentros en los conchales fueginos.* Jordi Estevez y Assumpció Vila, comp. Pp. 17-24. Argentina.

FABIANO, GRACIELA, and ORLANDO SANTANA. 2006. Las pesquerías en las lagunas costeras salobres del

Uruguay. In *Bases para la conservación y el manejo de la costa Uruguaya.* Rodrigo Menafra, Lorena Rodriguez, Fabricio Scarabino and Daniel Conde eds. Pp.557-565. Uruguay.

Farinati, Ester, Jorge Spagnulo, and Salvador Aliotta. 2006. Bioerosión en micromoluscos holocenos del estuario de Bahía Blanca, Argentina. *Ameghiniana* 43 (1): 45-54

Gandara, Manuel. 1990. La Analogía Etnográfica como Heurística: Lógica muestra, Dominios Ontológicos e Historicidad. In *Etnoarqueologia: primer Coloquio Bosch-Gimpera.* Yoko Sugiura and M. Serra eds. Pp. 44-84. México: UNAM.

Gaspar, María. 2000. Construcción de "Sambaquies" y ocupación del territorio brasileño por Pescadores, Recolectores y Cazadores. In *Arqueología de las Tierras Bajas.* Alicia Durán and Roberto Bracco, eds. Pp. 331-341. Uruguay.

Gaspar, María., and Paulo De Blasis. 2006. Balance de las Investigaciones Arqueológicas en Sambaquis. In *Problemáticas de la Arqueología Contemporánea.* Antonio Austral y Marcela Tamagnini, comp. 1: 281-283. Rio Cuarto, Argentina.

Gould, Richard. 1978. The Anthropology of human residues. *American Anthropologist* 80:815-835.

Hammer, Øyvind. 2010. PAST. PAleontological STatistics. User`s guide,Version 2.04.

Henderson, Wayne, Laurie Anderson, and Charles McGimsey. 2002. Distinguishing natural and archaeological deposits: stratigraphy, taxonomy, and taphonomy of Holocene Shell-Rich accumulations from the Lousiana Chenier Plain. *Palaios* 17:192-205.

Jorcín, Adriana. 1996. Distribución, Abundancia y Biomasa de *Erodona mactroide. Revista Brasileira de Zoología* 56(1): 115-162

Kidwell, Susan, Franz Fürsich, and Thomas Aigner. 1986. Conceptual framework for the analysis and classification of fossil concentrations. *Palaios* 1 (3): 228-238.

López, José, Eugenia Villarmarzo, and Laura Brum. 2009. Análisis de las plantas arqueológicas del sitio La Esmeralda, (Rocha, Uruguay). In *La Arqueología como profesión: los primeros 30 años*. Laura Beovide, Carina Erchini and Gonzalo Figueiro, comp. Pp.218-229. Uruguay.

Martínez, Sergio, Alejandra Rojas, Martin Ubilla, Mariano Verde, Daniel Perea and Gustavo Piñeiro. 2006. Molluscan assemblages from the marine Holocene of Uruguay: composition, geochronology, and paleoenvironmental signals. *Ameghiniana* 43 (2): 385-397.

Orquera, Luis., and Ernesto Piana. 2000. Composición de conchales de la costa del canal de Beagle ("Tierra del Fuego, República Argentina). *Relaciones de la Sociedad Argentina de Antropologia 25*: 249-274.

Piana, Ernesto., and Luis Orquera. 2010. Shellmidden Formation At The Beagle Channel (Tierra Del Fuego, Argentine). In *Monumental Question: Prehistoric Megaliths, Mounds, and Enclosures.* David Calado, Maxiliam Baldia and Matthew Boulanger, eds. *BAR International Series 2122.* Pp. 263-271. Hadrian Books, Oxford, England.

Politis, Gustavo. 2002. Acerca de la Etnoarqueología en América del Sur. *Horizontes Antropológicos* 8 (18): 61-91.

—. 2004. Tendencias de la Etnoarqueología en América Latina. In *Teoría Arqueológica de América del Sur.* Gustavo Politis and Roberto D. Peretti, eds. Pp.85-119. Argentina: INCUAPA, UNICEN.

Politis, Gustavo., and Arturo Jaimes. 2005. Patrones de descarte entre los Hoti del Amazonas venezolano. In *El contexto dinámico de la cultura material a través del tiempo.* Eduardo Williams ed. pp. 237-267. Mexico: El Colegio de Michoacán.

Prado, Flavia 2003. Organização e uso do espaço em duas aldeias xerente: uma abordagem etnoarqueológica. MS thesis. Universidade de São Paulo, Brasil

Velázquez, Adrián. 2007. La producción especializada de los objetos de concha del Templo Mayor de Tenochtitlan. *Colección Científica, Antropología,* México: INAH

Williams, Eduardo. 2005. La etnoarqueología, Arqueología como Antropología. Etnoaqueología. *In El contexto dinámico de la cultura material a través del tiempo.* Eduardo Williams, ed. Pp.13-33. México:El Colegio de Michoacán.

Zuschin, Martin, Michael Stachowitsch, and Robert Stanton. 2003. Patterns and processes of shell fragmentation in modern and ancient marine environments. *Earth-Science Reviews* 63 (2003): 33–82

11 - MARINE RESOURCE EXPLOITATION AT MERSA/WADI GAWASIS (RED SEA, EGYPT). THE HARBOUR OF THE PHARAOHS TO THE LAND OF PUNT

Alfredo CARANNANTE
Laboratorio di Scienze e Tecnologie applicate ai Beni Culturali, Università degli Studi "Suor Orsola Benincasa" di Napoli, Via Santa Caterina da Siena 37. Università degli Studi di Napoli "L'Orientale", Piazza San Domenico Maggiore 12, Napoli, Italy.
acarannante@unior.it

Rodolfo FATTOVICH
Università degli Studi di Napoli "L'Orientale", Piazza San Domenico Maggiore 12, Napoli, Italy.
rfattovich@liberto.it

Carla PEPE
Laboratorio di Scienze e Tecnologie applicate ai Beni Culturali, Università degli Studi "Suor Orsola Benincasa" di Napoli, Via Santa Caterina da Siena 37, Italy.
pepecar@liberto.it

Abstract : *Considerable evidence testifies to the use of the Mersa/Wadi Gawasis area as a harbour for seafaring expeditions in the Red Sea from the late 3rd to mid 2nd millennium BC.*
Several thousands of marine organism (fish, sea-turtle, molluscs, crustaceans and barnacles) remains have been found in the archaeological contexts of the site. Most of them represent food remains and assist in the reconstruction of the diet of ancient seafarers. Other remains testify to the use of shells as raw materials to produce ornaments such as different kinds of beads and household objects such as spoons and lamps. Some evidence suggests that tortoiseshell was also obtained from sea turtle carapace in the site.
Large heaps of hundreds of Lambis *shells have been found covering man-made structures overlooking the seashore of Mersa Gawasis. These structures are interpreted as cult places. Archaeomalacological analyses have revealed that the* Lambis *shells were neither used as food nor worked as raw materials. Their use was probably connected with ritual activities, as some evidence suggests.*

Keywords : *Egypt, Red Sea, Middle Kingdom, Shell, Fish.*

Introduction

Data concerning the exploitation of marine resources in ancient Egypt are very scarce. Aquatic resources were mainly obtained from the river Nile, whereas sea organisms had a less important role in the Egyptian economy and culture (Brewer and Friedman 1989; Darby, Ghalioungui and Grivetti 1977; Sahrhage 1998).

Archaeomalacological and archaeoichthyological remains found in Mersa Gawasis are a unique assemblage attesting to the exploitation of marine resources in ancient Egypt for food, sources of raw materials and for ornamental purposes, in addition to ritual activities which have never been identified before.

The aim of this work is to contribute to the knowledge of these aspects in the ancient Egyptian world, which have received scant attention.

The site of Mersa Gawasis (R.F. and C.P.)

The archaeological site of Mersa Gawasis is located in a bay on the Egyptian coast of the Red Sea where the Nile valley is nearest to the coast, about 25km south of

Figure 11-1. Location of Mersa Gawasis and Koptos with the route through the eastern desert evidenced.

the modern port of Safaga (fig. 11-1). The bay (*mersa*) is at the mouth of a dry river bed (*wadi*) whose terminal tract cuts into a fossil coral reef. The ancient settlement extends to the north onto a coral terrace and onto the slope of the *wadi,* and to the northeast to the edge of the cliffs bordering the bay.

The site was initially investigated in the late 1970s by Abdel Monem Sayed (University of Alexandria) who identified the coastal site with the Pharaonic harbour of *S3ww,* which was used for sea-faring expeditions to the land of Punt during the 12th Dynasty (Sayed 1979).

From the 5th Dynasty onward, the toponym of Punt frequently occurs in Pharaonic royal records, in private inscriptions, in religious and literary texts and in iconography. The land of Punt is always presented as a region close to the sea in east Africa from where luxurious goods such as gold, ebony, ivory, incense, myrrh, ostrich egg-shells and leopard skins were imported (Bard and Fattovich 2007). The "Tale of a Shipwrecked Sailor to Punt", a Middle Kingdom text, and the reliefs of Deir el-Bahri Temple of Hatshepsut (ca. 1450 BC) are among the most famous references to this important land.

The Mersa Gawasis excavation project of the "L'Orientale" University of Naples (UNO), the Italian Institute for Africa and the East (IsIAO), in collaboration with Boston University (BU), started in 2001, under the co-direction of Rodolfo Fattovich (UNO) and Kathryn Bard (BU).
Recent archaeological investigations have revealed different areas of use and occupation at the site and proved that it was an important harbour from which seafaring expeditions were sent to the land of Punt during the Middle and the New Kingdoms (Bard and Fattovich 2007). Several *stelae* mentioning voyages to Punt (some dedicated to the god Min of Koptos, the patron of explorers) were discovered at the site. Other relevant finds include potsherds of Eritrean and south Arabian ceramics and 22 inscribed shipping crates (one marked in hieroglyphic "wonderful things of Punt") (Bard and Fattovich 2007).

The excavated areas are distributed on the top of the fossil reef and at the foot of the cliff just on the top of the ledge where evidence of temporary shelters and many hearths have been found.

At the foot of the cliff a series of at least seven man-made caves excavated in the fossil reef have been discovered. They were used as storage rooms for the imported goods and for sails, ropes and disassembled ships (Bard and Fattovich 2007). Significant amounts of material with evident nautical associations were recovered inside these caves and from the surrounding areas, including ship timber, tenons, a hold, sail fragments, caches of thick ropes and several limestone anchors. Mersa/Wadi Gawasis is extraordinary, among other reasons, for being the first Egyptian site to yield unequivocal remains of ancient sea-going vessels. It opens up new lines of inquiry into Egypt's maritime technologies, with previously undocumented fastening systems and materials.

The coastal site functioned as the temporary staging point for a number of official expeditions under Pharaonic administration over the course of several centuries. Written accounts describe how ship timbers constructed in the Nile Valley were transported across the desert and then re-assembled on the Red Sea shore (Bard and Fattovich 2007).

The faunal assemblage (A.C.)

The excavations at Mersa Gawasis have revealed numerous bioarchaeological remains such as dried seeds and fruits, wood, mammal and fish bones, shells and other marine invertebrate remains. Bioarchaeological remains have been collected by hand picking in all strata where human activity was identified.

Archaeomalacological findings represent the most important elements of the faunal assemblage found in the site.

During the excavation seasons of 2006 and 2007 - on the basis of the agreement taken with those in charge of the project - the Bioarchaeology Laboratory team from the University of Naples "Suor Orsola Benincasa" prepared a field laboratory oriented to preliminary analyses of the aquatic vertebrate and invertebrate remains found at Mersa/Wadi Gawasis (Carannante and Pepe 2007).

Facies analyses on cores taken along the *wadi* as well as on stratigraphic sections facilitated the palaeoecological reconstruction of the evolution of the coast.

The lowest stratum excavated in the *wadi* consists of coral-reef blocks covered with a coarse terrigenous-calcareous sand with a lot of skeletal debris. Large fragments and complete shells (*Strombus* sp., *Lambis* sp., *Tridacna* sp., *Pinna* sp.) occur in this stratum. Different benthic foraminifera (*Paneroplis* spp., *Sorites* sp., *Elphidium* spp., Miliolidae.) and fragments of echinoids, bivalves and micro-gastropods occur in the sandy skeletal matrix. Taphocoenosis data suggest a shallow subtidal environment in an open embayment. In turn, the great quantity of epiphyte foraminifera (*Peneroplis* spp., *Elphidium* spp., *Sorites* sp.) suggests sea-grass covered bottoms.

This stratum is covered by a facies consisting of fining upward sands. The top of the bed shows some evidences of cross-lamination. The taphocoenosis is rich in benthic forams and small thin bivalves with closed valves (mainly *Diplodonta subrotunda*). These data suggest a shallow subtidal environment possibly evolving to a submerged beach.

The top of this stratum is covered with a mottled bed which shows clear evidence of bioturbation mainly due to the presence of roots. Benthic assemblages are characterized by gastropods (*Potamides conicus* and *Littoraria intermedia*), which are typical of a mangrove zone. The malacological data are consistent with the evidence of intensive bioturbation suggesting that a mangrove ecosystem was expanding in a coastal zone.

A few Middle Kingdom potsherds were found at the top of this bed, suggesting that when human habitation began at Mersa Gawasis the site overlooked an open bay that was partially colonized by mangroves.

The mottled bed is covered by strata consisting of fine sands with large porcelain foraminifers (*Sorites* sp. and *Amphisorus* sp.) typical of a shallow subtidal (tidal flat) environment covered with sea-grass. *Elphidium* sp., *Peneroplis* sp. and Miliolidae foraminifers are also present. Some *Ammonia* sp. foraminifers typical of more restricted conditions were also found. Shell, crustacean and fish remains occur too. A great quantity of ceramics was found in this facies and flat pebbles occur at the base. Hearths and ash layers are embedded locally in such strata.

The faunal data suggest an original biocoenosis typical of a more restricted embayment than the previous one. The occurrence of hearths suggests that this was a backshore with elements from the subtidal zone close to the occupation area.

The last facies displays a cyclic fining upward sequence (from fine sand to silt) with mud-cracks at the top in some profiles and fine sands containing small gravel and benthic foraminifera assemblages (*Peneroplis* spp., *Sorites* spp., Miliolidae. and *Ammonia* sp., all with very worn tests) in other profiles. These data attest the occurrence of depressed areas with ponds subject to periodic desiccation. Sandy covers with an intensely reworked bioclastic fraction occur marginally to the ponds area.

At their tops, all profiles are characterized by layers of colluvial / alluvial, sediments (coarse sand and fine conglomerates) with imbricated, angular clasts. Channel-like erosion-related surfaces cut locally into these deposits indicating hydraulic transport episodes in a wind-dominated setting. Several hearths and potsherds with a lot of food remains (mainly fish bones and shells) have been found at the bottom of this interval.

All these sedimentological and malacological data prove that the *wadi* was an open bay characterized by a mangrove ecosystem when the first Egyptian settlers arrived. The abundant sea-grass remains found during excavations also indicate that the beach stretched as far as the seafarer camp (fig. 11-2). This bay progressively evolved into a lagoon before being filled by prograding the continental sediments of the *wadi*. This particular palaeoecological situation represented an extra problem for archaeomalacological analyses, since thousands of shell and fish remains came from all the strata and most of them are not ascribable to anthropic activity. Only stratigraphic units containing artefacts and hearths have been considered for this work.

The subject of archaeozoological analyses presented in this work is limited to marine organism remains from stratigraphic units with evidence of human activity.

The aim of this work is to present the results of the archaezoological analyses carried out on such findings in order to understand: (i) the different uses of marine resources at Marsa Gawasis from the perspective of human ecology, (ii) the role of the marine world in the economy and culture of Egyptian explorers, and (iii) any subsequent traditions indicated by the working of ornamental shells.

The faunal assemblage examined consists of 3275 aquatic organism remains (table 11-1). Molluscs are generally the most well-represented *phylum* with 1989 gastropod specimens from 53 species, 912 bivalve specimens from 25 species and 28 scaphopod remains. Other marine invertebrate findings are represented by 16 coral fragments, 8 barnacles, 81 crab remains and 14 sea-urchin spines and

Figure 11-2. The fossil coral reef at Mersa/Wadi Gawasis with the locations of the explorers camp (a) and the shrines along the coastline (b) evidenced. The hatched line corresponds to the ancient shoreline. The white line evidences the present day shoreline.

Figure 11-3. A hearth excavated at Wadi Gawasis. In the box, a *Scarus* sp. premaxillar with a burnt edge suggesting the roasting of the fish.

Taxon	NISP	MNI	*Taxon*	NISP	MNI
Reptilia	**4**	**1**	**Gastropoda**	**1989**	**1975**
Cheloniidae	4	1	*Alvania* sp.	10	10
Osteichthyes	**223**	**189**	*Astrea* sp.	21	21
Scaridae indet.	115	69	*Bulla ampulla*	3	3
Scarus ghobban	3	3	*Cellana* sp.	1	1
Sparidae indet.	22	9	*Cerithium adansonii*	25	25
Acanthopagrus bifasciatus	3	3	*Cerithium caeruleum*	8	8
Serranidae indet.	5	4	*Cerithium* sp.	15	15
Mugilidae indet.	2	1	*Charonia tritonis*	4	2
Crenimugil crenilabus	1	1	*Clypeomorus consisus*	21	21
Scombridae	5	2	*Conus arenatus*	10	10
Chondrichthyes	1	1	*Conus lithoglyphus*	47	47
Dasyatidae (*Taeniura* sp.?)	1	1	*Conus magus*	1	1
Bivalvia	**912**	**874**	*Conus nussatella*	1	1
Anadara antiquata	43	34	*Conus pennaceus*	4	4
Anomia sp.	66	66	*Conus* sp.	122	122
Asaphis violacescens	17	15	*Conus textile*	1	1
Atactodea glabrata	286	286	*Cypraea arabica*	2	2
Brechites attrahens	1	1	*Cypraea grayana*	1	1
Chama sp.	1	1	*Cypraea nebrites*	11	11
Circe crocea	47	38	*Cypraea* sp.	42	42
Circenita callipyga	6	6	*Fusinus polygonoides*	1	1
Codakia tigerina	6	6	*Harpa amouretta*	2	2
Gafrarium pectinatum	2	2	*Lambis truncata*	73	65
Glycymeris pectunculus	2	2	*Littoraria intermedia*	1	1
Isognomon nucleus	1	1	*Mammilla melanostoma*	2	2
Leptomya subrostrata	108	105	*Mauritia* sp.	1	1
Loripes erythraeus	65	64	*Morula granulata*	2	2
Mactra olorina	25	13	*Nassarius gemmulatus*	126	126
Marcia flammea	1	1	*Nassarius obvelatus*	29	29
Modiolus sp.	25	25	*Nassarius* sp.	6	6
Pinctada margaritifera	6	6	*Nerita* sp.	903	903
Pinctada vulgaris	1	1	*Oliva* sp.	3	3
Plicatula sp.	119	119	*Patellidae*	2	2
Pteria sp.	12	12	*Phasianella solida*	150	150
Spondylus sp.	1	1	*Planaxis griseus*	6	6
Tapes sp.	3	3	*Polinices mammilla*	41	41
Tellina virgata	3	3	*Potamides conicus*	8	8
Tridacna maxima	65	63	*Pyramidella sulcata*	1	1
Scaphopoda	**28**	**28**	*Rhinoclavis* sp.	86	86
Dentalium longirostrum	28	28	*Strombus erythrinus*	11	11
Malacostraca	**81**	**80**	*Strombus fasciatus*	53	53
Coenobita scaevola	12	12	*Strombus gibberulus*	22	22
Ocypoda saratan	3	2	*Strombus mutabilis*	1	1
Thecostraca	8	8	*Strombus tricornis*	10	8
Cirripedia indetermined	8	8	*Subancilla annulata*	1	1
Echinoidea	**14**	**11**	*Terebra crenulata*	14	14
Cidaridae indetermined	3	2	*Tonna perdix*	1	1
Echinometra mathaei	1	1	*Trochus erythraeus*	1	1
Anthozoa	**16**	**10**	*Trochus maculatus*	1	1
Tubipora musica	4	4	*Turbo radiatus*	46	46
			Vermetus sp.	2	2
			Volema pyrum	26	26
			Xenoturris cingulifera	5	5

Table 11-1. Sea organisms taxa found in the anthropic strata of explorers camp at Wadi Gawasis with the corresponding Number of Identified Specimens (NISP) and the Minimum Number of Individuals (MNI).

theca (echinoid skeleton) fragments. Along with those of marine invertebrates, 223 fish remains and 4 sea-turtle bone fragments have been analyzed in order to define the exploitation of the marine environment at Mersa/Wadi Gawasis. Of such remains, 3094 were taxonomically identified beyond the class level.

Use of marine resources in the diet. (A.C.)

There is little evidence that marine resources were eaten in Egypt during the Dynastic Period in any quantity (Darby, Ghalioungui and Grivetti 1977:398). Nevertheless, this is probably due to the lack of coastal sites. In the delta area, not only freshwater fish, but also marine fish entering brackish water were consumed (von den Driesch 1986; Boessneck 1988).

Many of the Mersa Gawasis remains are ascribable to dietary use attesting to marine food consumption. Shells of large edible gastropods and bivalves, fish and mammal bone remains have been found in the layers of charcoal and ash where cooking took place. Several hearths associated with Middle Kingdom potsherds have been identified around the explorers' camp of Wadi Gawasis (fig. 11-3). They were often in proximity to the ancient beach. Some fish bones darkened by fire suggest that fish were roasted directly on the hearths (fig. 11-3). Two-hundred twenty-three fish remains come from such anthropic contexts.

Of these remains, 156 were taxonomically identified. Among them, parrotfish (Scaridae) (fig. 11-4) prevail with approximately 76% in terms of identified specimens (118 NISP), corresponding at least to 72 individuals. Sea-breams (Sparidae) were the second-most abundant, with 16% (25 NISP, 12 MNI), whereas the remains of groupers (Serranidae, 5 NISP, 4 MNI), mullets (Mugilidae 3 NISP, 2 MNI) and bonitos (Scombridae 5 NISP, 2 MNI) are very rare.

Among the 72 Scaridae remains, at least three were from the bluebarred parrotfish (*Scarus ghobban*), three of the 12 Sparidae remains were doublebar bream (*Acanthopagrus bifasciatus*) and one of the Mugilidae was a fringelip mullet (*Crenimugil crenilabis*).

Figure 11-4. *Scarus* sp. remains from Wadi Gawasis.

It is interesting to note that the identified species are commonly found in coral reefs which often penetrate silty environments (Randall 1983) such as that of Wadi Gawasis.

The dietary use of sea molluscs can also be seen. Seven-hundred fifteen bivalve and 592 gastropod specimens (44.6% of the whole mollusc remains assemblage in terms of MNI) from Wadi/Marsa Gawasis are compatible with such use if we consider only the edible species and exclude the remains which show marine erosion or bioerosion marks, internal bio-fouling, gastropod predation holes and hermit crab filling. They are mainly Neritidae winkles (389 MNI),

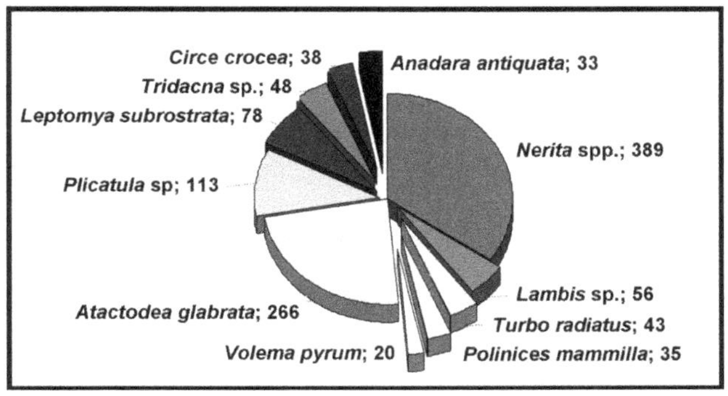

Figure 11-5. Quantitative composition of the molluscs alimentary species assemblage in terms of Minimum Number of Individuals.

smooth beach clam (*Atactodea glabrata*, 266 MNI), plicate oysters (*Plicatula* sp., 113 MNI) and *Leptomya subrostrata* (78 MNI). However, another three bivalve (*Tridacna* sp., *Circe crocea* and *Anadara antiquata*) and four gastropod species (*Lambis truncata, Turbo radiatus, Polinices mammilla* and *Volema pyrum*) might have been utilized as food at Mersa Gawasis (fig. 11-5) since their shells were found associated with hearths and other food remains.

Three of the *Lambis* spider conches found nearby the hearths are highly fragmented and show evidence of intentional percussion. In the same way, modern fishermen in the region smash the thick shell of this mollusc with a pebble in order to obtain the flesh to boil.

The ecology of these taxa (*see* Vine 1986) supplies other information about the environment where they were collected by the Egyptian seafarers.

Circe crocea lives in coral sands just beneath the surface whereas *Atactodea glabrata, Leptomya subrostrata, Anadara antiquata, Polinices mammilla* and *Volema pyrum* are common in mud and fine sediments, often among sea-grass meadows. The last species is also frequently associated with mangroves. Such ecological data confirms the palaeoecological reconstruction of Wadi Gawasis as an open bay with mangrove patches passing landward to a more restricted area (*e.g.* a silty lagoon) with sea-grass meadows.

The other mollusc taxa used as food at Wadi Gawasis were collected on the patch reef bordering the coast outside the bay. *Nerita* spp. live on rocks in the splash zones and *Plicatula* sp. lives attached to rocks in shallow water. *Lambis* spp. and *Turbo radiatus* frequent reef flats and coral rubble in shallow lagoons where *Tridacna* spp. are also present (Vine 1986).

Archaeomalacological interpretations often need further consideration in order to avoid excessive simplifications. Only several dozens of these shells can be definitely identified as food remains, such as the bivalves lying together with mammal and fish bones with evidence of fire exposure near the hearths (*e.g.* a concentration of 23 individuals of Venus clams, *Circe crocea*). The presence of other shells however, can not absolutely be attributed to anthropic activity. Their presence as well as the presence of the remaining mollusc shells with predation holes, bioerosion marks and internal bio-fouling (the 55.4% of the assemblage) may also be explained by such natural phenomena as wind transport or hermit crab roaming. For instance, the abundance of *Atactodea glabrata* at Mersa Gawasis may simply be a result of sedimentary transportation. The light valves of such intertidal species are easily transported by the wind onto emerged beaches.

The most abundant gastropod taxon at Mersa Gawasis may also be overrepresented due to another natural phenomenon. *Nerita* winkle shells are often inhabited by *Coenobita scaevola* land hermit crabs. Armies of this species move during the night from the Red Sea shoreline toward dead fish or other food on emerged beaches (Vine 1986). A Middle Kingdom jar found at Marsa Gawasis was full of *Nerita* and *Polinices* winkle shells filled with *Coenobita scaevola* remains. Such finding suggests that many of the *Nerita* shells found in proximity to the hearths with fish bones may be evidence for such a phenomenon. An analogous hypothesis is suggested by other crustacean remains such as some claws of the ghost crab *Ocypode saratan.* Crabs of this species emerge from their burrows where they are hidden during the day to scavenge along the tide line during the night (Vine 1986).

Unfortunately it is not possible to verify how much these natural phenomena influenced the presence of the Mersa Gawasis shell remains in the archaeological strata.

The ornamental use of marine resources. (A.C.)

Some of the archaeozoological remains from Marsa Gawasis are connected with ornamental use. About 2700 shells from the whole assemblage allow us to verify whether perforations are present. Of them, 186 show one or more holes. Most of the holes are natural and created by gastropod predators or marine abrasion but 37 shells are artificially holed by percussion, abrasion, sawing or drilling, suggesting that they were used as beads or pendants. It is not possible to verify if the remaining 149 shells with a natural bore were also used as pendants.

If we consider the artificially perforated shells, *Nerita* spp. winkles are the taxon most frequently utilized as ornaments with 30 shells (fig. 11-6), followed by cowries (Cypraeidae), cone shells (*Conus* spp.), *Polinices mammilla* and *Volema pyrum*.

Figure 11-6. Examples of artificially holed *Nerita* shells from Wadi Gawasis.

The Indo-Pacific/Red Sea genus *Nerita* perforated shells are reported in many Near Eastern sites from the Neolithic to Iron Age. Reese (1986) provides a census of similar pendants from Israel, in the West Bank and in Jordanian sites, such as Jericho, Abu Salem, Ashdod, Lachish, Giv'atayim, Tell Keisan, Beth Shan, Teleilat Ghassul, Buseirah, Beidha, Tawilan, Petra Jebel el-Jill and Wadi Jubaid. Cypriot Iron Age pottery has been found in some of these sites (Schreiber 2003) and an example of holed *Nerita* has been found in a *bothros* at Kition (Cyprus). Perforated *Nerita* shells have also been found in several northern Syrian and Iraqi sites including Tell Hadidi, Tell Brak, Tell al-Rimah, Nineveh, Nimrud and Umm Dabaghiyah (Reese 1986). There is also evidence for the ornamental use of *Nerita* shells in several Egyptian and Sudanese sites, such as El Omari, Maadi, Adaïma, Saggai, El Kadero, El Kadada and Esh Shaheinab (Arkell 1950:365; Gautier 1983; Boessneck, von den Driesch and Ziegler 1989; Van Neer 2002).

Other shells were also worked at Marsa Gawasis for ornamental purposes. Three small annular beads (a few millimetres in diameter) (fig. 11-7a) found during excavations were produced by using the pierced polished apexes of cone shells (*Conus* spp.). These were not the only kind of ornaments made from such species. Cone shells were also handcrafted on the spot to produce peculiar objects known by archaeomalacologists as "*Conus* whorl beads" or "*Conus* apex beads". Four specimens of these kinds of ornaments were found at Mersa/Wadi Gawasis (fig. 11-7c-e). Three of these beads were unfinished and allow us to reconstruct the sequence of the craft processes. The cone shell was initially bored on the apex by abrasion (fig. 11-7b). Afterwards, the last whorl and columella were removed by percussion and/or sawing (fig. 11-7c). Then the hole was enlarged by drilling the internal side (fig. 11-7d), and finally the bead was polished all over (fig. 11-7e).

Reese (1986, 1990) provides a census of what he first called "*Conus* whorl beads". Some examples of such kinds of ornaments have been found in the Turkish sites of Kurban Höyük, Korukutepe and Tell al-Judaidah in the plain of Antioch. There are many examples from Syrian sites such as Ras Shamra (Ugarit), Tell Fakhariya (11 examples), Mari (12) and Tell Hadidi (16). At Mari, six beads were found in the "*chambre des prêtes*" of the Ishtar temple and another six from three votive deposits at the temple of Dagan.

"*Conus* whorl beads" are reported in many Jordanian and West Bank sites (Reese 1986) such as Jericho, Ashdod, Lachish, 'Ain Shem, Megiddo, Tell Keisan, Beth Shan, Baq'ah, Teleilat Ghassul and Bab edh-Dhra. *Conus* beads of various types are also very common in the Early Neolithic and Late Bronze Age and in the Early Iron Age sites of Israel (Bar-Yosef Mayer 2002a).

The southern Sinai peninsula was a place where "*Conus* whorl beads" were produced (Bar-Yosef Mayer 2002a). Here, both the small disc beads from *Conus* and the "*Conus* whorl beads" have been found in many sites from the Pre-Pottery Neolithic B to the Early Bronze Age II (Bar-Yosef Mayer 2002*a*). "*Conus* whorl beads" seem to have been a particularly important component in EBII assemblages (Bar-Yosef Mayer 2002a).

According to Reese (1990) and Bar-Yosef Mayer (2002*a*), such beads represent a peculiar ornament of the Near Eastern (from Syria to Sinai) area from the Early Neolithic to the Iron Age. In fact "*Conus* whorl beads" have not been found in Cypriot and Aegean sites until now and are extremely rare in Egypt.

Hawara is the only Egyptian site where two very late (23rd Dynasty, ca. 800 BC) examples of such beads have been found (Reese 1986), although a *Conus* shell with its apex removed has been reported at Adaïma (Van Neer 2002) and other worked shells of this species are reported at Neolithic and Predynastic Maadi and El-Omari (Boessneck and von den Driesch, 1990; Boessneck, von den Driesch and Ziegler 1989).

Many of the better preserved ship timbers from Mersa Gawasis are carved in Lebanese cedar (*Cedrus libani*) wood obtained from Near Eastern sources (Ward and Zazzaro 2007). Is it possible that the archaeomalacological data suggest the presence of Near Eastern seafarers or shipwrights on the Egyptian vessels? Or did Egyptian explores produce the typical Near Eastern shell ornaments for Near East trade?

Figure 11-7. Examples of worked *Conus* shells from Wadi Gawasis. An annular disc (a); a perforated shell (b); a perforated shell missing the last whorl and the columella (c); a sketchy "*Conus* whorl bead" with the hole enlarged by internal drilling (d); a finished polished "*Conus* whorl bead" (e).

Some lithic tools from Mersa Gawasis might have been used to manufacture shell beads and other shell artefacts (Lucarini 2007:211). In particular, a group of microlithic perforators found in association with a concentration of shell remains may have been used to create striations that are visible on the artificial internal surfaces of the holes of certain Mersa Gawasis shell beads (Lucarini 2007:211).

Aquatic organisms as a source of raw materials. (A.C.)

Shell was also utilized at Mersa Gawasis to make diverse non-ornamental objects.

Eight spoon-shaped shell articles, whose actual function is uncertain, have been found in the site (fig. 11-8). These "spoons" were carved in the valve or last whorl of mollusc shells and carefully smoothed on the edges, which is confirmed by trace analysis. They might have been involved in some cosmetic activity rather than in meal consumption, although no trace of pigment has been observed. Many similar shell "spoons" as well as wood, alabaster and ivory replicas are exhibited in the Egyptian collections of several museums with attestation of their use as cosmetic palettes. Five of the shell "spoons" (7-12 cm length) from Mersa Gawasis were made from the nacreous valves of large freshwater bivalves of the Mutelidae family (fig. 11-8a). Such data demonstrates that freshwater shells were carried into the coastal site from the Nile or other East African rivers. This is not surprising. Reese (1988:264) reports that at least five valves of the Nilotic species *Aspatharia rubens* (Mutelidae) were found during excavations of the Egyptian temple of Timna near Aqaba, much more distant from the Nile Valley than Mersa Gawasis. The same author, in another paper (Reese 1986), provides a census of Mutelidae freshwater shells found in Near Eastern sites dating from Natufian Mesolithic to Islamic Age. Most of them have been found in Chalcolithic-Iron Age strata at Israeli and Jordanian sites. Such data indicate the importance of these freshwater shells beyond Egyptian world.

Three other spoons were made from marine shells. The first one is a small (~6 cm length) "teaspoon" carved in the nacreous valve of a pearl shell (Pteriidae.), probably *Pteria* sp. (fig. 11-8b). It is the only evidence for marine mother-of-pearl handicraft at Mersa Gawasis, even though six unworked *Pinctada* sp. shells have been found in the archaeological stratigraphic units.

A second "spoon" was carved in the last whorl and columellar lip of a large specimen of triton shell (*Charonia tritonis*) (fig. 11-8c). The third one was obtained from the last whorl of the rare Cassidae species *Phalium faurotis* and bored by sawing and drilling (fig. 11-8d).

One unique find is a large (~18 cm length) valve of a *Tridacna* shell (Giant Clam) showing clear evidence of burning only on the internal side (fig. 11-9). This shell was found near the entrance of an artificial cave. Traces of intense biofouling and bioerosional activities indicate that this shell was collected after the death of the mollusc and probably from a recent orictocoenosis, ruling out the hypothesis that it was cooked for dietary consumption. The centre of the internal side of the valve is characterized by a reddish burning colour surrounded by a black band. Some fringes, coinciding with undulations of the valve edge, show calcination in white, indicating where the highest temperatures were reached. Such different burning colorations suggest that the *Tridacna* valve was filled with fuel and used as a lamp, probably to illuminate the cave nearby.

The use of mollusc valves as lamps is also attested in Egypt in more recent times. Archaeometrical analyses on freshwater shell lamps from a Coptic church in Bawit (Egypt) revealed that Brassicaceae seed oil was utilized as a fuel (Romanus *et al.* 2008). Future analyses at Mersa Gawasis might also elucidate which kinds of fuels were used at the site.

Sea reptiles were also exploited as a source of raw materials at Mersa Gawasis.

Figure 11-8. Examples of spoon-shaped shell articles from Wadi Gawasis carved from a freshwater bivalve (a), from a Pteridae valve (b), from the last whorl and columellar lip of *Charonia tritonis* (c) and from the last whorl of *Phalium faurotis* (d).

Figure 11-9. *Tridacna* valve lamp from Wadi Gawasis.

Twenty large fragments of the flat bones of sea turtle have been found at the entrance of a cave. The fragments have been restored in four anatomical portions: two shields of the lower plastron, a marginal plate and a costal shield with two ribs from the carapace. No other turtle bone or tortoiseshell fragment has been found in the excavation. The *xiphiplastron* and *hypoplastron* show scarce corneous tortoise remains whereas only the external side of the dorsal costal shield shows a reddish coloration due to fire exposure and superficial cut marks oriented to remove the corneous tortoiseshell. Sahrhage (1998:77) affirms that many Egyptian bangles were produced with tortoiseshell and that such raw material was obtained by exposing the carapace of sea turtles caught on Red Sea coast to heat. The exploitation of tortoiseshell demonstrated at Mersa Gawasis represents an interesting archaeozoological counterpart of these kinds of ornaments.

Sea turtle exploitation in archaeological sites is rarely confirmed, but it is interesting here to recall the ethnographic account of Agatharchides of Cnidus who wrote about Red Sea in the 2nd century BC.

The Classical author describes the *chelenophagi* (turtle eaters) people of the Red Sea coasts as sea turtle hunters who lived by exploiting these reptiles (Burstein 1989).

The extensive exploitation of sea turtles demonstrated in the Bronze Age levels of Ra's al-Hadd in Oman seems to confirm the importance of such reptiles in the ancient cultures of the Arabian Peninsula and the Red Sea coasts as described by Agatharchides (Mosseri-Marlio 2000). They were used at Ra's al-Hadd, not only as food and as a source of tortoiseshell, but also as receptacles for bitumen used for sealing vessels and as a source of leather (Mosseri-Marlio 2000).

Shells as evidence of maritime activity (A.C.)

Many barnacles and bivalve shells (mainly Anomiidae-jingle shells and oysters) show evidence of wood impressions on the bottom side. Some of them were still attached to fragments of Lebanese cedar wood. These impressions testify that the shells grew upon flat wooden surfaces. Many of them were found in proximity to ship remains. Such data strongly suggest that the fouling shells were scraped away from the external surfaces of the timbers, suggesting that they were cleaned at the end of the voyage. A small barnacle was found still attached to a thin copper alloy band, attesting that metallic elements were also involved in the ship construction.

The impressive bioerosional damage of some hull timber fragments caused by shipworms (*Teredo* sp.) indicates that the vessel was in the water for a long time.

Red paint traces are often present on finished wood surfaces of timbers that bear evidence of shipworms. The same red paint characterizes much of the wood *debitage* at Mersa Gawasis (Ward and Zazzaro 2007). Red paint is present only on areas of extensive reworking or shipworm damage. Ward and Zazzaro (2007, 143) hypothesize that the paint was used to mark areas of the ship that needed to be removed. Lucarini (personal communication, May 2010) believes that some of the lithic scrapers found at Mersa Gawasis (Lucarini 2007) were utilized to remove marine fouling organisms and shipworm from the damaged parts of ship timbers.
Further analyses at Mersa Gawasis will investigate the degree of shipworm growth, barnacles and fouling shells in order to reconstruct the length of stay of the ships in sea water.

Shells in ritual activity (A.C. and R.F.)

Some peculiar archaeomalacological contexts at Mersa Gawsis merit specific analysis.

Twelve ceremonial structures stand along the seashore on the top of the fossil reef (fig. 11-2b). They are oval mounds of gravel and coral blocks, in some cases covering small chambers built with conglomerate slabs and blocks of fossil reef (Bard and Fattovich 2007). Limestone anchors, votive stelae and Middle Kingdom potsherds have been found in many of such structures together with many large shell remains, attesting to their use as shrines (Bard and Fattovich 2007). One of such mound (WG 29) is particularly relevant for archaeomalacological analysis since it was covered by more than a thousand large spider conch shell remains (fig. 11-10). Dozens of spider conch remains have also been found associated with each of the other ceremonial structures along the Mersa Gawasis coast. WG 29 consists of an oval platform constructed with conglomerate slabs and fossil reef blocks, extending approximately 9 m for 10 m and 1.2 m, with a ramp. It was probably utilized as an open air altar (Bard, Fattovich 2007, 244). Almost all the shells from WG 29 (1055 NISP) are identifiable as *Lambis truncata* remains corresponding to at least 703 specimens (MNI). Only 8 *Tridacna* sp. fragments

Figure 11-10. The ceremonial structure WG 29 with some of the *Lambis* shell remains.

Figure 11-11. The different *Lambis* shell fragmentation patterns at Mersa Gawasis. Flared outer lips (a); shells lacking of the last whorl (b); shells with a large hole on the last whorl (c) and discoidal fragments of the last whorl (d).

complete the archaeomalacological assemblage of the shrine. Such a peculiar context has been taphonomically analyzed in order to verify how the *Lambis* shells were utilized. The archaeomalacological assemblage has been considered separately from that excavated in the *wadi* area (fig. 11-2a).

Just four fragmentation pattern characterize all the *Lambis* shell remains from WG 29 (fig. 11-11). Four-hundred thirty-one *Lambis* shell fragments consist only of the flared outer lip ornamented by six hollow marginal digitations typical of the species (fig. 11-11a), 293 *Lambis* shells lack the last whorl (fig. 11-11b), 261 *Lambis* shells have a large hole on the last whorl (fig. 11-11c) and 70 are discoidal fragments of the last whorl (fig. 11-11d). Such constant fragmentation patterns may suggest that they were broken to extract the living mollusc. However, 72% of the *Lambis* remains from WG 29 show bioerosional marks and/or evidence of internal biofouling, which excludes a dietary use of the specimens, thus demonstrating that most of the shells were collected after the death of the mollusc.

Bar-Yosef Mayer (2002*a*) analyzed hundreds of *Lambis* remains and artefacts found in prehistoric sites of southern Sinai. Here shells of such species were collected after their death and used as raw material to produce different kinds of beads and bangles (Bar-Yosef Mayer 2002*a*). *Lambis* shell bangles have been found in some cases still on the arm bones of skeletons (Bar-Yosef Mayer 2002*a*). The works of Bar-Yosef Mayer (2002*a*, 2002*b*) have illuminated the production of *Lambis* shell bangles in Sinaitic sites such as Wadi Watir, and their trade in Chalcolithic and Bronze Age Egypt and the Near East. Bangles made from the bottom of the spire might have been made specifically for trade with Egypt, whereas another *Lambis* bangle type made from the last whorl of the shell has been found in several Caananite sites as far north as Bab edh-Dhra and Tell el-Far'ah (Bar-Yosef Mayer 2002*a*, 2002*b*).

It has been suggested that the fragmentation pattern of *Lambis* shells at Mersa Gawasis reflects their use as raw material. However, we may exclude such a hypothesis since neither the artefacts nor the working marks on *Lambis* shells are reflected in the site. Local fishermen maintain that such a fragmentation pattern is naturally produced on *Lambis* shells by the waves dragging them onto the reef. Indeed, recent *Lambis* shells with the same fragmentation pattern of that covering WG 29 structure have been found stranded along the coast of Mersa Gawasis, suggesting that people collected them on the beach.

Another non-dietary use of *Lambis* shells is found along the Red Sea. Local fishermen use these kinds of shell as light, readily available thermal-insulation "bricks" to build fishing huts along the coast (Vine 1986, 134). Several such shelters are still visible along the Egyptian coasts between Safaga and Marsa Alam. Is it possible that the shell mound covering WG 29 represents the remains of a modest fishing hut built upon the ancient structure? However, several *Lambis* shells lay beneath the conglomerate slabs and among the fossil coral blocks of WG 29. Such data confirm that the shell mound was related to shrine use in the Middle Kingdom, rather than to a recent hut.

Lambis shells seem to have been imbued with special symbolic and/or religious importance in the cultural world of the Mersa Gawasis explorers, if we relate them to the WG 29 shrine ritual activity and exclude their use as food, raw materials or "bricks".

Koptos in the Nile Valley is one of the nearest cities to Marsa Gawasis (fig. 11-1). It was rich and famous in antiquity and was a starting point for the expeditions in the eastern desert toward the Red Sea coast. The patron of Koptos was an ithyphallic god named Min whose attributes were emphasised in his iconography. He was a fertility and virility god but he was also lord of the eastern deserts and patron of the explorers (Kemp 2000:233). Some stelae at Marsa Gawasis mention Min of Koptos as a protector of the missions to Punt (Mahfouz, Manzo and Pirelli 2007).

Nineteenth century archaeological excavations at Koptos

discovered a head and parts of the legs and torsos of three gigantic statues (almost 4 m tall), currently exhibited in the Cairo Museum and in the Ashmolean Museum at Oxford (Williams 1988). The Koptos "Colossi" originally stood in the Min Predynastic temple of the city where they were impressive representations of the patron god (Williams 1988; Kemp 2000). All the statues represented naked human figures standing with their feet together and their right arm pressed against the side and the left one curved over the torso (fig. 11-12d). The left hand is represented as a ring in low relief around the erect phallus (Williams 1988). Several petroglyphs were carved on the legs of the three Min "Colossi" portraying sacral standards, gazelle heads, elephants, ostriches, shells etc. (fig. 11-12a-c). The compositions of inscribed reliefs are not identical but they are closely related. Shells are the only petroglyph that occurs on all the three statues (Williams 1988; Kemp 2000). Two specimens of *Lambis* shell are carefully represented on each statue (fig. 11-12a-c). They are easily identifiable by the six hollow marginal digitations plus the siphon channel that characterize the genus. Kemp (2000:233) hypothesizes that the animals portrayed on the Koptos "Colossi" (lions, gazelles, ostriches, Red Sea shells) represented the eastern regions reachable from the city whose lord was Min.

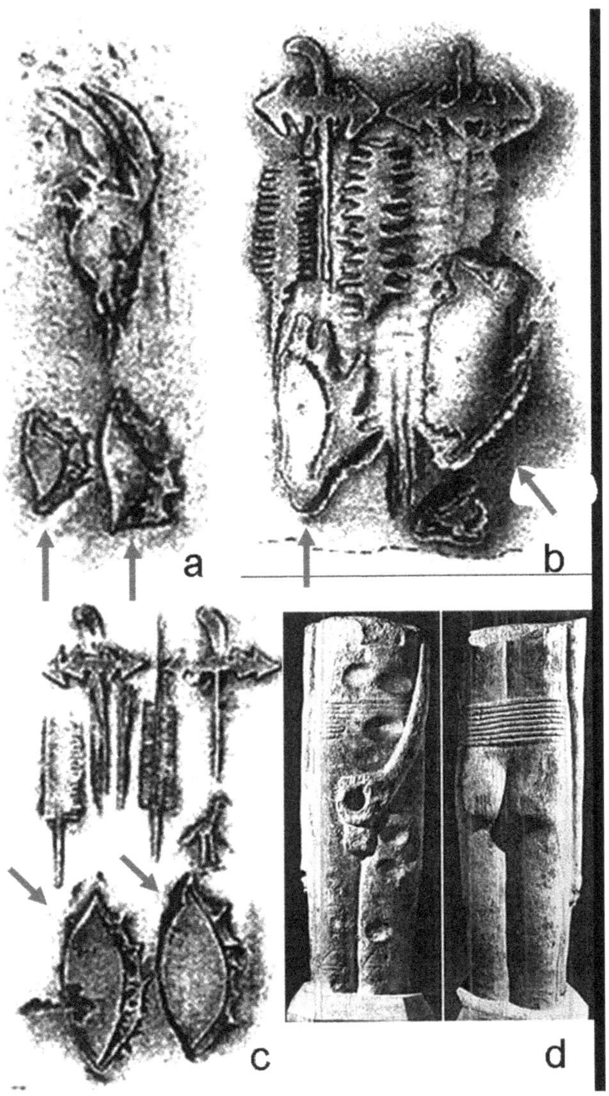

Figure 11-12. Drawings of the petroglyphs panels from the two Ashmolean colossi (a, b) and from the Cairo colossus (c) and the Ashmolean Min colossus 1894.105e (d) (from Kemp 2000 modified). Arrows point to the *Lambis* shell representations.

Kemp (2000:234, quoting Drouin 1989) recalls that the elephant is often associated with sexual demonstrations involving ithyphallic characters in Saharan prehistoric rock art. Curiously, the same authors do not correlate any phallic symbology with the *Lambis* shells portrayed on Min "Colossi", notwithstanding the six impressive digitations that stretch out from the external lip of this species.

Personal interviews with Red Sea fishermen revealed that nowadays men in Safaga collect living *Lambis* to consume their boiled meat as male aphrodisiac food. They also suggest that it is possible to guess the presumed itiphallic power of the mollusc by examining typical shell digitations.

As already hypothesized by one of the authors (Bard and Fattovich 2007:244), the votive deposition of hundreds of *Lambis* shells in the WG 29 shrine and in the other ceremonial structures along the Mersa Gawasis coast is related to a specific symbolism that involves the cult of the god Min, patron of Koptos, lord of the eastern desert and protector of explorers. Such a scenario would seem to reveal a never-before-known marine side of the god of Koptos and suggests that his prerogative was to protect not only travellers in the eastern desert, but also to protect seafarers. Moreover, it is one of the first substantiations of marine organisms in Egyptian worship world.

Conclusions

Archaeozoological data from the area of Mersa/Wadi Gawasis offer a unique picture of marine resource exploitation in ancient Egypt.

Despite the taphonomic complexity of the contexts, many mollusc shells and fish remains found in ash layers around hearths indicate their use as food by Egyptian seafarers. The ecology of the species suggests that they were fished or collected on the patch-reef along the coast as well as in the bay, which palaeocological analyses indicate was in the present-day *wadi*, just in front of the archaeological site. Some burnt mouth bones attest that some fish were roasted directly on the fire whereas the flesh of some large molluscs as *Lambis* spider conches were probably boiled after their shell was smashed. It is not possible to verify if smaller gastropods and bivalves were also cooked or eaten raw.

Egyptian seafarers also used mollusc shells for other purposes.

Shells of different gastropod species were collected to produce beads or pendants. Some of them were artificially perforated by percussion, abrasion, sawing or drilling and some lithic tools found in the site might have been involved in such handicraft production.

Most of the ornaments are simple holed shells. Other shells were worked on the spot to manufacture more complicated objects, such as small annular elements and peculiar objects known by archaeomalacologists as "*Conus* whorl beads", which represent a peculiar ornament of the Near East area from the Early Neolithic to the Iron Age. Such a datum suggests the presence of Near Eastern seafarers or shipwrights among the Egyptian explorers. Also supporting this hypothesis is that many components of the ships whose timbers have been discovered at Wadi Gawasis were procured in the Near East, carved in Lebanese cedar wood.

Many marine organism remains found at Wadi Gawasis are important indicators of maritime activity. Barnacles, oysters and jingle shells showing evidence of wood impressions on the bottom side have been found in close proximity to ship timber remains and attest that they were scraped away from the hull at the end of the voyage.
The abundance of barnacles and bio-fouling bivalves, their dimensions, together with the intensive shipworm (*Teredo* sp.) damage on several ship timbers, testify that the vessels were involved in long-duration voyages. Further analyses will be focused on the shipworm, barnacles and fouling shells' degree of growth in order to reconstruct how long the ships were in sea water.

Other shells were utilized as a source of raw material. Red Sea bivalves and gastropod shells, as well as freshwater valves imported from the Nile, were worked at Wadi Gawasis to produce fine spoon-shaped objects whose use is probably connected with cosmetics preparation. A large *Tridacna* valve was filled with fuel and utilized as lamp, as the different burning colours inside the shell attest.

Sea reptiles were also involved in Wadi Gawasis handicraft. Several elements from the carapace and the lower plastron of a large sea turtle show evidence of fire exposure and cut marks, suggesting they were worked to obtain the precious tortoiseshell that was utilized in ancient Egypt to manufacture bangles. In addition, classical literary sources and archaeological data from Egypt and the Arabian Peninsula confirm that sea turtles were an important resource for Red Sea and Aden Gulf peoples.

A last archaeomalacological assemblage from Mersa Gawasis has been considered separately due to its peculiar context. It consists of hundreds of *Lambis* shells amassed in large heaps covering man-made structures interpreted as cult places. Taphonomical analyses show that *Lambis* shells were not used as food, raw material, instruments or building elements. Sedimentological analyses verify their stratigraphic connection with the ancient cult structures. Thus *Lambis* shells seem to have been imbued with special symbolic and/or religious importance in the cultural world of the Egyptian explorers. This may also represent the first evidence of a ritual use of shells in ancient Egypt. Moreover, the Mersa Gawasis data strongly suggest a never-before-reported symbolic association between the god Min of Koptos, which some stelae at Mersa Gawasis remember as protector of the missions to Punt, and the *Lambis* spider conch shells, whose representation cover the legs of the three gigantic statues of the god discovered in his Predynastic temple at Koptos.

Acknowledgement

We would like to thank the reviewers, Catherine Dupont and Wim Van Neer, for their accurate proof-reading, for all the just advice, and for all the changes they suggested. They were very important to improving this paper.

We would like also to thank the four Red Sea fishermen who, offering tea on the beach between the sea and the desert, explained to us the different uses of *Lambis* molluscs and shells in recent Egyptian tradition.

References

ARKELL, A.J. 1950. The Use of *Nerita* Shells in Early Times. *Annales du Service des antiquités de l'Egypte* 50:365-366.

BAR-YOSEF MAYER, D.E. 2002*a*. The Shells of the Nawamis in Southern Sinai. In *Proceedings of the fifth international symposium on the archaeozoology of southwestern Asia and adjacent areas*. H. Buitenhuis, A.M. Choyke, M. Mashkour, and A.H. Al-Shiyab, eds. Pp.166-180. Groningen: ARC-Publicaties 62.

—. 2002*b*. Egyptian-Canaanite Interaction during the Fourth and Third Millennia B.C.: The Shell Connection. In *Egypt and the Levant: Interrelations from the 4th through the Early 3rd Millennium B.C.E.* E.C.M. van den Brink and T.E. Levy, eds. Pp. 129-135. London: Continuum.

BARD, K.A., and R. FATTOVICH, eds. 2007. *Harbor of the Pharaohs to the Land of Punt. Archaeological Investigations at Mersa/Wadi Gawasis Egypt, 2001-2005.* Naples: Università degli Studi di Napoli "L'Orientale".

BOESSNECK, J. 1988. *Die Tierwelt des Alten Ägypten.* München: Beck Verlag.

BOESSNECK, J., and A. VON DEN DRIESCH. 1990. Tierreste aus der vorgeschichtlichen Siedlung von El-Omari bei Heluan/Unterägypten. In *El Omari. A Neolithic settlement and other sites in the vicinity of Wadi Hof,*

Helwan. F. Debono and B. Mortensen, eds. Pp. 99-107. Mainz am Rhein: Verlag von Zabern.

Boessneck, J., A. von den Driesch, and R. Ziegler. 1989. Die Tierreste von Maadi und Wadi Digla. In *Maadi III. The non-lithic small finds and the structural remains of the Predynastic settlement.* I. Rizkana and J. Seeher, eds. Pp. 87-128. Mainz am Rhein: Verlag von Zabern.

Brewer, D.J. and R.F. Friedman. 1989. *Fish and Fishing in Ancient Egypt.* The Natural History of Egypt. Vol. II. Warminster: Aris & Phillips.

Burstein, S. 1989. *Agatharchides of Cnidus on the Erythraean Sea.* London: Hakluyt Society.

Carannante, A., and C. Pepe. 2007. Shells. In *Harbor of the Pharaoh to the Land of Punt. Archaeological Investigations at Mersa/Wadi Gawasis Egypt, 2001-2005.* K.A. Bard and R. Fattovich, eds. Pp. 212-216. Naples: Università degli Studi di Napoli "L'Orientale".

Darby, W.J., P. Ghalioungui, and L. Grivetti. 1977. *Food: the Gift of Osiris.* Vol.I. London: Academy Press.

Drouin, J. 1989. The Bestiary of Rupestrian and Literary Origin in the Sahara and the Sahel: an Essay in the Investigation of Correlation. In *Animals in to Art.* H. Morphy, ed. Pp. 343-356. London: Unwin Hyman.

Gautier, A. 1983. Animal life along the prehistoric Nile: the evidence from Saggai 1 and Geili (Sudan). *Origini* 12:50-115.

Kemp, B. 2000. The Colossi from the Early Shrine at Coptos in Egypt. *Cambridge Archaeological Journal* 10(2):211-242.

Lucarini, G. 2007. Lithics and grinding stones. In *Harbor of the Pharaoh to the Land of Punt. Archaeological Investigations at Mersa/Wadi Gawasis Egypt, 2001-2005.* K.A. Bard and R. Fattovich, eds. Pp. 196-212. Naples: Università degli Studi di Napoli "L'Orientale".

Mahfouz, E., A. Manzo, and R. Pirelli 2007. Textual Evidence. In *Harbor of the Pharaoh to the Land of Punt. Archaeological Investigations at Mersa/Wadi Gawasis Egypt, 2001-2005.* K.A. Bard and R. Fattovich, eds. Pp. 217-238. Naples: Università degli Studi di Napoli "L'Orientale".

Mosseri-Marlio, C. 2000. Sea Turtle and Dolphin Remains from Ra's al-Hadd, Oman. In *Archaeozoology of the Near East IV A: proceedings of the fourth International Symposium on the Archaeozoology of Southwestern Asia and Adjacent Areas.* M. Mashkour, A.M. Choyke, H. Buitenhuis and F. Poplin., eds. Pp. 94-103. Groningen: ARC Publicatie 32.

Randall, J.E. 1983. *Red Sea Reef Fishes.* London: Immel Publishing Limited.

Reese, D.S. 1986. The Marine and Freshwater Shells. In *The Late Bronze and Early Iron Ages of Central Transjordan. The Baq'ah Valley Project, 1977-81.* P.E. Mc Govern, ed. Pp. 320-332. Philadelphia: University of Pennsylvania.

—. 1988. Recent Invertebrates as Votive Gifts. In *The Egyptian Mining Temple at Timna. Researches in the Arabah 1959-1984: Vol I.* B. Rothenberg, ed. Pp. 260-268. Institute for Archaeo-Metallurgical Studies. London: Thames & Hudson.

—. 1990. Marine and Worked Shells. In *Town and Country in Southeastern Anatolia. Vol.II. The Stratigraphic Sequence at Kurban Höyük.* G. Algaze, ed. Pp. 410-416. Oriental Institute Publications 110. Chicago: University of Chicago.

Romanus, K., W. Van Neer, E. Marinova, K. Verbekele, A. Luypaerts, S. Accardo, I. Hermans, P. Jacobs, D. De Vos, and M. Waelkens. 2008. Brassicaceae Seed Oil Identified as Illuminant in Nilotic Shells from a First Millennium AD Coptic Church in Bawit, Egypt. *Analytical and Bioanalytical Chemistry* 390 (2):783-793.

Sahrhage, D. 1998. *Fischfang und Fischkult im Alten Ägypten.* Kulturgeschichte der antiken Welt 70. Mainz am Rhein: Verlag von Zabern.

Sayed, A.M.A.H. 1979. Discovery of the Site of the 12th Dynasty Port at Wadi Gawasis on the Red Sea Shore. In *Acts of the First International Conference of Egyptology.* W.F. Reineke, ed. Pp. 569-578. Berlin: Akademie Verlag.

Schreiber, N. 2003. *The Cypro-Phoenician pottery of the Iron Age.* Leiden-Boston: Brill.

Van Neer, W. 2002. Le matériel faunique. In *Adaïma. 1. Economie et Habitat.* B. Midant-Reynes and N. Buchez, eds. Pp. 521-565. Cairo: Institut français d'archéologie orientale 45.

Vine, P. 1986. *Red Sea Invertebrates.* London: Immel Publishing.

von den Driesch, A. 1986. Fische im alten Ägypten. *Documenta Naturae* 34:1-33.

Ward, C,. and C. Zazzaro. 2007. Finds: Ship evidence. In *Harbor of the Pharaoh to the Land of Punt. Archaeological Investigations at Mersa/Wadi Gawasis Egypt, 2001-2005.* K.A. Bard and R. Fattovich, eds.

Pp. 135-163. Naples: Università degli Studi di Napoli "L'Orientale".

Williams, B. 1988. Narmer and the Coptos Colossi. *Journal of the American Research Center in Egypt* 25:35-59.

12 - SHELLFISH GATHERING DURING THE IRON AGE AND ROMAN TIMES IN THE NORTHWEST OF THE IBERIAN PENINSULA

Carlos FERNÁNDEZ-RODRÍGUEZ, Víctor BEJEGA-GARCÍA, Eduardo GONZÁLEZ-GÓMEZ-DE-AGÜERO

Department of History, University of León, Campus de Vegazana s/n, 24071 León, Spain, cferr@unileon.es, arqueomalacoleon@hotmail.es

Abstract: *In the archaeological record of the northwest of the Iberian Peninsula there is virtually no reference to the use of marine resources before the Iron Age. In this period, the 1st millennium BC, the building of stable and fortified settlements known as 'Castros' suggests a more intensive use of environmental resources in order to meet supply needs. The rubbish of coastal settlements often includes large numbers of shells as a result of the gathering and consumption of marine shellfish, mixed with other types of remains including mammal and fish bones and other organic and inorganic materials. The study of these remains illustrates the existence of different strategies in the exploitation of the marine environment.*

The Romanisation process of this area caused major economic shifts that also affected these marine resources. The demand for these products, especially oysters, encouraged trade relationships between coastal towns and villages and those inland, a previously unknown activity between hill-fort settlements in the Iron Age and even in Roman times.

Keywords: *Zooarchaeology, Shellfish gathering, Iron Age, Roman Times, Northwest of the Iberian Peninsula*

Introduction

Zooarchaeological studies in the northwest of the Iberian Peninsula have developed slowly and belatedly, as a result of local lithological conditions that do not favour the preservation of these kinds of organic remains. In consequence, it was only in the final decades of the last century when progress began to be made in this facet of research (Fernández Rodríguez 2005).

Within this general framework, archaeomalacological studies have been no different even though the first references to the discovery and study of shell-middens were made in the early twentieth century. At that time, the study of macrolithic assemblages found on the southwest coast of Galicia led to the discovery of several deposits of shells which were tentatively associated with the lithic materials, which were attributed to the early Holocene (e.g. Jalhay 1925, 1929). The study of these middens was minimal, in most cases limited to a simple list of the species found and a visual estimate of their numbers. Recent reappraisals of some of the deposits have given them a completely different chronology and linked them with the Roman period.

The first references to shell-middens at hill-forts also appeared early in this century, led by the detailed study made by López Cuevillas and Bouza Brey (1926) of the deposits at the so-called O Neixón Pequeño Hill-fort (Boiro, A Coruña). They published a drawing of the stratigraphy and pointed out the importance that studying these sites would have for understanding subsistence at the settlements.

After this time only isolated, generic references were made to shell-middens in northwest Iberia until the 1970s, when Vázquez Varela (1975a, 1975b, 1975c) undertook the study of a few samples collected from shell-middens at a number of hill-forts. However, the lack of information about the sampling methodology used raises serious doubts about how representative the results are. Nonetheless, this line of research was pursued and in the 1990s it was Rodríguez López, together with the above-cited researcher, who began a more systematic study of the deposits, applying a rigorous, modern methodology (e.g. Rodríguez López 1992, 1995, 2001). This research has been continued in more recent years by the authors of the present paper.

The results obtained in the last 20 years allow us to put forward some general ideas about the shell-fishing carried out by the first stable, settled communities in the northwest of the Iberian Peninsula and the modifications that took place as a result of the Romanisation of this region.

Chronological and cultural background

The name 'Castro Culture' is used in the northwest of the Iberian Peninsula to define a period characterised by the presence of fortified settlements located on high ground (hill-forts). From the chronological viewpoint, this period extends beyond the limits of the Iron Age, as its beginnings go back to the Final Bronze Age. It lasted, at least for some authors, during the whole of the Roman period. These hill-forts were the first permanent settlements in the northwest, and it was therefore the first time that a population was fixed in the region (fig. 12-1).

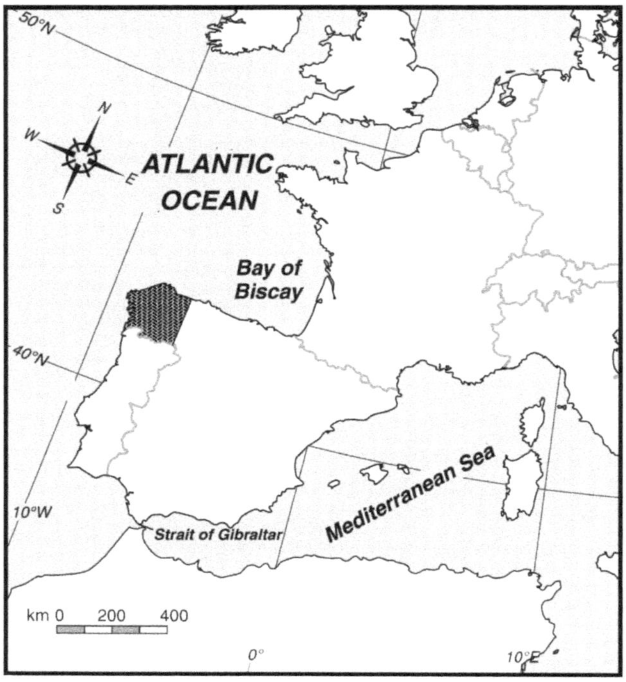

Figure 12-1. Location of the study area (Northwest of the Iberian Peninsula).

Several stages of development have been established within the Castro Culture (Carballo 1996; González Ruibal 2006/2007; Peña Santos 1996). As mentioned above, it began in the late Bronze Age with a first formation phase (tenth or eighth centuries to fifth century BC), characterised by the sedentism of the population in small fortified settlements. A second phase evolved between the fifth and second centuries BC, with an increase in the number and size of the hill-forts, the consolidation of their cultural characteristics, and evidence of contacts with the Mediterranean. The third phase (second century BC to first century AD) was characterised by an increase in social complexity and the cultural development of the settlements, as well as the first contacts with Rome. Some authors (Arias Vilas 2002; Peña Santos 1996) add another phase after the first century AD marked by the process of Romanisation. This involved progressive acculturation and modification of the characteristics of the culture, although some other earlier aspects were maintained. It gave rise to the so-called Galaico-Roman Phase.

Since research began into the proto-history of northwest Iberia, the 'castro', or hill-fort, has been regarded as the characterising element of this culture. Several types of castros can be differentiated depending on their location (Carballo Arceo 1996), including on hill tops, on hillsides, in valleys, and on coastal headlands. Among the marine settlements, a differentiation has been made (Rodríguez López and Fernández Rodríguez 1996) between littoral castros located in direct contact with the sea and sub-littoral castros situated a certain distance from the coast but which exhibit evidence of the use of marine resources.

The economy in the Castro Culture was based on cereal farming complemented with legumes and the gathering of wild fruit (Dopazo Martínez, Ramil Rego, and Fernández Rodríguez 1996; Rodríguez López, Fernández Rodríguez, and Ramil Rego 1993). The livestock consisted mainly of sheep, goats and cattle, while pigs were less important. Hunting was almost nonexistent and little evidence for this activity has been found (Fernández Rodríguez 2000). Fishing was concentrated on littoral species (*Sparus aurata*, *Labrus bergylta* and *Pagellus bogaraveo*) and the use of simple techniques such as lines of metal hooks (Ferré 2003).

The process of the conquest of the northwest of the Iberian Peninsula by the Romans took place during the first century BC, and culminated at the turn of the century, when the whole of northern Iberia was definitively integrated within the Roman economic-administrative system. The process of Romanisation caused the disappearance of some hill-forts. Other hill-forts survived and enlarged and some new ones were founded. In addition, other types of settlements were created (*villae*, *vicus*, factories, ports and towns) which implied a reorganisation of the region within the parameters of Roman administration (Pérez Losada 1996).

At Romanised hill-forts, stock-breeding and agriculture would not undergo any significant changes, although some techniques were introduced in the latter (Dopazo Martínez, Ramil Rego, and Fernández Rodríguez 1996). In fishing, catchment areas were increased and new fishing techniques were introduced which enabled pelagic species (such as *Sardina pillchardus*) to be caught (Ferré 2003).

In contrast, at the new settlements, especially in the large towns, changes took place in the products consumed. For example, the consumption of pork increased in place of ovicaprines. In addition, improvements were made to the livestock (an increase in the size of domestic animals for consumption) and some new species, such as geese, were introduced (Fernández Rodríguez 2003). This took place in the context of growth in the volume of trade, which also included marine produce.

The exploitation of molluscs in the Iron Age

The oldest evidence of the exploitation of marine molluscs in the northwest of the Iberian Peninsula dates back to a single site dated to the Chalcolithic period. This is the shell-midden located on Guidoiro Areoso, an islet in the Ria of Arousa (Vázquez Varela 1988; Vázquez Varela and Rodríguez López 1999), for which no results have been published. It is only in the Iron Age when enough sites have been found to affirm that the marine environment was being exploited habitually. It is possible that the absence of earlier sites is due to the characteristics of the lithology in northwest Iberia. The acidic pH and high rainfall causes the loss of organic matter in a relatively short time (Martínez Cortizas, Ramil Rego, and Llana Rodríguez 1993). We also cannot rule out the possibility that changes in sea level have resulted in at least some of the older coastal settlements being submerged under the sea (Rodríguez López and Fernández Rodríguez 1996).

In inland areas, however, the exploitation of river molluscs has been documented in the Bronze Age and throughout the Iron Age (Moreno Nuño 1995a, 1995b). For example, in the Province of León, valves of *Margaritifera margaritifera*, *Unio pictorum*, *Potomida littoralis* and *Anodonta* sp. have been recorded at sites such as the hill-forts of Sacaojos (Santiago de la Valduerna) (Driesch and Boessneck 1980), La Muela (Valencia de Don Juan) and Gusendos (Gusendos de los Oteros). The latter two sites remain unpublished.

Shellfish gathering at Early Iron Age hill-forts

In Galicia, the only shell-midden that can be attributed to this period is the one found at O Neixón Pequeño Hill-fort (Boiro, A Coruña) (Vázquez Varela and Rodríguez López 1999/2000). The most abundant species in this midden are *Cerastoderma edule* and *Littorina littorea*. Other species are much less frequent. Sandy and muddy substrates were exploited more than rocky shores, although the lack of information about the method and nature of the sampling raises some doubts about the value of the results (fig. 12-2).

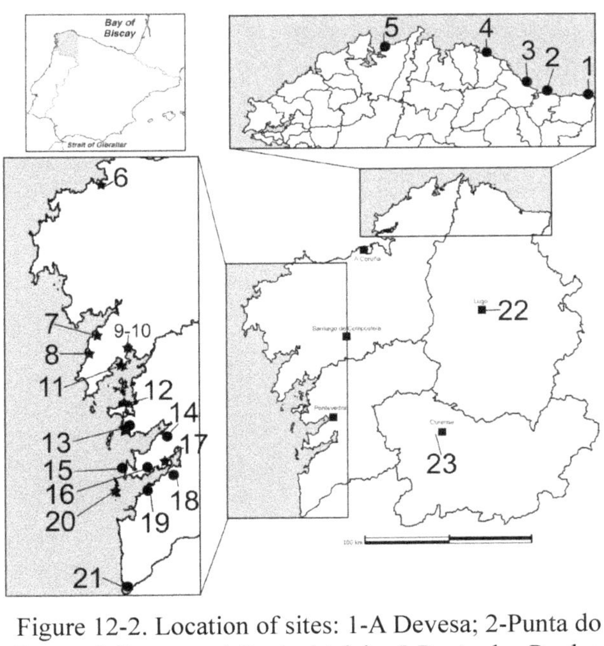

Figure 12-2. Location of sites: 1-A Devesa; 2-Punta do Castro; 3-Fazouro; 4-Punta Atalaia; 5-Punta dos Prados; 6-Borneiro; 7-Baroña; 8-Queiruga; 9-Neixón Pequeno; 10-Neixón Grande; 11-Achadizo; 12-Cantodorxo; 13-A Lanzada; 14-A Subidá; 15-Facho de Donón; 16-Castro de los Remedios; 17-Montealegre; 18-A Peneda; 19-Vigo; 20-Cíes; 21-Santa Tegra; 22-*Lucus Augusti*; 23-Santomé.

Although it is located outside modern Galicia, Campón Hill-fort (Villaviciosa, Asturias) provides more detailed information about shell-fishing at this time. The study (Rodríguez López, Vázquez Varela, and Camino Mayor 2005) reflects a predominant exploitation of sandy-muddy substrates, which reflects the surroundings of the site. Four species predominate (*Solen marginatus*, *Ostrea edulis*, *Stramonita haemastoma*, *Patella* sp.) and these are found in similar proportions, as are the molluscs of secondary importance (*Cerastoderma edule*, *Ruditapes decussatus*, *Venus verrucosa*, *Acanthocardia tuberculata*). This suggests that the shellfish were gathered in a sandy environment, with some use of rocky shores, with calm waters and little wave action. Equally, the exploitation of the species seems to have been balanced, with no kind of specialisation. The presence of specimens of *Stramonita haemastoma* with *Balanus perforatus* adhered to the shells, indicating a use of the lower intertidal zone or the upper low shore, seems to reflect the importance of the tides in the gathering process (Table 12-1).

	Castro del Campón (ss. VIII-VI BC)
	MNI
Solen marginatus	13
Ostrea edulis	10
Stramonita haemastoma	11
Patella sp.	15
Cerastoderma edule	5
Ruditapes decussatus	5
Venus verrucosa	3
Acanthocardia tuberculata	1
TOTAL	**63**

Table 12-1. Distribution of species (MNI) in the sites of Early Iron Age.

Shellfish gathering at Late Iron Age hill-forts

The molluscs found at Late Iron Age sites correspond to both rocky and sandy habitats, with rock species predominating. The distance of estuaries and beaches from sites means that species from that kind of substrate are a minority. However, at settlements located within the area of an estuary, such as Borneiro Hill-fort (Cabana, A Coruña) (Vázquez Varela and Rodríguez López 1997), species from a sandy-muddy shore, like *Cerastoderma lamarcki*, become more common. In contrast, in the area of the Ria of Arousa we have been able to show that the importance of rocky substrates increases further away from the back of the ria. This may be due to larger expanses of rock and more wave-beaten zones, which favoured the development of certain species preferring a hard substrate. Therefore, the exploitation strategies and the species gathered would have depended on the environment in the immediate surroundings of the settlements, with a catchment radius that would be no greater than 1500 m (Rodríguez López 2001).

The choice of rocky substrate species would also be linked to the environment, according to models of inter-species segregation. Hence, in rias like the Ria of Arousa, the sites nearest the mouth of the estuary will exhibit an increase in species such as *Patella* sp., which is due to a larger proportion of wave-beaten areas and rocky zones in the surroundings, whereas species preferring calmer waters, like *Mytilus* sp. decrease in importance. Similarly, in the case of *Osilinus lineatus* and *Littorina littorea* the former predominates at the mouth of the rias due to the greater salinity of the water compared with at the back of the ria, where the latter species predominates.

In addition to the influence of the environment, other factors come into play, such as the criterion of human selection owing to the greater ease in gathering certain species. It is clear that rock species are more visible than those in sandy substrates, which means that they can be selected and gathered with simpler methods. At the same time, it seems that other factors, such as taste and nutritional value may also be significant, as the two species with the highest yield in terms of the ratio between shell and food (*Patella* sp. 74.79%-25.21%; *Mytilus* sp. 50.80%-49.20%) are also exploited the most (Troncoso, Vázquez Varela and Urgorri 1995/96).

Despite the clear predominance of species from rocky substrates, sandy and muddy shores are represented at many sites. If we examine the percentage frequency of the species, two conclusions may be reached. In the first place, the species from sandy-muddy substrates play a secondary role, with the exception of Borneiro Hill fort, mentioned above. The settlements with few beaches in their surroundings exhibit a minimal percentage of species from that kind of environment, as compared with other sites where the surrounding area allows this substrate to be exploited more intensively.

In the second place, the settlements with a higher percentage of species from sandy and muddy substrates also display greater diversity of species, which may mean that the environment was exploited non-selectively. As it is impossible to see the mollusc before it is dug out, unlike the case on rocky shores, it cannot be identified until it has been gathered. Despite this, the exploitation concentrated on certain bivalves, such as *Ruditapes decussatus*, *Solen marginatus* and *Ensis* sp., which appear in greater numbers than other species from the same environment.

Although a wide range of species were gathered, it can be said that each site specialised in a group of two or three species from both sandy and rocky shores. This group is formed by the main species, which are not the same at all the sites but which depend on the physical environment around the settlement. After these species come a second group which tends to vary more in each case. These secondary species appear in low percentages which are sometimes under 10%. The main and secondary species together form the basic contribution of molluscs to the diet, although other species are found too. These may be regarded as unimportant due to the low frequencies with which they appear. They may be shells gathered on the coast after the animal had died or connected to the collection of other species.

The most common species in the studied samples are *Mytilus* sp., *Littorina* sp., *Patella* sp., *Solen/Ensis*, *Ruditapes decussatus* and *Ostrea* sp., of which the first three are present at most sites. They are evidence of a preferential exploitation of the middle-high intertidal zone, although some high shore species (*Patella vulgata*, *Littorina littorea*) are also common, together with more infrequent specimens characteristic of low shore species (*Paracentrotus lividus*, *Stramonita haemastoma*, *Charonia* sp.). These are the result of gathering dependent on tidal action, or the collection of dead specimens on the coast.

Through the use of multiple sampling strategies, both horizontally and vertically, in shell-middens that are apparently homogeneous, we have been able to appreciate significant variations in the importance of the species that form them. This phenomenon seems to correspond to the adaptation of shell-fishing to several factors, such as seasonality, the cycle of the tides or alternation in catchment areas to encourage the regeneration of resources.

For example, at Neixón Grande Hill fort, it was seen that in some levels, species belonging to a sandy-muddy substrate predominate (*Solen marginatus*, SU 38), whereas in others rock species are in the majority (*Mytilus* sp. SU 34) (Bejega García, Fernández Rodríguez, and Fuertes Prieto 2008; Bejega García and Fernández Rodríguez, 2008). Similarly, in one of the columns of samples from Cantodorxo Hill fort rocky substrate species are the most abundant while sandy substrate species predominate in the other (Fernández Rodríguez *et al.* 1998). These variations may be reflecting alternation in the exploitation of habitats or resource catchment areas to encourage the regeneration of the species. Biometric studies carried out at several sites, such as Neixón Grande Hill fort (Bejega García, Fernández Rodríguez, and Fuertes Prieto 2008), seem to support this hypothesis, as the individuals are all of a similar size, indicating that the larger specimens were selected within a balanced exploitation of the environment (Table 12-2).

In summary, it can be said that the use of these resources took place within a kind of subsistence economy that was closely tied to the physical conditions of the surrounding area, in which molluscs would have played a significant role as a complement to the diet, whilst avoiding an over-exploitation of the environment.

The exploitation of molluscs in the Roman Age

With the introduction of the Roman system in the northwest of the Iberian Peninsula, a series of administrative and

	Neixón Grande (s. IV-II BC)	Achadizo (s. V-II BC)	Cantodorxo (s. IV-II BC)	Borneiro (s. IV-I BC)	Queiruga (s. IV-I BC)	Facho de Donón (s II BC-I AD)
Patella sp.	4	272	650	1	2737	666
Gibbula sp.			35		4	4
Osilinus lineatus		1	56			
Littorina littorea	86	265	16	1		
Mytilus sp.	422	266	182	6	306	375
Chlamys varia	3					
Ostrea edulis	62	106		16		
Cerastoderma lamarcki				445		
Cerastoderma edule	4	22	10	1		
Dosinia exoleta		49				
Venus verrucosa		5	7			
Ruditapes decussatus	65	165	301	7		8
Venerupis rhomboideus		35				
Venerupis pullastra		7				
Lutraria lutraria		3				
Solen marginatus	294	6				
Ensis sp.		12	12			
Pollicipes cornucopia					86	376
Paracentrotus lividus					11	2
Other species	9	36	50		17	9
TOTAL	**949**	**1250**	**1319**	**477**	**3161**	**1440**

Table 12-2. Distribution of species (MNI) in the sites of Late Iron Age.

socio-economic changes occurred. Many of the hill-forts were abandoned, as new ones developed and pre-existing ones were enlarged. At the same time, new kinds of settlements appeared, such as villas, towns, army camps, fish-salting factories and ports, which took part in the trade networks that were established throughout the region. This new system was going to have a profound impact on socio-economic relationships and the exploitation of resources.

Shell-fishing at Romanised hill-forts

At indigenous settlements (hill-forts), the shell-fishing model would undergo significant changes in comparison with the previous period. The immediate surroundings of the settlement would still be exploited, but this area was enlarged. For example, at some sites (such as the hill-forts at A Devesa and Punta dos Prados on the Bay of Biscay) species have been found which held significant economic importance for Roman society, and whose catchment areas would have been located about 3000 m from the settlement (Vázquez Varela and Rodríguez López 1995/96, 1997/98) (Table 12-3).

The exploitation continued to be focussed mainly on the intertidal zone, although low shore species acquired certain importance. The presence of individuals corresponding to this level has been regarded as proof of sea bottom fishing in the Roman world (Vázquez Varela 1998; Vázquez Varela and Rodríguez López 1999/2000; Rodríguez López 2001). However, it is extremely difficult to determine for certain the existence and use of this type of technique, as we know of no tools that can be associated directly with it. In addition, low shore species with bromatological value are not very abundant. However, the variation in the frequencies of such species as *Stramonita haemastoma* and *Paracentrotus lividus* in the same shell-midden could be related to tide dynamics. At times of low tide, the catchment areas would increase in size considerably and the gathering of these species would become easier (Fernández Rodríguez, González Gómez de Agüero, and Bejega García 2008; Fernández Rodríguez and Rodriguez López, 1994,1996; González Gómez de Agüero 2009). In addition, the results of the study of the shell-midden at Castro de Punta Atalaia (Cervo, Lugo) have shown that at times when the low shore was being exploited, there is less pressure on the intertidal zone, which might reflect the use of strategies to avoid the over-exploitation of resources.

However, the duration of the period (first to fourth centuries A.D.) and the absence of absolute dates for these deposits are an obstacle to determining the use of sea-bottom fishing in the Roman Age. It is possible that both techniques could have co-existed, or that at a certain time the sea bottom began to be fished, perhaps as a consequence of over-

	Fazouro (s. I AD)	Punta do Castro (s. I-IV AD)	Punta Atalaia (s. I-III AD)	Punta dos Prados (s. I-II AD)	Devesa (s. I-IV AD)	Remedios (s. I-IV AD)	Santa Trega (s. I-IV AD)	Peneda (s. I-IV AD)	Suvidá (s. I-IV AD)
Patella sp.	48	43	6410	56	142	15	3123	1	110
Osilinus lineatus		17	44	14	16		69		1
Littorina littorea			1			9		1	217
Charonia lampas	23			1	2				
Stramonita haemastoma	89	31	73	4	23				
Mytilus sp.		8	184	22	2	1	898	7	679
Chlamys varia				1					
Ostrea edulis		10		7	50			35	5
Cerastoder-ma edule	2			19		25			1
Venus verru-cosa				4		17			
Ruditapes decussatus				5		11		189	138
Solen mar-ginatus								54	93
Pollicipes cornucopia			991	7			514		
Paracentro-tus lividus	5		15				34		
Other spe-cies	1		13	4	1	3	1		11
TOTAL	**168**	**109**	**7731**	**144**	**236**	**79**	**4639**	**287**	**1255**

Table 12-3. Distribution of species MNI) in the Romanised Hill forts.

exploitation that forced an enlargement of the resource catchment area.

At this time, species typical of rocky substrates are not only the most abundant but practically the only ones that are found, although a use of sandy shores can also be identified. The selection of species was still influenced by the surrounding environment, although the anthropic factor became increasingly important, as we shall see below. In addition to this general view two groups of sites can be recognised, one located in open areas where rock species were gathered almost exclusively and a another group located in the rias, with a larger proportion of species typical of sandy substrates.

Depending on the position of the sites, in one group or the other, differences can be seen in the species being gathered. However, a common exploitation pattern seems to exist, marked by certain specialisation in a predominant species. At sites located in open areas, *Stramonita haemastoma* and *Patella* sp. are the most frequent species. They are found in frequencies that are often over 50% of the total number of individuals. Among the sites located in rias, species found in calm water, such as *Mytilus* sp. and *Ruditapes decussatus* are the most common, with a significant presence of molluscs from sandy substrates. Despite this specialisation, we continue to find a variable group of species forming what we have called secondary resources. Once again we can see that the two or three species making up this group of secondary resources are closely linked to the environment. In open areas, *Mytilus* sp., *Pollicipes cornucopia* and *Osilinus lineatus* are usually quite common. In contrast, in the rias, *Littorina littorea* and *Ruditapes decussatus* are the most typical species. The group of species with an occasional presence is much less important.

This model is influenced by the surroundings of the hill-fort, but other factors are also acting on the variations in the species being gathered and the catchment areas, such as the time of year and the tides. Thus, it appears likely that

Paracentrotus lividus was gathered between autumn and spring, the time when the genital gonads are largest, and they are more easily accessible at times of low tide.

Biometric studies carried out for different species have shown that samples from these Romanised hill-forts exhibit a quite wide size-range, which suggests certain pressure on the environment. This was perhaps because of an increase in the population or a greater demand due to trade in inland areas with indigenous settlements, such as Santomé Hill-fort (Ourense) (Fernández Rodríguez and Rodríguez González 1999) where an *Ostrea edulis* valve was documented, and particularly with new settlements.

This model of exploitation would have existed beyond the northwest of the Iberian Peninsula, as a similar pattern has been identified at a number of hill-forts in northern Portugal, such as Cividade de Âncora, Côto da Pena and Terroso (Cabral and Silva 2003; Flores Gomes and Carneiro 2005; Silva 1996).

The use of molluscs at other types of settlements: fishing and trade

With the appearance of new types of settlements and new socio-economic roles, other models of exploitation arose that were linked more closely to anthropic variables than to environmental factors. Whereas in pre-Roman times molluscs were consumed in coastal areas, with the foundation of Roman settlements this consumption spread to inland regions. Mollusc remains have been documented at several Roman urban or military settlements, such as *Lucus Augusti* (Vázquez Varela 1996), *Asturica Augusta* (Fuertes Prieto and Fernández Rodríguez 2010) and *Legio* (Fernández Rodríguez and Fuertes Prieto 2003). The species found tend to be bivalves (e.g. *Pecten maximus*, *Ruditapes decussatus*) and from sandy substrates, although gastropods and echinoderms have also been documented in much smaller quantities. Despite the relative variety of the species usually found at this kind of site, it is *Ostrea edulis* which is the most abundant mollusc (Table 12-4).

It is therefore significant that these species which are the most abundant at the inland sites become scarcer at coastal hill-forts relative to earlier periods. This seems to indicate a double exploitation of the environment by these communities, some species were obtained for local consumption and other species were obtained for trade. The latter species would have been highly valued in gastronomy or craftsmanship and this is a factor that could explain the model of specialisation mentioned above. It is a well-known fact that some marine molluscs, mainly *Ostrea edulis*, held a high gastronomic value among the Roman elite, which generated a flourishing trade in this produce (Vázquez Varela 1998; Vázquez Varela and Rodríguez López 1997/98, 1999/2000).
Biometric studies made of oyster valves from several settlements have revealed significant size differences.

	Lucus Augusti (s. I-IV AD)	*Asturica Augusta* (s. I-IV AD)	*Legio* (s. II-III AD)
Patella sp.		4	
Osilinus lineatus			1
Astraea rugosa	1		
Stramonita haemastoma	3	4	
Mytilus sp.		2	
Chlamys opercularis		1	
Pecten maximus		15	
Ostrea edulis	184	467	9
Cerastoderma edule		2	
Acanthocardia tuberculata		2	
Laevicardium crassum		3	
Dosinia exoleta		1	
Ruditapes decussatus		1	1
Venerupis pullastra		1	
Spisula solida		1	
Paracentrotus lividus		1	
Unio pictorum		1	
Margaritifera margaritifera			1
TOTAL	**188**	**506**	**12**

Table 12-4. Distribution of species (MNI) in Roman settlements.

Thus, specimens from *Legio* (9.5 cm) and *Asturica Augusta* (9 cm) display similar sizes to coastal sites, such as Punta dos Prados (9 cm), whereas specimens from other sites, such as *Lucus Augusti* (7 cm) and the settlement at Lancia (5 cm) have a lower mean size (Fuertes Prieto and Fernández Rodríguez 2010). This may indicate a distribution of oysters related to the purchasing power of the consumers, that is, to the importance of the settlements, although more data is needed to confirm this hypothesis.

In addition, the use of molluscs for non-nutritional purposes has also been documented in the Roman world, for example as adornments (Fuertes Prieto and Fernández Rodríguez 2010) or as building materials (Vázquez Varela 1996).

River molluscs, such as *Unio pictorum*, were still being exploited within a tradition that, as we have seen, dates back to the Bronze Age.

Conclusions

After reviewing the characteristics of shell-fishing in northwest Iberia, we can describe a series of patterns that define this activity amongst coastal indigenous populations and the changes that took place during the Romanisation process.

Shell-fishing was carried out in the immediate surroundings of the settlements from the early Iron Age to the Roman period, although modifications occurred during this time. With the incorporation of local communities into the Roman system, the catchment areas were enlarged, both horizontally (along the coast) and vertically (to lower levels on the shore). In this way species of great socio-economic value were gathered at greater distance from the hill-forts, and a larger number of low shore species were collected (Table 12-5).

This fact, supported also by the biometric data, is an indication of greater pressure on the environment in Roman times. During the Iron Age the specimens tend to be similar in size, while the size range increases with Romanisation, which may be due to an increase in demand for trade and consumption amongst communities that had grown in size, together with specialisation in the exploitation of certain species.

At the same time, phenomena such as low tides played an important role, as the potential catchment area increased in size considerably. Together with diversifying the areas where shell-fish were gathered, this was a determining factor in the establishment of a balanced exploitation model to avoid the over-exploitation of species.

Since the shell-fishing took place in the immediate surroundings of the hill-forts, the environment was a determining factor in the kinds of species that were gathered. During the Iron Age, most molluscs were gathered on rocky shores although typical species of sandy coasts are also found at the sites. In addition, the exploitation appears to have been balanced, with three or four main species and a wide range of secondary species.

However, another important factor must be taken into account, namely anthropic selection. Where this selection is seen most clearly is at Romanised hill-forts, which tended to specialise in one main species, which was nearly always gathered on a rocky substrate. In the same way, selection is reflected in the increasing importance of certain species that were consumed in more Romanised contexts, and the shells of these species have been found in large numbers at inland settlements.

The exploitation of marine resources in the northwest of the Iberian Peninsula could be defined as characteristic of a subsistence economy until the Romanisation process began to modify this situation and exploitations were expanded to cater for trade with inland populations. The role of malacofauna in the diet of Roman urban and military centres seems to have been quite important, as this was a produce with social and commercial prestige. In turn, for coastal communities it became a produce with which they could trade with the hinterland.

The consumption of decapod crustaceans is another matter, as these remains are rarely found at either Iron Age sites or Roman centres. At the moment it is impossible to say whether the remains reflect an occasional consumption of these animals or, on the contrary, whether their consumption was more common than is reflected in the deposits.

In addition to the exploitation of marine molluscs, it has been seen that river molluscs were consumed at inland

Rocky Intertidal	Sandy Intertidal	Muddy Sand Intertidal	Infralittoral
Patella sp.	*Chlamys* sp	*Ostrea edulis*	*Astraea rugosa*
Osilinus lineatus	*Ruditapes decussatus*	*Cerastoderma lamarcki*	*Charonia lampas*
Littorina littorea	*Venerupis pullastra*	*Cerastoderma edule*	*Pecten maximus*
Mytilus sp.	*Venus verrucosa*		*Stramonita haemastoma*
Pollicipes cornucopia	*Solen marginatus*		*Paracentrotus lividus*
	Ensis sp		

Table 12-5. Most representative species of each ecosystem.

sites from the Bronze Age onwards. The presence of terrestrial gastropods is more problematic. Although some specimens have been found at Roman sites, their small numbers and occasionally difficult contextualisation raises doubts about whether this presence is due to contemporary or earlier intrusions.

References

Arias Vilas, F. 2002. Las fases de la Cultura Castreña en Galicia: un debate abierto. In *Los poblados fortificados del noroeste de la Península Ibérica: Formación y desarrollo de la Cultura Castreña.* M. A. de Blas Cortina, and A. Villa Valdés, eds. Pp. 127-138. Coloquios de Arqueología en la cuenca del Navia. Navia (Spain): Ayuntamiento de Navia.

Bejega García, V. and C. Fernández Rodríguez. 2008. O aproveitamiento dos recursos mariños no Castro Grande d'O Neixón. In *Os Castros de Neixón (Boiro, A Coruña) II. De espazo natural a paisaxe cultural.* X. Ayán Vila, ed. Pp. 195-222. Serie Keltia 40. Noia (Spain): Editorial Tosoxoutos.

Bejega García, V., C. Fernández Rodríguez, and N. Fuertes Prieto. 2008. Los restos faunísticos del Castro Grande de O Neixón (Boiro, A Coruña): Análisis zooarqueológico. In *I Congreso Internacional de Arqueoloxía de Vilalba.* E. Ramil Rego, ed. Special issue, *Férvedes* 5:33-42.

Cabral, J. P., and A. C. F. Silva. 2003. Morphometric analysis of limpets from an Iron-Age shell midden found in northwest Portugal. *Journal of Archaeological Science* 30:817-829.

Carballo Arceo, L.X. 1996. Os castros galegos: espacio e arquitectura. *Gallaecia*, 14/15: 309-357.

Dopazo Martínez, A., P. Ramil Rego, and C. Fernández Rodríguez. 1996. Arqueometría aplicada a yacimientos galaico-romanos del NW peninsular. Valoración de la actividad agrícola y ganadera. In *Biogeografía Pleistocena y Holocena de la Península Ibérica.* P. Ramil, C. Fernández, and M. Rodríguez, eds. Pp. 317-332. Santiago (Spain): Xunta de Galicia.

Driesch, A. von den and J. Boessneck. 1980. Tierknochenfunde aus Sacaojos bei La Bañeza (Prov. León). *Studien über frühe Tierknochenfunde von der Iberischen Halbinsel* 7:122-159.

Fernández Rodríguez, C. 2000. Los macromamíferos en los yacimientos arqueológicos del Noroeste peninsular: un estudio económico. In *Tesis Doctorales de la Universidad de Santiago de Compostela (año 2000): Humanidades.* Cd-rom edition. Santiago (Spain): University of Santiago.

Fernández Rodríguez, C. 2005. La arqueozoología en el noroeste de la Península Ibérica: historia de las investigaciones. *Munibe* 57(1):511-523.

Fernández Rodríguez, C., and C. Rodríguez López. 1994. *Análisis de la fauna del Castro de Fazouro. Campaña de 1992.* Research report. Vilalba (Spain): Museo de Prehistoria e Arqueoloxía de Villalba.

——. 1996. Nuevos datos acerca de los recursos económicos del poblado castreño de Santa Trega (A Guardia, Pontevedra). *Gallaecia* 14/15:359-392.

Fernández Rodríguez, C., C. Rodríguez López, C. Ferré, and J. M. Rey. 1998. Sondeos en el conchero del Castro de Punta de Cantodorxo (O Grove, Pontevedra): Análisis zooarqueológico. *Gallaecia* 17:177-197.

Fernández Rodríguez, C., E. González Gómez de Agüero, and V. Bejega García. 2008. Estudio del conchero del Castro de Punta Atalaia (San Cibrao, Lugo). In *I Congreso Internacional de Arqueoloxía de Vilalba.* E. Ramil Rego, ed. Special issue, *Férvedes* 5:43-52.

Fernández Rodríguez, C., and N. Fuertes Prieto. 2003. Análisis de la fauna de Maestro Copín y San Salvador del Nido (León). In *León I. La época romana en León: aspectos arqueológicos.* B.E. Fernández Freire, ed. Pp. 201-231. Arqueología Leonesa II. León (Spain): University of León.

Fernández Rodríguez, C., and X. Rodríguez González. 1999. Análisis de los restos faunísticos del conjunto arqueológico de Santomé (Ourense). *Boletín Auriense* 29:23-38.

Ferré, C. 2003. *Contribución al estudio de las ictiofaunas holocenas en Galicia.* PhD dissertation, University of Santiago, A Coruña (Spain).

Flores Gomes, J. M. and D. Carneiro. 2005. *Subtus Montis Terroso - Património Arqueológico no Concelho da Póvoa de Varzim.* Póvoa de Varzim: Câmara Municipal.

Fuertes Prieto, N. and C. Fernández Rodríguez. 2010. El comercio y consumo de moluscos en Época Romana en *Asturica Augusta* (León). In *I Reunión de Arqueomalacología de la Península Ibérica.* E. González Gómez de Agüero, V. Bejega, C. Fernández, and N. Fuertes Prieto, eds. Special issue, *Férvedes* 6:147-158.

González Gómez de A., E. 2009. El marisqueo en los castros galaico-romanos de la costa cantábrica gallega: el conchero de Punta Atalaya (San Cibrao, Cervo, Lugo). *University of León (Spain).* Dissertation Degree. Unpublished.

GONZÁLEZ RUIBAL, A. 2006/07. *Galaicos. Poder y comunidad en el Noroeste de la Península Ibérica (1200 a.C.-50 d.C.)*. Thematic issue, vol. 1, *Brigantium* 18-19.

JALHAY, E. 1925. El Asturiense en Galicia. *Boletín Arqueológico de la Comisión Provincial de Monumentos Históricos y Artísticos de Orense* VII (nº 160):341-352.

——. 1929. Un nuevo conchero prehistórico descubierto en Galicia. *Boletín de la Comisión Provincial de Monumentos Históricos y Artísticos de Orense* VIII (nº 189):425-431.

LÓPEZ Cuevillas, F., and F. BOUZA BREY. 1926. O Neixón. *Boletín de la Real Academia Gallega* 181:1-11.

MARTÍNEZ CORTIZAS, A., P. Ramil REGO, and C. LLANA RODRÍGUEZ. 1993. Edafología y palinología: aplicación al estudio de yacimientos al aire libre en Galicia. In *Actas Iº Congresso de Arqueologia Peninsular*, vol. I. Special issue, *Trabalhos de Antropologia e Etnologia*, XXXIII(1-2):449-469.

MORENO NUÑO, R. 1995*a*. Arqueofaunas de la Península Ibérica: Un ensayo de síntesis. *Complutum* 6:353-382.

——. 1995*b*. Catálogo de malacofaunas de la Península Ibérica. *Archaeofauna* 4:143-272.

PEÑA SANTOS, A. 1996. A secuencia cultural do mundo castrexo galaico. In *A Cultura Castrexa galega a debate*. J. M. Hidalgo Cuñarro, ed. Pp. 63-103. Tui (Spain): Instituto de Estudios Tudenses.

PÉREZ LOSADA, F. 1996. Hacia una definición de los asentamientos rurales en la *Gallaecia*: poblados (*vici*) y casas de campo (*villae*). In *Los Finisterres atlánticos en la antigüedad. Epoca prerromana y romana*. C. Fernández Ochoa, ed. Pp. 189-197. Madrid (Spain): Ayuntamiento de Gijón.

RODRÍGUEZ LÓPEZ, C. 1992. *Nuevas aportaciones al estudio del aprovechamiento de los recursos marinos en los castros gallegos: el Castro de Cabo de Cruz*. Trabajo de Investigación de Tercer Ciclo, Universidade de Santiago, A Coruña (Spain).

——. 1995. Dos modelos de explotación del mar en la Galicia Castreña: El Castro de Queiruga y el Castro de Facho de Donón. In *Actas del XXII Congreso nacional de Arqueología: Vigo 1993*, vol. 2. A. Beltrán, ed. Pp. 173-178. Vigo (Spain): Xunta de Galicia.

——. 2001. La explotación de los recursos marinos en la prehistoria del Noroeste peninsular. In *VIII Semana Galega de Historia. Galicia Mare Nostrum: a importancia do mar en Galicia*. Pp. 27-45. Colección Editorial da A.G.H. 15. Santiago (Spain): Asociación Galega de Historiadores.

RODRÍGUEZ LÓPEZ, C., C. FERNÁNDEZ RODRÍGUEZ. 1996. Una aproximación al estudio de los yacimientos castreños del litoral galaico: dimensiones ambientales y económicas. In *Biogeografía Pleistocena y Holocena de la Península Ibérica*. P. Ramil, C. Fernández, and M. Rodríguez, eds. Pp. 363-375. Santiago (Spain): Xunta de Galicia.

RODRÍGUEZ LÓPEZ, C., C. FERNÁNDEZ RODRÍGUEZ, and P. RAMIL REGO. 1993. El aprovechamiento del medio natural en la Cultura Castreña del Noroeste Peninsular. In *Actas Iº Congresso de Arqueologia Peninsular*, vol. I. Special issue, *Trabalhos de Antropologia e Etnologia*, XXXIII(1-2):285-305.

RODRÍGUEZ LÓPEZ, C., J. M. VÁZQUEZ VARELA, and J. CAMINO MAYOR. 2005. Concheros castreños y romanos del cantábrico occidental (Asturias y Galicia). *Gallaecia* 24:61-73.

SILVA, A. C. F. 1996. Malacofauna e arqueologia. *Al-Madan* 5:89-95.

TRONCOSO, J. S., J. M. VÁZQUEZ VARELA, and V. URGORRI. 1995/96. La malacofauna en la alimentación de la Cultura Castreña de la Edad del Hierro en Galicia. *Brigantium* 9:107-114.

VÁZQUEZ VARELA, J. M. 1975*a*. Estudio de los concheros de los castros de A Peneda y Montealegre (Pontevedra). *Gallaecia* 1:142-146.

——. 1975*b*. El conchero del Castro de Queiruga (La Coruña). *Boletín del Seminario de Estudios de Arte y Arqueología de la Universidad de Valladolid* 40/41:500-503.

——. 1975c. Estudio del conchero protohistórico de las Islas Cíes. In *Actas de las I Jornadas de Metodología aplicada a las Ciencias Históricas. I Prehistoria e Historia Antigua*. Pp. 291-295. Santiago (Spain): Universidade de Santiago.

——. 1988. El marisqueo en los castros de las rías gallegas. In *Coloquio de etnografía marítima (Santiago 1984)*. Pp. 209-219. Santiago (Spain): Museo do Pobo Galego and Xunta de Galicia.

——. 1996. Del mar al camino, del camino a la mesa: la fauna marina de las excavaciones arqueológicas de 1986, 1990 y 1991. In *Lucus Augusti. I. El amanecer de una ciudad*. A. Rodríguez Colmenero, ed. Pp. 107-122. A Coruña (Spain): Fundación Pedro Barrié de la

Maza.

——. 1998. O aproveitamento dos recursos mariños na prehistoria e antigüedade de Galicia. In *Historia da pesca en Galicia*. C. Fernández Casanova, ed. Pp. 13-49. Biblioteca de Divulgación, Serie Galicia 24. Santiago (Spain): Universidade de Santiago.

Vázquez Varela, J. M., and C. Rodríguez López. 1995/1996. La fauna marina del Castro de Punta dos Prados (Espasante, Ortigueira, La Coruña). *Brigantium* 9:75-106.

——. 1997. Nuevas perspectivas en el estudio del aprovechamiento de los recursos marinos. El Castro de Borneiro (Cabana, La Coruña, Galicia). *Lancia* 2:83-109.

——. 1997/1998. Nuevos datos acerca del aprovechamiento de los recursos marinos en los castros de la costa Cantábrica Gallega: El yacimiento de A Devesa (Ribadeo Lugo). *Boletín do Museo Provincial de Lugo* VIII(1):91-123.

——. 1999/2000. El aprovechamiento de los recursos marinos en la prehistoria y la antigüedad de Galicia. *Boletín do Museo Provincial de Lugo* IX:335-365.

13 - SHELLFISHING AND HORTICULTURE IN PREHISTORIC NORTHERN NEW ZEALAND

Tiffany JAMES-LEE
Department of Anthropology, Gender and Sociology, 1C13a, Richardson Building, University of Otago, PO Box 56, Dunedin, New Zealand,
tiffany.james-lee@otago.ac.nz

Abstract : *This paper examines the role of shellfish in prehistoric diet in the Coromandel region of New Zealand. Previous diet reconstructions in New Zealand have argued that fish, mammals and birds were significantly more important than shellfish. However these studies were generally biased towards southern New Zealand which differs significantly from the north because it lacked horticulture. The role that shellfish played in the horticultural economy of a selected region in northern New Zealand is examined, testing the hypothesis that they helped to sustain communities that were constrained in mobility, and therefore access to other resources by both the requirements of garden maintenance and the close territorial proximity of other horticultural communities.*

Keywords : *Shellfish, Horticulture, Prehistory, New Zealand, Subsistence*

Introduction

Most studies of shellfishing have been undertaken in the context of hunter-gatherer economies, with shellfishing generally seen as a standby resource, adding variety to diet (Álvarez *et al.* 2010). While there are some exceptions, there has been little consideration of the role of shellfishing amongst horticulturalists (although see Claassen 1986). This paper considers a case study of archaeological evidence for shellfish use in a horticultural context from northern New Zealand. It will also consider why shellfish dominate the late period prehistoric middens of this area, in contrast to other regions and periods, where other resources are dominant.

Compared to most other countries, New Zealand has a relatively short archaeological record. It was the last major landmass in the world to be colonised by people before the modern era (Anderson 2002). The earliest reliable radiocarbon dates point to human settlement during the thirteenth century AD (Newnham *et al.* 1998, Wilmshurst *et al.* 2008). Between 85 and 90% of the country's land area was covered in dense forest before Maori permanently settled New Zealand, with natural fire rare (Wilmshurst *et al.* 2004). The ancestors of Maori came from Eastern Polynesia, bringing with them plants such as *Ipomoea batatas* (sweet potato, kumara), *Dioscorea alata* (yam), *Lagenaria siceraria* (bottle gourd), *Colocasia esculenta* (taro), *Cordyline fruticosa* (Pacific Island cabbage tree) and *Broussonetia papyrifera* (paper mulberry), and fauna such as the Pacific rat (kiore, *Rattus exulans)* and the Polynesian dog (kuri, *Canis familiaris)* (Davidson 1984; Horrocks 2004; Walter, Smith, and Jacomb 2006; Wilmshurst *et al.* 2008). If chicken and pigs were taken on the migrations they did not survive the journey or leave evidence of their arrival (McGlone, Anderson, and Holdaway 1994). The early Polynesians found a temperate and varied environment, notable for its lack of native land mammals (except three species of bat), and the presence of large flightless birds, the moa. The various moa species ranged in weight from 25 to 250 kg (Anderson 2002). The first settlers quickly focused on these large flightless birds; it is now postulated that in most areas of New Zealand moa became extinct within 150 years of the arrival of humans, if not earlier (Holdaway and Jacomb 2000; Worthy and Holdaway 2002). As well as the avian megafauna, the first settlers relied upon fishing, agriculture and fern-root collection (bracken, *Pteridium esculentum*). The latter was the most important indigenous plant used by Maori at the time of European arrival, and involved repeated anthropogenic fires which prevented forest regeneration and encouraged the growth of the starchy rhizome (Horrocks 2004; Wilmshurst *et al.* 2004). Sustained predation drove between 30-40 bird species into extinction, and the increased levels of anthropogenic burning dramatically impacted on the vulnerable forest (Anderson 2002; McGlone 1989; McGlone and Wilmshurst 1999). After fish, marine mammals such as fur seals (*Arctocephalus fosteri*) and sea lions (*Phocarctus hookeri*) were the next most prevalent source of meat in prehistoric sites (Smith 2002). In the humid north-west, resource competition was more intense, and led to increased territoriality and warfare.

The first Polynesians to settle in New Zealand adapted their tropical horticulture to its temperate climate and distinct seasons (Davidson 1984). They modified their gardening techniques by adding heat-retaining gravel to soils, selecting fast maturing plant types and the storage of tubers in pits over winter (Anderson 2002; Davidson 1984). For successful germination and a reasonable crop of tubers, kumara requires temperatures of >15°C for five continuous months, and five frost-free months to grow. The North Island climate meets these requirements, but in the South Island the east coast is marginal for successful kumara horticulture, and is limited to north of the Banks Peninsula midway down the island (Bassett *et al.* 2004; Yen 1961). Early in the growing season is when young kumara plants are susceptible to frost damage, so kumara gardens are usually located close to the coast where warmer micro-climates are ensured (Bassett *et al.* 2004). Kumara was stored in subterranean pits over winter, which could be sealed off to provide warmth and control humidity (Yen 1961). Archaeological features associated with kumara gardening include storage pits of various dimensions, field systems of earth or stone garden walls, terraces, and the addition of charcoal, sand, gravel pebbles and shell fragments to garden soils (Davidson 1984; Horrocks and Barber 2005).

The study area for this paper, the Coromandel Peninsula is located on the north-eastern coast of the North Island, east of Auckland city (fig. 13-1). The peninsula is large, extending roughly 85 km from the western end of the Bay of Plenty, and its broadest point is nearly 40 km wide. Its coasts are characteristically rocky on the west, with sandy beaches and harbour estuaries punctuated by rocky outcrops on the east. The Coromandel Range forms the spine of the Peninsula, rising nearly 900 m (Allo 1972; Law 1982). Its rugged nature means that even today much of the area is isolated and uninhabited. Much of the steep and hilly peninsula is covered in temperate rainforest, and a forest park covers a large proportion of the interior. Many islands and island groups lie offshore of the peninsula, including the large Great Barrier Island to the north.

Sites from the Coromandel have been significant in New Zealand archaeology since the late 1950s, and were used to provide a basis for the separation of Maori prehistory

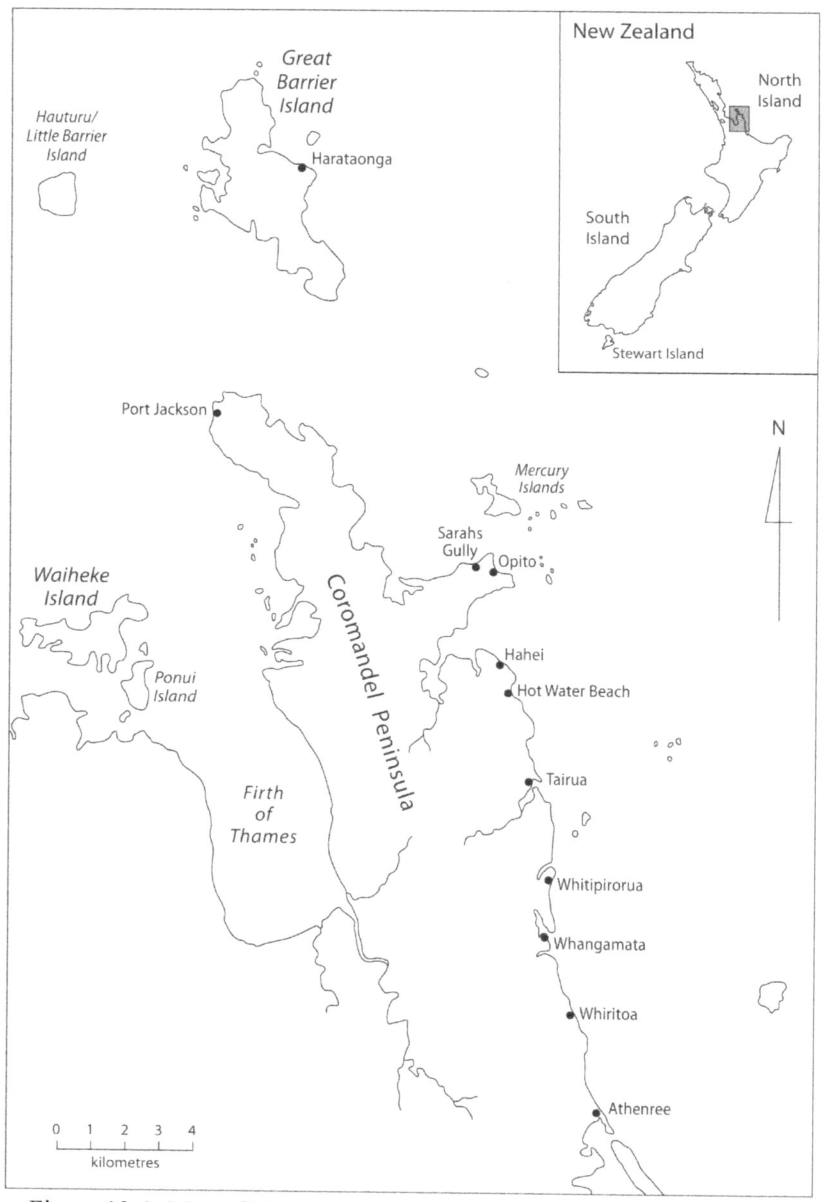

Figure 13-1. Map of New Zealand showing location of the Coromandel Peninsula and Whangamata.

into Archaic and Classic periods (Davidson 1979; Golson 1959; Law 1982). Initially most of the investigations on the Coromandel focused on early sites, which were typically quite small and unspecialised (Edson and Brown 1977; Green 1963; Jones 1973; Law 1972; 1982; Leahy 1974; Rowland 1975; Smart and Green 1962). They had evidence for production and maintenance of tools, underground storage of horticultural crops and the preparation and consumption of a wide variety of marine and terrestrial foods. A review of economic data from these sites up to the end of the 1970s proposed a pattern of generalised bird hunting, and indicated that fishing was of particular importance (Davidson 1979). Avifaunal assemblages from early sites investigated by this time included only small numbers from each species. However, a considerable variety of now extinct species, such as various moa, the Eyle's harrier (*Circus eylesi*), New Zealand coot (*Fulica prisca*), and New Zealand raven (*Corvus antipodum*) were identified. The early sites at Coromandel exhibited the importance of fishing through fishing material culture and substantial fish faunal remains. *Pagrus auratus* (snapper) seems to have been the most commonly caught species, with *Arripis trutta* (kahawai) occasionally appearing in noticeable numbers. With regards to shellfish gathering in the Coromandel, Davidson observed that local conditions are an important consideration in evaluating exploitation. At Tairua there was an apparent shift from rocky shore to estuarine and sandy beach species. Assemblages from early sites in the Coromandel showed general trends such as the decline and extinction of moa and an increase in the exploitation of estuarine shellfish. Most sites reflected generalised exploitation of a range of food resources, but with a strong marine emphasis. The sites can be interpreted as either seasonal camps or specialised activity areas within larger settlements (Davidson 1979).

In recent years much of the archaeology conducted on the Peninsula has been salvage work, driven by a housing and development boom and associated utilities installations, particularly in the first 10 years of the new century (e.g. Furey 2003; Gumbley 2002; Gumbley, Campbell, and Bacquie 2001; Gumbley and Hoffmann 2007; Hoffmann 2009; Mallows 2009; Sewell 2005; Simmons 2002). On-going forestry management on the Peninsula has also generated a large number of 'post-harvest' reports, which generally have not involved excavation of features or middens. Most of the recently recovered archaeological evidence has remained in the grey literature, and part of this research project has focused around using this new data to reassess our understanding of subsistence and settlement in the region. It includes fine-grained analysis of fauna from one site on the east coast of the Coromandel Peninsula – Cabana Lodge, at Whangamata – and is also contributing to a wider comparative study of prehistoric marine resource use being undertaken by Smith (Smith 2010b, 2011a, 2011b; Smith and James-Lee 2009).

The case study, Cabana Lodge, is an extensive midden site situated adjacent to the wharf at Whangamata (fig. 13-2). The site is so extensive that it was initially recorded as two separate sites (T12/2 and T12/3) (Gumbley 2003). Previous investigations (Allo 1972; Jolly 1978) report that the upper deposits include nineteenth century artefacts, while the lower layer includes material typical of early pre-European settlements. These include artefacts indicating adze and fishhook manufacture at the site, in particular one-piece hooks and worked bone from moa and sea mammal. Faunal remains suggest that seals were the main source of meat, followed by dogs and fish, while shellfish were few and unimportant. However sampling methods used in these early investigations were not clearly described, giving cause to doubt the reliability of such relative abundance estimates. An opportunity for more detailed analysis of the prehistoric occupation layer arose when development of a motel complex required further excavations (Gumbley and Hoffmann 2008). As well as providing the faunal remains used in the present study, these excavations yielded two radiocarbon dates on shellfish which indicate that the prehistoric layer was deposited in the mid- to late-14th century AD (Gumbley and Hoffmann 2008).

Method

Faunal remains from the 2008 excavations were identified using reference material in the Archaeological Laboratories at the University of Otago, and quantified in terms of numbers of identified specimens (NISP), minimum numbers of elements (MNE), and minimum numbers of individuals (MNI). In using Minimum Numbers of Individuals as a measure of abundance, there was the implicit assumption all animals in the assemblage were complete when they were arrived at the site, implying that all useable meat would have been available for consumption. This assumption is reasonable for shellfish, fish and small-to-medium bird species, but is uncertain for dogs, and not applicable to larger species such as seals, cetaceans and moa. The latter group is often represented in assemblages as incomplete skeletons, which have been butchered off-site. In these cases, Minimum Numbers of Butchery Units (MNBU) are calculated instead of MNI, i.e. "the smallest number of butchery units necessary to account for all of the remains of a taxon in an archaeological assemblage" (Smith 2011b:3). Meat yields for each taxon were then converted to energy yields (kcal) using taxa-specific kcal per kg values (Smith 2011b).

While archaeofaunal remains were usually identified to species level, on occasion coarser identification is only possible to the genus or family level, or to even more generic categories such as 'seal or dog' or 'fish ?sp'; if these are not subsumed by specific species during MNI calculations, it is necessary to define appropriate values for meatweight, nutritional and energy yields for these less precise taxonomic categories. The MNI for each

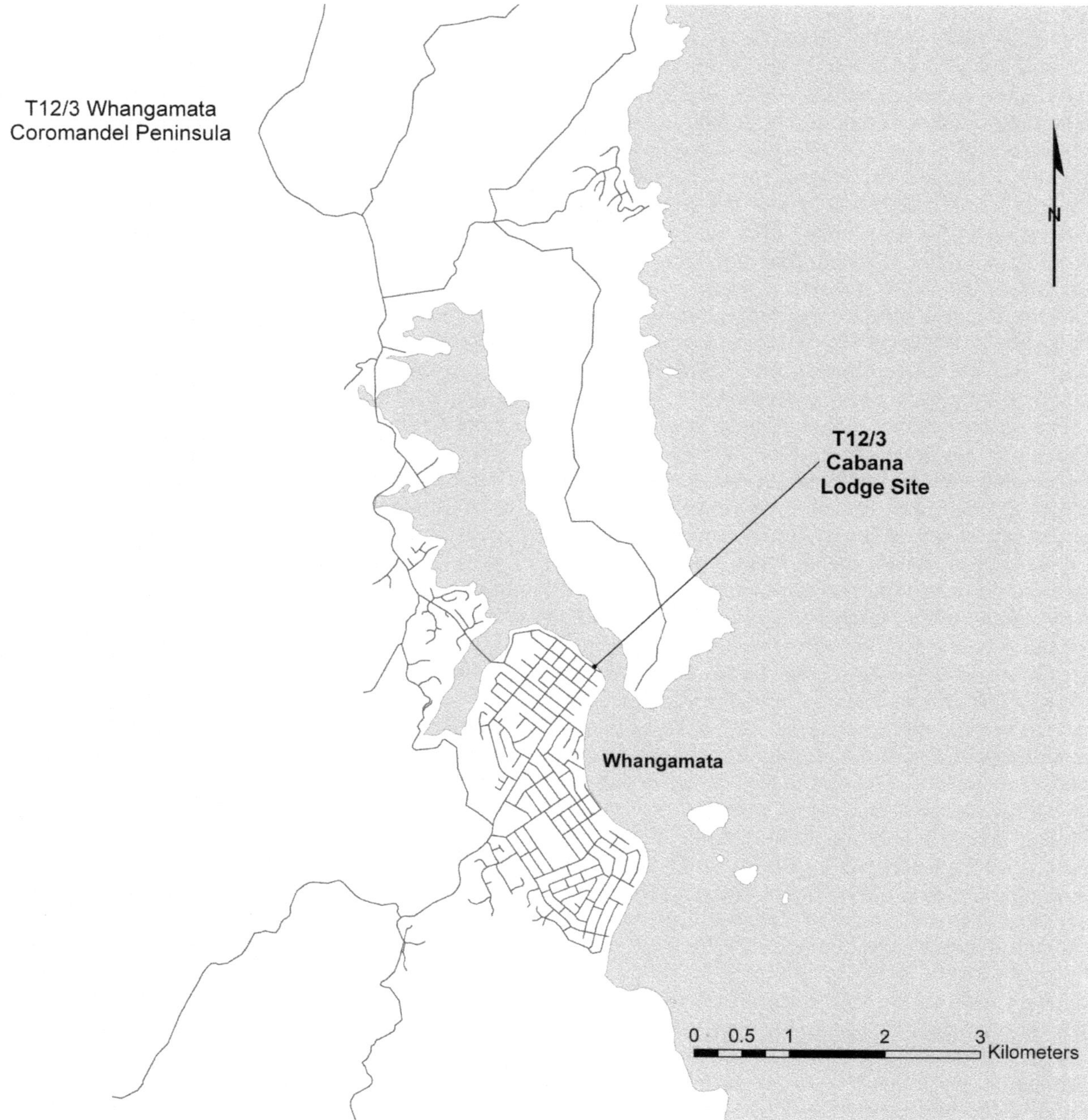

Figure 13-2. Map of Whangamata, Coromandel Peninsula, indicating location of Cabana Lodge site T12/3.

species or other taxonomic group was converted into estimates of the weight of usable meat that they represent, using standardised meat yield values for each taxon (Smith 2011b). The next step in the process relies upon mean weights of useable meatweight for each taxonomic group or species; conservative estimates for mean useable meatweights in the New Zealand context are used and listed in Smith (2011b). Values for finfish and a small number of shellfish were sourced from detailed proximate composition analyses (Vlieg 1988); the values for remaining taxa were derived from comparable species or best available sources (Alexander *et al.* 1979; Altman and Ditmar 1968; Bunce *et al.* 2003; Denniston 1972; Froese and Pauly 2010; Leach 2006; Maxwell 2010; Paul 2000; Raven and Bracegirdle 2010; Smith 1985; Stewart and Stahl 1977; White 1953; Worthy and Holdaway 2002; Worthy *et al.* 2005). To facilitate inter-site comparisons species were grouped into seven faunal classes (fish, shellfish, marine birds, marine mammals, terrestrial mammals, moa and smaller terrestrial birds) and the proportion of total energy harvest from each of these was calculated. Comparisons were made with energy yield data presented by Smith (2010c), along with energy yields calculated from faunal data reported by Furey (2009), Hoffman (2009) and Mallows (2009).

In describing the methods for calculating energy yields it is prudent to discuss cumulative error in the successive steps of calculation. White made the first attempt at calculating meat yields (1953); his work was extended by

Clark (1954), who converted these useable meat values for each taxon into energy yields. Later, nutritional values for protein, carbohydrate and fat yielded by meat yields were determined by Denniston (1972). This approach has seen further refinements with body part analysis to identify butchery units that were consumed and their corresponding meat yields, and increasingly precise nutrient and energy yield data (e.g. Grayson 1979; Lyman 1979; Smith 1975; Stewart and Stahl 1977). In the New Zealand setting Shawcross investigated nutritional components of prehistoric diet from two sites in the northern North Island (1967, 1970, 1972); his principal goal was to use calorie yield to estimate the size of the human populations and the duration of their occupations at the sites, rather than to address the relative importance of different faunal classes in the prehistoric diet. Shawcross' work involved attempts at calculating cumulative errors. For snapper (*Pagrus auratus*) he calculated total weight of individuals using element measurements and a 10% error range; he calculated *edible* meatweight to be 60% of whole fish weight with an error range of 10%; and finally a further error range of 10% for seasonal variations in fish weight was also included. Shawcross took into account variations in calorie values due to factors including fish age, variety, season caught, locality and quality by giving a degree of error of 29% (Shawcross 1967). He then allocated similar cumulative errors for the remaining fish, shellfish and mammal species. He calculated a total energy value for the entire excavated midden, and used this value to attempt to calculate the size of the human population and duration of occupation of the site. Although later disregarded, Shawcross' work mobilized further research in more precise calculations for determining meatweights for the archaeological remains of various classes of fauna in New Zealand (Leach *et al.* 1996; Leach *et al.* 2001; Nichol 1978; Nichol 1988; Smith 1985, 2004, 2011b). The method of calculating assemblage energy yields to evaluate the relative dietary importance of various taxa and classes of taxa has been further developed and described by Smith (1985, 2004, 2011b). He calculated cumulative errors by applying an error of ± 5% to the means of assemblages of variable MNI sizes; an error of 15% to energy yields to account for variations in age, size and condition of individual animals, variations in butchery practices and also consumption practices; and finally an error of 10% for final component energy proportions, because of judgements "about the composition of an appropriate regional data set" and included "several adjustments to calculated values for perceived biases in the archaeological record" (Smith 2011a:18). This resulted in a total cumulative error of ± 30%. Similarly, in this research project, cumulative errors of 15% are included for variations in age, size and condition of individual animals, variations in butchery practices and consumption practices, and a further 5% for the assemblage means that are calculated for each time period. This gives a total cumulative error of 20%.

Results

NISP, MNE and MNI totals for the Cabana Lodge assemblage are shown in table 13-1. Shellfish made up 98% of all identified individuals, with half of these being Tuangi cockles (*Austrovenus stutchburyi*), and about one third pipi (*Paphies australis*), both of which are soft-shore species. The large NISP value for shell sp. represents fragments that are almost certainly from species already represented in the MNE and MNI counts. Fish were the best represented class of vertebrates, and 70% of these were *Pagrus auratus* (snapper), with *Latridopsis ciliaris* (blue moki) and *Nemadactylus macropterus* (tarakihi) the only other species to make more than minor contributions. The high NISP and MNE counts for fish ?sp. are mostly vertebrae, spines and rays almost certainly from the species positively identified. The marine bird class is small, with *Eudytpula minor* (little blue penguin) accounting for two out of five individuals, and probably for most of the unidentifiable bones. Five *Arctocephalus forsteri* (New Zealand fur seal) are present in the marine mammal class, represented by two juveniles, two adult females and one adult male, various elements and body parts present. The terrestrial mammal class is represented by two *Canis familiaris* (kuri, Pacific dog) (one juvenile/subadult and one adult) and two *Rattus* sp. (rat, probably the kiore, Polynesian rat). Terrestrial birds include one indigenous species, *Halcyon sancta vagans* (New Zealand kingfisher), one historic introduced species - *Egretta novaehollandiae* (white-faced heron), and the bones from two *Gallus gallus* (domestic chicken). The latter two species are interpreted as intrusions from the historic layer, as chickens were not present in prehistoric New Zealand. These were excluded from meat weight and energy yield calculations. Although moa bone fragments were present in the assemblage, they were not able to be identified to element or species, and are likely to be debris from tool manufacture; thus they were also excluded from further analysis.

The meat weights and energy yields calculated for each species are also shown in table 13-1 and the relative contributions for each main faunal class to total meat and energy yields for the assemblage are summarised in table 13-2. Variations in meatweight due to age, sex and butchery units were taken into account where applicable (i.e. dogs and fur seals). Despite their numerical dominance in the assemblage, shellfish contributed less than 3% of edible meat and less than 2% of the energy harvested. Sea mammal (i.e. fur seal) contributed the overwhelming majority of the assemblage meat and energy yields at 48.92% and 56.95% respectively. Fish contributed a similar amount of meatweight and but less energy (45.01% and 37.75%). Terrestrial mammals provided only slightly greater proportions of total meatweight and energy than shellfish (2.80% and 2.30%). Marine and terrestrial birds were no more than marginal energy sources; as noted above, moa did not contribute to the energy yield.

To place Cabana Lodge in context it has been compared with data from thirty further assemblages from the Coromandel (See Table 13-3). These have been placed into three chronological groups, based upon carefully scrutinised and recalibrated radiocarbon dates (Smith 2010a). Table 13-4 shows the relative proportions of energy provided by each faunal class in the early period. At most sites either fish or marine mammal are the dominant sources of energy, with moa sometimes also figuring strongly. However, the overall impression is that there was considerable inter-site variation in the nature of the relative importance of fauna during the early period. This scenario changes dramatically in the middle period with sets of values that fall well outside the error ranges for the previous period (Table 13-5). Marine mammals and moa are no longer represented, and smaller birds and terrestrial mammals each occur in only one assemblage. Fish are represented in only three. Shellfish have undoubtedly become the major source of energy from fauna. Indeed, in over the half the assemblages from this period, they are the *only* animal food represented. This trend continues into the late period (Table 13-6), with every single assemblage comprised entirely of shellfish.

Discussion

Clearly these data indicate that there was a significant narrowing of the range of fauna exploited through time in the Coromandel. But one issue that should be addressed is just how realistic the pattern is. The extinction of moas and extirpation of seals from northern New Zealand are well attested (Holdaway and Jacomb 2000; Smith 2002), and obviously influence the results observed here. But, is it likely that fish, birds and terrestrial mammals all disappeared or dropped off the culturally preferred menu?

Taxon	Common Name	NISP	MNE	MNI	Meat Yield (kg)	Energy Yield (kCal)
Shellfish						
Austrovenus stutchburyi	cockle	9930	9554	4777	9.554	9362.92
Paphies australis	pipi	6455	6455	3228	3.228	3163.44
Turbo smaragdus	cat's eye	661	661	507	2.028	1987.44
Paphies sp.	paphies sp.	938	938	469	0.469	459.62
Paphies subtriangulata	tuatua	269	269	135	0.27	264.60
bivalvia indet.	bivalve sp.	3203	136	84	0.084	82.32
gastropoda indet.	gastropod sp.	242	66	66	0.066	64.68
Nerita melanotrogus	black nerita	29	29	29	0.029	28.42
whelk sp.	whelk sp.	25	25	25	0.025	24.50
Struthiolaria sp.	ostrich foot sp.	19	18	18	0.036	35.28
Amalda australis	southern olive shell	12	12	12	0.024	23.52
Penion sulcatus	siphon whelk	9	9	9	0.009	8.82
limpet sp.	limpet sp.	8	8	8	0.008	7.84
mussel sp.	mussel sp.	15	15	8	0.024	23.52
Crepidula costata	ribbed slipper shell	6	6	6	0.012	11.76
Cominella adspersa	speckled whelk	6	6	6	0.006	5.88
Amphibola crenata	mudsnail	4	4	4	0.004	3.92
Cookia sulcata	cook's turban	3	3	2	0.030	29.40
Austrofusus glans	knobbed whelk shell	2	2	2	0.002	1.96
Sigaptella novaezelandiae	circular slipper shell	1	1	1	0.001	0.98
Zeacumantus lutulentus?	horn shell?	1	1	1	0.001	0.98
Struthiolaria vermis	small ostrich foot	1	1	1	0.001	0.98
Pratulum pulchellum	strawberry cockle	2	2	1	0.001	0.98
shell sp.	shell sp.	20,836	0	0	0.000	0.00
Chiton sp.	chiton sp.	25	25	p	0.001	0.98
Shellfish Totals		**42702**	**18246**	**9399**	**15.913**	**15594.74**
Fish						
Pagrus auratus	snapper	1124	864	125	192.500	225995.00
Latridopsis ciliaris	blue moki	113	113	24	45.360	58831.92
Nemadactylus macropterus	tarakihi	74	72	13	7.280	12638.08
Meuschenia scaber	leatherjacket	5	4	4	6.300	11516.40
Arripis trutta	kahawai	20	20	3	3.150	3150.00
Notolabrus celidotus	spotty	3	3	3	1.050	1063.65
Chelidonichthys kumu	red gurnard	19	6	1	1.050	1018.50
Genyagnus monopterygius	spotted stargazer	2	2	1	1.050	1347.15
Helicolenus percoides	sea perch	1	1	1	1.050	1071.00
Zeus faber	john dory	1	1	1	0.000	0.00
Gempylidae indet.	snake mackerel sp.	1	1	1	1.050	729.75

Carangidae indet.	Carangidae ?sp	1	1	1	0.770	912.45
Pagrus auratus?	snapper?	1	1	0	0.000	0.00
Notolabrus celidotus?	spotty?	1	1	0	0.000	0.00
fish sp.	fish sp.	20507	1966	0	0.000	0.00
Fish Totals		**21873**	**3056**	**178**	**260.61**	**318273.9**
Terrestrial Bird						
Halcyon sancta vagans	New Zealand kingfisher	1	1	1	0.0455	80.08
Terrestrial Bird Totals		**1**	**1**	**1**	**0.0455**	**80.08**
Marine Bird						
Eudyptula minor	little blue penguin	15	15	2	1.540	4943.40
Larus novaehollandiae scopulinus	red-billed gull	3	3	1	0.182	584.22
Puffinus griesus	sooty shearwater	1	1	1	0.560	1797.60
Anas superciliosa	grey duck	1	1	1	0.700	2247.00
Eudyptula minor?	little blue penguin?	1	1	0		
Aves indet.	bird sp.	142	16	0		
Marine Bird Totals		**163**	**37**	**5**	**2.982**	**9572.22**
Terrestrial Mammal						
Canis familiaris	dog	40	29	2*	16.000	20152.00
Rattus sp.	rat	9	9	2	0.200	252.00
Rodentia indet.	rat?	4	4	0		
mammalia indet.	mammal sp.	123	0	0		
medium mammal	medium mammal sp.	7	4	0		
small mammal	small mammal sp.	11	5	0		
Terrestrial Mammal Totals		**194**	**51**	**4**	**16.200**	**19404.00**
Marine Mammal						
Arctocephalus fosteri	fur seal	22	21	5**	283.260	480141.20
Marine Mammal Totals		**22**	**21**	**5**	**283.260**	**480141.20**
Moa						
Moa sp.	moa ?sp.	26	0	0	0.000	0.00
Moa Totals		**26**	**0**	**0**	**0.000**	**0.00**
ASSEMBLAGE TOTALS		**64981**	**21411**	**9574**	**294.600**	**376066.10**

*MNI made up of 1 subadult and 1 adult, which have varying meatweight values.

**MNI made up of 2 juveniles, 2 female adults and 1 male adult

Table 13-1. Cabana Lodge prehistoric faunal assemblage frequencies, meat yields and energy yields.

Class	Meat Yield (kg)	%MTWT	Energy Yield (kCal)	% Energy Yield
Shell	15.91	2.75	15594.74	1.85
Fish	260.61	45.01	318273.90	37.75
Marine Bird	2.98	0.52	9572.22	1.14
Moa	0.00	0.00	0.00	0.00
Terrestrial Bird	0.05	0.01	80.08	0.01
Terrestrial Mammal	16.20	2.80	19404.00	2.30
Marine Mammal	283.26	48.92	480141.20	56.95
Total	**579.01**	**100.00**	**843066.14**	**100.00**

Table 13-2. Cabana Lodge percentage meat yield, percentage meatweight (MTWT) and energy harvest by animal class

The possible reasons behind the reduction in fish, bird and terrestrial mammals in the late prehistoric Coromandel require further exploration. If a wider view is taken, and data from nearby Auckland/Hauraki Gulf (Table 13-7) is looked at, it is apparent that fish, birds and terrestrial mammals were indeed still available and exploited near the Coromandel (Table 13-8). Fish are present in all but one of these late period assemblages from around Auckland. There appear to be major differences between the assemblage classes contributing energy in assemblages on the Coromandel Peninsula and around Auckland. The causes of this could be pre-or post-depositional, and it is important to consider factors that may be skewing the Coromandel data.

Sampling may well be an issue. Many of the early period Coromandel assemblages came from research-driven excavations, while most from the later periods are from salvage excavations where the choices about which parts of sites were investigated are driven by development rather than research considerations. Thus the later assemblages may be less representative of the *complete* archaeological record. However, this is a moot point, as salvage archaeology more often relies on random sampling than research excavations, some of which, it can be argued, tend to select 'richer' features or sites. Furthermore, sample sizes for most of the later period sites are generally smaller, and it is suspected that the midden analysis may not always have been as fine-grained or rigorous as might be the case in a research-driven context. While this might have contributed to smaller fish bones or otoliths being missed, it seems unlikely that this could explain all of the differences. Another potential set of reasons may reflect cultural practices. It is possible that some of the "missing" resources were being processed and preserved at other locations, possibly on a seasonal basis. Greater reliance could have been placed on food sources such as sharks which are poorly represented in the archaeological record, due to their cartilaginous elements. It is also possible that, in the late period, with meat sources from large taxa becoming scarcer, scavenging by dogs had a much greater impact on the survival of bones, particularly smaller species such as fish and birds, than had been the case in earlier periods.

Assemblage Code	Assemblage Name	Period	Source
CCK L3	Cross Creek Midden Layer 3 (T10/399)	E	Smith 2010c
CCK L5	Cross Creek Midden Layer 5 (T10/399)	E	Smith 2010c
CCK L7	Cross Creek Midden Layer 7 (T10/399)	E	Smith 2010c
CCK L9	Cross Creek Midden Layer 9 (T10/399)	E	Smith 2010c
CL T12/3	Cabana Lodge T12/3	E	*this manuscript*
HHB upper	Hahei Beach upper (T11/242)	E	Smith 2010c
HHB middle	Hahei Beach middle (T11/242)	E	Smith 2010c
HHB lower	Hahei Beach lower (T11/242)	E	Smith 2010c
HWB middle	Hot Water Beach middle (T11/115)	E	Smith 2010c
HWB lower	Hot Water Beach lower (T11/115)	E	Smith 2010c
PJK	Port Jackson (S09/53)	E	Smith 2010c
AHU-944 #3	Ahuahu (T10/944)	M	Furey 2009
ANA lower	Anatere Pa, Athenree (U13/78)	M	Smith 2010c
ANA upper	Anatere Pa, Athenree (U13/78)	M	Smith 2010c
BWT-874	Bowentown 874 (U13/874)	M	Smith 2010c
BWT-875	Bowentown 875 (U13/875)	M	Smith 2010c
BWT-876	Bowentown 876 (U13/876)	M	Smith 2010c
KTR-78	Koutunui Road, Athenree (U13/78)	M	Smith 2010c
MTR-990	Matarangi Beach (T10/990)	M	Smith 2010c
WHA-1035	Beach Road, Whangamata (T12/1035)	M	Hoffman 2009
WHA-1044 lower	Whangamata 1044 (T12/1044) lower	M	Smith 2010c
WHA-1044 upper	Whangamata 1044 (T12/1044) upper	M	Smith 2010c
WHI-624	Whitianga Waterways 624 (T11/624)	M	Smith 2010c
WHI-926	Whitianga Waterways 926 (T11/926)	M	Smith 2010c
WHI-927 B	Whitianga Club Marina Apartments Area B (T11/926)	M	Smith 2010c
NH-173	Ngarimu Heights, Thames Coast (T12/173)	L	Hoffman 2009
WHA-959	Moana Park Whangamata 959 (T12/959)	L	Smith and James-Lee (2009)
WHA-106	Whangamata Rising Main	L	Mallows (2009)
WHB-1246	The Crescent, Waihi Beach (U13/1246)	L	Smith 2010c
WHI-858	Whitianga Waterways 858 (T11/858)	L	Smith 2010c
WHI-927 A	Whitianga Club Marina Apartments Area A (T11/927)	L	Smith 2010c

Table 13-3. Archaeozoological assemblages and sources from the Coromandel Peninsula.

Assemblage	Period	Energy Shellfish (%kCal)	Energy Fish (%kCal)	Energy Marine Birds (%kCal)	Energy Marine Mammals (%kCal)	Energy Moa (%kCal)	Energy Terrestrial Small Birds (%kCal)	Energy Terrestrial Mammals (%kCal)	Total
CCK L3	E	21.44	52.22	3.22	15.46	0.00	1.06	6.61	100.00
CCK L5	E	0.55	24.41	3.48	60.93	6.73	0.00	3.89	100.00
CCK L7	E	0.60	11.05	0.38	82.36	2.71	0.54	2.36	100.00
CCK L9	E	0.09	4.16	5.14	77.02	11.73	1.74	0.11	100.00
CL T12/3	E	4.15	84.17	1.95	4.63	0.00	0.02	5.09	100.00
HHB upper	E	3.73	92.63	3.49	0.00	0.00	0.00	0.14	100.00
HHB middle	E	1.89	89.34	2.91	0.00	0.00	5.76	0.09	100.00
HHB lower	E	2.89	40.79	4.10	0.00	42.20	0.00	10.02	100.00
HWB middle	E	0.91	23.05	11.34	23.81	31.41	3.17	6.31	100.00
HWB lower	E	0.54	14.66	5.28	50.27	22.34	2.68	4.23	100.00
PJK	E	0.41	0.00	1.66	88.48	7.36	2.09	0.00	100.00
Averages		**3.38**	**39.68**	**3.90**	**36.63**	**11.32**	**1.55**	**3.53**	**100.00**
± 20% cumulative error		0.68	7.94	0.78	7.33	2.26	0.31	0.71	
Range		**2.71 - 4.06**	**31.74 - 47.62**	**3.12 - 4.69**	**29.31 - 43.96**	**9.05 - 13.58**	**1.24 - 1.86**	**2.83 - 4.24**	

Table 13-4. Early period Coromandel assemblages showing energy proportions by animal class.

Assemblage	Period	Energy Shellfish (%kCal)	Energy Fish (%kCal)	Energy Marine Birds (%kCal)	Energy Marine Mammals (%kCal)	Energy Moa (%kCal)	Energy Terrestrial Small Birds (%kCal)	Energy Terrestrial Mammals (%kCal)	Total
AHU-944 #3	M	24.68	0.00	0.00	0.00	0.00	0.00	75.32	100.00
ANA upper	M	2.70	97.30	0.00	0.00	0.00	0.00	0.00	100.00
ANA lower	M	100.00	0.00	0.00	0.00	0.00	0.00	0.00	100.00
BWT-874	M	19.13	80.87	0.00	0.00	0.00	0.00	0.00	100.00
BWT-875	M	2.17	63.47	34.37	0.00	0.00	0.00	0.00	100.00
BWT-876	M	100.00	0.00	0.00	0.00	0.00	0.00	0.00	100.00
KTR-78	M	100.00	0.00	0.00	0.00	0.00	0.00	0.00	100.00
MTR-990	M	100.00	0.00	0.00	0.00	0.00	0.00	0.00	100.00
WHA-1035	M	100.00	0.00	0.00	0.00	0.00	0.00	0.00	100.00
WHA-1044 lower	M	100.00	0.00	0.00	0.00	0.00	0.00	0.00	100.00
WHA-1044 upper	M	100.00	0.00	0.00	0.00	0.00	0.00	0.00	100.00
WHI-624	M	100.00	0.00	0.00	0.00	0.00	0.00	0.00	100.00
WHI-926	M	100.00	0.00	0.00	0.00	0.00	0.00	0.00	100.00
WHI-927 B	M	100.00	0.00	0.00	0.00	0.00	0.00	0.00	100.00
Averages		**74.91**	**17.26**	**2.45**	**0.00**	**0.00**	**0.00**	**5.38**	**100.00**
± 20% cumulative error		14.98	3.45	0.49	0.00	0.00	0.00	1.08	
Range		**59.92 - 89.89**	**13.81 - 20.71**	**1.96 - 2.95**	**0.00 - 0.00**	**0.00 - 0.00**	**0.00 - 0.00**	**4.30 - 6.46**	

Table 13-5. Middle period Coromandel assemblages showing energy proportions by animal class.

Assemblage	Period	Energy Shellfish (%kCal)	Energy Fish (%kCal)	Energy Marine Birds (%kCal)	Energy Marine Mammals (%kCal)	Energy Moa (%kCal)	Energy Terrestrial Small Birds (%kCal)	Energy Terrestrial Mammals (%kCal)	Total
NH-173	L	100.00	0.00	0.00	0.00	0.00	0.00	0.00	100.00
WHA-959	L	100.00	0.00	0.00	0.00	0.00	0.00	0.00	100.00
WHA-106	L	100.00	0.00	0.00	0.00	0.00	0.00	0.00	100.00
WHB-1246	L	100.00	0.00	0.00	0.00	0.00	0.00	0.00	100.00
WHI-858	L	100.00	0.00	0.00	0.00	0.00	0.00	0.00	100.00
WHI-927 A	L	100.00	0.00	0.00	0.00	0.00	0.00	0.00	100.00
Averages		**100.00**	**0.00**	**0.00**	**0.00**	**0.00**	**0.00**	**0.00**	**100.00**
± 20% cumulative error		20.00	0.00	0.00	0.00	0.00	0.00	0.00	
Range		**80.00 - 120.00**	**0.00 - 0.00**	**0.00 - 0.00**	**0.00 - 0.00**	**0.00 - 0.00**	**0.00 - 0.00**	**0.00 - 0.00**	

Table 13-6. Late period Coromandel assemblages showing energy proportions by animal class.

Assemblage Code	Assemblage Name	Period	Source
BTF 132	Bream Tail Farm (R08/132)	L	Smith 2010c
OMH 035	Omaha Beach (R09/887)	L	Smith 2010c
OMH 042	Omaha Beach (R09/887)	L	Smith 2010c
SUN upper	Sunde upper, Motutapu Island (R10/25)	L	Smith 2010c
CRD West	Cryer's Road West (R11/1519)	L	Smith 2010c
CRD A41	Cryer's Road A41 (R11/1519)	L	Smith 2010c
PUR 86	Puriri 86 (T12/882)	L	Smith 2010c
PUR 91 340	Puriri 91 340 (T12/340)	L	Smith 2010c
PUR 91 886	Puriri 91 886 (T12/886)	L	Smith 2010c
PUR 91 885	Puriri 91 885 (T12/885)	L	Smith 2010c

Table 13-7. Additional late period assemblages from Auckland/Greater Hauraki.

Assemblage	Period	Energy Shellfish (%kCal)	Energy Fish (%kCal)	Energy Marine Birds (%kCal)	Energy Marine Mammals (%kCal)	Energy Moa (%kCal)	Energy Terrestrial Small Birds (%kCal)	Energy Terrestrial Mammals (%kCal)	Total
BTF 132	L	3.06	96.94	0.00	0.00	0.00	0.00	0.00	100
OMH 035	L	55.15	44.85	0.00	0.00	0.00	0.00	0.00	100
OMH 042	L	100.00	0.00	0.00	0.00	0.00	0.00	0.00	100
SUN upper	L	2.05	92.85	1.90	0.00	0.00	0.30	2.90	100
CRD west	L	95.52	4.48	0.00	0.00	0.00	0.00	0.00	100
CRD A41	L	58.87	13.66	15.28	0.00	0.00	0.00	12.19	100
PUR-86	L	7.42	92.58	0.00	0.00	0.00	0.00	0.00	100
PUR-91 340	L	15.91	84.09	0.00	0.00	0.00	0.00	0.00	100
PUR-91 886	L	19.01	80.99	0.00	0.00	0.00	0.00	0.00	100
PUR-91 885	L	23.38	76.62	0.00	0.00	0.00	0.00	0.00	100
Average	**L**	**38.04**	**58.71**	**1.72**	**0.00**	**0.00**	**0.03**	**1.51**	**100**
± 20% cumulative error		7.61	11.74	0.34	0.00	0.00	0.01	0.30	
Range		**30.43 - 45.65**	**46.97 - 7.45**	**1.38 - 2.06**	**0.00 - 0.00**	**0.00 - 0.00**	**0.02 - 0.04**	**1.21 - 1.81**	

Table 13-8. Late period Auckland/Greater Hauraki assemblages showing energy proportions by animal class.

These considerations suggest that the archaeological data do not quite represent reality (fig. 13-3). Accounting for dog scavenging (and the non-preservation of dog coprolites), if 15% of the energy harvest calculated for shellfish is shifted to fish, and 1 or 2% of the total is also allowed to come from each of the mammal and bird components (excluding moas) we would get something like the pattern shown in the lower graph of fig. 13-3. horticulture was possible, while the southern region was beyond its limits (fig. 13-5). In both areas plant foods will have supplemented the diets reconstructed from animal foods represented in the archaeological record. Both had a range of wild plant foods that would have contributed to this, but the availability of horticultural crops in the north means that this was possible to a far greater extent than was the case in the south. Elsewhere, Smith has noted

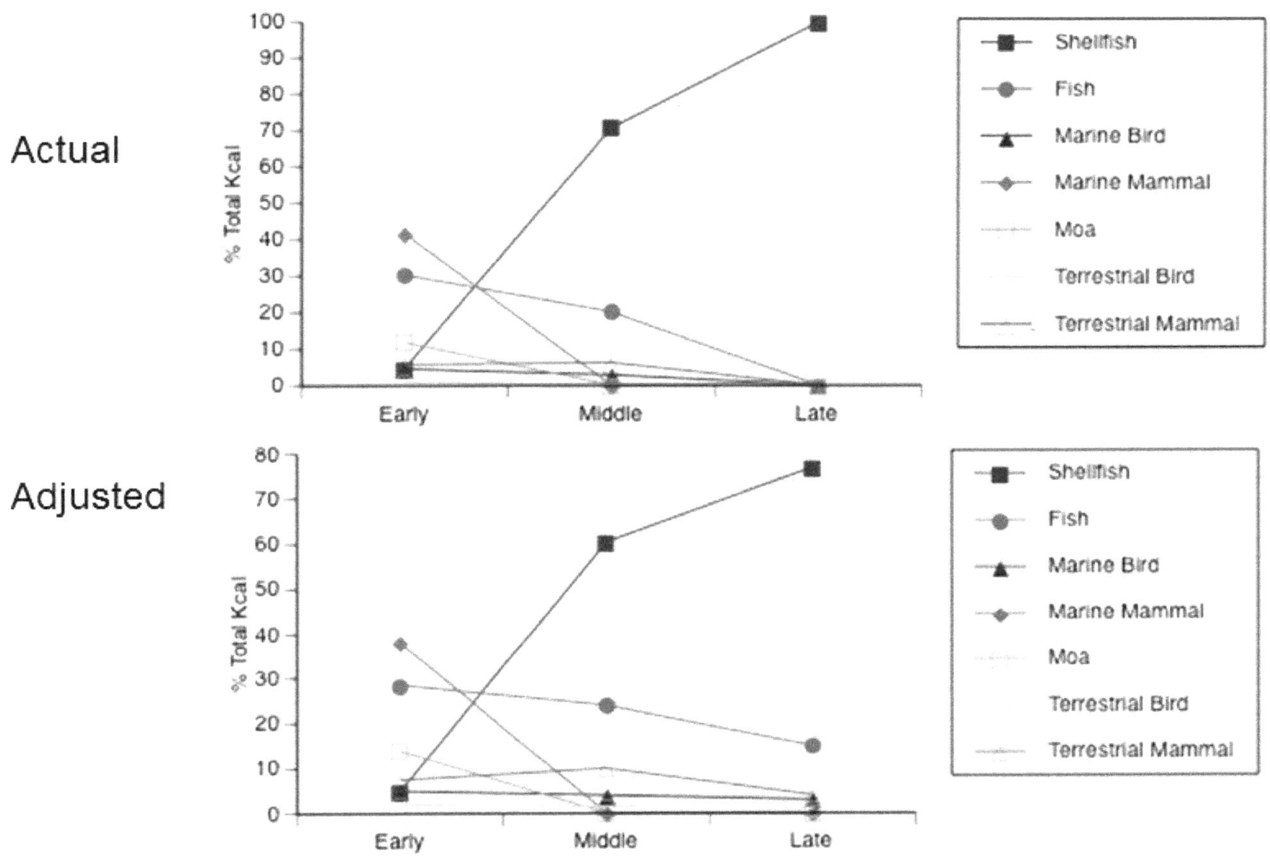

Figure 13-3. Graphs of averaged %kCal by animal class for Coromandel assemblages with actual (upper) and adjusted (lower) values.

Although there is no way of really knowing, it is more likely this is what really happened than strictly following the data would suggest. Whatever the case, it seems clear that shellfish became the mainstay of the animal component of the diet for people in the Coromandel region. To get a better estimate of what reality might have been, a couple of adjustments to period averages can be made. If our adjusted averages are compared with the same sort of information that has been compiled for a study area near the southern end of the South Island (Smith 2011a) we see a vastly different pattern (fig. 13-4). There is the same loss over time of the large animals – moas and marine mammals – although this happens much more slowly in the south. Most of the slack is taken up by fish. Although shellfish do increase in importance, it is nothing like what happens in the Coromandel.

One of the significant differences between the two areas is that the northern study region is within the area where that the nutritional composition of the late period northern faunal assemblages indicates that a substantial proportion of energy must have been derived from plant foods (Smith 2002). The limited evidence that we have from isotope studies on human bone is consistent with this (Leach *et al.* 2003).

What the horticultural economy allowed was sustained population growth. The estimates shown in fig. 13-6 for Otago-Catlins and the Greater Hauraki region (including the Coromandel) are based on the analysis of radiocarbon dates and historical data (Smith 2011a). It shows rapid growth of the human population around the beginning of the Middle period in the horticultural zone (for discussion see Groube 1970). This coincides with the time at which moas and fur seals were exterminated from northern New Zealand, and the transition from *generalised* assemblages of fauna in Coromandel sites to *specialised* assemblages

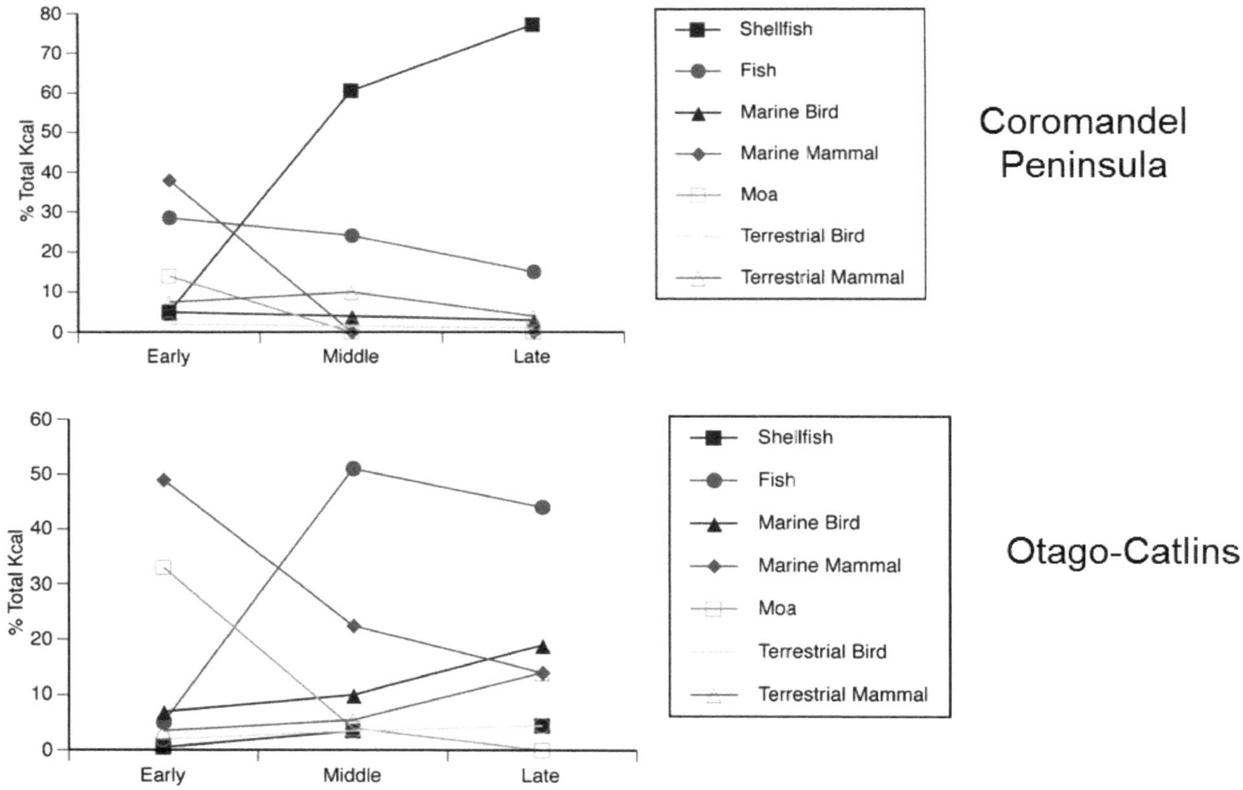

Figure 13-4. Graphs of averaged %kCal by animal class for Coromandel assemblages (upper) and Otago-Catlins assemblages (lower).

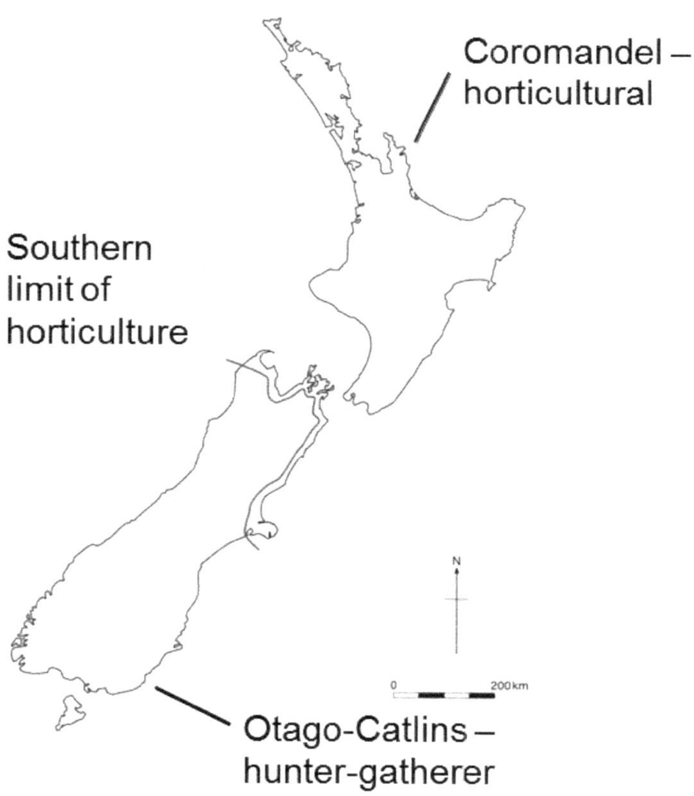

Figure 13-5. Map of New Zealand indicating the Coromandel and Otago-Catlins areas, and the southern limit for horticulture.

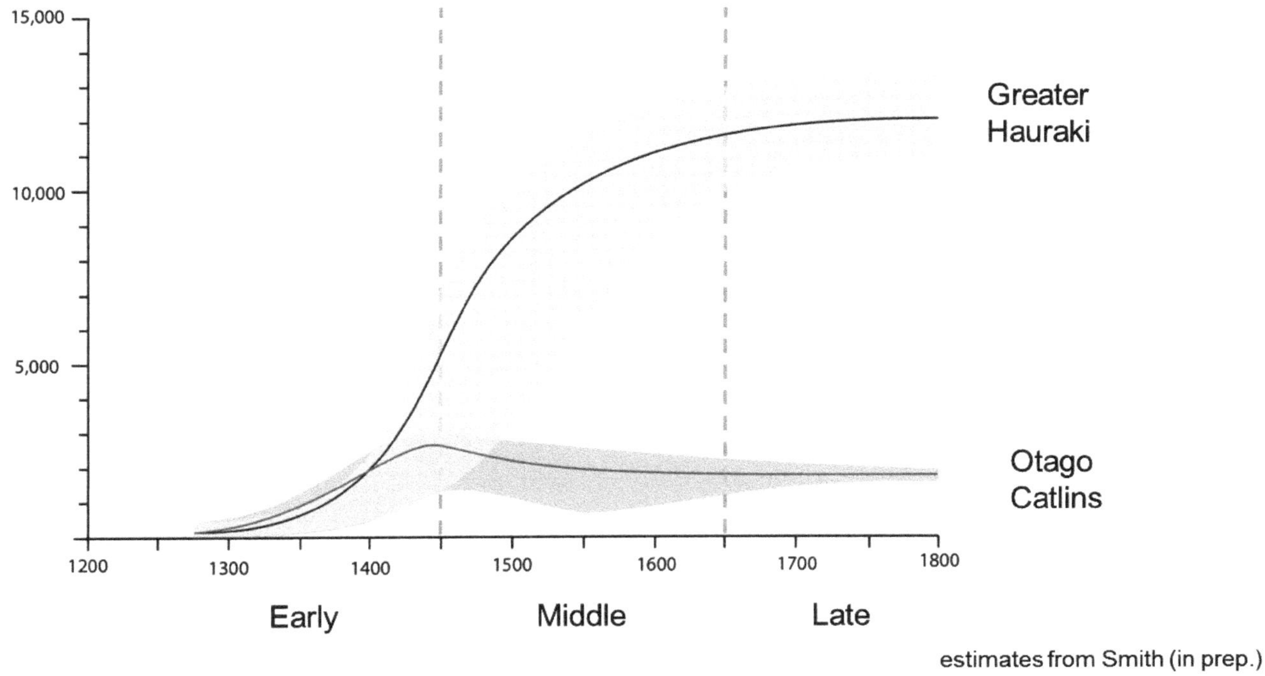

Figure 13-6. Population growth estimates for Otago-Catlins and Greater Hauraki regions (Smith 2011a).

of shellfish and fish. Other changes that appear to have occurred between the early and late periods include the "...the decentered efflorescence of paa" (Barber 1996) and the large numbers of larger kuri (dog) skeletons found in early sites compared to lower numbers and smaller mean bodyweights of kuri in late sites (Clark 1997).

Many of the middle and late period sites on the Coromandel Peninsula are clustered around harbours and estuaries in which there are large beds of shellfish such as cockles (*Austrovenus stutchburyi*) and pipi (*Paphies australis*). The shores around these harbours also have soils suitable for growing kumara and other cultigens. Tending gardens of kumara would have required much time spent preparing the soil and tending the plants. Travelling away from home base would have left crops vulnerable to raids on kumara stored in pits. Less time would have been available for gathering resources further away – especially as they became depleted – leading groups to forage for food in the areas closest to their gardens. The importance of shellfish in this context is that they would have provided a stable resource, capable of sustained harvesting. They were locally abundant, did not require specialist skills or travelling long distances to exploit. The meat from shellfish was able to be dried and stored for later use, and the shells used in tool making, such as scrapers for flax weaving (Paulin 2007).

Conclusion

In conclusion, then, what this case study has shown is that shellfish played a significant role in the horticultural economy of the prehistoric inhabitants of the Coromandel Peninsula, and that they helped to sustain a rapidly growing population there. To what extent this might have occurred elsewhere in the world remains to be seen, but this case study suggests that closer attention to this aspect of the archaeological record can be revealing.

Acknowledgements

Many thanks to Ian Smith for support and critical feedback in the development of this paper; Heather Sadler for IT support; to Les O'Neil for illustrations; and to the reviewers for their comments.

References

Alexander, R. M., G. M. O. Maloiy, R. Njan, and A. S. Jayes. 1979. Mechanics of running in the ostrich (*Struthio camelus*). *Journal of Zoology* 201:363-76.

Allo, J. 1972. The Whangamata Wharf site (N49/2): excavations on a Coromandel coastal midden. *Records of the Auckland Institute and Museum* 9:61-79.

Altman, R. A., and D. S. Ditmar. 1968. *Metabolism*. Bethseda: Biological Handbooks.

Álvarez, M., I. B. Godino, A. Balbo, and M. Madella. 2010. Shell middens as archives of past environments, human dispersal and specialized resource management. *Quaternary International* doi: 10.1016/j.quaint.2010.10.025.

Anderson, A. J. 2002. "A fragile plenty: pre-European Maori and the New Zealand environment " in

Environmental Histories of New Zealand. Edited by E. Pawson and T. Brooking, pp. 19-34, 297-298. Auckland: Oxford University Press.

BARBER, I. G. 1996. Loss, change, and monumental landscaping: towards a new interpretation of the "Classic" Maaori emergence. *Current Anthropology* 37:868-880.

BASSETT, K. N., H. W. GORDON, D. C. NOBES, and C. JACOMB. 2004. Gardening at the edge: documenting the limits of tropical Polynesian kumara horticulture in southern New Zealand. *Geoarchaeology: An International Journal* 19:185-218.

BUNCE, M., T. H. WORTHY, T. FORD, W. HOPPITT, E. WILLERSLEV, A. DRUMMOND, and A. COOPER. 2003. Extreme reverse sexual size dimorphism in the extinct New Zealand moa Dinornis. *Nature* 425:172-5.

CLAASSEN, C. 1986. Shellfishing seasons in the prehistoric southeastern United States. *American Antiquity* 51:21-37.

CLARK, G. 1954. *Excavations at Star Carr: an early Mesolithic site at Seamer near Scarborough, Yorkshire*. Cambridge: University Press.

CLARK, G. R. 1997. Maori subsistence change: zooarchaeological evidence from the prehistoric dog of New Zealand. *Asian Perspectives* 36:200-219.

DAVIDSON, J. M. 1979. "Archaic middens of the Coromandel region: a review," in *Birds of a Feather: Osteological and Archaeological papers from the South Pacific in Honour of R.J. Scarlett*, vol. 62. Edited by A. Anderson, pp. 183-202. Oxford: British Archaeological Reports International Series.

—. 1984. "Subsistence economics," in *The Prehistory of New Zealand*, pp. 115-148. Auckland: Longman Paul.

DENNISTON, G. B. 1972. *Ashishik Point: an economic analysis of a prehistoric Aleutian community*. PhD Thesis, University of Madison,

EDSON, S. C., and D. BROWN. 1977. Salvage Excavation of an Archaic Burial Context, N44/97, Hahei. *Records of the Auckland Institute and Museum* 14:25-36.

FROESE, R., and D. PAULY. 2010. "FishBase," vol. 2010.

FUREY, L. 2003. *Investigation of T11/926, Cook Drive, Whitianga*. Unpublished report.

—. 2009. *Description of archaeological deposits in dunes, White Beach, Great Mercury Island (Ahuahu)*. Unpublished report.

GOLSON, J. 1959. "Culture change in prehistoric New Zealand," in *Anthropology in the South Seas*. Edited by J. D. Freeman and W. R. Geddes. New Plymouth: Avery.

GRAYSON, D. K. 1979. "On the quantification of vertebrate archaeofaunas," in *Advances in Archaeological Method and Theory*. Edited by M. B. Schiffer, pp. 199-237. New York: Academic Press.

GREEN, R. C. 1963. Summaries of sites at Opito, Sarah's Gully and Great Mercury Island. *New Zealand Archaeological Association Newsletter* 6:57-69.

GROUBE, L. M. 1970. "The origin and development of earthwork fortifications in the Pacific," in *Studies in Oceanic Culture History vol. 1 Pacific Anthropological Records No.11*. Edited by R. C. Green and M. Kelly, pp. 133-164. Honolulu: B.P. Bishop Museum.

GUMBLEY, W. 2002. *Assessment of Archaeological Values for T12/991, at Whiritoa, Coromandel Peninsula*. Unpublished report.

—. 2003. *Cabana Lodge, Whangamata: Assessment of archaeological deposits and their values*. Unpublished report.

GUMBLEY, W., M. CAMPBELL, and B. BACQUIE. 2001. *Pauanui Sewerage Treatment Plant, Settlement Road, Pauanui: Archaeological Survey and Assessment*. Unpublished report.

GUMBLEY, W., and A. HOFFMANN. 2007. *Assessment of Archaeological Values Relating to the Proposed Residential Subdivision, Skipper's Road, Opito Bay*. Unpublished report.

—. *2008. Archaeological Investigation Preliminary Report: T12/3, Cabana Lodge, Whangamata*. Unpublished report.

HOFFMANN, A. 2009. *T10/824, Kuaotunu, Coromandel Peninsula: Archaeological Investigation*. Unpublished report.

HOLDAWAY, R. N., and C. JACOMB. 2000. Rapid extinction of the Moas (Aves: Dinornithormes): Model, Test and Implications. *Science* 287:2250-2254.

HORROCKS, M. 2004. Polynesian plant subsistence in prehistoric New Zealand: a summary of microfossil evidence. *New Zealand Journal of Botany* 42:321-334.

HORROCKS, M., and I. G. BARBER. 2005. Microfossils of introduced starch cultigens from an early wetland ditch in New Zealand. *Archaeology in Oceania* 40:106-114.

Jolly, R. G. W. 1978. The East or Cabana Lodge site, Whangamata. *New Zealand Archaeological Association Newsletter* 21:135-137.

Jones, K. L. 1973. Excavations at Tairua (N44/2), 1958-1964: A synthesis. *New Zealand Archaeological Association Newsletter* 16:143-150.

Law, R. G. 1972. Archaeology at Harataonga Bay, Great Barrier Island. *Records of the Auckland Institute and Museum* 9:81-123.

——. 1982. "Coromandel Peninsula and Great Barrier Island," in *The First Thousand Years* Edited by N. Prickett, pp. 49-61. Palmerston North: Dunmore Press Limited.

Leach, B. F. 2006. *Fishing in Pre-European New Zealand*: New Zealand Journal of Archaeology Special Publication / Archaeofauna

Leach, B. F., J. M. Davidson, L. M. Horwood, and A. J. Anderson. 1996. *The estimation of live fish size from archaeological cranial bones of the New Zealand barracouta Thyrsites atun*. Wellington, N.Z: Museum of New Zealand.

Leach, B. F., J. M. Davidson, M. Robertshawe, and P. C. Leach. 2001. Identification, Nutritional Yield, and Economic Role of Tuatua Shellfish, *Paphies* spp., in New Zealand Archaeological Sites Leach, B. F., C. Quinn, J. Morrison, and G. Lyon. 2003. The use of multiple isotope signatures in reconstructing prehistoric human diet from archaeological bone from the Pacific and New Zealand. *New Zealand Journal of Archaeology* 23(2001):31-98.

Leahy, A. 1974. Excavations at Hot Water Beach (N44/69), Coromandel Peninsula. *Records of the Auckland Institute and Museum* 11:23-76.

Lyman, R. L. 1979. Available meat from faunal remains: a consideration of techniques. *American Antiquity* 44:536-546.

Mallows, C. 2009. *Archaeological Excavation: T12/1243 - Whangamata Waste Water Treatment Plant.* Unpublished report.

Maxwell, K. 2010. "One hundred years of the Otago groper fishery: combining multiple data sources," in *Paper presented to the N.Z. Marine Sciences Society Conference, Wellington, July 2010.*

McGlone, M. S. 1989. "The Polynesian settlement of New Zealand in relation to environmental and biotic change," in *Moas, mammals and climate in the ecological history of New Zealand* Edited by M. R. Rudge, pp. 115-130: New Zealand Journal of Ecology 12 (Supplement).

McGlone, M. S., A. J. Anderson, and R. N. Holdaway. 1994. "An Ecological Approach to the Polynesian Settlement of New Zealand," in *The Origins of the First New Zealanders*. Edited by D. G. Sutton, pp. 136-163. Auckland: Auckland University Press.

McGlone, M. S., and J. M. Wilmshurst. 1999. Dating initial Maori environmental impact in New Zealand. *Quarternary International* 59:5-16.

Newnham, R. M., D. J. Lowe, M. S. McGlone, J. M. Wilmshurst, and T. F. G. Higham. 1998. The Kaharoa Tephra as a critical datum for earliest human impact in northern New Zealand. *Journal of Archaeological Science* 25:533-544.

Nichol, R. K. 1978. *Fish and shellfish in New Zealand prehistory.* MA thesis, University of Auckland,

——. 1988. *Tipping the Feather Against a Scale: Archaeozoology from the Tail of the Fish.* Thesis, University of Auckland, Auckland.

Paul, L. J. 2000. *New Zealand Fishes: Identification, Natural History and Fisheries*. Auckland: Reed Books.

Paulin, C. D. 2007. Perspectives of Maori fishing history and techniques. *Tuhinga* 18:11-47.

Raven, J., and S. Bracegirdle. 2010. *New Zealand SeaShells: Visual Guide*. Wellington: Creatus.

Rowland, M. J. 1975. *Tairua and Offshore Islands in Early New Zealand Prehistory.* MA Thesis, University of Auckland, Auckland.

Sewell, B. 2005. *Archaeological monitoring on the Anarake Subdivision,* Whangapoua. Unpublished report.

Shawcross, F. W. 1967. An investigation of Prehistoric Diet and Economy on a Coastal Site at Galatea Bay, New Zealand. *Proceedings of the Prehistoric Society* 33:107-131.

——. 1970. Ethnographic Economics and the Study of Population in Prehistoric New Zealand: Viewed through Archaeology. *Mankind* 7:279-291.

——. 1972. "Energy and ecology: thermodynamic models in archaeology," in *Models in archaeology*. Edited by D. L. Clarke, pp. 577-622. London: Methuen.

Simmons, A. 2002. *Archaeological Report on a Pit Located in the Coastal Lifestyles Subdivision at Hotwater Beach, Coromandel Peninsula.* Unpublished report.

Smart, C. D., and R. C. Green. 1962. A stratified dune site at Tairua, Coromandel. *Dominion Museum Records in Ethnology* 1:243-266.

Smith, B. D. 1975. "Toward a more accurate estimation of meat yield of animal species at archaeological sites.," in *Archaeozoological Studies*. Edited by D. Clason, pp. 99-108. Amsterdam: North-Holland Publishing.

Smith, I. W. G. 1985. Sea Mammal Hunting and Prehistoric Subsistence in New Zealand. PhD Thesis, University of Otago, Dunedin.

—. 2002. "Retreat and resilience: fur seals and human settlement in New Zealand." *9th ICAZ Conference, Durham 2002, Durham, 2002.*

—. 2004. Nutritional perspectives on prehistoric marine fishing in New Zealand. *New Zealand Journal of Archaeology* 24(2002):5-31.

—. 2010a. Protocols for organising radiocarbon dated assemblages from New Zealand archaeological sites for comparative analysis. *Journal of Pacific Archaeology* 1:184-187.

—. 2010b. "Regional and temporal variations in energy harvests from prehistoric fauna in New Zealand," in *Paper presented to the New Zealand Archaeological Association Conference 2010*. Westport, New Zealand.

—. 2011a. *Estimating the magnitude of pre-European Maori marine harvest in two New Zealand study areas*. New Zealand Aquatic Environemnt and Biodiversity. Report No. XX.

—. 2011b. *Meat Weights and Nutritional Yield Values for New Zealand Archaeofauna*. Anthropology, University of Otago.

Smith, I. W. G., and T. James-Lee. 2009. *Data for an archaeozoological analysis of marine resource use in two New Zealand study areas*. Anthropology, University of Otago. Otago Archaeological Laboratory Report: Number 6.

Stewart, F. L., and P. W. Stahl. 1977. Cautionary Note on Edible Meat Poundage Figures. *American Antiquity* 42:267-270.

Vlieg, P. 1988. *Proximate Composition of New Zealand Marine Finfish and Shellfish*. Department of Scientific and Industrial Research.

Walter, R., I. W. G. Smith, and C. Jacomb. 2006. Sedentism, subsistence and socio-political organisation in prehistoric New Zealand. *World Archaeology* 38:274-290.

White, T. E. 1953. A Method of Calculating the Dietary Percentage of Various Food Animals Utilized by Aboriginal Peoples. *American Antiquity* 18:396-398.

Wilmshurst, J. M., A. J. Anderson, T. F. G. Higham, and T. H. Worthy. 2008. Dating the late prehistoric dispersal of Polynesians using the commensal Pacific rat. *Periodical of the National Academy of Sciences* 105:7676-7680.

Wilmshurst, J. M., T. F. G. Higham, H. Allen, D. Johns, and C. Phillips. 2004. Early Maori settlement impacts in northern coastal Taranaki, New Zealand. *New Zealand Journal of Ecology* 28:167-179.

Worthy, T. A., and R. N. Holdaway. 2002. *The Lost World of the Moa: Prehistoric Life of New Zealand.* Bloomington and Indianapolis: Indiana University Press.

Worthy, T. H., M. Bunce, A. Cooper, and R. P. Scofield. 2005. "Dinornis – an insular oddity, a taxonomic conundrum reviewed." *Proceedings of the International Symposium "Insular Vertebrate Evolution: the Palaeontological Approach", 2005*, pp. 377-390 12.

Yen, D. E. 1961. The adaptation of kumara by the New Zealand Maori. *Journal of the Polynesian Society* 70:338-348.

14 - FISHER-GATHERERS OF THE RED SEA: RESULTS OF THE FARASAN ARCHIPELAGO SHELL SITES PROJECT

Matt Gregory Meredith WILLIAMS
University of York, Archaeology Department, The King's Manor, York, YO1 7EP
matthew.williams@york.ac.uk

Abstract: *This paper presents the results of the Farasan Archipelago Shell Sites Project (FArSSite) – part of the Southern Red Sea Project. The discovery of over 1,500 shell sites on the Farasan Archipelago (Red Sea) represents one of the highest concentrations of such sites in the world. These have been extensively mapped and surveyed, allowing a picture of complex landscape change to be reconstructed.*

Two at risk sites were excavated in order to investigate spatial and temporal shell mound evolution. Analysis of the excavated material has revealed a varied pattern of coastal resource exploitation, over relatively short time periods. The two shell mounds show markedly different composition and structure, indicating that although similar marine habitats were being exploited, different activities were taking place resulting in contrasting assemblages.

Bayesian techniques have been applied to radiocarbon and amino-acid racemization data, in order to calculate the age and accumulation rates at the sites, showing that the sites may have been contemporaneous. The fisher-gatherers who built these mounds were without pottery; however the use of boats is implied given the island setting and distance of c.40 km across open water to the mainland where similar sites have been found.

Keywords: *Shell midden, Farasan islands, Red Sea, Formation processes, Dating*

Introduction

The FArSSite Project arose as part of the *Southern Red Sea Project: Farasan Islands*, tasked specifically with investigating the shell midden sites of the islands in greater detail. The Farasan Islands are an archipelago of up to 250 islands, centred on the two largest islands of Farasan Kabir and Saqid. The archipelago is over 40 km from the mainland, located in Saudi Arabian waters, off shore from Gizan (fig. 14-1a).

The research theme was to better understand the provenance of shell middens on the Farasan Islands, both spatially and temporally. That shell middens exist on the islands had been recorded in the early 1980's during archaeological survey, with a number of subsequent projects noting their presence and obtaining further dates from selected sites. Unfortunately the location of these sites and the origin of the dating samples was often unspecific (Bantan 1999; Deputy Ministry of Antiquities and Museums 1990; Zarins, Al-Jawad Murad, and Al-Yish 1980). However,

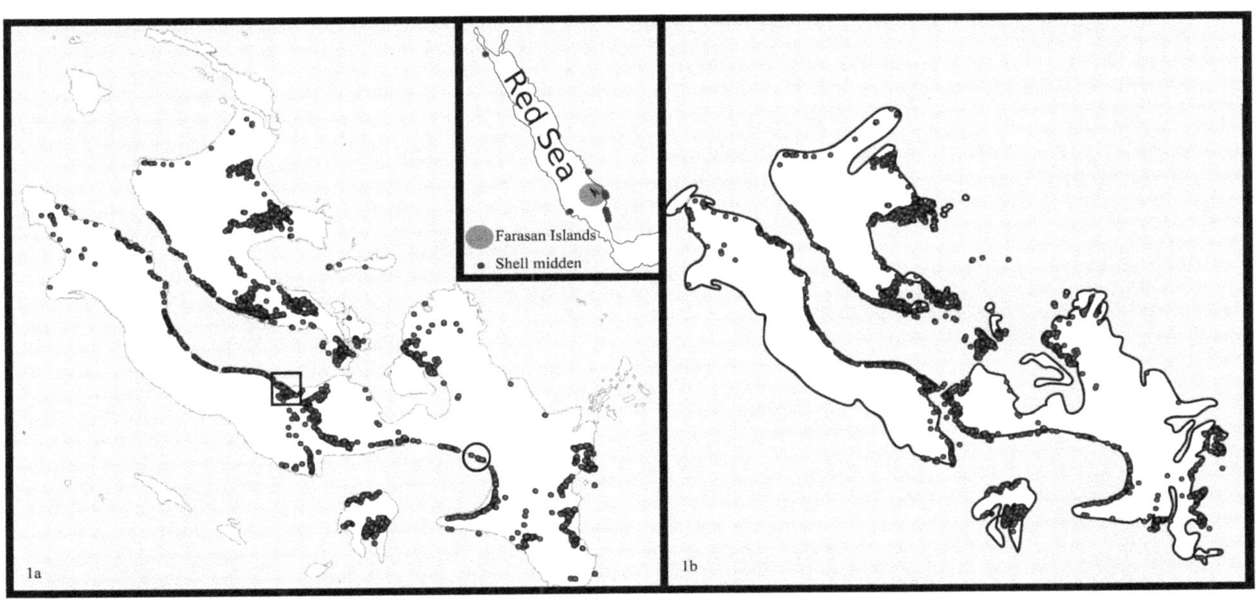

Figure 14-1. Insert (of 1a) shows location both of Farasan Islands, and other known shell middens in the Red Sea. 1a shows map of central islands of the Farasan Archipelago, with known shell midden sites; Khur Maadi sites are boxed, Janaba East sites are circled. 1b is a reconstruction of palaeoshorelines of Farasan Islands c.5000BP.

these reports demonstrated the presence of such sites on the islands, and gave an indication of the period during which some of these mounds may have formed.

It was not until the first survey of the islands by the Southern Red Sea Project in 2006 that the full extent of the shell sites of anthropogenic origin became apparent (see Bailey *et al.* 2007 a; Bailey *et al.* 2007 b; Bailey *et al.* in press). During these seasons of field work over 1,000 shell mounds were mapped across the islands.

This paper serves to outline the key results of the study, addressing in particular the distribution of sites across the islands, the excavations, and what these tell us about the archaeological environment and archaeological food resources on the islands.

Methods

The methodology of this project is composed of three main parts: survey, excavation, and dating. Surveying the islands encompassed mapping and measuring shell sites, as well as documenting the composition, nature and location of palaeocoastline features. Satellite image interpretation added another layer of data to these findings; shell sites and palaeocoastline features can be identified on the images, and by manipulating the different bandwidths which make up a satellite image it was possible to produce false colour composite images. These make it possible to distinguish between different sediment types and works particularly well on the Farasan Islands where vegetation cover is sparse.

Two sites were selected for excavation – both are the largest and most centrally located sites within a group of smaller shell sites. These were both 'at risk' sites which were under threat of imminent destruction. The objective of the excavations was to expose sections through the mounds, from which the internal structure of the sites could be assessed. These would be used in an attempt to reconstruct the evolution of the shell mounds and estimate the volume of shell within each depositional event.

Test pits were also excavated into shell sites across the islands, in order to obtain dating samples and assess the composition of the surface layers of the sites.

The final stage was the use of radiocarbon and amino-acid racemization dating (AAR). These had dual objectives: to investigate the accumulation rates of the two excavated shell mounds, and to date the samples obtained from test pits on other shell midden sites across the islands. Further details regarding the methodology used for AAR dating can be found in Penkman *et al.* (2008) and initial results are published in Demarchi *et al.* (2010).

Results and Discussion

Survey

The survey of the islands located 2,811 shell sites (fig. 14-1a), the vast majority of which are located along palaeocoastlines; these features, where associated with shell middens, were also mapped. When combined with false colour composite satellite images this allowed for a reconstruction of the palaeoshorelines at the time the shell middens were accumulating (fig. 14-1b). The reconstruction shows that at the time of shell mound accumulation the coastlines of the islands were much more convoluted with embayments and channels breaking up the larger islands.

The survey found that the size of the shell sites range from scatters as small as 0.5 m^2 to mounds up to 6m high and stretching up to 100m along the coastlines. The smallest scatters perhaps represent one meal or processing event, whilst the larger sites indicate reoccupation of the site over a longer period. The average size of site on the islands is 1 m high with a 20 m diameter, with the majority of sites being dome shaped.

The question of whether the shell mounds are of anthropogenic origins has already been addressed (see Bailey *et al.* 2007a); briefly the mounds are generally composed of well stratified deposits with archaeological artefacts within them. Some shell middens were found to be located on shallow banks of natural shell, with a clear distinction between them.

Excavation

Excavation of two shell mounds (KM1057 and JE0004) suggests that shellfish was not the only food resource exploited at shell midden sites. However these sites show different accumulation histories; the results of the 2008 field season excavation have been reported in Williams 2010 and Alsharekh *et al.* in press; for a detailed methodology and results please refer to these papers. The reports for the results of the 2009 field season excavations are currently in preparation.

Site KM1057 is part of a large group of shell mounds at the mouth of the former Khur Maadi bay (fig. 14-1); it is a 3 m high conical shell mound with a diameter of 30 m; over 33% has been destroyed through extraction by the building industry (fig. 14-2a). The site is located 400 m inland at the mouth of an in-filled and uplifted palaeobay, on a palaeoshoreline composed of a 0.5-1 m high cliff. In 2008 a 1 m wide stepped section was excavated through the deepest part of the mound (fig. 14-2b), this was composed predominantly of *Strombus fasciatus*, interrupted by two layers of *Chama reflexa* and *Spondylus marisrubri*. Fishbone was found in limited quantities in the upper layers of the mound, corresponding to an ash matrix.

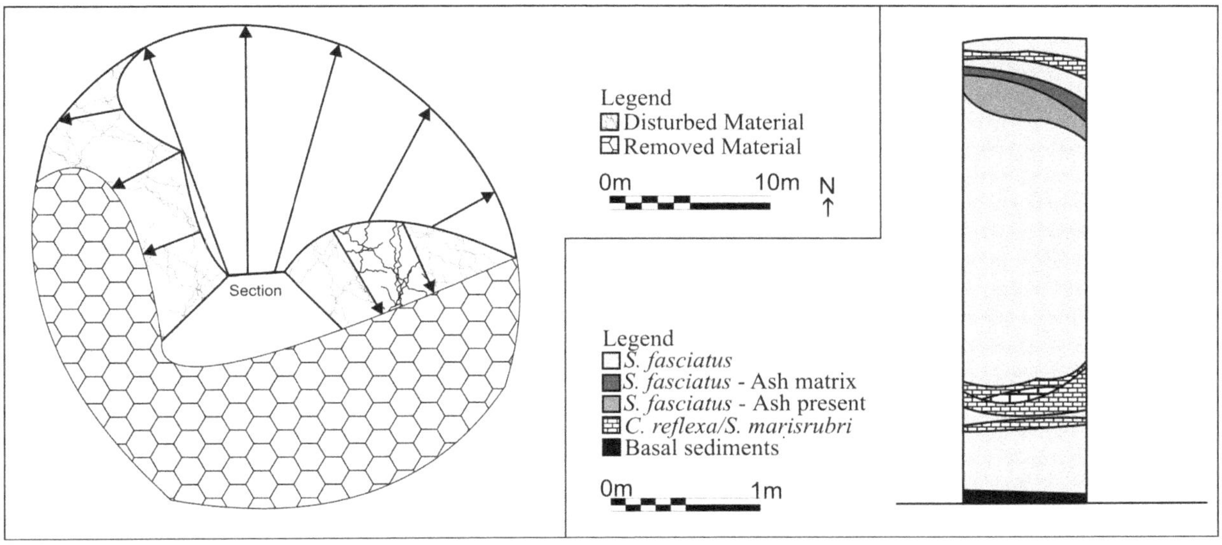

Figure 14-2. Plan of KM1057 (a) and South facing section through KM1057 (b). (After Alsharekh *et al.* in press).

It appears that KM1057 was predominantly used as a shellfish processing site, with little evidence for any other activity other than occasional fish processing. Based on current evidence it seems that habitation did not occur at this location, but more likely on one of the smaller satellite sites, since hearths were not found within the site, but were found at the smaller sites.

The estimated volume of the site (*c.*820 m^3) can be used to infer the potential for the amount of meat procured from the shell (assuming an average *c.*10% meat to shell weight yield – e.g. Erlandson 1988; Meehan 1977; Osborn 1977). Previous studies have suggested that 5 kg of shellfish meat a day would be needed to meet an adult's energy requirement for a day (*ibid.*). This gives a total of *c.*36,000 days worth of shellfish food contained within the mound. Whether the shellfish was used exclusively for food or whether it had other purposes is unknown; it may even be that shellfish was being processed for storage for consumption at a later time (consuming large quantities of protein for longer than a few weeks can result in protein poisoning, see Noli and Avery 1988).

Dating of KM1057 and adjacent sites used two different techniques, radiocarbon dating and AAR. Radiocarbon dating was initially used in order to define the temporal extent of KM1057 by taking radiocarbon dates from the base and top of the mound (Alsharekh *et al.* in press). These dates proved to be almost identical suggesting that the mound had accumulated rapidly over a maximum period of 400 years (Table 14-1). AAR dating was tested at this site and showed promising results, giving the possibility to determine if sites were from the same period (Demarchi *et al.* 2010). Two peripheral sites were dated using this method, suggesting that they accumulated during the same period as KM1057.

Assuming that the shellfish were being processed for food we can estimate that over a maximum 400 year period (Table 14-1) it would be enough to sustain five people for just under twenty days a year, or ten people for a little under ten days per year. Given that people were likely to have been exploiting and consuming other food resources as part of a wider diet (as evidenced by the fishbones in the upper levels) it is likely that the above figures are minimum estimates. It should be stressed that these figures are estimates, and that uncertainty is inherent.

A number of the smaller surrounding sites were test pitted, revealing diversity in their composition. Whilst many sites appeared to be dominated by *S. fasciatus* in the same way as KM1057, an equal number appear to be dominated by

Lab No.	Provenance	Sample Material	13C/12C Ratio	Conventional radiocarbon age BP	Calibrated age cal BC	2σ range cal BC
Beta-255383	JE0004 Top	shell	+1.6 ‰	5010±50	3310	3380–3080
OxA-19587	JE0004 Base	charcoal	-24.53 ‰	4709±31	3503	3373–3561
Beta-255385	KM1057 Top	shell	+2.4 ‰	4880±50	3070	3300–2900
Beta-255384	KM1057 Base	shell	+1.3 ‰	4850±50	3020	3270–2880

Table 14.1. Radiocarbon dates (after Alsharekh *et al.* in press). Calibrated dates are those supplied by the laboratory using the INTCAL04 dataset (Reimer *et al.* 2004), and taking account of the available regional offset for the marine reservoir effect (Hughen *et al.* 2004)

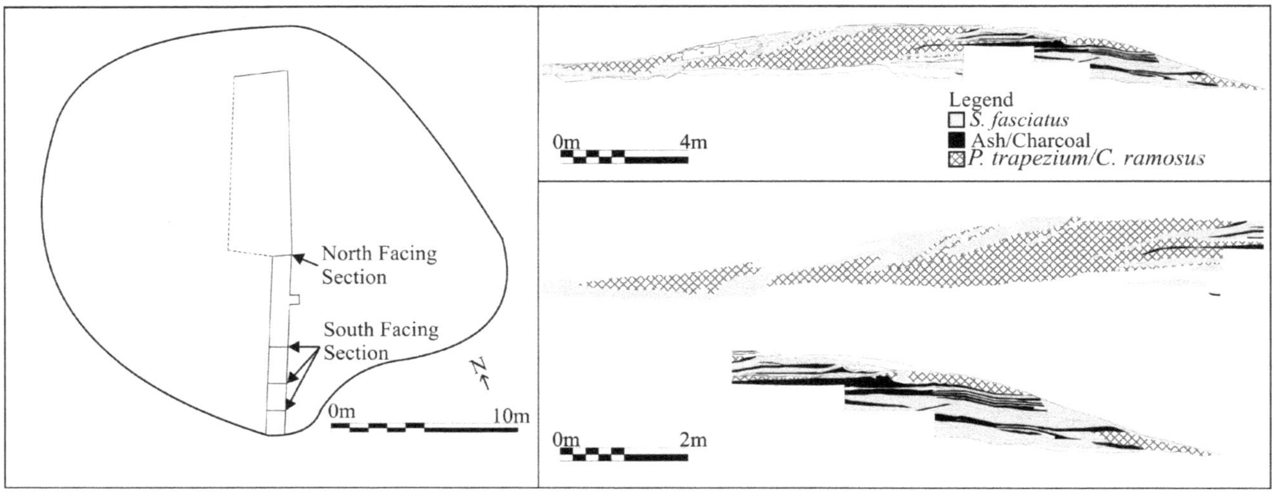

Figure 14-3. Plan of site JE0004 (a) and West facing section through site (above) with zoom in on each half (below) (b) (after Alsharekh *et al.* in press).

C. reflexa and *S. marisrubri* suggesting that different sites were associated with different activities. That this is not a hard rule is demonstrated by the five layers of *C. reflexa* and *S. marisrubri* within the stratigraphy of KM1057. However it does suggest that to some extent different shellfish species were processed at different sites, perhaps relating to the different habitats from which they were gathered.

Site JE0004 is part of a group of eight sites along a section of *c.*600 m of coastline. Of the seven sites peripheral to JE0004, three are low shell mounds, up to 1 m in height, the rest are scatters. Test pitting of the low mounds indicates that their internal structure and composition is similar to JE0004, which could be used to argue for a continuous exploitation strategy along this section of coastline. Further dating samples will test this hypothesis.

JE0004, located on a small but prominent headland in Janaba bay east (fig. 14-1), is a conical mound with a maximum depth of 1.5 m and a 20 m diameter (fig. 14-3a); it is located on a 3 m high cliff on the present day shoreline. Excavations were first started in 2006 with a series of step trenches down the southern flank (Bailey *et al.* 2007b). In 2008 a transverse trench was started across the centre of the mound as a continuation of the step trenches, reaching bedrock through the northern half of the mound (Alsharekh *et al.* in press; Williams 2010). In 2009 work began on excavating the southern half of the trench down to bedrock, in order to expose a full profile through the mound (fig. 14-3b). The excavations show that the mound is composed of two halves: on the southern (coastal) side the sequence is dominated by discrete layers of clean crushed *Strombus fasciatus* alternating with layers of ash and charcoal. The northern (inland) side of the mound is dominated by thick layers of larger shells (*Pleuroploca trapezium*, *Chicoreus ramosus*, *Chama reflexa* and *Spondylus marisrubri*) with smaller pockets of *Strombus fasciatus*. This configuration has been interpreted as two modes of deposition, with processing taking place on the southern side of the mound around hearths, and larger shells being dumped on the northern side of the mound away from the processing site (Alsharekh *et al.* in press; Williams 2010). The 2009 excavations reinforced this hypothesis, but also showed that the northern side of the mound postdates the southern side. This indicates a change in strategy at the site, with emphasis changing to larger species during the accumulation of the northern side. This may have been in response to geomorphological factors, or simply a change in preference. Given the nature of the substrate offshore from the site, it seems possible that changing ecological conditions may have played a role – the cliff drops into a subtidal fossil coral shelf *c.*0.5 m deep, which extends *c.*200 m offshore where it drops off into a series of deeper coral terraces. This is covered by patches of sand on which *S. fasciatus* thrives; where there is no sand cover species which favour a rocky substrate, such as *P. trapezium* and *C. ramosus* thrive. Deep water species such as *C. reflexa* and *S. marisrubri* can be found on the deeper coral terraces offshore. The sandy patches are not sheltered, and are exposed to adverse currents or storms. It is possible that the changes in species composition through the stratigraphy reflect variations in the size of the sand patches and thus the availability of *S. fasciatus*.

Fishbone was found throughout the site, with some poorly preserved mammal bones also present in the southern half of the mound. This indicates that the site was not used exclusively for shellfish processing, and that the site may have been used as a homebase from which resources could be gathered from a wider catchment – this is perhaps one reason why the site grew to be larger than the surrounding small shell mounds. The presence of features resembling post holes needs further investigating, but suggest that this was a more permanent site.

Calculating the amount of food represented by the volume of shellfish at JE0004 was also undertaken. The volume of the site is approximately *c.*200 m^3, equating to a little over 10,000 days worth of food for an adult. The radiocarbon dates suggest that this site accumulated over a period of between 400-600 years (Table 14-1), meaning that the shellfish would support five people for five days a year over 400 years or three days per year over 600 years, and ten people for two-three days a year over 400 years or one to two days over 600 years. Although these figures suggest only brief visits every year, this does not take into account the high concentrations of fishbone, and presence of mammal bone at the site which likely supplemented the diet. The fact that there are close to 3000 shell-bearing sites from this period across the islands suggests that these sites were part of a wider landscape exploitation strategy of the fisher-gatherer communities.

Using the site dimension data for all shell mounds collected from the survey, and extrapolating this for the sites not fully surveyed, it has been possible to estimate the total volume of shell contained within the sites. This was done by modelling each shell mound as a truncated cone. There is approximately 300,000 tons of shell contained in deposits across the islands, equating to enough to feed 60 people every day for 600 years. This is a minimum estimate, since it is unlikely that people subsisted entirely on shellfish.

There is a significant difference between these two groups of shell mounds and it is likely that a combination of both the differing local environmental settings (subtidal at JE0004, shallow intertidal and subtidal at KM1057) and exploitation strategies at each site affected this. Certainly the JE0004 deposits suggest a site more akin to a homebase, whereby the site was used as a base from which to exploit the wider environment. Whereas the excavations at KM1057 revealed little evidence of activities other than shellfish processing of a single species, with many surrounding sites also suggesting specialisation.

Perhaps these two areas were used for different purposes or by different fisher-gatherer groups, with differing exploitation strategies. It may even be that these two sites were occupied during slightly different periods, with exploitation strategies varying between the two. The dating of the sites (see Alsharekh *et al.* in press; Demarchi *et al.* 2010) suggests that the sites accumulated during overlapping or closely successive periods of shell mound building activity. This could be used to argue for exploitation strategies changing over time, by a single group, or different strategies by different groups.

Within the wider context of the area these shell mounds correspond in date to shell midden sites located on the mainland near to the coastal port of Gizan (fig. 14-1a insert). The composition of these sites is slightly different, having a greater component of *Terebralia palustris* – a mangrove dwelling gastropod (Zarins, Al-Jawad Murad, and Al-Yish 1981; Zarins and Al-Badr 1986; Zarins and Zahrani 1985). The Farasan Islands shell middens date to the final stages of shell mound building within this area, and are amongst the youngest yet found. However, older sites may still be found on the islands.

The climate was becoming more arid after a humid phase, with this transition occurring around 6000 BP in the region of the Farasan Islands (e.g. Fleitmann *et al.* 2007). In addition farming is thought to have reached the region somewhere around 4000 BP (Durani 2005). It may be that deteriorating climate forced coastal populations to intensify shellfish exploitation, especially during the dry season when other food resources may have been harder to access. With the arrival of farming, access to another stable source in food was introduced, reducing the reliance on other resources such as shellfish. However shellfish exploitation did not terminate with the arrival of farming and a number of sites have been dated to much later periods either through radiocarbon dating or their association with pottery. Indeed sites are still accumulating today, with some of the older sites being reused. Complicating the picture, an anonymous shell mound from Janaba bay was test excavated and dated by in a previous study, and three radiocarbon dates obtained, two from "Level 3", and the other form "Level 2". The Level 3 dates correspond to the dates from JE0004 and KM1057, however the Level 2 date is reported as 2410 ± 100 uncal. BP (Deputy Ministry of Antiquities and Museums, 1990). Unfortunately no further information is available, and the location, stratigraphy, provenance of the dating samples, and what material was dated remains unknown. It is therefore impossible to determine whether this is a case of continuous site use, or later reoccupation.

There have been several hypotheses as to why intensive shellfish exploitation suddenly appears at 6000 BP (Bailey & Milner 2002; Erlandson 2001). One suggests that intensive shellfish exploitation occurred before 6000 BP, but these sites were submerged by rising sea levels and are now underwater. Compounding this is that as sea levels rose the coastline was never static long enough for large mounds to accumulate. Alternatively, prior to 6000 BP only low level shellfish exploitation occurred; once sea level stabilised a unique window of ecological opportunity arose with large productive shell beds forming. Coastal populations took advantage of this, resulting in intensive shellfish exploitation.

This phenomenon is not restricted to the Red Sea, with many other regions experiencing a similar flourishing of shell midden sites at *c.*6000 BP. Earlier sites are known, mainly in areas where local tectonics and/or isostatic effects have resulted in uplift, but not exclusively so (Mayer and Beyin 2009). Indeed many of the pre-6000 BP sites along the Red Sea are found inland, some as a result of tectonic activity (eg Zarins, Al-Jawad Murad, and Al-Yish 1981; Zarins and Al-Badr 1986; Zarins and Zahrani 1985). This raises the question of whether earlier sites once existed in areas where tectonic uplift was not so dramatic, as these sites would now be submerged (Bailey *et al.* in press; Alsharekh *et al.* in press).

Conclusions

The shell midden sites of the Farasan Islands are of an unusually high density, unique for the Red Sea and the world. Dating suggests that this shell midden accumulation phase lasted at least 600 years, but could even extend between 5600-4500 years BP; future research may well extend this range. There is still uncertainty in some areas, evidenced both by the young dates produced both by Demarchi's preliminary study for a shell mound in the Hareed Bay area (Demarchi *et al.* 2010) and the Deputy Ministry of Antiquities and Museums results. These samples may well be evidence for reoccupation of sites, however further research is needed.

The volume of shell involved in the shell midden sites, and time depth at the two excavated sites suggests a broader shellfish exploitation strategy with sites occupied for short periods. Ethnographic studies suggest that this behaviour is often to fill a shortfall in food resources during the dry season, indeed the climatic data suggest increased aridity around this time (e.g. Meehan 1977). Based on information from the excavated sites a minimum population size of sixty people is inferred, however it is likely to have been more than this or represent a longer period of time.

Using the location of shell sites and their associated palaeoshorelines it has been possible to reconstruct the palaeoshorelines of the central islands of the Farasan Archipelago. These suggest that there has been tectonic uplift since the period of shell mound building, and that the shell mounds may have accumulated as a result of a window of ecological opportunity.

The excavations of two shell mounds show two very different patterns of accumulation, perhaps associated with the different environmental settings at each site. Certainly the activities at both sites differed: JE0004 was probably used as a homebase and has a wider range of activities present, associated with the processing of both marine and terrestrial resources, whilst KM1057 appears to only indicate shellfish processing.

The Farasan Island sites fit into the broader chronology for coastal resource exploitation, both regionally and globally. However the density of shell sites on the islands is unique for the Red Sea, with an unusually high level of intensification. Whether this was a seasonal exploitation, made by coastal groups from the mainland, or by a permanent community on the islands is at present impossible to say. It may have been due to deteriorating climate and the high productivity of the coastlines of the islands which attracted fisher-gatherers and led to such high levels of exploitation.

Acknowledgements

This was supervised by Geoff Bailey, who is thanked for his invaluable support on this project. Kirsty Penkman was instrumental in providing expertise and resources for the amino-acid racemization dating in this project, aided by Beatrice Demarchi who undertook the preliminary study and Richard Allen who helped with practical matters. The fieldwork and radiocarbon dating was made possible through grants form the British Academy and Leverhulme Trust. The Supreme Commission for Tourism and Antiquities, Riyadh, is thanked for granting a permit for fieldwork and for the participation of their staff. Thanks are also due to Graham Oliver, National Museum of Wales, for help with identification of marine molluscs, and Mark Beech, Abu Dhabi Authority for Culture and Heritage and University of York, for identification of fish bone material. Harry Robson is also thanked for his help and support during this project.

Bibliography

ALSHAREKH, ABDULLAH, BAILEY, GEOFF N., EVA M. LAURIE, GARRY MOMBER, LAWRENCE J. MORAN, ANTHONY SINCLAIR, MATTHEW G. M. WILLIAMS, NABIEL ALSHAIKH, ALMA'MARY, A., and SAUD ALGHAMDI. In press. Coastal archaeology in the Farasan Islands: report on the 2008 fieldwork of the joint Saudi-UK Southern Red Sea Project. *Atlal: Journal of Saudi Arabian Archaeology*

BAILEY, GEOFF N., ABDULLAH ALSHAREKH, NIC FLEMMING, KURT LAMBECK, GARRY MOMBER, ANTHONY SINCLAIR, CLAUDIO VITA-FINZI. 2007a. Coastal prehistory in the southern Red Sea Basin: underwater archaeology and the Farasan Islands. *Proceedings of the Seminar for Arabian Studies* 37:1–16.

BAILEY, GEOFF N., NIC FLEMMING, GEOFFREY C.P. KING, KURT LAMBECK, GARRY MOMBER, LAWRENCE J. MORAN, ABDULLAH AL-SHAREKH, CLAUDIO VITA-FINZI, C. 2007b. Coastlines, submerged landscapes and human evolution: the Red Sea Basin and the Farasan Islands. *Journal of Island and Coastal Archaeology* 2(2):127–60.

BAILEY, GEOFF N., ABDULLAH ALSHAREKH, NIC FLEMMING, GARRY MOMBER, LAWRENCE J. MORAN, ANTHONY SINCLAIR, GEOFFREY C.P. KING, CLAUDIO VITA-FINZI, ALMA'MARY, A., NABIEL Y. ALSHAIKH, and SAUD ALGHAMDI. In press. Coastal archaeology and prehistory in the Southwest Region of Saudi Arabia and the Farasan Islands: report on the 2004 and 2006 surveys of the joint Saudi-UK Southern Red Sea

Project. *Atlal: Journal of Saudi Arabian Archaeology*

Bailey, Geoff, and Nicky Milner. 2002. Coastal hunters and gatherers and social evolution: marginal or central? *Before farming: the archaeology of Old World hunter-gatherers* 3–4 (1):1–15.

Bantan, Rashad. 1999. *Geology and sedimentary environment of Farasan Bank (Saudi Arabia) Southern Red Sea: a combined remote sensing and field study.* PhD thesis, University of London Royal Holloway, London.

Demarchi, Beatrice, Matt G Williams, Nicky Milner, Nicola Russel, Geoff Bailey, and Kirsty Penkman. 2010. Amino acid racemization dating of marine shells: a mound of possibilities. *Quaternary International* 239 (1-2): 114–124.

Deputy Ministry of Antiquities and Museums. 1990. Radiocarbon dating and results. *Atlal, the Journal of Saudi Arabian Archaeology* 13: 74–75 (in Arabic).

Durrani, Nadia. 2005. The Tihamah coastal plain of South-West Arabia in its regional context c. 6000 BC–AD 600. BAR Int. Series. *Society for Arabian Studies Monographs* 4. Oxford: BAR Publishing,.

Erlandson, Jon M. 2001. The archaeology of aquatic adaptations: paradigms for a new millennium. *Journal of Archaeological Research* 9:287–350.

——.1988. The role of shellfish in prehistoric economies: a protein perspective. *American Antiquity*, 53(1):102–109.

Fleitmann, Dominik, Stephen J. Burns, Augusto Mangini, Manfred Mudelsee, Jan Kramers, Igor Villa, Ulrich Neff, Abdulkarim A. Al-Subbary, Annett Buettner, Dorothea Hippler, Albert Matter. 2007. Holocene ITCZ and Indian monsoon dynamics recorded in stalagmites from Oman and Yemen (Socotra). *Quaternary Science Reviews* 26(1):170–188.

Mayer, Daniella E. and Amanuel Beyin. 2009. Late Stone Age shell middens on the Red Sea coast of Eritrea. *The Journal of Island and Coastal Archaeology* 4(1):108–124.

Meehan, Betty. 1977. Man does not live by calories alone: the role of shellfish in a coastal cuisine. In *Sunda and Sahul*, J. Allen, J. Golson, and R. Jones eds., Pp. 493–531, Prehistoric Studies in Southeast Asia, Melanesia and Australia. New York: Academic Press.

Noli, Dieter and Graham Avery. 1988. Protein poisoning and coastal subsistence. *Journal of archaeological science* 15(4):395–401.

Osborn, Alan J. 1977. Strandloopers, mermaids, and other fairy tales: ecological determinants of marine resource utilization, the Peruvian case. In *For theory building in archaeology*, L.R. Binford, ed., Pp. 157–205, New York: Academic Press.

Penkman, Kirsty E.H., Darell S. Kaufman, Darell Maddy, Matthew J. Collins, 2008. Closed-system behaviour of the intra-crystalline fraction of amino acids in mollusc shells. *Quaternary Geochronology* 3(1-2):2–25.

Williams, Matt G.M. 2010. Shell mounds of the Farasan Islands, Saudi Arabia. *Proceedings of the Seminar for Arabian Studies* 40:357-366.

Zarins, Juris, Norman Whalen, Mohammed Ibrahim, Mursi Abd al Jawad, and Majid Khan. 1980. The Comprehensive Archaeological Survey program. Preliminary report on the Central and Southwestern provinces. *Atlal, the Journal of Saudi Arabian Archaeology* 4:9–36.

Zarins Juris, Abd Al-Jawad Murad, and Khalid S. Al-Yish, 1981. The Comprehensive Archaeological Survey Program, a. The second preliminary report on the southwestern province. *Atlal, the Journal of Saudi Arabian Archaeology* 5:9–42.

Zarins, Juris, and Hamad Al-Badr. 1986. Archaeological investigations in the southern Tihama Plain, II, *Atlal, the Journal of Saudi Arabian Archaeology* 10: 36-57.

Zarins, Juris, and Awad Zahrani. 1985. Recent archaeological investigations in the Southern Tihama Plain (The sites of Athar, and Sihi, 1404/1984). *Atlal, the Journal of Saudi Arabian Archaeology* 9:65-107.

15 - SHELL EXPLOITATION AT PLAYA DEL TESORO AND BANDERAS MEXICAN PACIFIC COAST

José BELTRÁN M.
Centro INAH NayaritLerdo de Tejada No. 76 ote. Centro, C.P. 63000 Tepic, Nayarit, Mexique,
jbcol51@hotmail.com

Abstract: *Archaeological studies about maritime exploitation along Mexican Pacific coast reveal the important place that mollusks had in the cultural development of coastal societies in west Mexican coast (2000 B.C. to 1530 A.D.), Playa del Tesoro and Punta Mita in Manzanillo and Banderas Bay show clearly different aspects of coastal subsistence.*
These bays functioned as dive centers, as well as showing evidence for the storing, processing and distribution of diverse ocean products. Mollusks were the largest and the best preserved zoological group, with one hundred and sixty species found during the archaeological excavations in both bays.
This research focuses on the species of mollusks that were collected for subsistence, such as clam shells and oysters, which supplied people along the coast. Other species were selected and transformed into personal ornaments and utilitarian items of material culture. Some of these materials were traded long distances.
Taxonomic identification provides information on the ecological niches that were exploited as well as the mobility patterns of different communities. It also gives us insights into the different ways in which mollusks were captured as well as the technology used for this purpose.
The presence of exogenous items indicates the importance of trade between western Middleamerica and other continental regions that were involved in long distance resource procurement and exchange. Shell is an archaeological material of great significance to the understanding of the processes of cultural interaction that took place between cultural regions. Shell materials such as Spondylus, Strombus, Pinctada *and* Murex, *as well as metal artifacts, turquoise and other luxury items, were exchanged and distributed along the Pacific coast and inland.*

Keywords: *Mexican Coast, Food, Ornament, Dye, Lime.*

Introduction

Archaeological excavations undertaken along the west Mexican coast, have allowed us to appreciate the importance of marine and estuary exploitation in the development of the pre-Columbian societies. Playa del Tesoro in the middle of Manzanillo and Santiago bay and Punta Mita and other sites in Bahía de Banderas (fig. 15-1), are clear examples of the diversity of this importance, since they functioned as centers for deep diving, storing, processing and distribution of diverse marine products. They stand as key examples of the longstanding tradition of marine exploitation of tropical coastal habitats.

The most well represented zoological group was the mollusks with 160 species found during the excavations in both (Villanueva 1990, 1998). Some artifacts are completely finished while others are at various stages of production. There is also some industrial waste relating to shell artifact production.

In addition to mollusk material, other taxa such as sponges, corals, sea urchin, fish, whale bones, crabs and marine turtles, as well as bird bones and terrestrial mammals were also recovered.

The taxonomical identification to the level of genus and species provides information on the different ecological niches were exploited as well as the level of mobility of the different communities. In this way we could achieve an approximation of the human activities undertaken in the capture of mollusks as well as the technology that was used for this purpose. In terms of the malacofauna, these bays belong to the Panamic Malacological Province which stretches from the Gulf of California to Cabo Blanco Peru, and includes around 2000 species (Abbot 1974). In Western Mesoamerica there are also specimens which belong to the California and the Caribbean Provinces showing the exchange that took place between different cultural areas.

Banderas and Manzanillo bays belong to one of the richest ecosystems on the planet, where the estuary in particular contains optimum conditions for the development of life. This wealth of life is due to the influence of the cold stream of California and of the mass of subtropical waters in the Gulf of California, as well as the Central America Stream coming from the south.

Bahía de Banderas and Manzanillo functioned as very important ports, and were key nodes in long distance mercantile routes as revealed by the presence of foreign and exotic materials. Both cultural and mercantile trade engaged with far-off continental regions, especially in the Mesoamerican Plateau, north of Mexico, the southwest

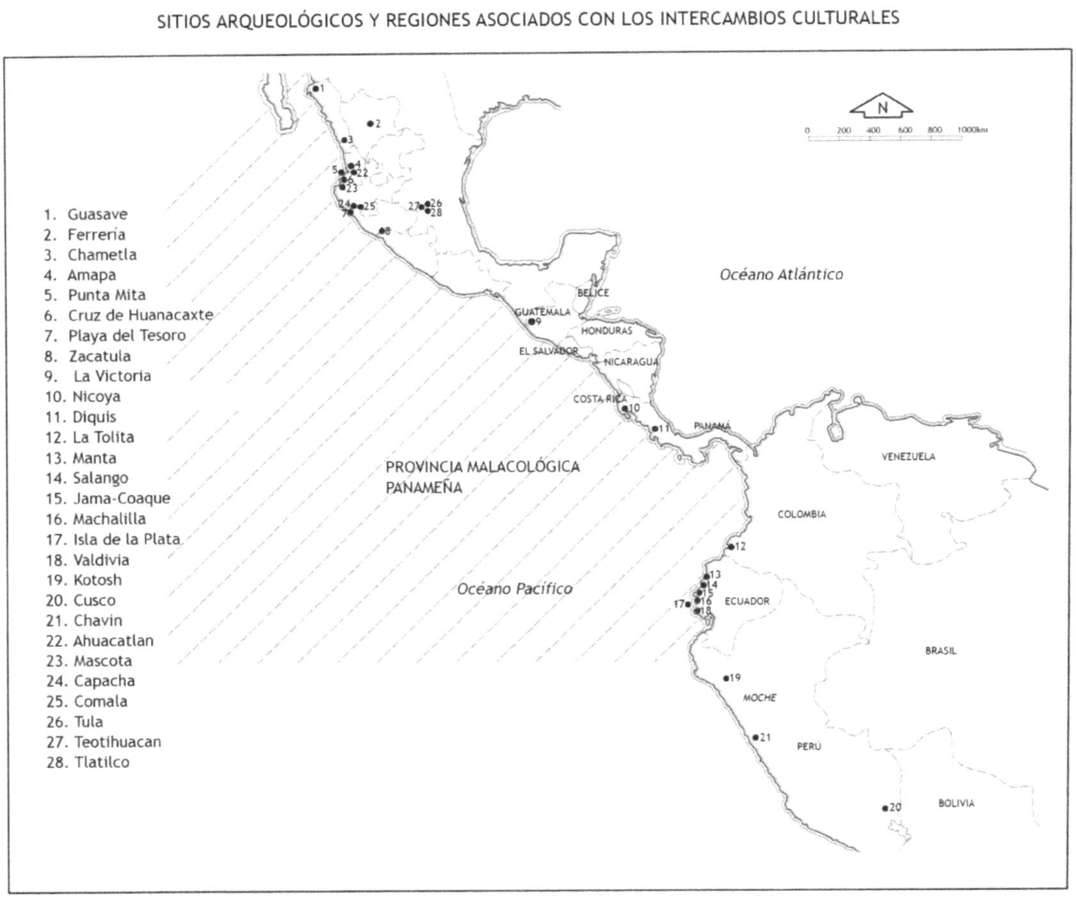

Figure 15-1. Playa del Tesoro, Banderas bay and sites along the Pacific coast.

United States, the Central American coast and Ecuador (Beltrán 2009).

As an important trade item, shell is a very important archaeological material for understanding the different stages of cultural interaction that existed across several regions of the American continent.

Playa del Tesoro and archaeological sites at the Banderas bay

Playa del Tesoro is the only archaeological site that has been excavated in Manzanillo Bay (Beltrán 1991), as opposed to the north shore of Banderas Bay, where forty sites have been recorded and test excavations were conducted at five. These coastal sites provide relevant information about shell exploitation (González and Beltrán 2007).

Playa del Tesoro is located on the Santiago peninsula which divides Manzanillo and Santiago bays (fig. 15-2). There are habitation and production areas located on the hill, near the port and in Salagua valley. There is a rich cemetery located near the sea in Playa del Tesoro. Cultural materials include a red on buff ceramic, figurines and a rich material inventory with local and exogenous manufacture. There is also evidence of regenerated cranial intervention, tooth work and cranial deformation.

There are three occupation phases at Playa del Tesoro: Early Morett (300 B.C. to 100 A.D.). Tesoro (A.D. 150 to 750), and a late occupation with metal artifacts and Toltec materials, such as Mazapa figurines, seals, and tripod vessels (A.D. 900 to 1100). The most intense occupation at Playa del Tesoro was A.D. 200 to 670 (Beltrán 2001).

Figure 15-2. Santiago Bay and Playa del Tesoro

Five sites at Banderas Bay were located at north shore, and represented different moments in the cultural sequence, such as Cruz de Huanacaxte (1200 B.C. to 1530 A.D.), Punta Mita, Higuera Blanca, Litibú and Pontoque (300 B.C. to 1530 A.D.). The architecture is composed of a ceremonial center with mounds and platforms, habitation and production areas associated with shell middens comprised of different mollusk species, as well as basalt and granite hammerstones (González and Beltrán, 2007).

Early occupation materials include local Capacha ceramics and figurines, and San Blas materials (B.C. 1200 to 500). Subsequently ceramics, figurines and another material from the shaft tomb tradition (300 B.C. to 600 A.D.) appear. In the final phases, Aztatlan was the main cultural tradition with a rapid spread in Banderas Bay (A.D. 900-1350).

Shell materials

The earliest evidence for marine exploitation on the west Mexican coast is at Matanchén, just south of San Blas, with people harvesting mollusks from B.C. 2200 to 1730 (Mountjoy 2000).

From the beginning of the formative period in western Mesoamerica, around 1500 B.C. the exploitation of marine and estuarine resources generated an important shell industry that encompassed two categories. The first category relates to satisfying basic food needs, and this in turn is related to the recollection and capture of mollusks. The second activity involves shell transformations and includes several discrete activities such as the exploitation of shell dye, lime preparation and the capture of several precious species which were widely valued. All of these were used in the making of diverse artifacts.
From this activity a whole craft industry centered on shells developed. Artifacts included elegant necklaces, bracelets, noserings, earrings and beads, as well as religious objects like rattles, big rings of nacreous shell, figurines, trumpets, and buttons. Utilitarian objects such as chisels, hooks, horns, awls and shuttles were also produced. All these were in great demand throughout the continent (Beltrán 2001).

Food resources

From earliest times, shells as well as seafood represented one of the more stable food resources, as it allowed coastal people to satisfy basic subsistence needs. Therefore, throughout time, a very important communal activity developed which centered on the capture of food species, both in the bays and estuaries (fig. 15-3). This has been shown through the presence of archaeological sample in Playa del Tesoro, Banderas Bay and isla Panales, in the classic and postclassic periods, including the remains of species *Ostrea corteziensis, O. columbiensis, O. angelica, O. iridescens* and *O. finsumosheri*, as well as *Anadara grandis* and *A. multicostata, Protothaca columbinsis, Chama mexicana, Trachycardium procerum* and *Conus princeps.* Species such as *Choromytilus palliopunctatus, Arca pacífica, Megapitaria squalida* and *Aequipecten circularis* have also been recovered. The small gastropod *Strombus gracilior* has been found in great quantities near big cooking pots (Beltrán 1991).

Figure 15-3. Panales Island mound shell, posclassic period

Shell dye

At Litibú and Cruz de Huanacaxte in the Mesoamerican postclassic period (900 A.D. to 1521 A.D.) an important activity related to the exploitation of certain mollusks was the procurement of dye for cloth, feathers and other materials (González and Beltrán 2007). All species related to dyeing came from family Muricidae, like *Hexaplex erythrostomus* and *H. regius*, or from family Thaididae such as *Purpura pansa, Thais biserialis, Acanthina brevidentata, Neorapana muricata* and *N. tuberculata* (fig. 15-4). Dyed cloth was used only by members of the upper class.

Figure 15-4. *Hexaplex regius,* Huanacaxte port

Species within family Muricidae are shells with a very complex structure including numerous spines; some have a pink mouth and some others a black or purple mouth. They are found in subtidal beds living in big colonies all along the rocky coast, feeding on oysters, mussels and other mollusks.

Some shells such as *Purpura pansa* live on stones where waves crash, which makes it very easy to catch them when the waves are low, therefore it is possible to obtain the dye without sacrificing the animal and this in turn helps to increase production.

When these mollusks are in danger they have the property of expelling a viscous liquid (made by oxidation with the environment) which produces a purple indigo dye of very high quality. This dye was used in dyeing cloth, feathers and other materials and was widely esteemed. The dye was collected in big shells in which pieces of cotton were soaked. Several genera in family Muricidae had to be collected from beyond the intertidal zone and it was thus necessary to sacrifice the mollusk. The intensity, quality and quantity of the liquid depended on the species selected (Nutall 1971).

Shell lime

At Playa del Tesoro in Tesoro phase, shell lime was produced as a raw material to be utilized in construction, production processes, food conservation and ceremonies (fig. 15-5 and 15-6).

Shells are made up of several layers of calcium carbonate (CaCo3). When burned they lose their structural integrity and become a very fine powder used for preparing a high quality lime.

During excavations at Playa del Tesoro a big shell lime deposit of 25 kg. was undercovered under a very hard floor made up of cemented shell lime. Underneath it a variety of plain wares, including a vase, seven vessels and two miniature bowls was found.

Figure 15-5. Shell lime deposit at Playa del Tesoro

Figure 15-6. Small jars with shell lime, Tesoro phase

Some shells which contained cemented lime inside were also found in species including *Choromytilus palliopunctatus, Megaptaria squalida* and *Anadara gigantea.* In the Andean region shell lime was frequently used as a utilitarian and ceremonial product; even nowadays it is used for chewing the sacred leaves of the coca plant (*Erythroxilum coca*) (Elera 1987).

Spondylus princeps, *the sacred oyster*

The capture of *S. princeps* was of great importance at Playa del Tesoro and Banderas Bay in the Mesoamerican classic (200 A.D. to 750 A.D.) and postclassic periods (900 A.D. to 1521 A.D.). Is a shell with long thorns of different hues of red. They are found attached to subtidal rocks and coral reef, and so specialized divers are needed for their capture.

It was called *Teochipolli* in Nahuatl and *Mullu* in Qechua languages. It was considered sacred so as a result it was highly sought after. There existed a vast trade from early times amongst the peoples of Mesoamerica, Ecuador, Peru and Bolivia. This species is appreciated for its exquisite flavor, but its real value was in its use in early fertility rites and also for calling rain (Elera 1987). It became a non-replaceable offering and was laden with symbolism as it was the favorite food of the gods. It was associated with very important deities in many cultural areas. It was also used as raw material for the making of bracelets, beads, and diverse objects. *Spondylus calcifer* was also found, a species with no thorns but a bigger and heavier form (fig. 15-7 and 15-8).

Figure 15-7. *Spondylus* necklage, burial 12-2 Tesoro phase

Figure 15-8. *Spondylus* necklace, burial 12-2 Tesoro phase

Strombus, *the shell trumpet*

Family Strombidae includes species of big, solid shells. They are found in sandy intertidal and subtidal zones. They can be eaten but their real importance was due to the shell which, when the apex was removed, was used as a trumpet in religious rites, ceremonies, wars and celebrations of new governors and death festivities. This is the reason why they were in such great demand in Mesoamerica and in the Andean region. They were also used in the fabrication of bracelets, awls, and chisels, as well as artifacts shaped as half moons, and once finished they were used as noserings or lances. Only the nobles, priests and warriors could use them. Some such artifacts were found in Playa del Tesoro and Punta Mita worked in the following species: *Strombus galeatus* (fig. 15-9) *S. peruvianus*. Two smaller species, *Strombus gracilior* and *S. granulatus*, were used mainly as food. Two other species not in family Strombidae were also used as trumpets: *Fasciolaria princeps*, one of the biggest and most beautiful is found in the Panamic Province, and *Hexaplex regius* which comes from the port of Huanacaxtle. The main importance of these shell trumpet species is that several examples were found which derive from the Caribbean Sea, such as *Strombus gigas*, *S. costatus* and *Turbinella angulata*, and these were found next to the great shells from the Pacific. This shows a commercial trade between the two coasts during the Classic Period.

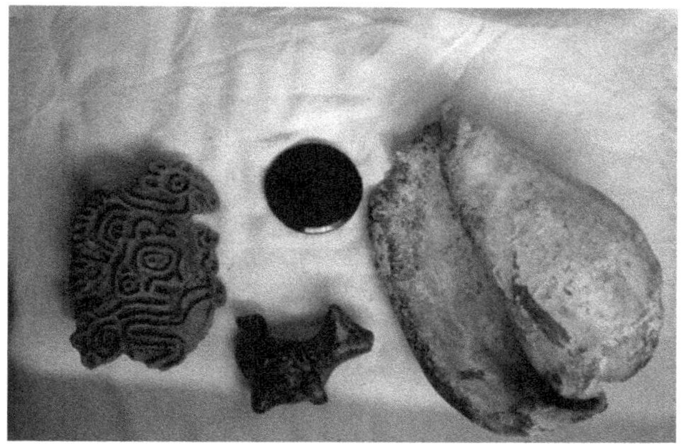

Figure 15-9. *Strombus galeatus* trumpet, whistle and stamp, Playa del Tesoro

Pinctada mazatlanica

The greatest producer of pearls is the oyster *Pinctada mazatlanica* greatly appreciated by the ancient people, and heavily exploited. These oysters are found on subtidal rocks, so diving is necessary for their capture. Plaques of the pearl oyster were widely used engraved with other materials (fig. 15-10).

Figure 15-10. *Pinctada mazatlanica,* burial 2 Tesoro phase

The edges of this shell were used as ornaments called *anteojera Tlaloc*, as well as fishhooks of 3cm to 7cm long, which are associated with fish bones from shores, reefs and estuaries. Examples of these fishhooks were found in Ecuador, Punta Mita and Playa del Tesoro (fig. 15-11), with a dive weight (fig. 15-12) which indicates a cultural trade between these people in the classic and postclassic periods (Marcos 1995).

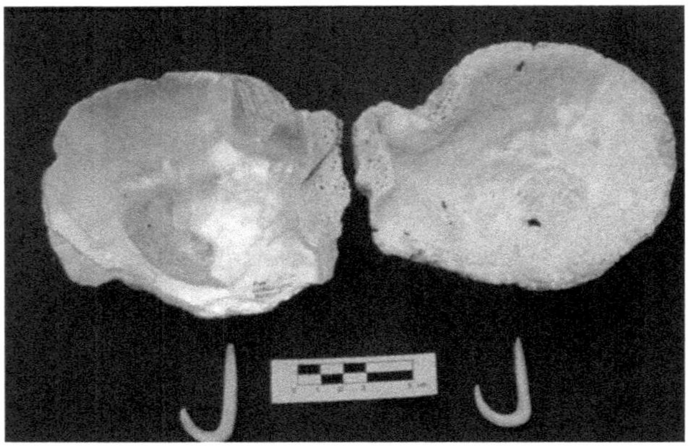

Figure 15-11. Hooks of *Pinctada mazatlanica,* Punta Mita

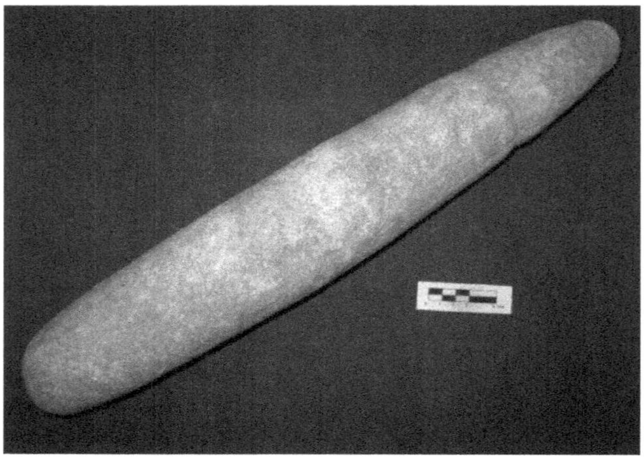

Figure 15-12. Dive weight, Punta Mita, posclassic

Glycymeris gigantea *bracelets*

During archaeological excavations at Playa del Tesoro, numerous bracelets made of *Glycymeris gigantea* were found, each carved from a single shell (fig. 15-13). The manufacturing technique began with drilling a hole in the center of the shell, and gradually enlarging it until it reached the border of the valve. Once the bracelet ring was roughly shaped, the exterior was carefully buffed thus obtaining a glossy finish and a beautiful plain bracelet of great quality. Some were shaped into animals like bats at the umbo (fig. 15-14).

Figure 15-13. *Glycymeris gigantea* bracelet, Burial 12- 3 Tesoro phase

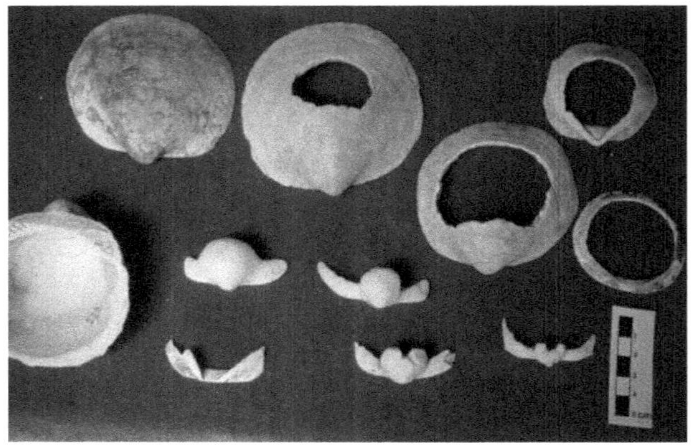

Figure 15-14. Work process *Glycymeris gigantea* shells, Playa del Tesoro

These artifacts have been found along the coast of Guerrero all the way up to Sonora, indicating a shared cultural tradition across this area. The shells are found in sandy substrates from 7 m. to 13 m. depth from California down to Guerrero (Keen 1971). Further big bracelets made of *Ancistromesus mexicanus* were also found in both bays. The same manufacturing technique was used in the making of the bracelets. These limpets live attached to rocks.

Pecten *pectorals*

In Playa del Tesoro beautifully carved Pectinidae scallop shells were found, utilizing species including *Argopecten circularis* and *Lyropecten subnodosus,* the largest and most attractive shell of purple, magenta and orange shades. *Argopecten circularis* is a smaller shell of great beauty and red hues. There are several worked pieces showing a small hole in the umbo to be used as a necklace, while others have a larger perforation of the same technique, and were destined to be worn as bracelets, or arm ornaments. There were also valves of *Pecten vogdesi, P. sericius* and *Chlamys lowei* which showed no evidence of carving. All of these are found from Ecuador to California in both deep and shallow waters. In the Panamic province there are less than twenty different species (Keen 1971).

Oliva *and* Conus *rattles and bells (fig. 15-15)*

These artifacts were used in the ceremonial dances as percussion instruments, attached to the legs of the dancers for keeping rhythm. They were found in Playa del Tesoro but mostly in Punta Mita deposits in classic and postclassic period, forming a collection of seventy artifacts, all made out of *Oliva porphyria,* the largest of the olives. Other species were also found including *O. incrassata* and *Conus vittatus,* one of the most beautiful species. These species come from deeper waters, but are also found in the subtidal were they feed on sea stars. They are predators of other mollusks, and some of them are highly poisonous. A large coneshell displayed at the Museum of the University of Colima (MAG) showing Figure 18. Hooks of *Pinctada mazatlanica,* Punta Mita hues of pink and blue, was found in a pillage shaft tomb. At the Quetzalcoatl Pyramid in Teotihuacan strings of *Oliva* were recovered.

Figure 15-15. *Oliva* rattles, Punta Mita

Beads and carved shells

Anadara and *Spondylus* shells were found in both bays showing rectangular or linear carvings for plaques and beads in the Tesoro phase, as well big round and square beads for necklaces in the reoccupation phase (fig. 15-16). Some had a perforation in the shell for suspension. Fragments of shell carvings were found along with *Trachycardium procerum* valves displaying rectangular cuts.

Figure 15-16. Necklace and beads made of *Spondylus*

Pendants

A wide collection of pendants was found in the excavations at both bays in the classic and postclassic periods. They can be classified into two groups: the first remain in their natural state and the others show some degree of working. Most of these pendants were found as parts of necklaces and bracelets, but others were also found separately. Two kinds of perforation can be observed: a very thin and fine hole through which a thread can be passed and the other has a wider piercing for a thicker string. These perforations can be conical or parallel. Examples of pendants with a fine drill-hole are manufactured from the following species: *Polymesoda mexicana, Chione californiensis, Argopecten circularis, Pitar lupanaria, Turritella leucostoma, Cassis centiquadrata* and *Chiton* sp. Rarer specimens such as *Turritella leucostoma* have a conical perforation as the only carving. Other species with this larger perforation include *Pseudochama inermis, Argopecten circularis, Placunanomia cumingii, Anadara formosa, Trachycardium consors, T. procerum* and *Anadara grandis*. A rockshell, *Morum tuberculosum,* with a large perforation and thin and fragile walls, is believed to be used for something other than a rattle. A snail shell, *Ortalichus,* has only one hole and was used as a whistle. The pendants and earrings show a great variety of designs: zoomorphic, phytomorphic and geometric. The most common shells were *Spondylus* followed by smaller amounts of *Pinctada mazatlanica* and *Lyropecten subnodosus.*

Mercantile Trade

One of the most important discoveries at the excavations along Playa del Tesoro and Punta Mita was the exotic materials coming from faraway parts of the continent. The exchange with the Andean world and others regions was focused on the demand for *Spondylus* the sacred oyster and the great *Strombus* which were obtained along the tropical coast of Ecuador. These products became scarce for overexploitation about 900 A.D. thus their shells were highly valued along the west coast of Mesoamerica and Andean region.

The main products exchanged were shells (fig. 15-11, fig. 15-17), metals (fig. 15-18), turquoise and precious stones, ceramics, obsidian (fig. 15-19) and fine textiles (Paulsen, 1977).

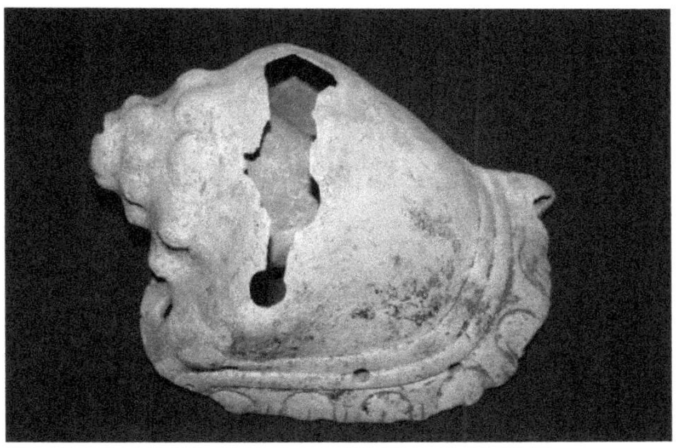

Figure 15-17. Caribbean *Strombus gigas,* Nayarit museum

Figure 15-18. Bronze axe, Punta Mita, posclassic period

Figure 15-19. Beads of amazonita necklace and obsidian arrow in cardiac zone, burial 7 Tesoro phase

During the Tesoro phase in Playa del Tesoro, exogenous materials like earrings with iridescent paint, "asientos platón", ceramics with incised decoration and metal fish hooks with South American influence are present. Also present were ceramics with fresco paint, earrings, candlesticks and obsidian flakes similar to Teotihuacan materials. Other artifacts include Mazapa figurines, shell beads (fig. 15-16, fig. 15-20), and tripod ceramics from the Toltec tradition (Beltran 1991).

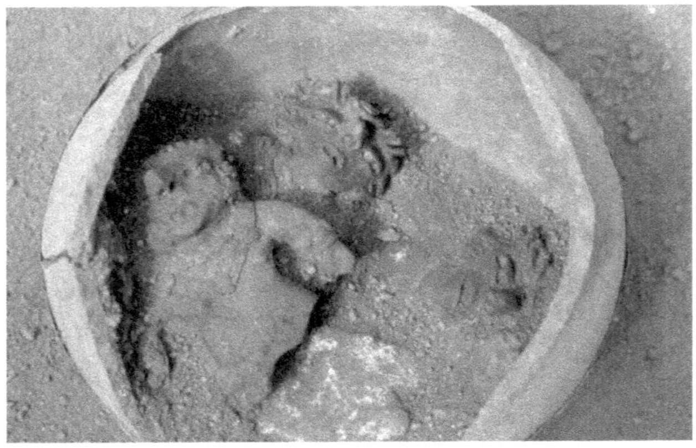

Figure 15-20. Figurines from Tesoro phase, burial 12-3

In Punta Mita materials include shell and metal fishhooks (fig. 15-11), dive weights (fig. 15-12), plumbate sherds and anthropomorphic stone sculptures similar to south American and Central American materials (Mountjoy and Beltrán 2004).

Conclusion

Shell is an archaeological material of great significance to the understanding of the processes of cultural interaction that took place between cultural regions. The exploitation of the coast by the ancient inhabitants of Banderas and Manzanillo bays provided great wealth and established a large market for ocean products, which in time developed into a well settled system of seaborne trade along the American tropical Pacific coast.

Without question there was a great deal of diversity and complexity in Playa del Tesoro and Bandera Bay cultural behavior. The presence of exogenous items indicate the importance of trade between west Mexican coast and other continental regions that were involved in long distance resource procurement and exchange. Shell materials such as *Spondylus, Strombus, Pinctada* and *Murex*, were exchanged and distributed along the Pacific coast and inland, as well as metal artifacts, turquoise and other luxury items (fig. 15-1).

These materials show the existence of commerce with the Mesoamerican highlands (Teotihuacan and Tula), the American Southwest, the Central American coast, Ecuador and north Peru. This demonstrates the diffusion and transmission of materials, technology and cultural elements which, to date, has not been evaluated thoroughly.

Such goods circulated widely throughout the Americas, thus showing the great enterprise executed in order to provide large quantities of shellfish for a market. Using the present data, the navigation routes have started to be reconstructed, with the presence of exogenous materials and the ports and sites involved in the cultural dynamic as a starting point.

References

Abbott, R. Tucker. 1974. *American Seashells*. Nueva York: Van Nostrand Reinhold Co.

Beltrán, José. 1991. Los concheros del puerto de Salagua (Playa del Tesoro). Thesis. México : Escuela Nacional de Antropología e Historia

——.2001. *La explotación de la costa del Pacífico en el Occidente de Mesoamérica y los contactos con Sudamérica y con otras regiones culturales*. Tepic, México : Universidad autónoma de Nayarit

——.2009. Playa del Tesoro, Colima, y los intercambios con Ecuador. In *Relaciones intra e interregionales en el occidente de México*. Daniel Barragán and Rafael Martínez, eds. Pp. 387- 412. Guadalajara México: Universidad de Guadalajara

Elera, Carlos. 1987. Inferencias socioeconómicas e ideológicas en torno a una tumba disturbada de la cultura Taicantin, valle del Viru, costa norte del Perú. *Cuicuilco* 18:62-78. INAH: México.

González Lourdes and José Beltrán. 2007. Arqueología de la bahía de Banderas. In *El occidente de México, Perspectivas multidisciplinarias*. Rosa Yáñez, ed. Pp. 312-324. Guadalajara México. Universidad de Guadalajara

Keen, Myra. 1973 (1958). *Sea Shells of Tropical west America. Maritime molluscs from Baja California to Peru*. 2° edition. Stanford CA: Stanford University Press.

Marcos, Jorge. 1995. El Mullu y el Pututo. La articulación de la ideología y el tráfico a larga distancia en la formación del estado Huancavilca. In *Primer encuentro de Investigadores de la costa Ecuatoriana en Europa: Arqueología, Etnohistoria, Antropología Sociocultural*. Quito, Ecuador. ABYA YALA

Mountjoy, Joseph B. 2000. Prehispanic Cultural Development along the Southern Coast of West Mexico. *In Greater Mesoamerica*. Michael S. Foster and Shirley Gorenstein, eds. Pp.81-106. University of Utah Press. Salt Lake City

Mountjoy Joseph B. and José Beltrán M. 2004. Anthropomorphic Peg Based Sculptures from the Banderas Valley of coastal west México, *Ancient Mesoamérica* 16: Pp. 155-168. Cambridge University Press

Nutall, Zelia. 1971. *Una curiosa supervivencia del caracol de Púrpura en Oaxaca*. México: Gobierno del Estado de Oaxaca.

Paulsen, Allison C. 1977. Patterns of Maritime Trade between South Coastal Ecuador and Western Mesoamerica 1500 B.C.-A.D 600. In *The Sea in the Pre-Columbian World*. Elizabeth P. Benson, ed. Washington. Dumbarton Oaks

Villanueva, Gerardo. 1990. Informe malacológico de Playa del Tesoro. Laboratorio de Bioarqueología, DSA. Mexico INAH

——.1998. Informe malacológico de Punta de Mita Laboratorio de Bioarqueología, DSA. Mexico INA

16 - OYSTERS, PHEASANTS AND FINE FOODS. "HIGH CLASS" PRODUCTS IN ALIFE (CAMPANIA, ITALY) DURING AND AFTER THE ROMAN EMPIRE

Alfredo CARANNANTE
Laboratorio di Scienze e Tecnologie applicate ai Beni Culturali, Università degli Studi "Suor Orsola Benincasa" di Napoli, Via Santa Caterina da Siena 37, Napoli, Italy,
alcarann@yahoo.it

Salvatore CHILARDI
Laboratorio di Scienze e Tecnologie applicate ai Beni Culturali, Università degli Studi "Suor Orsola Benincasa" di Napoli, Via Santa Caterina da Siena 37, Napoli, Italy,
schilar@tin.it

Daniela REBBECCHI
Laboratorio di Scienze e Tecnologie applicate ai Beni Culturali, Università degli Studi "Suor Orsola Benincasa" di Napoli, Via Santa Caterina da Siena 37, Napoli, Italy,
danielarebbecchi@virgilio.it

Annalisa DEL SANTO
Università degli Studi di Napoli "L'Orientale",
annalisa-ds@hotmail.it

Roberto VEDOVELLI
Laboratorio di Scienze e Tecnologie applicate ai Beni Culturali, Università degli Studi "Suor Orsola Benincasa" di Napoli, Via Santa Caterina da Siena 37, Napoli, Italy,
robertoind1981@libero.it

Abstract: *This work deals with the archaeozoological remains which were found in the* criptoporticus *of the little town of Alife in the northern part of the territory of Caserta. The faunal assemblage covers a period starting from the 2nd century up to the 7th century. Data from the* criptoporticus *area shows the presence of fine food items such as flat oysters, thorny oysters, moray eels, piglets, pheasants and partridges suggesting an aristocratic villa, probably owned by a very rich family.*
Large quantities of oysters attest to the different techniques of oyster-breeding during the Roman Imperial Age and features of the shells allow us to allocate a provenance of the Phlegraean coastal area.
Other shell and bird remains seem to be related to ritual activities in some particular contexts.

Keywords: *Roman Imperial Age, Oyster breeding, Fish, Mammals, Birds.*

Introduction

Several classical sources (Petronius, Juvenal, Martial, Pliny, Apicius, among others) refer that luxurious banquets represented one of the most important opportunities of wealth display in the Imperial Roman world. The same ancient sources state that marine molluscs (mainly oysters) and fish were among the most important resources used to this end.

Ancient Campania, and the Phlegraean Area in particular, became the heart of economic, political and military power of the Roman Empire. The Phlegraean town of Baiae was the habitual residence of the emperor, the main military Roman fleet was stationed in Miseno harbour nearby Baiae and the Phlegraean town of Puteolis was the most important commercial port of the whole Empire. Such a concentration of power in Campania meant that this region had a cornucopia of luxurious aristocratic banquets that contemporary authors then described in their works. Petronius' *Satyricon*, Juvenal's *Satires* as well as Martial's *Epigrams* present examples from the Campanian region. The most famous gourmands of Roman history, such as Lucullus, Lucius Murena, Sergius Orata and Apicius, had villas where they presented lavish banquets along the Phlegraean coast.
The aim of this work is to show the importance of the production and distribution of desirable marine resources

in a complex network which reached even the inner towns of southern Italy and to verify if this finding could be used as an indicator of the existence of rich purchasers that used such products to affirm their social status.

The town of Alife

Alife is a small town located (fig. 16-1) in the inner part of the administrative province of Caserta (Campania,

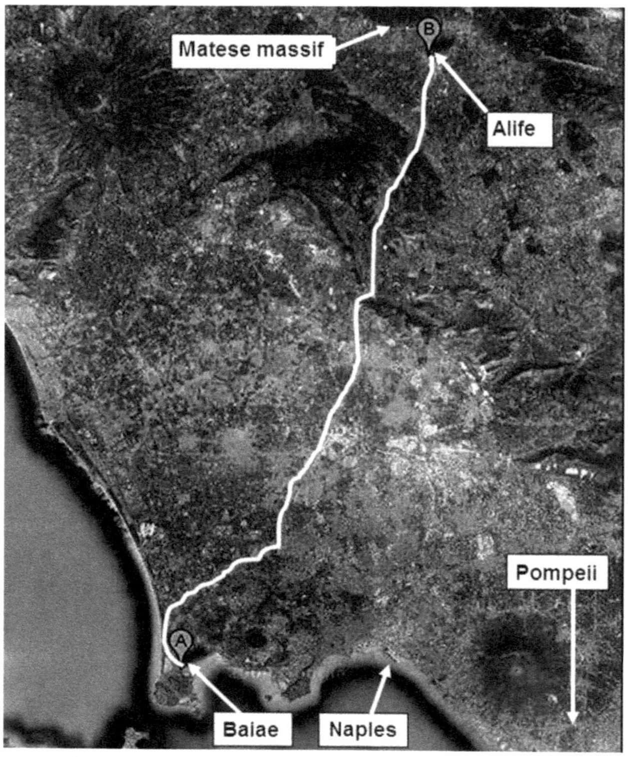

Figure 16-1. Satellite view of the central part of Campania with the location of the roman towns of Naples, Baiae, Pompeii and Alife, the shortest route from Baiae to Alife and the location of the Matese Massif.

southern Italy). It lies at the foot of the Matese Massif, 45 km as the crow flies from the coastline and 110 m above sea level. The small river Torano flows close to the ancient south-eastern walls of the town before flowing into the larger river Volturno 6 km downstream. Small settlements had existed since the Iron Age on the hills that surrounded the modern town, but the actual village was probably settled by Osco-Samnitic people before the 4th century BC (Miele and Sirano 2004). The ancient settlement took part in the war waged by the Samnites against Rome from 343 BC to 290 BC. It was destroyed by the Romans but it was rebuilt as an *oppidum* (a fortified village) and became a *municipium Romanorum* with the name of *Allifae* or *Alliphae* (Trutta 1776). The town plan reflects the typical features of a Roman military camp with two orthogonal main streets and a surrounding wall (fig. 16-2). During the 1st century BC, Alife was transformed into a military colony and the town noticeably grew in terms of wealth and prosperity. An amphitheatre was built in this phase (fig. 16-2b). An earthquake caused serious damage to the town in 369 AD, but its fortifications were quickly rebuilt. A bishopric was present in Alife during the 5th century, but it disappeared in the next century and the town quickly lost importance following the decline of the Roman Empire. Alife was then conquered by the Lombards and was destroyed during the 10th century by a Saracen raid. The town was rebuilt again and had a partial phase of economic recovery during the Norman age when, in the 12th century, it was ruled by Count Ranulf who began the construction of the Cathedral (Gambella 2000).

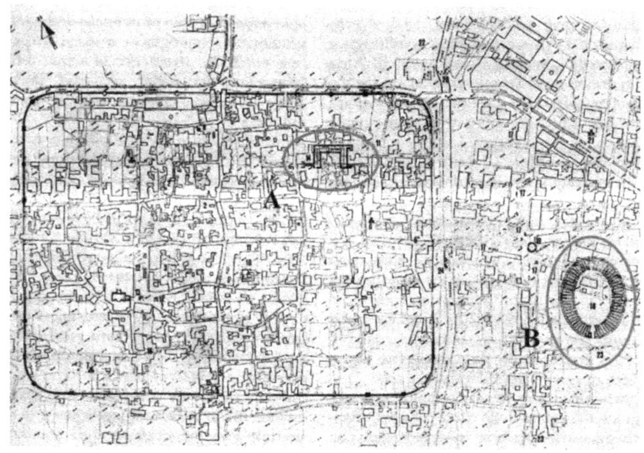

Figure 16-2. Plan of Alife showing the walls of the ancient town, the *criptoporticus* (A) and the amphitheatre (B). Note the orthogonal main streets that correspond to the *cardus* and the *decumanus* of an ancient roman military camp.

The *criptoporticus*

In Roman architecture a *criptoporticus* is a vaulted corridor or arcade below ground level. *Criptoportici* are present in Spain (Cadiz), Portugal (Coimbra and Lisbon), France (Arles, Narbonne and Reims) and even in some villas in England (Bignor and Woodchester) as well as in the eastern Mediterranean (e.g. Corynthum) (Corporate author 1973).

The Alife *criptoporticus* is U-shaped in plan (fig. 16-2a and 16-3) and can be divided into three principal "branches". The central one (i.e. the U base) is 44 m long, while both lateral branches have a length of 27.5 m each. The whole structure is longitudinally divided by a central row of pillars connected by arcades and on the inner alleys there are 21 chutes from which large accumulations of heterogeneous materials were discharged (fig. 16-3 and 16-4). The *criptoporticus* was excavated into the bedrock and the interior structure was formed using *opus caementicium*. The inner walls were then covered with a simple grey mortar plaster.

Most of the archaeozoological remains found in the *criptoporticus* come from heaps of rubble and waste which lay below the chutes. The *criptoporticus* was built during

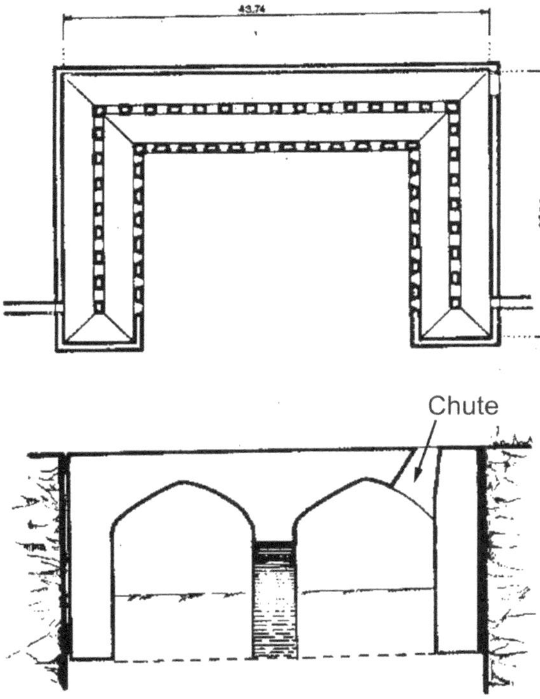

Figure 16-3. Alife, plan (A) and section (B) of the *criptoporticus*. Note the position of the niche at the centre of the horizontal branch and the chutes opening in the inner alleys.

Figure 16-4. View of part of the *criptoporticus* during the excavation. Some chutes (A) and the pillar row (B) are easily identifiable. An accumulation of discharged materials (C) showing its typical conoid shape is visible on the left. A detail of a rubble and shells heap is shown in the box.

the reign of Augustus Caesar and was abandoned during the 2nd century AD (phase 2), large quantities of pottery fragments and building waste were discharged from the chutes covering the older floor. These waste strata were later levelled (phase 3, dated to the first half of 3rd century AD) in order to obtain a new floor that was in use (phase 4, dated to the 3rd-4th century) until the beginning of 5th century, when the *criptoporticus* was filled again by large amounts of waste from the surface, perhaps related to a partial abandonment of the building above (phase 5, dated to the 5th-7th century). A large flood hit the area between the 8th and the 10th century causing the final decline of the structure.

Archaeological excavations started in May 2007 and ended in February 2008. They were carried out by the Medieval Archaeology department of the University of Naples Suor Orsola Benincasa in cooperation and on behalf of the Soprintendenza per i Beni Archeologici of Caserta and Benevento (Marazzi and Olivieri 2009).

The faunal assemblage

A large number of animal remains or archaeozoological material were recovered, chronologically spanning from the Roman Imperial age to the Middle Ages. The best archaeozoological records are, however, those that cover the time span from the 2nd century AD to the 7th century (i.e. phases 2-5). We focused our attention on these remains and their related strata in order to understand the economic and social transformations occurring in the town during those centuries.

Along with domestic mammals we found both wild and domestic bird remains, fish remains (both freshwater and marine species) and abundant mollusc shells. Shells represent an important element of the faunal assemblage found in the Alife *criptoporticus*.

Molluscs (A.C.)

The *criptoporticus* archaeomalacological assemblage represents a unique opportunity to investigate the exploitation of marine molluscs in Campania during Roman Imperial Age. Unfortunately published shell assemblages in the region contemporary with the occupation of Alife are quite scarce. Even though lots of shells were found in the Roman towns destroyed by the 79 AD Vesuvius eruption they are still to be analysed in detail. Moreover shells from Pompeii, Herculaneum and Oplonti provide evidence of mollusc use in the 1st century AD, hundreds of years before the deposition of a large part of the archaeozoological assemblages found in the *criptoporticus*.

The Alife archaeomalacological assemblage consists of 1404 fragments taking in a time span from the 2nd century to the 5th-6th century AD. The remains derive from five bivalve and two gastropod marine taxa. No scaphopod or cephalopod remains have been found in the *criptoporticus* nor other aquatic invertebrate remains. The archaeomalacological assemblage is completed by scarce terrestrial gastropod remains.

Flat oyster (*Ostrea edulis*) shells are the most abundant with 743 fragments mostly found in the heaps of phase 2 (447 fragments). None of them show evidence of marine erosion and internal biofouling or predation holes. Such evidence suggests the dietary use of the species. Forty-four oyster valves are characterized by one or more artificial holes passing through the shell edge (fig.16-5).

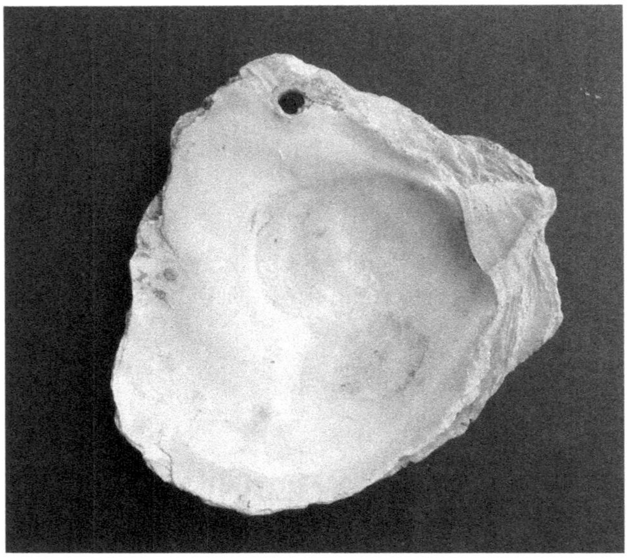

Figure 16-5. Oyster valve with artificial hole, indicating that young oysters were suspended with ropes to poles in order to favour their growth in protected areas of lagoons.

Such modifications match well with those from another important contemporary Campanian site. The imperial palace of Baiae stood looking onto a coastal lagoon, the *Lacus Baianus*, which has now vanished due to the typical bradyseism phenomenon of the Phlegraean Field area (Carannante 2002). Here dozens of *Ostrea edulis, Spondylus gaederopus* and *Lutraria oblonga* valves were found during underwater excavations in the gulf of Baiae where the *Lacus* was situated in the Roman Age (Carannante 2002). These shells were meal remains reutilized as filling in the creation of a quay placed on the shore of the lagoon at the end of 1st to the beginning of the 2nd century AD. Most of the oyster valves from Baiae showed the same artificial perforations as noted at Alife (Carannante 2002).

Many of the artificial holes in the Alife and Baiae oyster shells retain the casts of fine ropes used to suspend the living molluscs from perch systems embedded in the coastal lagoons. This was a typical oyster breeding technique common in the imperial Roman world (Günther 1897). Small oysters were perforated on the edge of the shells and hung in protected areas of lagoons to favour their free growth up to the largest sizes. The shells progressively grew around the ropes thus shaping the holes.

This oyster culture technique is also documented by classical sources and iconography. Ausonius (4th century AD) wrote that oysters raised at Baiae "...*pendent fluitantia palis...* " "hang floating from poles" and Pliny (*Naturalis Historia* IX, 79), among others, tells us that Sergius Orata (1st century BC) implanted the first oyster breeding ground in a coastal lagoon next to Baiae. In the same paragraph it is reported that "He (Sergius Orata) was the first to judge the oysters of Lake Lucrinus in Baian territory as being superior in delicacy of flavour."

Several glass flasks produced in Campania dating to the 3rd-4th century AD were illustrated with the most famous resorts along the Phlegraean coast from Baiae to Pozzuoli (Günther 1897; Kolendo 1977). One of the resorts is described on all the flasks by the inscription "*Ostriaria*" (place devoted to oyster breeding) (fig. 16-6). The associated illustration is easily interpreted as ropes with oysters dangling from a complex system of perpendicular poles.

This was not the only oyster breeding technique attested in Alife. Twenty lower oyster valves found in phase 2 and 3 stratigraphic units are still attached to large amphora fragments or show the concave cast of their surface (fig. 16-7). These valves show that pottery fragments were used as artificial oyster bed substrata – probably where it was not possible to utilize poles.

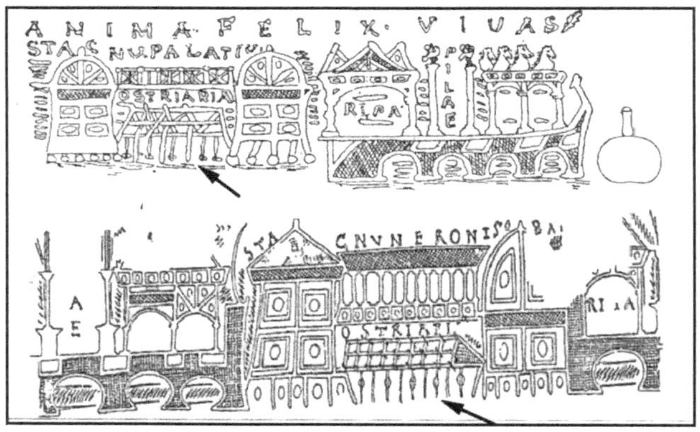

Figure 16-6. Illustrations of some glass flasks showing the inscription "Ostriaria". Arrows indicate these oyster breeding areas with their pole made structures (from Dubois 1907; Garcia y Bellido 1954 modified).

Figure 16-7. A pottery fragment with a fragment of an oyster valve attached. Pottery fragments were used as an artificial substratum for oyster breeding. *Cripteporticus* of Alife.

Two hundred and twenty-eight *Spondylus gaederopus* (thorny oyster) valves were found in association with the remains of *Ostrea edulis* at Alife. This is an additional connection with the Baiae imperial palace archaeomalacological evidence even though the stratigraphic association between both species cannot be explained by a common provenance from lagoon oyster breeding centres. Thorny oysters have never been bred and they were collected in open waters on rocky bottoms characterized by moving currents.

Nevertheless, the association between very large specimens of *Ostrea* and *Spondylus* which we found both in Alife and Baiae has a connection to ancient Roman culinary art. Apicius, the most famous gastronomer of Roman Age, describes a recipe typical of Baiae made from these two delectable marine molluscs. The *Embractum Baianum* (Baian stew) was made by stewing the minced meat of both flat and thorny oysters (and also "*urticas marinas*", "sea nettle", the delicious but urticant sea anemone) in a saucepan with oil, *garum* (fish brine), pine-kernels and several spices.

Some of the thorny oyster lower valves from Alife are still cemented onto pyroclastic rock fragments (fig. 16-8) which petrographic analysis has revealed to be the typical yellow tuff which characterizes the Phlegraean area.
All the data strongly suggest that both flat oysters and thorny oysters were brought to Alife from the Phlegraean coasts about 60 kms away as the crow flies (fig. 16-1). The distance between Phlegraean coast and Alife grows to 80 kms if we gauge distance following the Roman roads.

Figure 16-8. *Spondylus* lower valve with pyroclastic rock fragment cemented on the outer surface. Analyses show that these fragments come from the Yellow Tuff of the Phlegraean area.

Donax trunculus, the common tellina or bean-clam, is the second most common species in the archaeomalacological assemblage with 291 fragments.

Venus verrucosa remains are very scarce with only 31 fragments. None of the shells have predation holes, internal biofouling or marine erosion marks and this suggests the dietary use of such molluscs. In contrast, an ornamental use can be argued for three eroded valves of *Glycymeris glycymeris,* one of which appears to have been artificially perforated at the umbo.

Murex shells (30 *Hexaplex trunculus* fragments and 4 *Bolinus brandaris* fragments) are the only marine gastropod taxa attested in the Alife *criptoporticus*. The shells do not show marine erosion or bioerosion marks, internal bio-fouling or gastropod predation holes. Such evidence suggests they were used as food.

A detailed examination of the archaeomalacological assemblages from the different stratigraphic units and each chronological phase shows interesting patterns and changes.

Ostrea edulis is the dominant taxon during phase 2. Oyster remains form 82.5% (MNI) of the marine archaeomalacological assemblage (fig. 16-9a). They are followed in abundance by *Spondylus gaederopus* remains (15% MNI). The shells are clustered within the deposits. For example stratigraphic unit 293 (phase 2) is a concentrated heap of flat oyster and thorny oyster valves beneath one of the *criptoporticus* chutes. Several similar oyster heaps were found in the different areas of the *criptoporticus.*

The archaeomalacological assemblage of stratigraphic unit 50 dating to phase 3 (2nd-3rd century AD) is characterized by the prevalence of *Donax trunculus* remains (fig. 16-9b) constituting 82% (MNI), with oysters falling to 3.3%.

These data from stratigraphic unit 50 contrast not only with that from preceding phases but also with that of other contemporary stratigraphic units. The archaeomalacological assemblage of the remaining phase 3 stratigraphic units is very similar to that from phase 2 confirming the dominant role of *Ostrea edulis* (63.2%) followed by *Spondylus* (20.5%) and murex shells (8.5%), whereas bean-clams do not exceed 4.3% (MNI) (fig. 16-9c). Such data suggests that a different interpretation is required for stratigraphic unit 50 supported also by data obtainable from other faunal and non faunal archaeological remains.

An analogous anomaly is seen in the archaeomalacological assemblage from stratigraphic unit 195 dating to phase

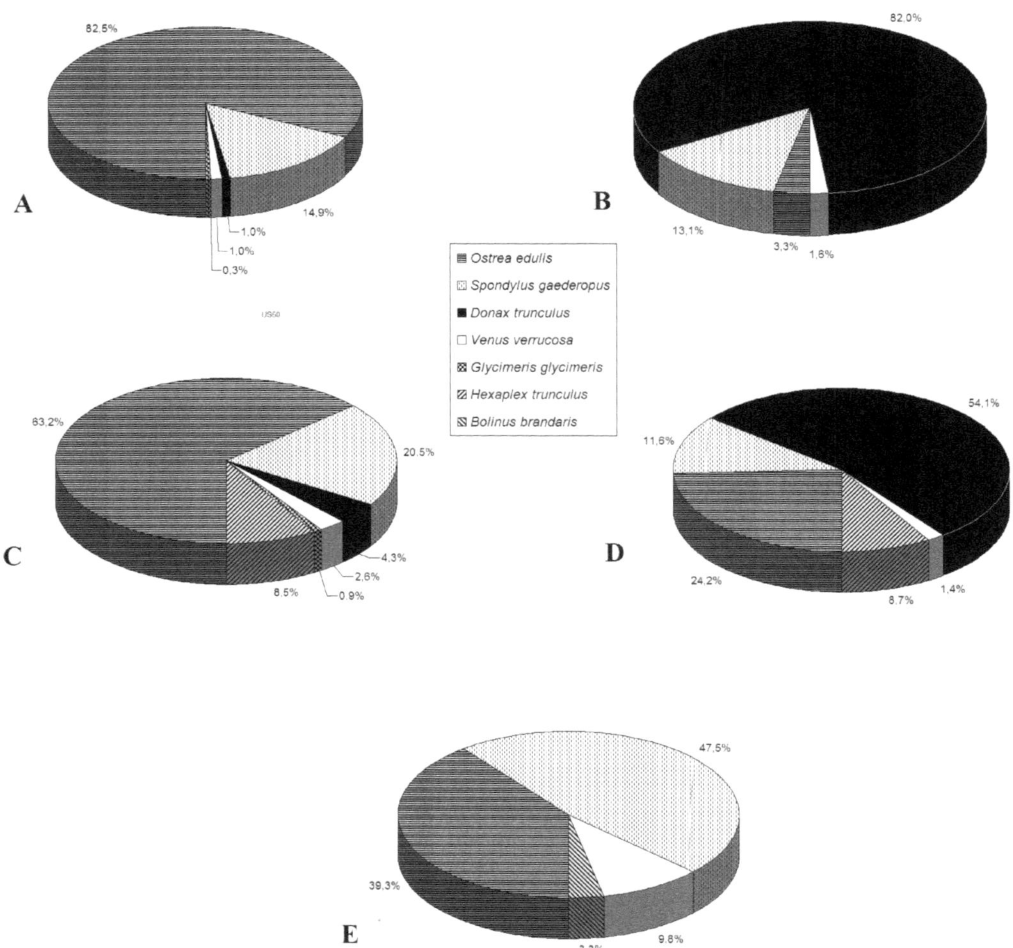

Figure 16-9. Quantitative composition of the archaeomalacological assemblage calculated in terms of MNI%. Each single set of data refers to a different phase or stratigraphi unit; A) Phase 2; B) Phase 3 (Stratigraphic Unit 50 only); C) Phase 3 (without Stratigraphic Unit 50 remains); D) Phase 4 (Stratigraphic Unit 195 only); E) Phase 4 (without Stratigraphic Unit 195 remains).

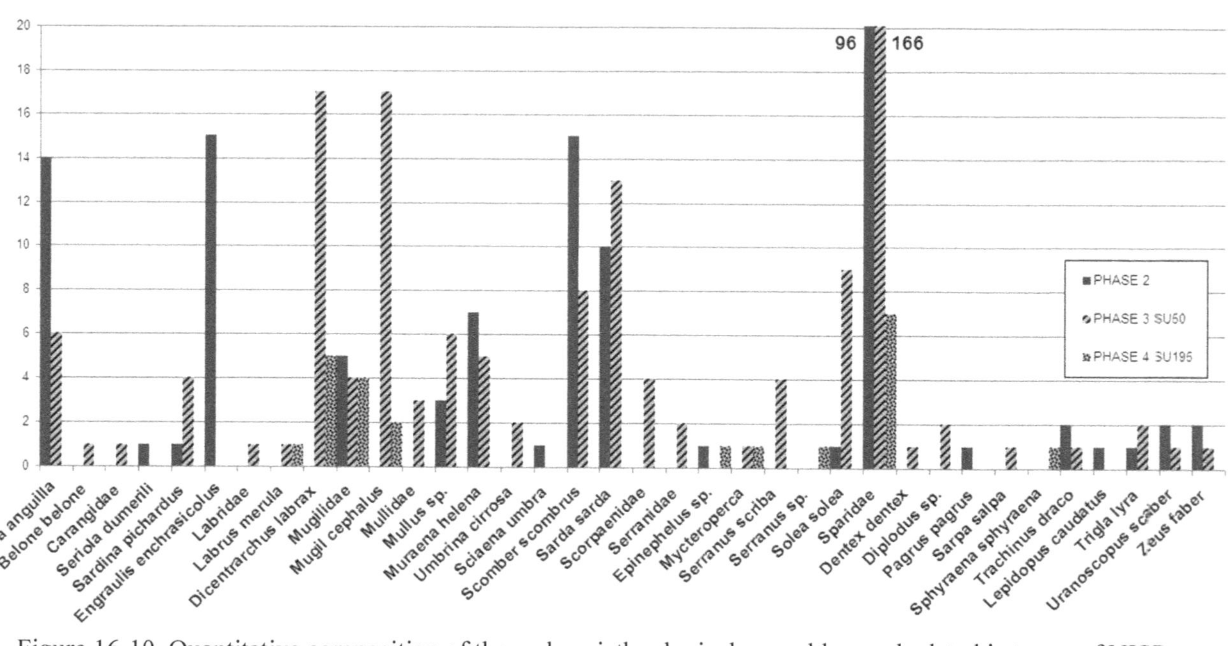

Figure 16-10. Quantitative composition of the archaeoicthyological assemblage calculated in terms of NISP.

4 (3rd/4th century AD). This is characterized by more than 54% *Donax trunculus* (MNI), followed by 24.2% *Ostrea*, 11.6% *Spondylus* and 8.7% murex shells (fig. 16-9d). As with stratigraphic unit 50, the data from stratigraphic unit 195 contrast with those from other contemporary stratigraphic units of phase 4. These other units are characterized by the prevalence of *Spondylus* (47.5% MNI) and *Ostrea* (39.3%) remains followed by *Venus verrucosa* (10%) and murex shells (fig. 16-9e), whereas *Donax* remains are completely absent. Indeed, all the *Donax* shells from phase 4 come from this single stratigraphic unit, 195.

Mollusc remains from stratigraphic units dating to phase 5 (5th-7th century AD) are very scarce. There are just 54 fragments which are largely assignable to reworked older strata.
Thirteen pulmonate gastropod shells in the Helicidae complete the archaeomalacological assemblage from phase 5. The stratigraphic units from 1 to 4 lack terrestrial mollusc remains, whereas pulmonate snail shells increase in abundance from phase 5 to the most recent levels of the *criptoporticus*. The presence of a small artificial hole on the body whorl of many of the shells suggests a dietary use. This is confirmed by black burn marks on many of the shells. The stratigraphic units dating to the medieval period (phases 5-9) are completely lacking in marine mollusc remains, but the presence of few freshwater mussel (Unionidae) valve fragments attests to the sporadic exploitation of freshwater molluscs.

Many classical authors —such as *Petronius* in his *Satyricon*, Martial in the *Epigrams*, Juvenal in his *Satire* – highlight the social importance of offering Phlegraean bred oysters during Roman Imperial Age banquets.

Given this documentary evidence, the abundance of marine mollusc remains of Phlegraean provenance at Alife during the period from the 2st to the 4th century AD certainly reflects an elitist exploitation of marine resources already documented in other Campanian Imperial contexts such as at Baiae, where an intensive consumption of bred oysters and thorny oysters matches well with the Alifan datum (Carannante 2002).

Fish (A.C. and A.D.S.)

Marine molluscs were not the only aquatic organisms exploited in Alife. Nine hundred thirty-six fish remains were found during the excavation of the *criptoporticus*. All of them were collected from the sieves (1 mm mesh) from stratigraphic units rich in other archaeozoological remains. Eighty-three percent (778 remains) of the archaeoichthyological assemblage has been identified, with the remaining 158 fish bones (small fragments, scales and spines) remaining unidentified.

Only 30% of the identified fish specimens are ascribable to freshwater species. A surprising result if we consider that the geographic location of Alife is in close proximity to the rivers Volturno and Torano but 45 kms away from the coast.

The trout (*Salmo trutta*) is the most well represented (250 remains) freshwater species in the archaeoichthyological assemblage (fig. 16-10) followed by specimens in the Cyprinidae (38 NISP). A few Cyprinidae cranial bones were identifiable as tenches (*Tinca tinca*). Twenty eel (*Anguilla anguilla*) remains attest the use of this catadromous species probably captured in the rivers.

Sparidae remains (sea bream) are the most abundant of the marine and brackish water fish with 274 bones. However most are vertebrae which do not enable us to determine the species. Cranial sparid bones provide identifications of two sargos (*Diplodus sp.*), one dentex (*Dentex dentex*), one common seabream (*Pagrus pagrus*) and one salema (*Sarpa salpa*).
Transitional water fish seem to have been important in the Alifan diet. Thirty-two grey mullet remains (only identifiable to Mugilidae family) and 25 sea bass (*Dicentrarchus labrax*) remains have been found in the *criptoporticus*. One very large vertebra attests that sea bass of more than a metre in length arrived at Alife.

The dominance of brackish water fish remains within the *criptoporticus* archaeoichthyological assemblage suggests that resources from coastal lagoons or river mouths were important sources of supply at Alife.

Other fish remains also demonstrate that open sea fishing contributed to Alifan diet.

In the archaeoichthyological assemblage at Alife the following species have also been identified; 23 bonito (*Sarda sarda*) remains, 23 mackerel (*Scomber scombrus*) remains, 15 anchovy remains (*Engraulis encrasicholus*), 14 Serranidae remains (three at least from the groupers *Epinephelus guaza* and *Mycteroperca rubra* and four remains of *Serranus scriba*), 12 red mullet remains (*Mullus* sp.), 12 Mediterranean moray eel (*Muraena helena*) remains (fig. 16-11), 10 common sole (*Solea solea*) remains, 5 of *Sardina pilchardus*, 4 of *Scorpaena sp.*, 3 of John Dory (*Zeus faber*), 3 of the *Labrus* sp. (probably *Labrus merula*), 3 of the greater weever (*Trachinus draco*), 3 of the sea robin *Trigla lyra*, 3 of the stargazer (*Uranoscopus scaber*), 2 of greater amberjack (*Seriola dumerili*) and 2 of bearded umbrine (*Umbrina cirrosa*). Single bones of other species such as the brown meagre (*Sciaena umbra*), the garfish (*Belone belone*), the silver scabbardfish (*Lepidopus caudatus*) and the European barracuda (*Sphyraena sphyraena*) were also identified.
Apicius, in the fourth book of his *De Re Coquinaria*, describes several recipes prepared with the species identified for Alife including Mediterranean barracuda, sole, red mullet, anchovy, mackerel, grey mullet and dentex. The ninth book of Apicius is entirely devoted to

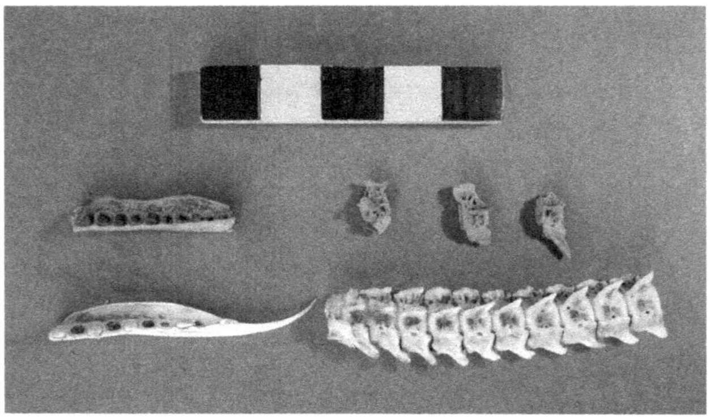

Figure 16-11. Mediterranean moray eel remains. Lighter specimens at the bottom are modern ones showed as comparison.

marine resources and reports many recipes for red mullet, bonito, mackerel and grey mullet, whereas the tenth book is centred on sauces to dress boiled or grilled fish. An entire paragraph of the tenth book describes different ways to serve the Mediterranean moray, but several sauce recipes are also suggested for eel, scorpion fish, red mullet, bonito, mackerel, dentex and other fish in the Sparidae. It is interesting to note that the remains of twelve of the seventeen marine fish species mentioned in the *De Re Coquinaria* have been found in the Alife *criptoporticus*.

Literary sources suggest that many of the fish species identified at Alife were an important status indicator. The social value of red mullet in imperial times was made famous by the fourth Juvenal satire and in Pliny's *Naturalis Historia* (IX, 30), where its very expensive cost is reported, and in Petronius' *Satyricon* (XXXV), where a pair of red mullets are served during the Trimalchio banquet to symbolize the constellation of Pisces in a dish evoking the zodiac.

The Mediterranean moray was a much appreciated fish in the Imperial Age. Pliny reports that Caius Hirrus invented preserves for moray as it was a very expensive species (IX, 81), that the orator Hortensius had several fish preserves in Bauli in the territory of Baiae and that Licinius Murena was the first to create preserves for other fish (IX, 80).

Bonito (IX, 18), mackerel (IX, 19) sea bass or Mediterranean barracuda (IX, 28), John Dory at Gades (Cadiz) (IX, 32) and salema at Ebusus (Ibiza) "which is considered elsewhere an unclean fish" (IX, 32) are also reported in Pliny's *Naturalis Historia* as valuable species. Trout and Sparidae are the dominant fish taxa in all the phases of the *criptoporticus*. All archaeoichthyological assemblages from the different phases show high taxonomic diversity with a NISP/number of taxa ratio of 13:3 for phase 2 where 253 fish bones were found, 16:6 for phase 3 where 497 remains were found, and a minimum value of 3:1 for phase 4 where we have only 28 fish bones. Archaeoichthyological remains from phases 3 and 4 all derive from stratigraphic units 50 and 195: units also noted above for the distinctiveness of their shell assemblages. Phases following phase 4 lack any fish remains.

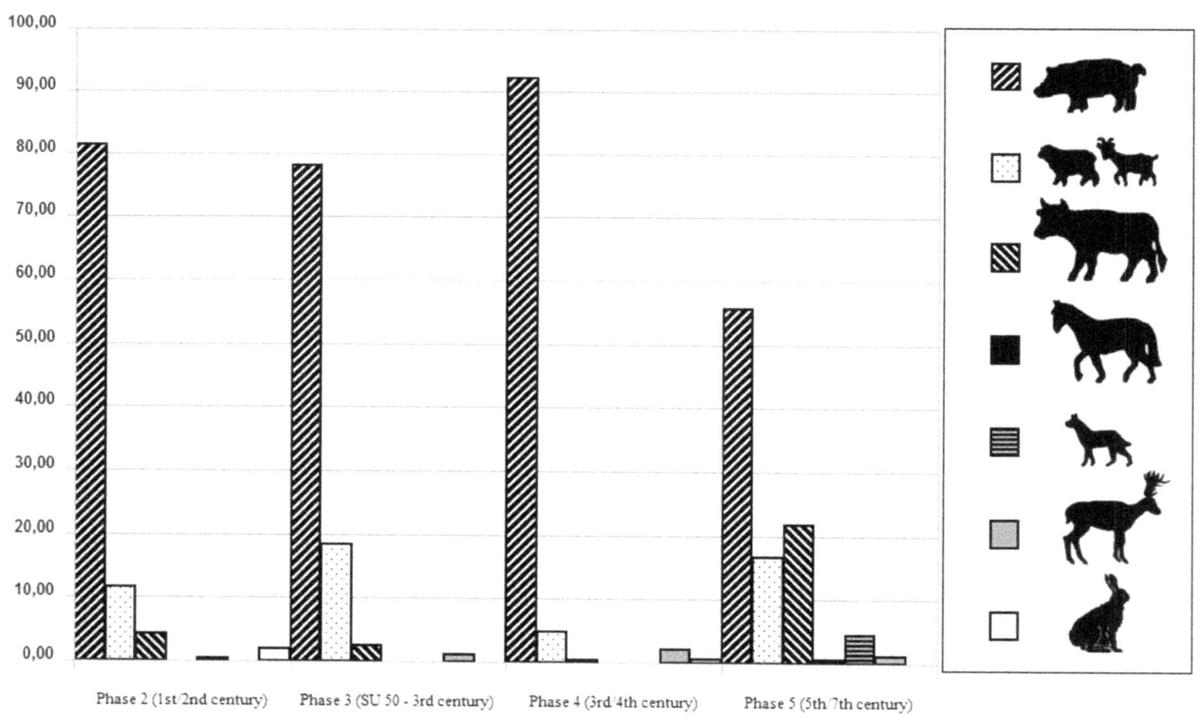

Figure 16-12. *Criptoporticus* of Alife . Mammal remains. Quantitative composition of the faunal assemblage calculated in terms of NISP%. Note the decrease of pig remains in phase 5.

Mammals (S.C. and R.V.)

Pigs are the most abundant taxon in all the four examined phases (fig. 16-12), with a NISP percentage that ranges from 81% to 90% between the 1st-2nd century and the 4th century decreasing to 55% during the 5th-8th century when bovine remains increase and became the second mammal taxon in terms of NISP%.

Sheep and goats are present in variable quantities, ranging from less than 7% NISP during the 3rd-4th century (when pigs reach their peak in terms of frequency) to a maximum value of 14% in the last phase.

Bovines are always the third most important taxon in terms of frequency except in the 5th/6th century but this could easily be explained as in stratigraphic unit 8 (a level dating to this chronological phase) a high number of cattle bone fragments were recovered, several of them showing traces of anthropic modifications such as some sawed horn cores (fig.16-13). These data could be related to handicraft activities that were carried out in the above *sub divo* building(s), from which the waste was discharged using *criptoporticus* disposal chutes.

These data are totally consistent with what we know about the exploitation and production of pork meat during the Roman Imperial Age and the Medieval Age in southern and central Italy.

Pigs were a fundamental resource in the Roman world and their breeding increased from the Republican to the Imperial Age in line with pork consumption. Data coming from the urban area of Rome (Minniti 2005) clearly shows this trend, which is also well known in Campania where bovines were the principal domestic mammals bred in the Gulf of Naples area before the 2nd century BC, with pigs becoming more important from the 1st century BC (King 1999).

After the fall of the Roman Empire this picture changes dramatically. The percentage of pigs decreases in Rome from 80% reached during the Imperial Age (De Grossi Mazzorin 1989, 1995, 1998; Tagliacozzo 1993) to a lower value of 55% in 5th century strata which come from the *Schola Praeconum* (Barker 1982) and the *Meta Sudans* (De Grossi Mazzorin 1995). This trend continues in the following centuries as attested by the faunal assemblages of *Cripta Balbi*, where the percentage of pigs decreases from 60% to 47% between the 7th and the 10th century (Minniti 2005), or in some contemporaneous assemblages which come from several central Italian sites such as Monte Gelato, Farfa and San Donato (Clark 1997) or the kitchen complex of the San Vincenzo al Volturno abbey (Chilardi 2007).

Figure 16-13. Sawed horn core of bovine from SU 8.

Birds. (D.R.)

The bird assemblage consists of 246 fragments deriving 11 taxa. Two hundred twenty-five of them were identified to anatomical element with 140 then being taxonomically identified.

Domestic species predominate: chickens (*Gallus gallus*), geese (*Anser anser*) and brents (*Branta bernicla*) are the most common species, but the rock dove (*Columba livia*) and the common wood pigeon (*Columba palumbus*) are also present. These species were generally important sources of meat, eggs and feathers. Wild (or semi-wild) taxa, such as pheasants (*Phasianus* sp.), and partridges (*Alectoris* sp.) are also present in the faunal assemblage. These species lived in the Matese area until the first half of the 20th century.

Aquatic avifauna is represented by the common teal (*Anas crecca*) and the grey heron (*Ardea cinerea*). Their presence could be related to the proximity of lakes and rivers as such as Lake Matese, the river Volturno and the torrents Albento and Torano.

The avifaunal assemblage is completed by Passeriformes remains, and indicates the presence of a cultivated area for their diet of grain.

Lower limbs are the most common anatomical elements, mainly tibiotarsi (probably because of their higher mechanical resistance), followed by upper limbs among which the humerus prevails. Skulls, mandibles, carpometacarpi, tarsometatarsi, lower and upper phalanxes are rare.

Smaller elements may have been lost during the sieving and washing processes but there is no evident explanation for the lack of larger elements such as carpometacarpi and tarsometatarsi. Our hypothesis is that these bones were eliminated before consumption as heads and limb extremities have a low value in terms of meat content.

The bird remains from stratigraphic units related to phase 2 (1st-2nd century) are very scarce and not well preserved. They comprise 8 humeri fragments of *Gallus gallus* and a humerus and coracoid of the rock pigeon.

Stratigraphic unit 50 (phase 3), dating to the 3rd century, is characterized by a higher variety of species and by an abundance of wild taxa (fig. 16-14a): fowl (*Gallus gallus*), brent (*Branta bernicla*), goose (*Anser anser*), common teal (*Anas crecca*), passerine remains, partridge (*Alectoris* sp.) and pheasant (*Phasianus* sp.). Pheasant remains represent

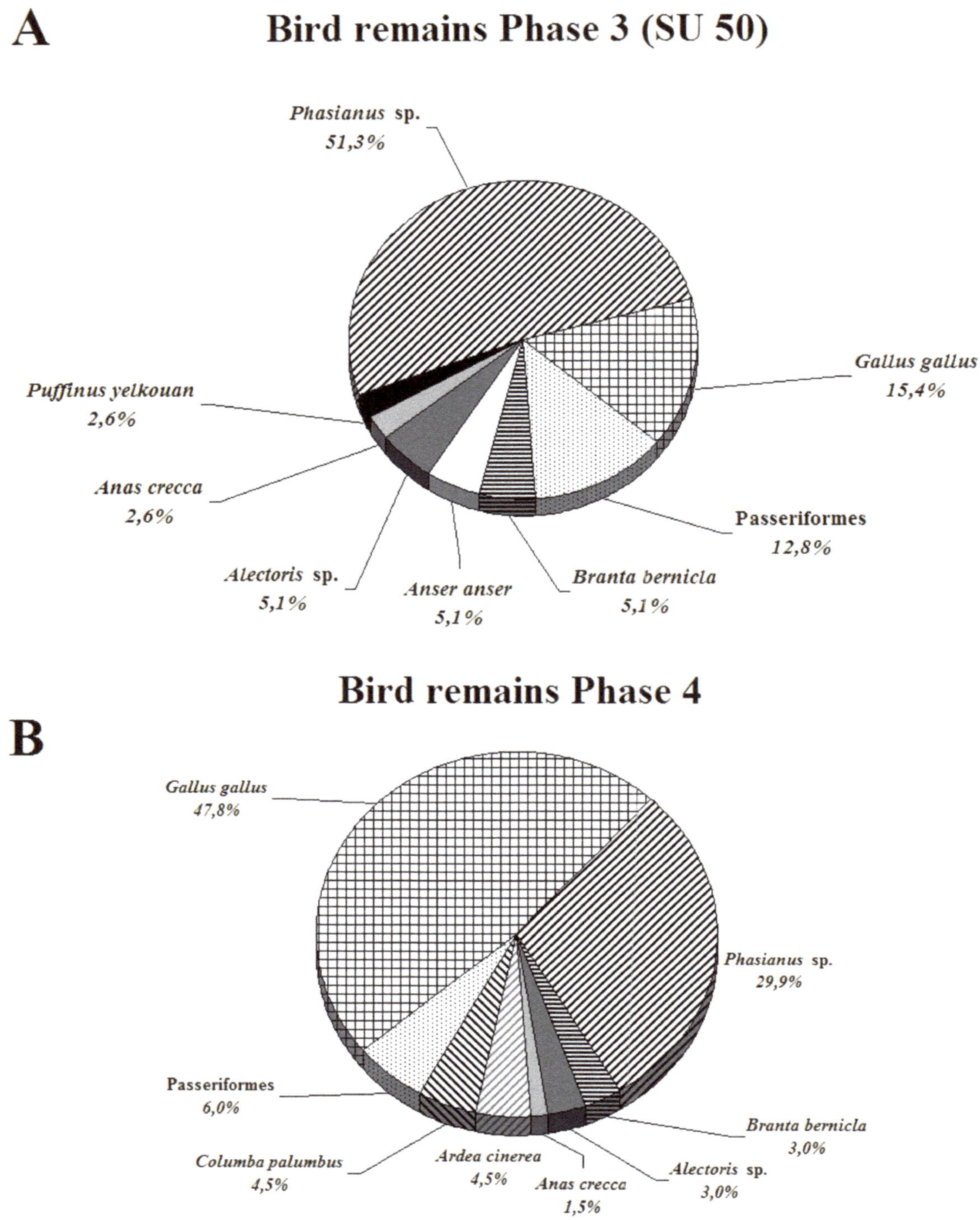

Figure 16-14. Quantitative composition of avifaunal assemblage calculated in terms of NISP%.

more than 50%NISP of the identified bird bones (fig. 16-15), with several of them showing cut marks.

A peculiar finding is a single yelkouan shearwater (*Puffinus yelkouan*) humerus. This typical coastal species living on rock cliffs and in the open sea can be hardly related to an alimentary use unlike marine fish and molluscs found at Alife.

An increase in poultry consumption is seen in phase 4 dating to the 3rd-4th century (fig. 16-14b). The remains of fowl become the most abundant in the assemblage displacing the pheasant as the predominant taxon. This trend is comparable with data from the urban area of Rome, where chicken farming was widespread during the Roman Imperial Age (De Grossi Mazzorin, 2000).

Fowl remains from these stratigraphic units belong to small chickens below the average size recorded for the Roman

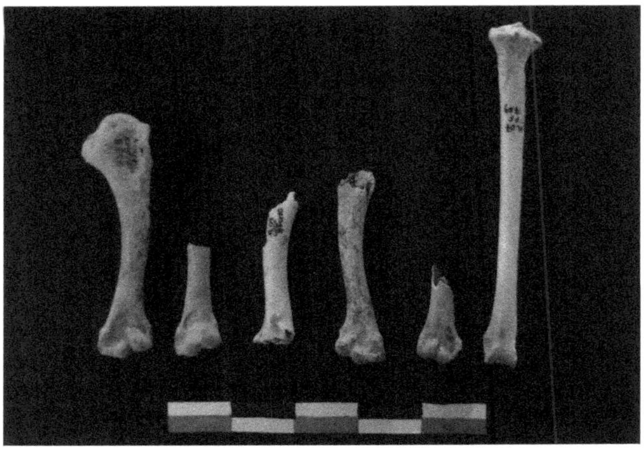

Figure 16-15. *Phasianus* sp. remains from Stratigraphic Unit 50 (phase 3).

Figure 16-16.The vaulted niche (A) at the centre of the central branch with the altar remains (B) right at its foot surrounded by the Stratigraphic Unit 50 with fire traces. A gold headed hairpin (C) found in such Stratigraphic Unit is shown in the box.

Imperial Age and more comparable with those from the Medieval period (Riedel and Rizzi, 1997).

Three pheasant humeri from this phase show cut marks next to the distal epiphysis related to the disarticulation of the humerus from the radius-ulna.

Anser anser and *Branta bernicla* were also important resources during phase 4. The identification of the anatomical parts shows a predominance of wing and leg remains.
Stratigraphic unit 195 related to the same phase 4 yielded peculiar data concerning non-dietary bird exploitation. Here the remains of an almost complete fowl skeleton were found. Taphonomic evidence suggests only partial consumption of the meat. This anomalous pattern could be related to ritual activities which other anomalous findings from the same stratum indicate.

The avifauna from phase 5 (5th-7th century) are represented by fowl remains, fragments of the upper limbs of *Anser anser*, *Phasianus*, *Alectoris* and a tibiotarsus of *Branta bernicla*. *Gallus gallus* is the most frequent taxon and the only one whose remains belong to almost the whole skeleton but the total number of identified specimens for this phase is very low (only 18 fragments) and cannot be used for statistical studies.

Conclusions

Archaeozoological remains found in the excavation of the Alife *criptoporticus* are a useful example of luxury food production and distribution. In this faunal assemblage marine resources are the most important social indicator.
Archaeozoological data provide an economic and social context, and point to a wealthy social class which was able to transport high value marine resources towards the inner part of the Campanian territory from the 1st to the 4th century AD. Archaeomalacological data indicate that marine molluscs were carried from the Flegrean area (see fig. 16-1) where the imperial palace of Baiae stood and where the most famous oyster breeding and fish preserving centres were located to supply the Roman aristocracy and even the imperial court with their products.

The study of the oyster remains has highlighted at least two different breeding techniques based on hanging the living molluscs to poles by ropes and onto artificial substrates made of amphora fragments.

The presence of game birds (pheasant, partridge, heron, and teal) as well as the consumption of piglets confirms the wealth of part of the Alifan community.

The Alifan economic situation appears to change dramatically after the 4th century when the absence of marine species and the significant reduction of bird remains reflect a social decline of the local community.

Two stratigraphic units require further explanation. Stratigraphic unit 50 dating to phase 3 is the only level with fire traces that was not located beneath a chute but just in front of a vaulted niche (*arcosolium*) at the centre of the central branch (the base of the U shaped structure). This niche (fig. 16-16) is the only one in the whole *criptoporticus* and traces of a small structure (probably an altar) have been found right at its foot. Several bronze coins, pottery lamps and a gold headed hairpin (see fig. 16-16) have been found in this stratigraphic unit around the presumed altar. The other unusual unit, stratigraphic unit 195 dating to phase 4, contained scattered human remains belonging to at least two individuals. Such archaeological findings strongly suggest that these deposits are not simple waste dumps.

Additionally, both stratigraphic units 50 and 195 are the only deposits of their phases containing fish remains and both are characterized by a predominance of *Donax trunculus* in the archaeomalacological assemblages. These data are noteworthy as stratigraphic units contemporary to 50 and 195 have *Ostrea* and *Spondylus* as predominant taxa as in all the other strata.

How can we explain such anomalies? An almost complete fowl skeleton with evidence of partial meat consumption was found in stratigraphic unit 195 suggesting some kind of ritual activity. Fowls were frequently used as sacrificial items in divination rites, and Juvenal (*Satires* XIII, 230) notes that they were sacrificed to the *Lares*, offering part of them to the deities (...*Laribus cristam promittere galli*...). The *Lares* were household gods in the Roman religion connected to the underworld and were often considered to be the spirits of ancestors.

Fowl were also sacrificed during the Mithraic Mysteries and their remains have been found in the excavations of the *mithraeum* area in the *Crypta Balbi* in Rome (De Grossi Mazzorin 2000). The *criptoporticus arcosolium* with the small altar suggests a strong connection to the Roman *lararia*, domestic shrines where *Lares* rites were attended. Typical *lararia* consisted of a wall vaulted niche that hosted small *Lares* statuettes. They were widespread throughout the Roman world and they were located virtually in any room of the house and even in kitchens and stores.

A structure interpreted as a *sacellum* consecrated to *Lares compitales* dating to the 2nd-1st century BC has been found in Pompeii (VIII, 4, 24) (Anniboletti 2008). This consisted of a small altar and an *arcosolium* that recalls the *criptoporticus* context. Many archaeological remains were found around the altar and filling two small votive pits located behind. These pits contained bronze coins, lamp fragments, fowl remains and many fish bones (Anniboletti 2008) and thus correspond closely to the Alifan data. The faunal assemblage of this *lararium* includes some mollusc taxa such as *Patella* spp., species in the Cardiidae and Veneridae and, particularly, dozens of valves identified as *Tellina planata* within the text (Anniboletti 2008). This last identification seems to be incorrect. Looking at the illustration (Anniboletti 2008, fig. 16-8 and 16-9) it is as apparent that the shells shown are *Donax trunculus* rather than *Tellina planata*. The misidentification may be due to the fact that in Italian "tellina" is the common name for *Donax* spp.

"*Tellina planata*" (actually *Donax trunculus*) valves associated with *Patella* shells were also found in at least two other *lararia* in Pompeii (many shells in a pit next to a niche consecrated to *Lares Viales* on the façade of building VI 2, 16-21 and in a pit next to a *lararium* in the *domus* VII 15, 8) (Anniboletti, 2004).

All of this evidence allows us to hypothesize a role of *Donax trunculus* in ritual activities connected to the *Lares* cult in the Roman world.

References

ANNIBOLETTI, LARA. 2004. Domus VI 2, 16-21. Saggio F. In *Il progetto Regio VI. Campagna di scavo 2003*. F. Coarelli and Fabrizio Pesando, eds. Pp. 149-152. Rivista di Studi Pompeiani, no. 15. Rome: L'Erma di Bretschneider.

——. 2008. Il sacello VIII 4, 24: un culto collegiale a Pompei. *Fold&R Fasti On Line Documents and Research. The Journal of Fasti Online*. http://www.fastionline.org/docs/FOLDER-it-2008-104.pdf.

APICIUS. *De Re Coquinaria* (Bompiani edition, with Italian translation by G. Carazzali, 3rd edition, 2003).

AUSONIUS. *Epistolae* (Il Cardo edition with Italian text and comments by L. Mondin, 1995).

BARKER, GRAEME W. 1982. The animal bones. In *The Schola Praeconum, I*. David Whitehouse, Graeme Barker, David Reese, eds. Pp. 53-73. Papers of the British School at Rome, 50.

CARANNANTE, ALFREDO. 2002. Relazione archeo-malacologica sui campioni di fauna marina provenienti dal saggio di scavo nel lotto d'acqua Nautica Fusaro di Baia. Appendice a F. Maniscalco, N. Severino, Recenti ipotesi sulla conformazione del Lacus Baianus, *Ostraka*, XI (1):175-176.

Centre national de la recherche scientifique (France). Colloques internationaux. Sciences humaines (Corporate author). 1973. *Les cryptoportiques dans l'architecture romaine: Ecole française de Rome, 19-23 avril 1972* Rome: Ecole Française de Rome.

CHILARDI, SALVATORE. 2007. I resti di mammiferi dall'area delle cucine dell'abbazia altomedievale di San Vincenzo al Volturno (IS): risultati preliminary. In *Atti del 4° Convegno Nazionale di Archeozoologia. Pordenone 13-15 novembre 2003. Quaderni del Museo Archeologico del Friuli Occidentale*, 6:355-359.

CLARK, GILLIAN. 1997. Monastic economies? Aspects of production and consumption in early medieval central Italy. *Archeologia Medievale* XXIV:31-54.

DE GROSSI MAZZORIN, JACOPO. 1989. Nota preliminare sulla fauna. In *Curia, Forum Iuliu, Forum Transitorium*. C. Morselli and Edoardo Tortorici eds. Pp. 340-347. Lavori Soprintendenza Archeologica, no. 14. Rome.

——. 1995. La fauna rinvenuta nell'area della *Meta Sudans* nel quadro evolutivo degli animali domestici in Italia. In *Atti del 1° Convegno Nazionale di Archeozoologia. Rovigo 5-7 marzo 1993. Padusa Quaderni*, 1:309-318.

——. 1998. L'analisi dei resti ossei animali. In *Roma – Saggio di scavo nell'aula di S. Isidoro in Thermis, IV*. D. Candillo *et al.* Eds. Pp. 416-425. Notizie Scavi dell'Antichità, anni CCCXCIII-CCCXCIV, s. IX, voll. VII-VIII, 1996-1997.

——. 2000. Introduzione e diffusione del pollame in Italia ed evoluzione delle sue forme di allevamento fino al Medioevo. In *Atti del 3° Convegno Nazionale di Archeozoologia. Siracusa, 3-5 Novembre 2000*. Ivana Fiore, Giancarla Malerba and Salvatore Chilardi eds. Pp. 357-360. Rome: Istituto Poligrafico e Zecca dello Stato.

Dubois, Charles. 1907. *Pouzzoles antique (histoire et topographie)*. Paris: Fontemoing.

Gambella, Angelo. 2000. *Medioevo alifano. Potere e popolo nello stato normanno di Alife*. Rome: Drengo.

Garcia y Bellido Antonio. 1954. El vaso puteolano de Ampurias. *Archivo Español de Arqueologìa*, 27:212-226.

Günther Robert Theodore. 1897. The Oyster Culture of the Ancient Romans. *Journal of the Marine Biological Association of the United Kingdom (New Series)*, 4: 360-365.

Juvenal. *Satirae* (Mondadori edition, with Italian translation by B. Santorelli, 2011).

King, Anthony. 1999. Diet in the roman world: a regional inter-site comparison of the mammal bones. *Journal of Roman Archaeology*, 12:168-202.

Kolendo, Jerzy. 1977. Parc à huitres et viviers à Baiae sur un flacon en verre du Musée National de Varsovie. *Puteoli. Studi di storia antica,* 1:108-125.

Marazzi, Federico and Donatina Olivieri, 2009. *Il criptoportico romano di Alife - Il monumento e la sua esplorazione.* Piedimonte Matese: Arti Grafiche Grillo.

Martial. *Epigrammata* (Garzanti edition, with Italian translation by A. Carbonetto, 2008).

Miele, Floriana and Francesco Sirano, eds. 2004. *Ager Allifanus. La piana alifana alla luce delle recenti ricerche archeologiche.* Piedimonte Matese: Ikon.

Minniti, Claudia. 2005. L'approvvigionamento alimentare a Roma nel Medioevo: analisi dei resti faunistici dalle aree di scavo della *Crypta Balbi* e di Santa Cecilia. In *Atti del 3° Convegno Nazionale di Archeozoologia. Siracusa, 3-5 Novembre 2000.* Ivana Fiore, Giancarla Malerba and Salvatore Chilardi eds. Pp. 469-492. Rome: Istituto Poligrafico e Zecca dello Stato.

Petronius. *Satyricon* (Rizzoli edition with Italian translation by A. Aragosti, 1995)

Pliny. *Naturalis Historia (VIII-XI)* (Rizzoli edition with Italian translation by F. Maspero, 2011).

Riedel, Alfredo and Rizzi, Jasmine. 1997. Studio preliminare dell'avifauna di Verona medioevale (Piazza del Mercato Vecchio e del Tribunale). In *Atti del 2° Convegno Nazionale di Archeozoologia. Asti, 14-16 Novembre 1997*. Pp. 343-344. Forlì: ABACO.

Tagliacozzo, Antonio. 1993. I reperti faunistici. In *Caput Africae I, Indagini archeologiche a Piazza Celimontana (1984-1988). La storia, lo scavo, l'ambiente*. Pavolini Carlo, ed. Pp. 251-278. Rome: Istituto Poligrafico e Zecca dello Stato.

Trutta, Gianfrancesco. 1776. *Dissertazioni istoriche delle antichità alifane*. Naples: A. Forni.

17 - ARCHAEOZOOLOGICAL ANALYSIS OF MOLLUSCAN FAUNA FROM THE LATE BRONZE AGE STRATUM OF SITE 4 OF TELL JENIN (NORTHERN WEST BANK, PALESTINE)

Ademar EZZUGHAYYAR,
Department of Biology and Biochemistry, Birzeit University, P. O. Box 14, West Bank, Palestine,
ademar@birzeit.edu

Khalid M. SWAILEH
Department of Biology and Biochemistry, Birzeit University, P. O. Box 14, West Bank, Palestine,
kswaileh@birzeit.edu

Abstract: *Molluscan shells were collected and analyzed from the Late Bronze Age stratum of Tell Jenin (West Bank, Palestine). A total of 2922 (MNI) shells were identified from the stratum and found to be belonging to 44 species of terrestrial, freshwater and marine molluscs. Additionally, a total of 6200 unknown shell fragments were collected from the same stratum. Landsnail shells belonging to 18 species and having a total of 1974 specimens were clearly dominant over freshwater molluscs (17 species and 916 specimens) and marine Mediterranean molluscs (9 species and 32 specimens). Among terrestrial molluscs,* Calaxis hierosolymarum *and* Cecilioides genezarethensis *(family Ferrussaciidae) were the most dominant landsnail of that Age. Freshwater molluscs were dominated by* Semisalsa contempta *and* Pseudamnicola solitaria, *whereas Mediterranean snails were dominated by* Glycymeris violascens *and* Glycymeris pilosus. *The construction phase of the site (occupation phase) contained more specimens of molluscs than the destructive (abandonment) phase. The molluscan study and analysis revealed paleoclimatic variations during the Late Bronze Age. It also reflected different patterns of trade exchange, food sources, and the use of molluscan shells as artifacts, including their use as ornaments, in traditional and ritual activities by the inhabitants of the site.*

Keywords: *Tell Jenin, Late Bronze Age, Molluscan Fauna, Paleoclimatic Variations.*

Introduction

The Tell Jenin (Jenin city, West Bank-Palestine) (UTM 1785-2075) is located about 40 km east of the Mediterranean coast and 100 km north of Jerusalem (Ezzughayyar, Al-Zawahra, and Salem 1996; Glock 1987). This site is found at 150 m altitude (fig. 17-1). Its annual precipitation is about 500 mm/year, with an average temperature of 20° C (10° C in January and 30° C in August). According to reports from field inspectors of the Department of Antiquities during the British Mandate, Tell Jenin was first identified by Albright (1926) as the site of an ancient town on top of which was a modern cemetery and a threshing floor. The Tell is also known as Tell el-Nawar (Arabic word for gypsies), because of annual nomad encampments on the mound prior to 1948. Precise stratigraphic knowledge of the archaeological history of the site and the region of Jenin began in 1977 with Birzeit University salvage excavations at four sites of the Tell (Glock 1979; 1987). The excavations and surveys of site 4 (1980- 1983) indicate that the most ancient occupation of the Tell dated to Late Neolithic to early Chalcolithic period (7000-4000 BC).

Shells of mollusks from marine and freshwater environments have been used by humans since about 50,000 BP for food (Volman 1978), ornaments (Bar-Yosef 1989, 1991; Bar-Yosef Mayer 2005a, 2008; Biggs 1963; Claassen 1991; Goring-Morris 1989) and sometimes as

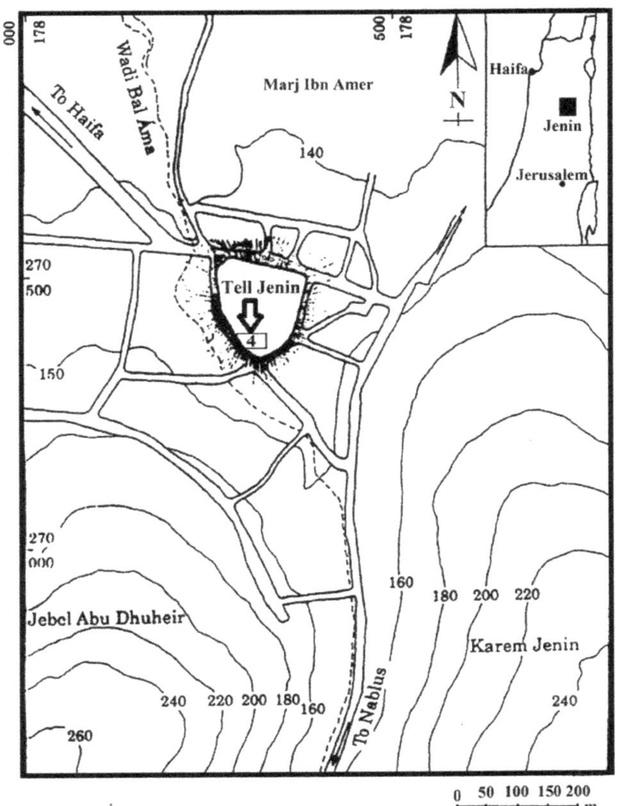

Figure 17-1. Sketch map showing the location ofTell Jenin and the archaeological Site 4 excavated by Birzeit University.

currency in trade (Reese 1991a). Shells are found in a great many archaeological sites, in many contexts, such as burials, food preparation and craft working areas, storage installations and refuse heaps. As ornaments they were used for necklaces, bracelets, clothing decorations or inlay in artifacts of other materials. Shells were probably collected directly from the sea or river or by exchange, or as gifts (Bar-Yosef Mayer 2008; Reese 1991a).

The Late Bronze Age stratum was chosen for analysis as it contains the majority of molluscan faunal remains in the site. Moreover, this stratum was deserted and faced severe erosion cycles (the destruction phase), separating the two occupation phases where ash and seeds were recovered.

The purpose of this study is to present a systematic and quantitative analysis of marine, freshwater shells and land snails, as each group may represent a different paleoclimatic line of evidence in prehistoric times. Following from this, the reconstruction of ecological or environmental parameters within the Late Bronze Age can be isolated. In addition, study of the molluscs could reveal the patterns of trade and exchange, food procurement, and traditional activities of the inhabitants of the site (1550-1200 BC).

Materials and Methods

Samples of the snail shell-bearing beds of the Late Bronze Age (stratum IV) from Site 4 were sieved under water with fine-meshed sieves (2 mm) and left to dry. About 300 samples of soil (8 liters each) were floated to retrieve imaterial for microanalysis. Shells (complete, broken, and fragments) obtained by sieving were picked out by hand for later analysis. The minimum number of individuals (MNI) was calculated by using the "state of preservation" and "shell part" factors. That is, the number of complete shells of each species was counted and the number of the most frequent shell part (apex, siphonal canal or umbo) was added to it.

Molluscan specimens were then identified and classified by taxa (genus and species). Statistical analysis, as well as provenance including area, locus, basket and period, were entered into database programs (D-base IV and Excel). Ordinarily the pottery excavated was washed, sorted and recorded each day. The outline of the occupational history of Tell Jenin is based on these field readings of the ceramic material. Nomenclature of the marine specimens was done according to Abbot and Dance (1982), Sharabati (1984), Tornaritis (1987) and Barash and Danin (1982). The nomenclature of the terrestrial and freshwater molluscs followed Mienis (1982, 1983 and 1986a), Schütt (1983), and Heller (1993). The identification of specimens to species level was finalized by comparing them to the molluscan collection at the Department of Biology (Birzeit University).

Results

The molluscan specimens derived from the Late Bronze Age stratum at Tell Jenin consisted of 2922 specimens (MNI) and 6200 unidentified fragments that were too small to be categorized

Molluscs	No. of specimen
Freshwater Gastropoda	
Family Thiaridae	
Melanoides tuberculata (Müller, 1774)	29
Family Melanopsidae	
Melanopsis buccinoidea (Olivier, 1801)	133
Melanopsis costata (Olivier, 1804)	74
Melanopsis cerithiopsis Bourguignat, 1856	13
Family Neritidae	
Theodoxus jordani jordani (Sowerby, 1836)	123
Family Cochliopidae	
Semisalsa contempta (Dautzenberg, 1894)	103
Semisalsa galilaea (Preston, 1913)	10
Family Hydrobiidae	
Pseudamnicola solitaria Tchernov, 1971	148
Orientalina gaillardoti (Bourguignat, 1856)	178
Islamia mienisi (Schütt, 1991)	3
Opercula of unspecified species	10
Family Valvatidae	
Valvata saulcyi Bourguignat, 1853	1
Family Planorbidae	
Bulinus truncatus (Audouin, 1826)	18
Planorbis planorbis antiochianus Locard, 1883	16
Gyraulus piscinarum (Bourguignat, 1852)	40
Freshwater Bivalvia	
Family Unionidae	
Unio terminais Bourgnignat, 1852	14
Total	**916**

Table 17-1. Taxonomic composition of freshwater mollusc shells from Site 4 in Tell Jenin from the late Bronze Age stratum.

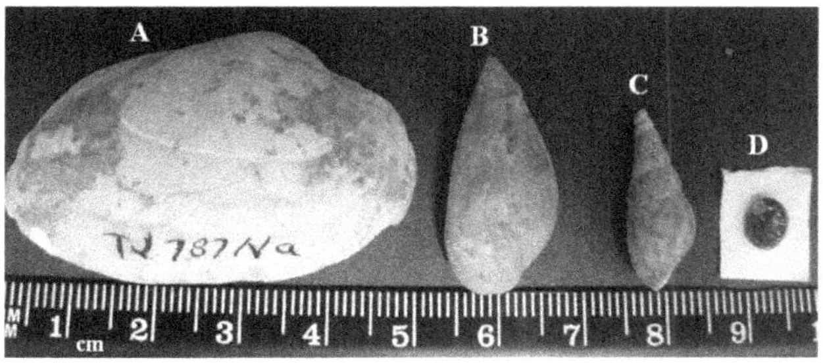

Figure 17-2. Some examples of freshwater mollusc shells: A: *Unio terminalis*; B; *Melanopsis buccinoidea*; C: *Melanoides tuberculata*; D: *Theodoxus jordani jordani*.

Molluscs	No. of specimen
Terrestrial Gastropoda (Land snails)	
Family Enidae	
Euchondrus septemdentatus (Roth, 1839)	126
Euchondrus sp.	41
Family Ferrussaciidae	
Calaxis hierosolymarum (Roth,1855)	791
Cecilioides genezarethensis Forcart, 1981	663
Cecilioides judaica (Mousson, 1861)	4
Family Pristolomatidae	
Vitrea contracta (Westerlund, 1871)	68
Family Sphincterochilidae	
Sphincterochila cariosa (Olivier, 1804)	1
Family Hygromiidae	
Trochoidea simulata (Ehrenberg, 1831)	12
Trochoidea tuberculosa (Conrad, 1852)	5
Xeropicta carmelensis Forcart, 1976)	7
Xeropicta vestalis (Pfeiffer, 1841)	50
Monacha syriaca (Ehrenberg, 1939)	30
Monacha obstructa (Pfeiffer, 1842)	32
Family Pyramidulidae	
Pyramidula rupestris hierosolymitana (Bourguignat, 1852)	30
Family Trisexodontidae	
Caracollina lenticula (Michaud, 1831)	51
Family Helicidae	
Levatina spiriplana caesareana (Mousson, 1854)	36
Helix engaddensis Bourguignat, 1852	27
Total	**1974**

Table 17-2. Systematic analysis of terrestrial mollusc shells from Site 4 in Tell Jenin from the late Bronze Age stratum.

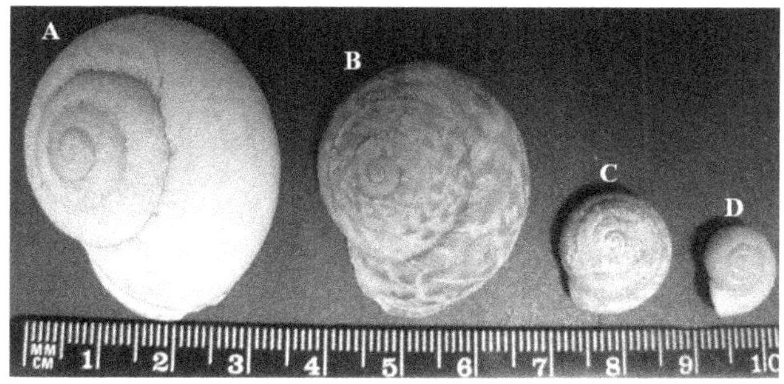

Figure 17-3. Some examples of landsnails. A: *Levantina spiriplana caesareana*; B: *Helix engeddensis*; C: *Sphincterochila cariosa*; D: *Monacha syriaca.*

Figure 17-4. Some examples of marine mollusc shells. A: *Cerastoderma glaucum*; B: *Conus mediterraneus* C: *Nassarius gibbosulus* D: *Phalium granulatum*; E: *Bolinus brandaris*; F: *Glycymeris violascens.*

Systematic analysis

In order to classify specimens to the species levels, shells were examined carefully by eye or with the help of a compound microscope (magnification 2x-4x). Results obtained are summarized in tables 17-1 to 17-3. In addition, some shells representing the three groups of molluscs were photographed and are shown in fig. 17-2 to 17-4. The distribution of specimens by habitat is shown in fig. 17-5. Manipulated shells are shown in fig.17- 6.

Molluscs	No. of specimen
Marine Bivalvia (Pelecypoda)	
Family Glycymerididae	
Glycymeris pilosus (Linnaeus, 1767)	6
Glycymeris violascens (Lamarck, 1819)	9
Family Cardiidae	
Cerastoderma glaucum (Poiret, 1789)	4
Marine Gastropoda	
Family Conidae	
Conus mediterraneus Hwass, 1792	3
Family Cassidae	
Phalium granulatum undulatum (Gmelin, 1791)	3
Family Nassariidae	
Nassarius gibbosulus (Linnaeus, 1758)	2
Family Muricidae	
Bolinus brandaris (Linnaeus, 1758)	2
Scaphopoda	
Family Dentaliidae	
Dentalium sp.	3
Total	**32**
Unidentified Fragments	6200

Table 17-3. Taxonomic composition of marine mollusc shells from Site 4 in Tell Jenin from the late Bronze Age stratum.

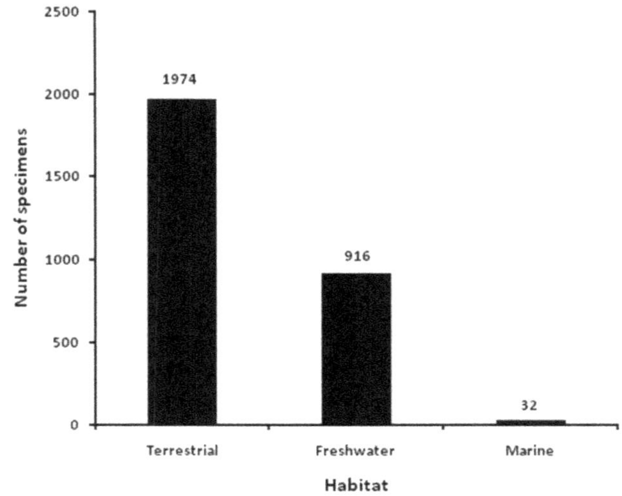

Figure 17-5. Distribution by habitat of molluscs found in the Late Bronze Age stratum of Tell Jenin.

Discussion

Land snails are the only group of molluscs which might have entered the site by themselves, and were not necessarily brought in by humans. Their dominance in this stratum may reflect a very short human occupation, since they are typical of deserted sites while marine and freshwater shells were more frequent in the construction phase (fig. 17-5). These observations are in accordance with those of Bar-Yosef and Heller (1987) and Ezzughayyar, Al-Zawahra, and Salem (1996). *Monacha* and *Xeropicta* species might have entered the site after it was deserted and covered by weeds. *Monacha* species are common today throughout Mediterranean regions, and where the climate is cooler and receives more rain in the hilly region, the shells become larger. *Monacha* (fig. 17-3D) is an annual species that lives only for one year and its size depends largely on the amount of rainfall during the winter period. Thus, it could reflect paleoclimatic variation in the studied prehistoric sites (Heller and Tchernov 1978). For example, the average diameter of *Monacha* shells from Tell Jenin archaeological site studied was 6.2 mm, whereas the present mean diameter of the snail's shell is 9.2 mm corresponding to 400 mm annual rainfall. This means that the climate of the Late Bronze Period was less moist.

Biggs (1963) recovered some fragments of *Monacha*, *Sphincterochila* and *Trochoidea* (*Xerocrassa*) from the Alter of the Sanctuary at En-Gedi and dated them to the 4th Millennium BC. Schütt (1983) has described a closely related species, *M. syriaca,* from various sites in Jordan like north Shuna, Ajlun and Irbid.

Pyramidula hierosolymitana has been found in Cyprus, Syria, Lebanon, Palestine, Israel and Jordan. It is widely distributed in the mountains, especially lichen and algae covered sides of the rocks, where the snails are camouflaged by a thin layer of mud covering their shells (Mienis 1986a).

Calaxis and *Ceciliodes* species are tiny subterranean land snails which were concentrated in specific loci (116, 189, 213, 220 and 240) of abandonment phases of the site. These loci are tombs that contain a large quantity of bone fragments. Mienis (1992a) had noticed that, *Calaxis* species were attracted to the fungus growing on decaying bones and were collected from a human skull at Khirbet Amirat (Israel). *Caracollina lenticula* is a small lentiform species found in the site. Its general distribution is circum-Mediterranean from the Cape Verde Islands to Palmyra. It is a relict of the Pleistocene and post-pluvial past, when the Mediterranean climatic zone extended further inland from the Levantine Coast (Schütt 1987). However, the activities of rodents (like the white-toothed shrew) and birds cannot be excluded; since predation on freshwater molluscs, like *Melanopsis* and land snails, like *Monacha*, by birds and rodents has been frequently noted (Mienis 1992b; 1996).

Euchondrus species have a large, discontinuous geographical distribution. They seem to be confined primarily to sandy areas in the coastal plains of Israel and

Egypt (Bar 1992). *E. spetemdentatus* was abundant in the destruction phases of Tell Jenin. This species could prevent or exclude other species like *E. sulcidens* or *E. ovularis* from inhabiting the site, since it has come down from the hills and occupied the "Qurqar" hills in the northern of Palestine.

Xerocrassa species are subterranean snails occurring in the drier parts of the deserts in Israel, like the coastal area of the Dead Sea, the Negev and the adjoining areas in Sinai. Many species of *Xerocrassa* were described in archeological sites of the Levant, and found in the destruction and abandonment phases the Late Bronze Age stratum of Tell Jenin (Ezzughayyar, Al-Zawahra, and Salem 1996).

Melanopsis species are abundant in the site; *M. buccinoidea* (fig. 17-2, B) is distributed all over the eastern Mediterranean region and Mesopotamian basin where there are several subspecies. *M. certhiopsis* is considered to be a relict of the *Melanopsis* inhabiting the Pliocene connection between Euphrates and Jordan Valley (Schütt 1987). *Melanopsis cerithiopsis* is sympatric with *M. buccinoidea* in most of the streams and sources in the lower Jordan valley and Emeq Bet Shean, (except for the Jordan River, which is inhabited by *M. costata*), without showing any interbreeding (Mienis 1983).

Since the *Melanopsis* species of Tell Jenin were concentrated in the construction phase of the Late Bronze Age stratum, it is reasonable to conclude that these freshwater snails were brought within the mud from the river banks (Al-Muqatta' River) used for mudbrick manufacturing or building. The idea that freshwater shells were brought into the site with mudbricks was first suggested by Bar-Yosef and Heller (1987). This coincides with the observations of Ezzughayyar, Al-Zawahra, and Salem (1996). *Theodoxus (Neritaea) jordani* is a very abundant freshwater prosobranch in our flotation samples, especially in the construction phases of the site (fig. 17-2D). This species occurs mainly in flowing or turbulent water bodies, springs, rivers and lakes. *Theodoxus* is a suitable object for studies in paleozoo-geography of the Middle Eastern countries (Roth 1987). The faunal history of the genus *Theodoxus* suggests that its origin lies on the Eurasian plate, whereas the subgenus *Neritaea* originated from the northern part of the African plate, which in Jurassic times included the so-called Apulian microplate (Biju-Duval, Dercourt, and Le Pichon 1976). The distribution of *Theodoxus* populations could be explained by derivation from snails inhabiting brackish sections of the coastal belt of the Aegean and Marmara basins. In this way brackish water basins serve as pathway from regions at the former southern border of the Eurasian plate into regions of the Apulian plate (Roth 1987).

The presence of considerable number of hydrobiid snails in the site reflects the richness of brackish springs in the area, since these very small snails inhabit fresh water springs and probably live in the stenothermic parts of the outflow of subterranean waters. So, it was reasonable to find these snails in the construction phases of Tell Jenin (mudbrick construction). The opercula that were found in the construction phases of the site were related to the hydrobiid species. *Levantina* and *Helix* species (fig. 17-3A & B, respectively) were the largest in size of the land snails recovered at this site, perhaps indicating that these species could be dietary items for the people of that time. The importance of these snails is related to their sensitivity to climatic conditions which is reflected in the dimensions of their shells. Bar-Yosef Mayer (2005b) had explained that climate and seasonality reconstructions are typically undertaken when studying middens. In many archaeological sites, land and freshwater snails are useful in reconstructing past environments.

The presence of marine shells (fig. 17-4) in the Late Bronze Age stratum supports the idea that shellfishing was practiced by the inhabitants of the site. Since shellfish perish very rapidly and have to be consumed shortly after gathering, most of shell debris would probably not be taken back to the habitation site (40 km from the Mediterranean shore), but rather consumed on the sea shore. This might explain the small quantity of edible shellfish in the site. It is not clear whether they were collected directly by the Tell inhabitants or by exchange with other groups who lived closer to the shores. Of more central importance to local diet were the domestic animals (sheep, goat and cattle) that were an important source of food and raw material in the Bronze Age economy of the southern Levant (Bar-Yosef Mayer 2005a; Horwitz and Tchernov 1989). Marine shells have been used for cultural purposes in the southern Levant since Paleolithic times (Bar-Yosef 1989; Reese 1991b).

The common Mediterranean marine bivalve *Glycymeris* was more abundant than other bivalves at the site. All specimens were perforated at umbo region (fig. 17-6B). These bivalves were perforated either naturally or by people for wearing as necklaces, since they are strong shells (Bar-Yosef Mayer, Vandermeersh, and Bar-Yosef 2009). Biggs (1963; 1969) suggested that *Glycymeris* symbolize the moon or the moon deity, while *Cardium (Cerastoderma)* with the ridges represent the sunrays (fig. 17-4A).

At Tell Ta'annek (10 km north of Tell Jenin), there is a sharp increase in number of perforated *Glycymeris* in the Bronze Age, which might indicate a rise in the use of shell beads as a social status symbols (Ezzughayyar and Al-Zawahra 1996). *Unio terminalis* was the only freshwater bivalve found at Tell Jenin (fig. 17-2A). The specimens were fragile and tend to break readily, which explain the large number of fragments. These bivalves might have been embedded in mud brought in for brick manufacture or building. The activities of rodents or birds should also be considered (Mienis 1992a).

The presence of Red Sea *Dentalium* species at Tell Jenin

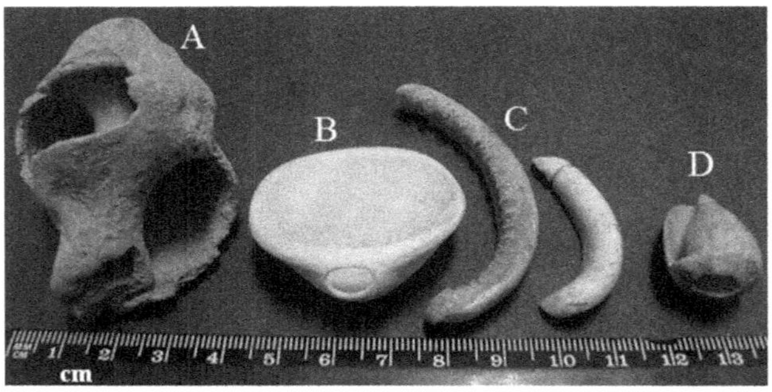

Figure 17-6. Modified shells. A: *Bolinus brandaris*; B: *Glycymeris violascens*; C: *Cassid lips*; D: *Conus mediterraneus*.

supports the idea that these shells were used as exchange items between southern and northern communities. The *Dentalium* specimens are found in burials decorating more males than females (Bar-Yosef Mayer 2005a; Belfer-Cohen 1988).

Muricid shells include the Mediterranean Basin gastropod *Bolinus brandaris* (fig. 17-4E), known for its association with purple dye production in Roman times. This ancient industry dates back to 1600-1500 BC in Northern Syria and Greece (Reese 1987). This is not the case of Tell Jenin, since thousands of *Murex* specimens would be needed to extract a small amount of dye (Spanier 1986).

The shells of *Nassarius* (fig. 17-4C) species have been used since the Neolithic by men and women as an expression of beauty or for some magical and ritual purposes (Bar-Yosef Mayer 2005a; Reese 1991a).

The lips of *Phalium* shells (fig. 17-6C) are known from a number of Near Eastern and Mediterranean archaeological sites; most of these lips can be seen as personal ornaments or offerings in graves or sanctuaries (Reese 1989).

Conus mediterraneus specimens found in the site had holes at the apical region, so we suggest their use as necklaces (fig. 17-6D). *Conus* shells with apical holes are found at a number of sites in Israel, including Beth Shan, Ain Shems and Megiddo (Bar-Yosef Mayer 2005a; Reese 1986).

It is customary to assume that naturally accumulated paleontological assemblages reflect the fauna that once existed in an area, whereas archaeological assemblages reflect human activities, which are not necessarily local, and therefore cannot be used for environmental reconstruction. However, if one can separate species that naturally accumulate at a site from those brought in by humans, then, one can use those species in environmental reconstruction and compare the conditions under which species exists today (Brewer 1992; Claassen1991).

The Southern Levant exhibits remarkable spatial variability in climatic and vegetation zonation (Zohary 1981). This mosaic environment is clearly reflected by the regional faunal diversity (Tchernov 1979). The various groups of marine, freshwater, and terrestrial snails and their distribution represent changing ecological and environmental parameters thus shedding light on human social and economic activities within the Late Bronze period of Tell Jenin.

Conclusions

The molluscan fauna of Tell Jenin accumulated in the Late Bronze Age stratum during the human occupation period of the site. The dominance of landsnails may reflect a very short human occupation or the desertion of the site. However, the dominance of the freshwater snails during the construction phase is thought to be due to their presence in the mud used for mudbrick building. The presence of marine shells in Tell Jenin supports the idea that collection of dead shells was familiar to the inhabitants of the site rather than shellfishing as an economic practice. The presence of perforated shells in the site indicates their use as beads or pendants. Finally, the small shell size of some annual molluscs, compared to present size of these molluscs, indicates less humid climatic conditions during the Late Bronze Age.

Acknowledgements

We are very grateful to Dr. L. Glock from the Palestinian Institute of Archaeology for making the faunal material of Tell Jenin available. We also thank Mr. Ibrahim Iqtait for drawing fig. 17-1.

References

ABBOT, R.T. DANCE S.P. 1982. *Compendium of Seashells.* Dutton Inc, New York.

ALBRIGHT, W.F. 1926. The topography of the tribe of Issachar. *Zeitschrift für die Alttestestamenialische Wissenschaft* 44:225-236.

BARASH, A. and Z. DANIN 1982. Mediterranean mollusca of Israel and Sinai: Composition and distribution. Israel *Journal of Zoology* 31:86-118.

BAR, Z.M. 1992. The geographical distribution of *Euchondrus sulcidens* (Mousson, 1861). *Argamon* 13:1-8.

BAR-YOSEF, D.E. 1989. Late Paleolithic and Neolithic marine shells in the Southern Levant as cultural markers. *Proceeding of the 1986 shell bead Conference* C.F. Hayes (ed.), Rochester Museum and Science Center, Research Records, 20:169-174.

——. 1991. Changes in the selection of marine shells from the Natufian to the Neolithic. In: Bar-Yosef,O.& Valla, F.R.(eds.): *The Natufian Culture in the Levant:* 629-636. International Monographs in Prehistory. Ann Arbor, Michigan.

Bar-Yosef, D.E. and J. Heller. 1987. Mollusca from Yiftahel lower Galilee, Israel. *Paleorient* 13:131-135.

Bar-Yosef Mayer, D.E. 2005a. The exploitation of shells as beads in the Paleolithic and Neolithic of the Levant. *Paleorient.* 31(1):176-185.

——. 2005b. Archaeomalacology: molluscs in former environments of human behavior. *Proceedings of the 9th Conference of the International Council of Archaeology, Durham, 2002.* Bar-Yosef, D.E. (Ed), Oxford Books, Oxford:132-147.

——. 2008. Invertebrate Analysis In: *Encyclopedia of Archaeology*, ed. by Deborah M. Pearsall. Academic Press, New York.

Bar-Yosef Mayer, D.E., B. Vandermeersh, and O. Bar-Yosef. 2009. Shells and ochre in Middle Paleolithic Qafzah Cave, Israel: Indications for modern behavior. *Journal of Human Evolution* 56 (3):307-314.

Belfer-Cohen, A. 1988. *The Natufian settlement at Hayonim Cave.* Unpublished Ph.D. thesis, The Hebrew University, Jerusalem.

Biggs, H.E.J. 1963. On the mollusca collected during the excavations at Jericho 1952-1958, and their archaeological significance. *Man* 153:125-128.

——. 1969. Molluscs from human habitation sites, and the problem of ethnological interpretation: In: Brothwell, D. & Higgs, E. (eds.): *Science in Archaeology:*423-427. Thames and Hudson, London.

Biju-Duval, B., J. Dercourt, and X. Le Pichon. 1976. La genese de la Mediterranee. *La recherche* 71:811-822.

Brewer, D.J. 1992. Zooarchaeology, method, theory and goals. *Archeological Method and Theory* 4:195-244.

Claassen, C. 1991. Normative thinking and shell-bearing sites. *Archaeological Method and Theory* 3:249-298.

Ezzughayyar, A. and M. Al-Zawahra. 1996. Molluscan fauna from Tell Ta'annek (Northern West Bank-Palestine). *Archaeozoologia* VIII/1.2:71-88.

Ezzughayyar, A., M. Al-Zawahra, and H. Salem. 1996. Molluscan fauna from site 4 of Tell Jenin (northern West Bank-Palestine). *Journal of Archaeological Science* 23:1-6.

Glock, A.E. 1979. Tell Jenin. *Revue Biblique* 86:110-112.

——. 1987. Tell Jenin excavations 1977-1983. Birzeit *Research Review* 4:4-30.

Goring-Morris, A. N. 1989. Sociocultural aspects of marine mollusk use in the terminal Pleistocene in the Negev and Sinai regions of the Southern Levant. In *Proceeding of the 1986 Shell Bead Conference,* C.F. Hayes (Ed), Rochester Museum and Science center, *Research Records*, 20:175-187.

Heller, J. and E. Tchernov. 1978. Pleistocene land snails from coastal plain of Israel. *Israel Journal of Zoology* 27:11-19.

Heller, J. 1993. *Land snails of the land of Israel: Natural history and a field guide.* The Ministry of Defense, Israel. 271pp.

Horwitz L.K. and E. Tchernov. 1989. Animal exploitation in the Early Bronze Ages of the Southern Levant, in: *L' urbanisation de la Palestine à l'age du Bronze ancien.* P.D. Miroscedji (Ed.). Oxford, BAR International Series, 527:279-296.

Mienis, H.K. 1982. A checklist of the land molluscs of Israel and the Administered Areas. *Levantina* 39:457-465.

Mienis, H.K. 1983. A preliminary checklist of freshwater molluscs of Israel and the Administered Areas. *Levantina* 47:543-450.

——.1986a. A revised checklist of the brackish and freshwater molluscs from Israel and the Administered Areas. *Levantina* 63:675-682.

——.1986b. Notes on molluscs from Cyprus: A brief note concerning the family Pyramidulidae. *Levantina* 62:673-674.

——.1992a. Land snails in a human skull from Khirbet Amarit, Israel. *The Papustyla* 6(4):46.

——.1992b. Additional notes concerning predation on landsnails by white-toothed shrews in Israel. *Malacological Newsletters* 11:47-50.

——.1996. Selective predation on *Monacha haifaensis* (Mollusca: Gastropoda) by *Crocidura suaveolens* (Mammalia: Insectivora). *Nachrichtenblatt der Ersten Vorarlberger Malakologischen Gesellschaft* 4:43-46.

REESE, D.S. 1986. *The marine and freshwater shells, in: The Late Bronze and Iron Ages of Central Transjordan:* The Baq'ah valley project, 1977-1981 P. Mc Govern (ed.), Monograph 65 Philadelphia, The University Museum XIII:320-332

——. 1987. The Mediterranean shell purple–due industry. *American Journal of Archaeology* 90(2):183.

——. 1989. On Cassid lips and Helmet shells. *Bulletin of the American Schools of Oriental. Research* 257:33-39.

——. 1991a. The trade of Indo-Pacific shells into the Mediterranean Basin and Europe. *Oxford Journal of Archaeology* 10(2):159-196.

——. 1991b. Marine shells in the Levant: upper Paleolithic, Epipaleolithic and Neolithic, in: *The Natufian Culture in the Levant,* O. Baryosef and F.R. Valla (Eds), International Monographs in Prehistory, Ann Arbor Archaeological Series 1, 613-628

ROTH, J.R. 1987. Data on the distribution and faunal history of the Genus *Theodoxus* in the Middle East (Gastropoda: Neritidae). In: F. Krupp, W. Schneider & R. Kinzelbach (eds.). *Proceedings of the Symposium on the Fauna and Zoogeography of the Middle East Mainz 1985.* Wiesbaden. Beihefte zum TAVO. A 28, 73-79.

SCHÜTT, H. 1983. Die Molluskenfauna der SuBwasser in Einzugsgebiet des Orontes unter Berucksightigung benachbarter FluBsysteme. *Archiv fur Mollusken funde* 113:Pp 17- 91, 225-228.

SCHÜTT, H. 1987. The Molluscs of the Oasis Palmyra. In: F. Krupp, W. Schneider and R. kinzelbach (eds.), *Proceedings of the Symposium on the Fauna and Zoogeography of the Middle East,* Mainz 1985. Wiesbaden. Beihefte zum TABO, A 29. Pp 62-72.

SHARABATI, D. 1984. *Red Sea Shells*. KPI, London, Boston, Melbourne, Hentey, 128 pp.

SPANIER, E. 1986. Cannabilism in Muricid snails as a possible explanation for archaeological findings. *Journal of Archaeological Science* 13:463-468.

TCHERNOV, E. 1979. Quaternary Fauna. In: A. Horowitz (Ed.). *The Quaternary of Israel.* New York, Academic Press. Pp 259-290.

TORNARITIS, G. 1987. *Mediterranean Sea shells*, Nicosia, G. Tornaritis, 190 pp.

VOLMAN, T.P. 1978. Early archaeological evidence for shellfish collecting, *Science* 201:911-913.

ZOHARY, M. 1981. On the flora and vegetation of Middle East structure and evolution. In: W. Frey & H.-P. Uerpmann (eds.), *Beitrage zur Umweltgeschichte des vorderen orients.* Wiesbaden. Ludwig Reichert. Pp. 1-25.

18 - ACQUISITION AND MANAGEMENT OF MARINE INVERTEBRATE RESOURCES AT A PRE-ROMAN COASTAL SETTLEMENT : THE SITE OF DOSSEN ROUZ (LOCQUÉMEAU-TRÉDREZ, BRITTANY, FRANCE)

Caroline MOUGNE,
UMR 6566 CReAAH (Centre de Recherche en Archéologie, Archéosciences, Histoire, bâtiment 24-25, Université de Rennes 1-CS74205, 35042 Rennes Cedex, France,
caroline.mougne@free.fr

Catherine DUPONT,
CNRS, UMR 6566 CReAAH,
catherine.dupont@univ-rennes1.fr

Anna BAUDRY,
INRAP «Institut national de recherches archéologique préventives», UMR 6566 CReAAH
(Base Archéologique de Poitiers, 122 rue de la Bugellerie, 86000 Poitiers, France,
anna.baudry-dautry@inrap.fr

Laurent QUESNEL,
CNRS, UMR 6566 CReAAH,
laurent.quesnel@univ-rennes1.fr

Marie-Yvane DAIRE.
CNRS, UMR 6566 CReAAH,
marie-yvane.daire@univ-rennes1.fr

Abstract: *The Dossen Rouz settlement, located on the seashore of the commune of Locquémeau-Trédrez, (Brittany, North western France) was dedicated to the production of sea salt during the Late Iron Age (3rd-2nd cent. BC). This archaeological site has revealed the exploitation of a large spectrum of animal species (shells, crustaceans, echinoderms, fishes and mammals). The archaeological deposit has been modified by modern incursions of seaborne shell. Distinguishing between the natural and anthropic accumulations is essential to understanding how the deposit was created. In this paper we attempt to distinguish the natural deposits with the aim of obtaining some reliable data on the Iron Age diet at Dossen Rouz. The variety and quantity of archaeological marine invertebrate remains encountered is significant for the Iron Age at Dossen Rouz. For shells and especially sea urchins, rarely identified in archaeological contexts, metric quantitative and taphonomic studies, have allowed us to develop new approaches, such as acquisition methods.*

Keywords: *Natural Deposit, Iron Age, Shellfish, Sea Urchin, Subsistence Strategies.*

Introduction

The subsistence economy of populations settled in the western part of France during the Iron Age remains poorly understood, particularly as regards marine resource use on the coast of the Channel (Bizien-Jaglin and Laperas 2004; Campbell 2005; Campbell 2007, 2008-a; Dupont 2006-a) and of French Atlantic Ocean (Dupont 2009; Dupont 2011; Germinet 2009:294-319; Gruet 1998; Méniel 2008). This subject is a little better known on the French Mediterranean coast (Bardot 2010; Brien 1988, Brien-Poitevin 1992, 1993) and some studies have been undertaken in other European countries (for example: Bejega García *et al.* 2010; Berghe, Gleba and Mannering 2009; Cabral and Silva 2003; Cunliffe and Hawkins 1988; Mallory and Mc Cormick 1988; Mc Cormick *et al.* 1996; Theodoropoulou 2008). Thus, the study of Dossen Rouz, a pre-Roman site located on the north coast of Brittany in the commune of Locquémeau-Trédrez (Côtes d'Armor), presents an opportunity to better understand the way of life of people living during the Iron Age.

This site had a dual function, indicated by the presence of both domestic structures and others dedicated to the production of sea salt, dated to 3rd-2nd cent. BC (fig. 18-1). Over many centuries, the Dossen Rouz site has been within the reach of high tides, causing erosion and damage to part of the remains (Daire 2011). The presence of marine invertebrates (serpulids and cirripeds like barnacles) on artefacts (ceramics, briquetage (ceramic saltpans) and attached to the inside of some shells), confirms that the site was temporarily submerged by the sea. Several modern activities are known on the coast of Locquémeau. Late 19th and early 20th century photographs illustrate the burning

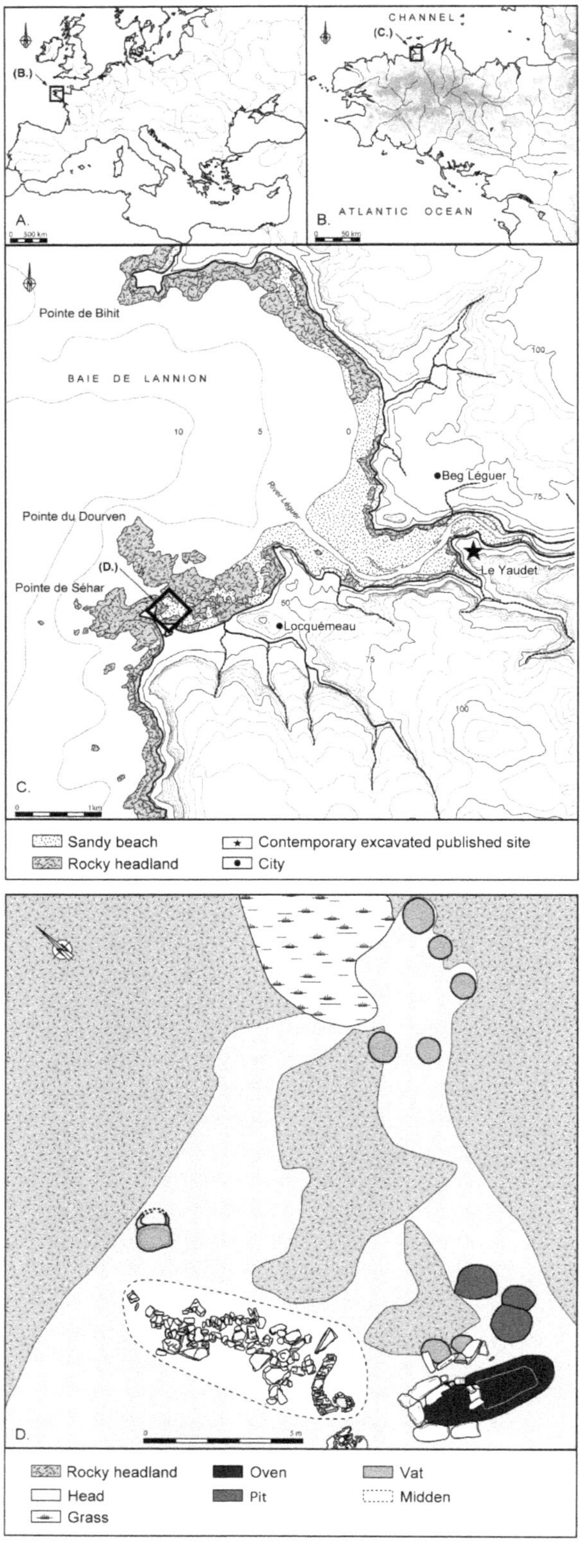

Figure 18-1. Location of the archaeological site of Dossen Rouz (CAD L. Quesnel).

of seaweed to produce soda and the drying of fishing nets on the Pointe de Séhar (Daire 2011). To distinguish the Late Iron Age (LIA) horizon and obtain reliable data on the subsistence strategies of the Dossen Rouz community, and also to determine the stratigraphic sequence of the site, a study of the modern shells was first undertaken. For the Iron Age period, the aim of these analyses is mainly to understand the difference in content between the shell assemblages of the occupation and potential post-abandonment phases to define the modes of subsistence of people and activities associated with the exploitation of marine resources (fishing and the gathering of shells and echinoderms). The results contribute to an understanding of the site and of its economy. The archaeological site of Dossen Rouz is, from this perspective, not only interesting in terms of the number and the variety of shells present but also for the presence of sea urchins, rarely discovered in such large quantities on the coastal assemblages (Mougne and Dupont 2011).

Materials and Methods

During the limited period of excavation of the Dossen Rouz site, carried out under the direction of M.-Y. Daire, a fieldwork protocol was established for the recovery of archaeozoological remains: collection was visual whilst excavating for larger and better preserved specimens and in the case of ecofact concentrations sieving with fine meshes (2 mm) and sorting of sedimentary samples in the laboratory.

Two sedimentary samples from the Iron Age horizon (20 litres in total), one from the midden and one from the modern horizon (10 litres in total) were taken. This protocol allowed us to identify species that have been subject to significant fragmentation or which are too small to be easily seen. It showed that the sediments found inside the larger shells, such as limpets (*Patella* sp.) and abalone (*Haliotis tubercula tubercula*), contained small invertebrate species as well as archaeozoological remains. The data revealed a wide faunal spectrum, including shells, sea urchins, mammals, fishes and crabs. Identification of the marine invertebrates (gastropods, bivalves, crustaceans and sea urchins) was carried out with the comparative collection from the Archéosciences Laboratory of Rennes (Comparative collection Gruet & Dupont, UMR 6566-CReAAH). Identifications were confirmed using reference works on molluscs (gastropods and bivalves) and crustaceans (Audibert and Delemarre 2009; Fretter and Graham 1977, 1984; Poppe and Goto 1991, 1993; Quéro and Vayne 1998), and on echinoderms (Hayward and Ryland 1995; Mortensen 1943; Southward and Campbell 2006).

Scientific names used for the marine molluscs identified correspond to the international standards of the CLEMAM (2011). Several methods of counting were used to calculate the relative proportions of each species. The NISP (Number of Identified Specimens; Grayson 1984) corresponds to all the identified remains greater than 1 mm. Minimum Number of Individuals (MNI) was used for shells and sea urchins. For the spiral gastropods calculating the MNI depended on the presence of the peristome (Dupont 2006-b). For the conical gastropods, an individual was counted if examination showing an apex and a part of the muscle scar located inside the shell (Dupont 2006-b). For bivalves, a MNI was obtained after assigning valves to left or right according to the position of the teeth in the hinge, the ligament and the pallial sinus line (Dupont 2006-b; McCarthy, Finlay and Mc Clean 1999). Echinoderms are represented by only one species of sea urchin (*Paracentrotus lividus*). Every complete test contains ten hemipyramids and ten auricles (5 left and 5 right for each) (fig. 18-2) (Hayward and Ryland 1995). MNI was calculated using these two anatomical parts. The *rotula* (five per test) and the epiphyses (ten per test) could have been used for the MNI calculations, but the overall numbers were too small (Campbell 2008-a, 2008-b; Gruet 2003; Mougne and Dupont 2011). Next, the remains of each species were weighed (in grams). Finally, shells were measured with a digital calliper graduated in millimetres (0.01) according to the procedures established by C. Dupont (2006-b) and sea urchins to those established by Y. Gruet (2003). At least 60 individuals were selected randomly and measured for each square and horizon when possible.

Identified species

Thirty four species of mollusc, one species of echinoderm and one species of crustacean were identified in the Late Iron Age and modern horizons (table 18-1). Precision is necessary for echinoderm determination. Regular standard sea urchin test is characterized by a particular number of pores pairs. The number, the shape and the placement of pores pairs constitute a key to identify echinoderm species. *Paracentrotus lividus* and *Strongylocentrotus droebachiensis* are two sea urchins that are present on the Channel coast and have five or six pores pairs (Hayward and Ryland 1995; Mortensen 1943; Southward and Campbell 2006) (fig. 18-2-A). Nowadays, *Strongylocentrotus droebachiensis* does not exist on the seashore next Locquémeau. This species lives in more northerly areas of Europe. However, we took it into account in case of climatic variations and of modification its biogeographic distribution. Some authors (Campbell 2008-a, 2008-b; Mortensen 1943) use a specific anatomical part, the auricle (fig. 18-2-C), to distinguish these two species. Unfortunately auricles found on the site of Dossen Rouz are poorly preserved and the two sea urchins species cannot be differentiated. Thus, the distinction was made by considering the characters of the ambulacra and the median interambulacran line and the global shape of the test (fig. 18-2-B).

Natural and cultural accumulations

Some phenomena create accumulations of shells particularly at coastal sites (Claassen 1998). It is not uncommon for archaeologists to report difficulty in disentangling the natural or anthropic origin of such

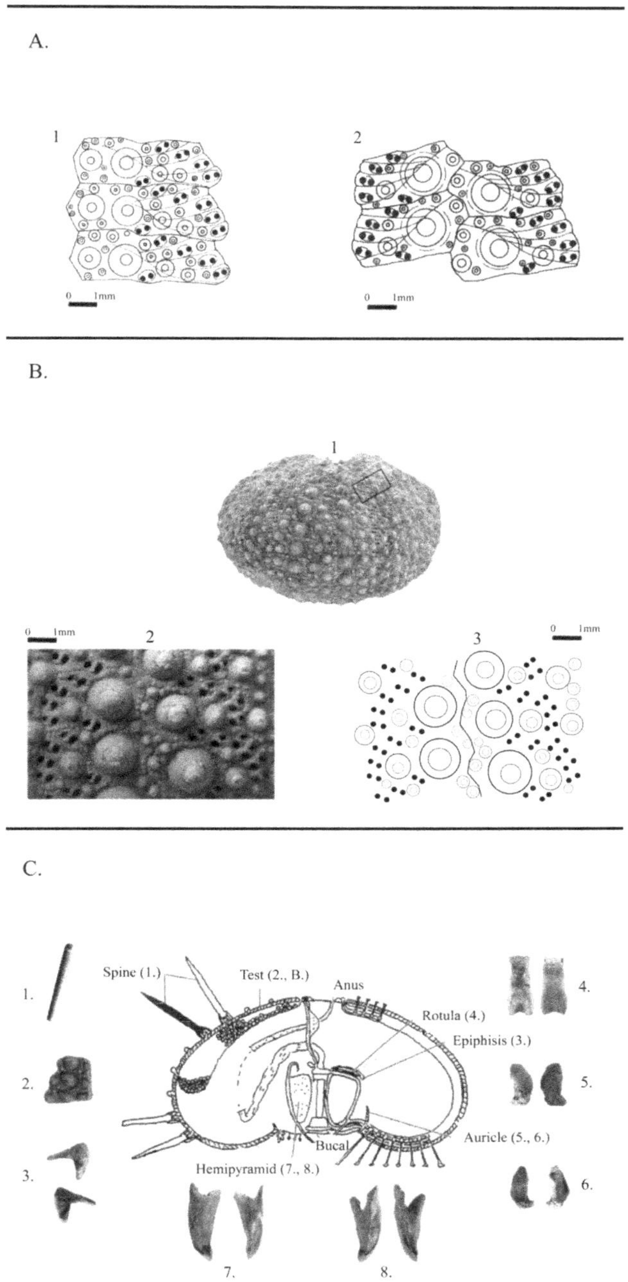

Figure 18-2. Determination and description of the violet sea urchin (*Paracentrotus lividus*).
A : Test plates of *Strongylocentrus droebachiensis* (1) and of *Paracentrotus lividus* (2); A.1 and A.2 : (after Mortensen, 1943; Southward and Campbell, 2006); **B** : Test of *Paracentrotus lividus* of the Dossen-Rouz site at Late Iron Age horizon; B.1 : Test of *Paracentrotus lividus* of the Dossen Rouz site at Late Iron Age horizon (L=48mm); B.2 : Close-up of the test in order to identify the species; B.3 : Schematic drawing of close-up B.2; **C** : Anatomical parts description of *Paracentrotus lividus*; C.1 : L=11mm; C.2 : L=7mm; C.3 : L: 6mm; C.4 : L=6mm; C.5 : right L=8mm; C.6 : left L=8mm; C.7 : left L= 14mm; C.8 : right L = 14mm (after Jammes 1904) (CAD C. Mougne).

		Modern horizon						Late Iron Age horizon					
Latin name	**Common name**	**NISP**	**% NISP**	**MNI**	**% MNI**	**Weight (g)**	**% Weight**	**NISP**	**% NISP**	**MNI**	**% MNI**	**Weight (g)**	**% Weight**
Gastropods:													
Patella sp.	limpet	17	1.33	3	0.29	13.32	2.64	20028	88.11	4992	74.27	15990.06	90.93
Osilinus lineatus	thick topshell							6	0.02	6	0.09	15.53	0.09
Gibbula umbilicalis	flat top shell	308	24.19	308	30.28	133.03	26.30	388	1.70	385	5.72	190.09	1.08
Gibbula pennanti	top shell	241	18.93	241	23.7	157.82	31.22	414	1.82	412	6.13	288.51	1.65
Gibbula cineraria	grey top shell	84	6.60	84	8.26	28.60	5.66	63	0.28	63	0.94	30.46	0.17
Gibbula magus	giant top shell							1	0.01	1	0.02	2.73	0.02
Gibbula spp.	top shell	216	16.97	46	4.52	21.34	4.22	249	1.09	103	1.53	28.97	0.16
Jujubinus exasperatus								1	0.01	1	0.02	0.14	0.01
Jujubinus sp.								1	0.01	1	0.02	0.13	0.01
Littorina obtusata	flat periwinkle	232	18.22	219	21.53	90.40	17.89	293	1.29	284	4.22	123.82	0.70
Littorina littorea	common periwinkle	20	1.57	16	1.57	36.86	7.29	89	0.39	74	1.10	128.82	0.73
Littorina saxatilis	rough periwinkle	1	0.08	1	0.10	0.94	0.19	1	0.01	1	0.02	0.22	0.01
Littorina sp.	periwinkle	4	0.31	2	0.20	0.58	0.11	8	0.03	5	0.07	0.18	0.01
Lacuna pallidula		22	1.73	22	2.16	3.03	0.60	17	0.07	16	0.24	2.23	0.01
Nucella lapillus	dogwhelk	5	0.39	4	0.39	2.42	0.48	20	0.09	18	0.27	18.05	0.10
Trivia monacha	European cowrie							1	0.01	1	0.02	0.01	0,00
Nassarius incrassatus	thick-lipped dog	2	0.16	2	0.20	0.17	0.03	18	0.08	16	0.23	2.58	0.01
Nassarius reticulatus	netted dog whelk	2	0.16	2	0.20	1.40	0.28	16	0.07	13	0.19	10.99	0.06
Nassarius sp.		1	0.08	1	0.10	0.06	0.01	8	0.03	7	0.10	0.21	0,00
Haliotis tuberculata tuberculata	abalone	3	0.23	1	0.10	3.94	0.78	169	0.74	41	0.61	505.45	2.87
Calliostoma zizyphinum	painted top shell	2	0.16	2	0.20	1.39	0.27	8	0.03	8	0.12	12.01	0.07
Calliostoma sp.		2	0.16	2	0.20	0.41	0.08	2	0.01	2	0.03	0.05	0.01
Ocenebra erinaceus	sting winkle	1	0.08	1	0.10	0.09	0.02	14	0.06	12	0.18	9.10	0.05
Bittium reticulatum	needle whelk	30	2.36	27	2.65	0.12	0.02	114	0.5	113	1.68	1.55	0.01
Rissoa parva		2	0.16	2	0.20	0.01	0.01	15	0.07	15	0.22	0.08	0.01
Cingula trifasciata								2	0.01	2	0.03	0.01	0.01
Mangelia coarctata								2	0.01	2	0.03	0.17	0.01
Tricolia pullus pullus	pheasant shell	20	1.57	20	1.96	1.55	0.30	16	0.07	16	0.23	1.06	0.01
Trochidae								1	0.01	1	0.02	0.02	0.01
Unspecified gastropods								55	0.24	26	0.38	1.57	0.01
Bivalves:													
Ostrea edulis	flat oyster	1	0.08	1	0.10	2.65	0.52						
Mytilus edulis	mussel	45	3.53	5	0.49	4.35	0.86	668	2.90	65	0.94	171.26	0.98
Pecten maximus	great scallop							2	0.01	1	0.02	21.32	0.13
Cerastoderma edule	common cockle							5	0.02	1	0.07	1.97	0.01
Cerastoderma sp.	cockle							6	0.02	1	0.03	0.36	0,00
Ruditapes decussatus	European carpet clam	2	0.16	1	0.10	0.20	0.04	20	0.09	1	0.07	7.44	0.05
Venus verrucosa	warty venus							1	0.01	1	0.02	5.59	0.03
Venerupis aurea	golden carpet	3	0.23	1	0.10	0.31	0.06	1	0.01	1	0.02	0.19	0.01
Tapes sp.	carpet shell							1	0.01	1	0.02	0.01	0.01
Anomia ephippium	saddle oyster	2	0.16	1	0.10	0.32	0.06						
Dosinia sp.								1	0.01	1	0.02	1.55	0.01
Glycymeris sp.	bittersweet							1	0.01	1	0.02	9.67	0.05
Veneridae		1	0.08		0.10	0.10	0.02	2	0.01	1	0.02	0.03	0.01
Unspecified bivalve		4	0.32		0.10	0.23	0.04						

Molluscs total		1273		1015		505.60		22728		6712		17584.20	
Cancer pagurus	edible crab							1		1		0.02	
Balanus sp.		13		1		1.02		64				1.33	
Unspecified Crab								6		1		0.05	
Crustacean total		13		1		1.02		71		1		1.40	
Paracentrotus lividus	violet sea urchin	2		1		0.01		4925		3		42.89	
Echinoderm total		2		1		0,01		4925		3		42.89	
Total		**1288**		**1017**		**506.63**		**27724**		**6716**		**17628.49**	

Table 18-1. Spectrum of marine invertebrates on the modern and Late Iron Age (LIA) horizons at Dossen Rouz.

deposits (Bernáldez Sánchez *et al.* 2010; Heinrich 1994). At Dossen Rouz, the shelly modern horizon (just at the top of the Late Iron Age one) has been studied in order to determine its origin. The malacological spectrum of the "modern" horizon (19th-20th cent. AD) is constituted of 22 species among which there are 17 gastropods and 5 bivalves (table 18-1).

Gibbula spp. (*G. umbilicalis*, *G. pennanti* and *G. cineraria*) dominate the corpus at 67 % of the NISP and the MNI (table 18-1). The next largest in the corpus is *Littorina obtusata*, which represents 18 % of the NISP and 22 % of the MNI. Measurements were taken on *Gibbula* spp. and *Littorina obtusata* (fig. 18-3). The histograms of the lengths of both species show that all size classes are present, from juveniles to adults. This profile corresponds to a natural distribution without human selection. These species of gastropods are often defined as associated shells in shell middens because of their small sizes (Bizien-Jaglin and Laperas 2004; Dupont 2006-b; Gruet and Dupont 2009). "Associated shells" are those living in the same biotope as other favoured species. Other small gastropods occurring in the modern horizons, such as *Lacuna pallidula*, *Bittium reticulatum* and *Tricolia pullus pullus,* are represented by a MNI of between 1.5 and 3% and are also defined as associated shells (table 18-1). The larger species such as *Patella* spp., *Mytilus edulis*, *Ostrea edulis*, *Haliotis tuberculata tuberculata* and *Ruditapes decussatus* are poorly represented (between 1 and 5 individuals) and present some taphonomic indicators. Several limpets and mussels have perforations, potentially due to the action of predatory gastropods like the dogwhelk or murex (*Nucella lapillus* or *Ocenebra erinaceus*) (Henri-Martin 1932). These perforations at Locquémeau penetrate the full thickness of the test. This indicates that these shells were empty and dead when they were brought onto the site. Moreover, a number of larger shells have been damaged by marine erosion.

To understand the process of creation of the shell midden on the modern horizon, we had to discern the biotope of each shell present (table 18-2). Most derive from the rocky seashore, with the vast majority being small gastropods. The natural distribution of the sizes

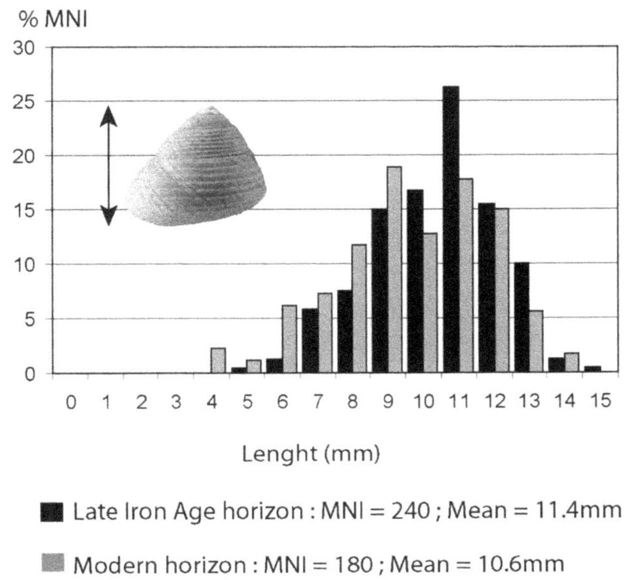

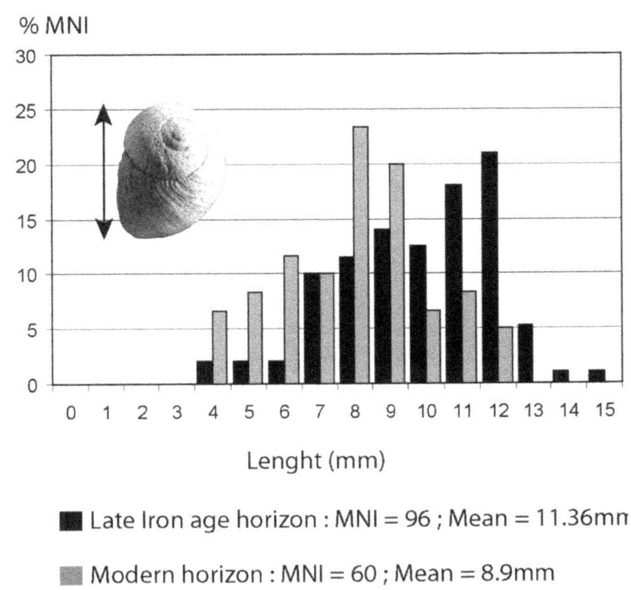

Figure 18-3. Distribution of length classes (mm) of *Gibbula* spp. (*G. cineraria*, *G. umbilicalis*, *G. pennanti*) and *Littorina obtusata* in the modern and Late Iron Age horizons.

Species	Substrate	Tidal range			
		Intertidal			Eulittoral
		H	M	L	
Littorina saxatilis	rock		■	■	
Patella sp.			■	■	
Patella vulgata			■		
Patella intermedia			■		
Nucella lapillus			■	■	
Gibbula pennanti			■	■	
Gibbula umbilicalis			■	■	
Littorina obtusata			■		
Osilinus lineatus			■		
Lacuna pallidula				■	■
Gibbula cineraria				■	■
Jujubinus exasperatus				■	■
Ocenebra erinaceus				■	■
Calliostoma zizyphinum			■	■	■
Trivia monacha				■	■
Rissoa parva			■	■	■
Tricolia pullus pullus				■	■
Nassarius incrassatus				■	■
Haliotis tuberculata tuberculata					■
Bittium reticulatum	rock and sand		■	■	■
Cingula trifasciata			■	■	■
Paracentrotus lividus				■	■
Mytilus edulis	rock and mud		■	■	■
Littorina littorea			■	■	
Ostrea edulis			■	■	■
Venerupis aurea	sand			■	■
Dosinia sp.			■	■	■
Venus verrucosa				■	■
Pecten maximus					■
Cerastoderma edule			■	■	■
Mangelia costata	sand and gravel			■	■
Glycymeris sp.					■
Ruditapes decussatus	sand and mud		■	■	■
Nassarius reticulatus			■	■	■
Gibbula magus				■	■

Table 18-2. Tide range and substrates of marine species (grey cell: potential presence; H: High tide level; M: Middle tide level; L: Low tide level; after Dupont 2006-b).

is in opposition with the location of the archaeological site, on the tidal range. The lack of marine sediments at the top of the site is also an indicator that these small shells did not live here. This whole population did not live at the top of the site but may have been transported by another process. Most of these shell species live on the rocky seashore (table 18-2) and inside or associated with vegetable species (seaweeds and marine plants), and some in more sandy environments. Although environmental origins are diverse, in table 3 we only summarize their association with seaweeds (Fretter and Graham 1976, 1977, 1978, 1980, 1981, 1984; Gruet 1989; Hawkins and Jones 1992; Hayward, Nelson-Smith, and Schields 1998). Further study of these associations between shellfish and algae was therefore undertaken (table 18-3).

The tidal range of seaweeds changes with geographic location, exposure to waves, temperature, tides and salinity of the water (Gruet 1989; Lewis 1964). Table 18-3 presents a model distribution of shellfish and seaweed zonation at the rough wave-beaten coast of the western Channel. *Gibbula* spp., *Littorina obtusata*, *Littorina littorea*, *Rissoa parva*, *Tricolia pullus pullus*, *Nassarius reticulatus* and *Bittium reticulatum* are associated with several seaweeds (table 18-3). The seaweeds encountered in the intertidal zone are entirely associated with the small gastropods and bivalves present at the site of Dossen Rouz (table 18-3). The majority of small gastropods live in association to *Fucus serratus* and the other seaweeds which occupy the same zone. The zonation of the seaweed *Fucus vesiculosus* (intertidal range) and of *Himanthalia elongata* (intertidal/sublittoral range) also corresponds with numerous shell species. However, the location of the archaeological site does not correspond with areas where seaweeds grow. Instead, the Dossen Rouz site is located in an intertidal zone with a regular flow of high tides across it and these tides are likely to have brought these seaweeds onto the site with the associated fauna (table 18-3). Modern maritime activities (production of soda and drying of nets) could also have contributed to these natural deposits. These processes could explain the accumulation of these abundant small gastropods in the stratigraphic sequences of the site.

The distribution histogram of the most representative species makes it possible to compare the malacological spectrum of the modern and LIA horizons (fig. 18-4-A).

The most obvious difference between them is the proportion of *Patella*. This species represents only 0.3 % of the MNI for the modern period, whereas it dominates the LIA period with 74 % of the MNI. *Gibbula* spp. (*G. umbilicalis*, *G. pennanti* and *G. cineraria*) are in first position for the modern horizon and second for the LIA horizon (table 18-1 and fig. 18-4-B). In spite of this difference between the two horizons, several similarities can be observed. In both modern and LIA horizons the MNI of each species of *Gibbula* is nearly the same (fig.

Tidal range	Shell species / Algae and sea grass species	*Littorina saxatilis*	*Osilinus lineatus*	*Littorina obtusata*	*Gibbula umbilicalis*	*Gibbula pennanti*	*Gibbula cineraria*	*Littorina littorea*	*Lacuna pallidula*	*Rissoa parva*	*Tricolia pullus pullus*	*Jujubinus exasperatus*	*Nassarius reticulatus*	*Bittium reticulatum*	*Anomia epphipium*
Intertidal	*Pelvetia canaliculata*	xxx	xx		x			x							
	Fucus spiralis	xxx	xxx	x	x			xxx						xx	
	Ascophyllum nodosum	x	x	xxx	xx			xx						xx	
	Fucus vesiculosus *Ulva* sp. *Cladophora*		x	xx	xx	x		xx	x				x	xx	
	Fucus serratus *Chondrus crispus* *Ulva* *Cladophora* *Lomentaria* *Gigartina* *Laurencia pinnatifida* *Codium*			x	x	xxx	xxx	xx	xx	xxx	xx		xxx	x	x
Intertidal/ Sublittoral	*Himanthalia elongata* *Zostera marina* *Corallina officinalis* *Cystoseira* *Codium*						xx	x		xxx	xxx	xx	xx	x	xx
	Laminaria saccharina *Zostera marina*						xx	x		xx	xx	xx	xx		xx
Sublittoral	*Laminaria digitata* *Zostera marina* *Corallina officinalis* *Callithamnion tetricum* *Plumularia plumosa* *Nitophyllum punctatum*						x			xx	xx	xx	xx		xx

Table 18-3. Seaweeds environments and potential association with some molluscs (Fretter and Graham 1976, 1977, 1978, 1980, 1981, 1984; Gruet 1989; Hawkins and Jones 1992; Hayward and Ryland 1995; Hayward, Nelson-Smith and Schields 1998; Lewis 1964). xxx : important association; xx : potential association; x : rare association.

18-4-B). *G. cineraria* forms only a small proportion in both horizons, doubtless because it lives on the lower seashore (table 18-2). *G. umbilicalis* and *G. pennanti,* however, have a high and almost identical percentage in both horizons (35-45 % of the MNI). *G. umbilicalis* numbers are lower in the LIA horizons and *G. pennanti* numbers are lower in the modern horizons. A combination of factors may explain the differences in the proportions of *G. pennanti* and *G. umbilicalis,* including higher sea levels in the modern context favouring the presence of deeper species, the biotope of the species and the contribution of modern maritime activities. It is difficult, however, to favour one of these factors. The distributions of the lengths of *Gibbula* in the LIA and modern horizons are similar and correspond to the natural size-class distribution of the species (fig. 18-3). The shells from the LIA horizons do not seem to have been sorted in the manner described for the modern horizon. It seems likely that other small species such as *Littorina obtusata* or *Bittium reticulatum* were introduced the same way and correspond to an accidental natural incursion. With the exception of limpets, edible shell species are represented only by rare individuals and by individuals of small size in the LIA horizon. The European carpet clam (*Ruditapes decussatus*) is represented only by 20 remains or 1 individual. *Venus verrucosa*, *Venerupis aurea*, *Cerastoderma edule* were also identified with between 1 and 10 fragments.

Nevertheless, one should bear in mind that some of the smaller species of shells in the LIA horizons may be "accessory" species, the presence of which is an intended consequence of human activity (Dupont 2006-b: 45). These smaller species settle on seaweeds, in mud and sand, or on larger shells which may have been collected by people. They are evidence of activities using sea products. This argument would be supported by a high percentage of burnt remains in the case of seaweeds used as fuel. However, in the LIA horizons only 0.3 % of the complete assemblage of marine resources bears

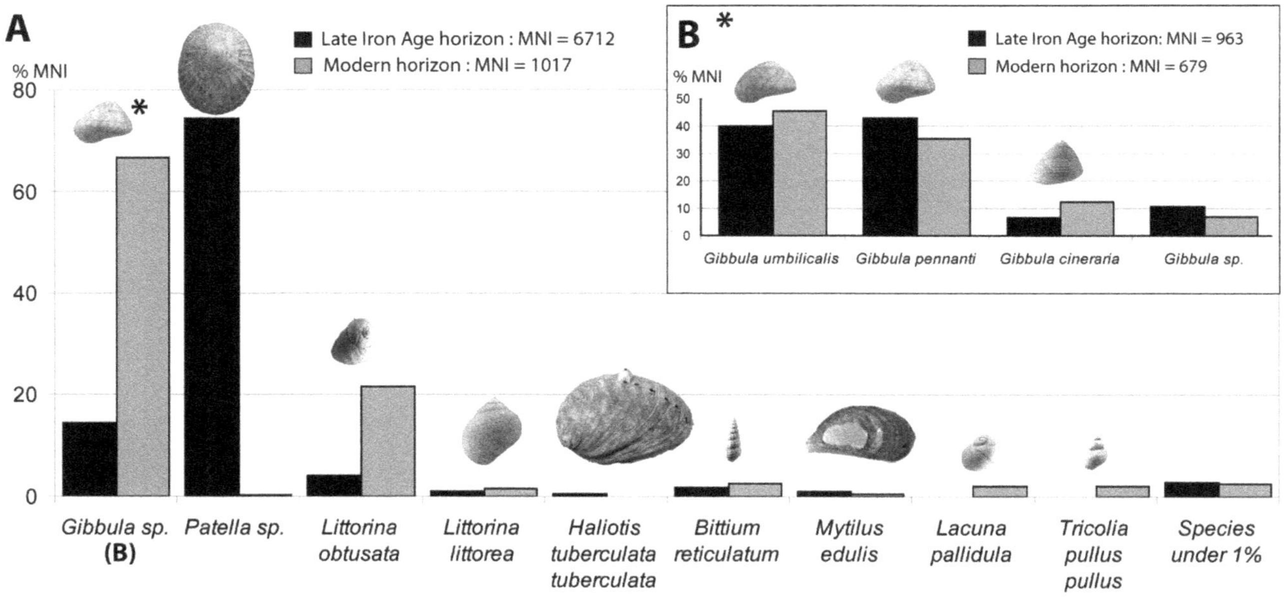

Figure 18-4. Comparison of Minimum Number of Individuals percentage of the majority of dominant seashell species (A) and of the *Gibbula* species (B) in the modern and LIA horizons.

indications of burning on the small species such as the *Gibbula* and the *Bittium*. This extremely low percentage does not appear to confirm the hypothesis. Of the thirty four species of shells identified from both horizons, four can be eaten (*Patella* sp., *Haliotis tuberculata tuberculata*, *Mytilus edulis* and *Littorina littorea*), with the remaining 30 determinable as "accessories" (fig. 18-5). The accessory species represent 98 % of MNI in the modern horizon and 23 % of MNI in the LIA horizon (table 18-1).

Exploitation of marine invertebrates during the Late Iron Age

Faunal spectrum

After demonstrating that 23 % of shell species in the Iron Age horizons were intrusive species, it is concluded that only 4 species of molluscs among the 32 counted could have been consumed by humans (table 18-4). As emphasized above, the marine resources are dominated by *Patella* spp., whether we use the Number of Identified Specimens (95.5 %), the Minimum Number of Individuals (96.5 %) or the quantification by weights (95.2 %) (table 18-4). The term *Patella* spp. usually includes three species present along the French maritime façade. First, *P. vulgata* lives on high intertidal rocks because it has high desiccation tolerance (Fretter and Graham 1976). It is abundant on all rocky shores from the most exposed to the most sheltered. At Dossen Rouz, it represents 99 % of the total MNI of limpets. By contrast, *P. intermedia* is less tolerant to continuous wetting and drying (Fretter and Graham 1976). It occurs on exposed rocky coasts only and frequents middle and lower stretches of the intertidal zone. This explains its low percentage, only 1 %. Finally, *P. ulyssiponensis* lives in the low intertidal zone and is accessible during spring high tides (Fretter and Graham 1976). We have not identified it at Dossen Rouz. These results indicate that at Dossen Rouz people preferentially gathered limpets which were easily and consistently accessible. Aside from being the major species at the Dossen Rouz site it is also common in many other coastal archaeological sites in Brittany and elsewhere in Europe for all periods (Bizien-Jaglin and Lapéras 2004; Campbell 2007; Cuenca Solana, Clemente Conte and Gutiérrez Zugasti 2010; Dupont 2006-b, 2009; Gutiérrez Zugasti 2008). The goal in gathering it may have been to feed not only people but perhaps also animals (Levasseur 1999). Some edible species with remains recovered in smaller quantities were also identified. The periwinkle (*Littorina littorea*) was one of these, with 74 individuals. The abalone (*Haliotis tuberculata tuberculata*) is represented by 41 individuals which correspond to 3 % of the weight of the remains in the sample. Regarding the mussel (*Mytilus edulis*), the number of remains represented 3.2 % totalling only 65 individuals. The mussel shell is more fragile than other marine molluscs (limpets or common periwinkles) and its fragmentation is much more important (Dupont 2006-b; Light 2005).

The single species of echinoderm identified, *Paracentrotus lividus* (fig. 18-2-B and 2-C), represents 17.7 % of the NISP of the species found in the LIA horizons (NISP = 4925) and 19 % of the NISP of edible species. The very low MNI of the violet sea urchin (5 individuals) could result on the one hand from the low proportion of the sampled and sieved sediment and, on the other, from the fact that the anatomical parts of the skeleton used to establish this quantification (the MNI) are extremely fragile and small. The presence of this species is very interesting because

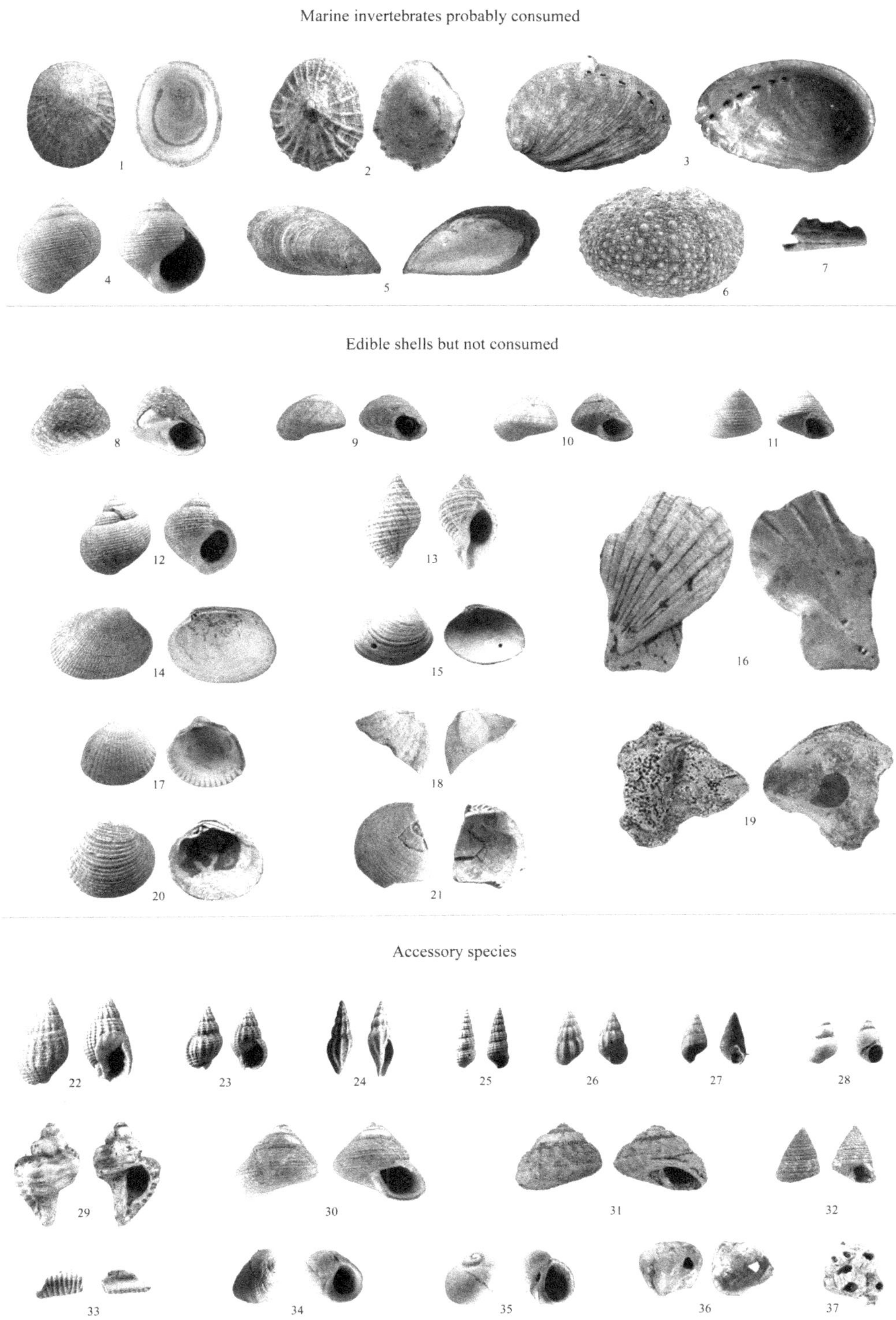

Figure 18-5. Marine invertebrates identified at Dossen Rouz (CAD. C. Mougne ; C. Dupont and L. Quesnel) **1** : *Patella vulgata* (L=38mm), **2** : *Patella intermedia* (L=34mm), **3** : *Haliotis tuberculata tuberculata* (L=90mm), **4** : *Littorina littorea* (L=18mm), **5** : *Mytilus edulis* (L=72mm), **6** : *Paracentrotus lividus* (L=48mm), **7** : *Cancer pagurus* (L=7mm), **8** : *Osilinus lineatus* (L=19mm), **9** : *Gibbula umbilicalis* (L=11mm), **10** : *Gibbula pennanti* (L=13mm), **11** : *Gibbula cineraria* (L=11mm), **12** : *Littorina saxatilis* (L=14mm), **13** : *Nucella lapillus* (L=19mm), **14** : *Ruditapes decussatus* (L=35mm), **15** : *Venerupis aurea* (L=27mm), **16** : *Pecten maximus* (L=61mm), **17** : *Cerastoderma edule* (L=23mm), **18** : *Dosinia* sp. (L=29mm), **19** : *Ostrea edulis* (L=58mm), **20** : *Venus verrucosa* (L=31mm), **21** : *Glycymeris* sp. (L=37mm), **22** : *Nassarius reticulatus* (L=21mm), **23** : *Nassarius incrassatus* (L=11mm), **24** : *Mangelia coarctata* (L=11mm), **25** : *Bittium reticulatum* (L=5mm), **26** : *Rissoa parva* (L=3mm), **27** : *Cingula trifasciata* (L=3mm), **28** : *Tricolia pullus pullus* (L=8mm), **29** : *Ocenebra erinaceus* (L=33mm), **30** : *Calliostoma zyzyphinum* (L=21mm), **31** : *Gibbula magus* (L=21mm), **32** : *Jujubinus exasperatus* (L=8mm), **33** : *Trivia monacha* (L=5mm), **34** : *Littorina obtusata* (L=12mm), **35** : *Lacuna pallidula* (L=9mm), **36** : *Anomia ephippium* (L=10mm), **37** : *Balanus* sp. (L=21mm).

Species	NISP	% NISP	MNI	% MNI	Weight (g)	% Weight
Patella sp.	20028	95.5	4992	96.5	15990	95.2
Littorina littorea	89	0.4	74	1.4	128.8	0.8
Haliotis tuberculata tuberculata	169	0.9	41	0.8	505.5	3
Mytilus sp.	668	3.2	65	1.3	171.2	1
Total	20954	100	5172	100	16795.5	100

Table 18-4. Consumed shell spectrum on the Late Iron Age (LIA) horizons of the Dossen Rouz site.

it is rarely identified on coastal archaeological sites (Campbell 2008-a, 2008-b; Gruet 2003; Gutiérrez Zugasti 2011; Lawrence 2007). Lastly, seven crustacean remains, with one identified as *Cancer pagurus* (edible crab), have been found but there is no evidence as to whether crab was consumed (table 18-1).

The histogram of the lengths of *Patella* spp. shows the dominance of shells more than 20 mm long (fig. 18-6), indicating that individuals of middle and large size were preferentially exploited for food by the population of Dossen Rouz (Dupont 2006-b). For common periwinkles, the data show a size concentration between 18 to 22 mm with some rare small individuals which may have been deposited by high tides. At the Neolithic coastal site of Ponthezières (Charente-Maritime, France), the distribution histogram of the lengths of common periwinkles is approximately the same (Gruet and Dupont 2009:563). It seems that selective gathering was practised by the inhabitants of Dossen Rouz. Abalones (*Haliotis tubercula tubercula*) present at the Dossen Rouz site are generally large (between 60 and 100 mm in length). Therefore, people favoured the gathering of large individuals. Finally, the size of the violet sea urchins (*Paracentrotus lividus*) could be approximately determined using the lengths of the hemypiramid. Their diameter is estimated between 40 and 60 mm (Gruet 2003; Mougne and Dupont in press). These sea urchins were also selected for their large dimensions. A biometric analysis for each edible shell and sea urchin species shows a distribution of gathering by selection, with a concentration of large sizes.

Gathering Strategies

Humans appear to have exclusively exploited the rocky seashore (table 18-2). As for the violet sea urchin, it populated the rocky seashore and sometimes the sand (Quéro and Vayne 1998). The extent of the coastal territory used by people in search of marine food seems to have been quite restricted, concentrated on this single environment, which corresponds with the present surroundings of the site. Most of the edible species in the LIA horizons were available in the intertidal zone and thus could have been gathered on foot when the tides permitted (table 18-2). The level of the tides is especially important for obtaining particular species such as the abalone (*Haliotis tuberculata tuberculata*) and violet sea urchin (*Paracentrotus lividus*), for which collecting requires very low tides only occurring during spring or autumn equinoxes (Poppe and Gotto 1993). Conversely, the limpets found at the archaeological site are strictly intertidal species and people would have gathered them on the seashore, as also was the case for common periwinkles and mussels.

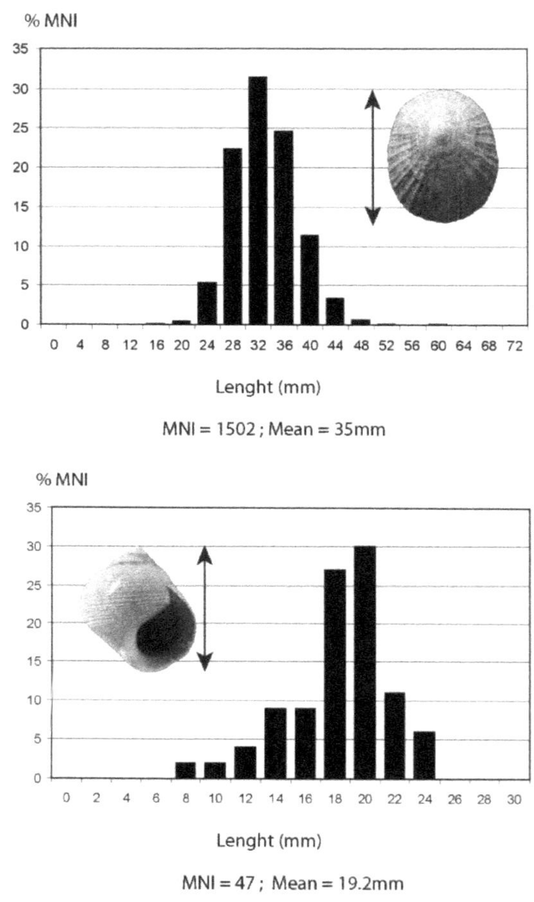

Figure 18-6. Distribution of length classes (mm) of *Patella* sp. and *Littorina littorea* in the Late Iron Age horizon.

Selection of species, therefore, seems mainly to have been linked to their accessibility. The most exploited shells required little investment – they were visible to the naked eye – and limited effort was required in collecting them because they were obtainable from both high and middle shore levels. Shell gathering could have been carried out on a daily basis for species such as the limpet, the common periwinkle and the mussel which were abundant and common on the coast of Dossen Rouz. Only the abalone and the violet sea urchin are limited to a specific opportunity for collection. Sea urchins are consumable only during specific periods: gonads or female sexual glands are the edible parts of the sea urchin and are particularly fleshy just after the reproduction period, during winter or spring (Quéro and Vayne 1998). These data allow us to state that the people of Dossen Rouz had access to coastal resources for most of the year.

Conclusion

Some coastal archaeological sites are endangered by rising sea levels. In response to this threat, the ALERT project (Archaeology, Coasts and Global Warming) was set up in 2007 and the aim of indexing and sometimes studying such sites before their disappearance (Lopez-Romero and Daire 2008).

The malacological study of Dossen Rouz is one of the rare studies applied to Iron Age contexts along the French maritime façade (Atlantic Ocean and English Channel). This study shows interesting methodological developments linked to the original position of this archaeological site, i.e. the seashore. Assessment of the assemblage of species, their size, their taphonomic characteristics and their composition relative to modern horizons, makes it possible to attribute their presence mainly to the accumulation of seaweed at the high water mark. These incursions that reached the Late Iron Age horizon have been eliminated from the archaeological assemblage. These results are dependent on the use of sieving. Sieving some sedimentary samples with fine meshes (2mm), also makes it possible to isolate some species rarely described in archeological context such as sea urchins.

The consumption of domestic mammals (cattle, sheep / goat, pigs) is attested at Dossen Rouz by evidence of butchery on bone remains. However, the subsistence economy of this community during the Iron Age seems to have been marked by a significant degree of exploitation of marine resources: shellfishes, echinoderms and crustaceans. Among these, molluscs seem to represent a substantial majority.

The anthropic malacological spectrum indicates that the limpet was the most abundant and consumed shellfish during the Iron Age. Its accessibility and abundance in the natural environment are adequate to explain this predominance. Other species such as abalones, mussels, periwinkles and clams, and also the violet sea urchin, were consumed secondarily. The pre-Roman inhabitants of Dossen Rouz mainly exploited the rocky seashore and selected middle- to large-sized individuals. Collection may have occurred during various seasons and a daily presence of humans gathering marine resources at Locquémeau-Trédrez is not excluded.

The Dossen Rouz site illustrates the development of Iron Age maritime communities characterized by a high density of settlements, especially those dedicated to the production of salt, during the Late Iron Age along the Atlantic façade. Marine resources probably contributed to this development, particularly by underpinning the subsistence economy. Further malacological studies could provide a better understanding of the importance of marine resources in the diet, and a better knowledge of the species exploited in relation to the nature of coastal occupations and exchange. Another line of research concerns the rhythms (seasonal, annual?) and duration (generation, century?) of these coastal settlements. An approach incorporating sclerochronological analysis could deepen our understanding.

Acknowledgements

We would like to thank Greg Campbell and Yves Gruet for their comments on the first version of this paper.

References

Bardot, Anne. 2010. *Les coquillages en Gaule romaine, entre Méditerranée et Rhin. Approche socio-économique et socio-culturelle*. PhD dissertation, University of Bordeaux 3, Bordeaux, inédit.

Bejega García, Víctor, Eduardo González Gómez de Agüero, Carlos Fernández Rodríguez and Juan Carlos Álvarez García. 2010. Los concheros de O Neixón (Boiro, A Coruña) y Punta Atalaia (San Cibrao, Cervo, Lugo): Una propuesta de muestreo y excavación de depósitos de la edad del Hierro y época Romana en Galicia. In I Reunión de arqueomalacología de la Península Ibérica. González Gómez de Agüero, Eduardo, Víctor Bejega García, Fernández Rodríguez and Natividad Fuertes Prieto, eds. *Férvedes Revista de Investigación* 6: 33-42.

Berghe, Vanden I., Margarita Gleba, and Ulla Mannering. 2009. Towards the identification of dyestuffs in Early Iron Age Scandinavian peat bog textiles. *Journal of Archaeological Science* 36:1910-1921.

Bernáldez Sánchez Eloísa, Esteban García-Viñas, Carmen Pozo Cabello, M. Carmen Lozano-Francisco and José Luis Vera-Peláez. 2010. Inferencias bioestratinómicas en la interpretacíon arqueológica de yacimientos del Holoceno. In I Reunión de arqueomalacología de la Península Ibérica. González Gómez de Agüero, Eduardo, Víctor Bejega García, Fernández Rodríguez and Natividad Fuertes Prieto, eds. *Férvedes Revista de Investigación* 6: 85-94.

BIZIEN-JAGLIN, CATHERINE, and JACQUELINE LAPERAS. 2004. Les formations coquillières des Haches. In *Les Haches en Saint-Jacut-de-la-Mer (Côtes d'Armor)*. Bizien-Jaglin, Catherine, ed. Pp. 69-76. Alet: Centre Régional d'Archéologie.

BRIEN FRANÇOISE. 1988. Les Coquillages. In Le village gaulois de Martigues. Chausserie-Laprée, Jean, éd. *Dossiers Histoire et Archéologie* 28: 96-97

BRIEN-POITEVIN, FRANÇOISE. 1992. Collecte, consommation et réutilisation des coquillages marins sur le site de Lattes (IVe s. av.n.é.-IIe s. de n.é.). In : Recherche sur l'économie vivrière des Lattarenses. Py, Michel, ed. *Lattara* 5: 125-138

——. 1993. Etudes conchyliologiques de quelques sites, l'étang de Berre et la vallée de l'Arc. In *Archéologie et environnement de la Sainte-Victoire aux Alpilles*. Leveau Philippe and Mireille Provansal, eds. Pp. 285-300. Aix-en-Provence: Travaux du Centre Camille Julian.

CABRAL, JOÃO PAULO, and ARMANDO COELHO F. DA SILVA. 2003. Morphometric analysis of limpets from an Iron-age shell midden found in northwest Portugal. *Journal of Archaeological Science* 30: 817-829.

CAMPBELL, GREG. 2005. L'économie de subsistance. In *Les Fouilles du Yaudet en Ploulec'h, Côtes d'Armor*, vol. 2: *Le Site de la Préhistoire à la fin de l'Empire gaulois*. Cunliffe, Barry, and Patrick Galliou, eds. Monograph 63. Pp. 355-356. Oxford: Oxford University School of Archaeology.

——. 2007. Appendix G: The marine molluscs. In *Les Fouilles du Yaudet en Ploulec'h, Côtes-d'Armor*, vol. 3. Cunliffe, Barry, and Patrick Galliou, eds. http://www.arch.ox.ac.uk/research/research_projects/le_yaudet/appendices (accessed April 8, 2011)

——. 2008a. Sorry, wrong phylum: a neophyte archaeomalacologist's experiences in analyzing a European Atlantic sea urchin assemblage. *Archaeofauna* 17: 77-90.

——. 2008b. A preliminary study of methods for identifying archaeological sea urchin remains in the Pacific Northwest. *Canadian Zooarchaeology* 25:15-35.

CLAASSEN, CHERYL. 1998. Taphonomy of shell and shell assemblages. In *Shells*. Claassen, Cheryl, ed. Pp. 53-98. Cambridge Manuels in Archaeology. Cambridge: Cambridge University Press.

CLEMAM. 2010. Check List of European Marine Mollusca, http://www.mnhn.fr/base/malaco.html. (accessed April 15, 2010)

CUENCA SOLANA, DAVID, IGNACIO CLEMENTE CONTE and IGOR GUTIERREZ ZUGASTI. 2010. Utilización de instrumentos de concha durante el Mesolítico y Neolítico inicial en contextos litorales de la región Cantábrica: programa experimental para el análisis de huellas de uso en materiales malacológicos. *Trabajos de Prehistoria* 67: 211-225.

CUNLIFFE BARRY and STEVE HAWKINS. 1988. The shell midden deposits. In *Mount Batten Plymouth. A Prehistoric and Roman Port*. Cunliffe, Bary, éd. Pp. 35-38. Monograph 26. Oxford: Oxford University Committee for Archaeology.

DAIRE, MARIE-YVANE ed. 2011. Sur les rivages de la Manche... Le site de l'Âge du Fer de Dossen Rouz à Locquémeau-Trédrez (Côtes d'Armor). Etude pluridisciplinaire. Alet : Centre Régional d'archéologie, Rennes.

DUPONT, CATHERINE. 2006a. Etude archéozoologique: la faune marine, coquillages et crustacés de la phase 3. In *Grains de sel. Sel et salines du littoral bas-normand (Préhistoire – XIXe siècle), Entre Archéologie et Histoire. Carpentier, Vincent, Emmanuel Ghesquière and Cyril Marcigny, eds.* Pp.111-116. *Rennes: Centre Régional d'Archéologie d'Alet and A.M.A.R.A.I.*

——. 2006b. *La malacofaune de sites mésolithiques et néolithiques de la façade atlantique de la France: Contribution à l'économie et à l'identité culturelle des groupes concernés*. BAR International Series 1571. Oxford: BAR Publishing.

——. 2009. La consommation des coquillages marins. In L'habitat littoral des Ormeaux à Angoulins (Charente-Maritime): activités vivrières et salicoles entre marais et océan. Maguer, Patrick *et al.* In *Les Gaulois entre Loire et Gironde, Actes du XXXIe Colloque international de l'Association Française pour l'Étude de l'Age du Fer, 17 – 20 mai 2007, Chauvigny, Vienne, France, Tome I. Bertrand, Isabelle, Alain Duval, José Gomez de Soto and Patrick Maguer,* eds. Pp.89-98. Chauvigny: Association des publications chauvinoises.

——. 2011. The Dog Whelk *Nucella lapillus* and dye extraction activities from the Iron Age to the Middle Ages along the Atlantic coast of France. *The Journal of Island and Coastal Archaeology* 6: 3-23.

FRETTER, VERA, and ALAISTAIR GRAHAM. 1976. The prosobranch molluscs of Britain and Denmark, Part. 1- Pleurotomariacea, Fissurelacea and Patellacea. *The journal of molluscan studies*, suppl. 1: 1-38.

——. 1977. The prosobranch molluscs of Britain and Denmark, Part. 2- Tronchacea. *The journal of molluscan* studies, suppl. 3. Pp. 39-100. Reading:

Angus Graham Associates.

——. 1978. The prosobranch molluscs of Britain and Denmark, Part. 3- Neritacea,Viviparacea, Valvatacea, terrestrial and freshwater Littorinacea and Rissoacea. *The journal of molluscan studies*, suppl. 5: 101-152.

——. 1980. The prosobranch molluscs of Britain and Denmark, Part. 5- Marine Littorinacea. *The journal of molluscan studies*, suppl. 7: 243-284.

——. 1981. The prosobranch molluscs of Britain and Denmark, Part. 6- Cerithiacea, Stombacea, Hipponicacea, Lamellariacea, Cyperaeacea, Naticacea, Tonnacea, Heteropoda. *The journal of molluscan studies*, suppl. 9: 285-362.

——. 1984. The prosobranch molluscs of Britain and Denmark, Part. 8- Neogasteropoda. *The journal of molluscan* studies, suppl. 15: 435-556.

Germinet, David. 2009. *Homme et Animal dans les fermes du Poitou à la transition de l'âge du Fer à la période romaine*. PhD dissertation, University of François-Rabelais, Tours, inédit.

Grayson, Donald K. 1984. *Quantitative Zooarchaeology*. London: Academic Press.

Gruet, Yves. 1989. *Algues. Loire-Atlantiques Vendée*. Nantes: Bulletin de la Société des Sciences naturelles de l'Ouest de la France.

——. 1998. Un exemple d'occupation dans la plaine charentaise (fin de l'âge du Fer et Moyen-Âge), le site de Mortantambe à Cabariot. In *L'estuaire de la Charente de la Protohistoire au Moyen-Âge*. Laporte, Luc, ed. Pp. 129-134. Documents d'Archéologie Française, 72. Paris: La Maison des Sciences de l'Homme.

——. 2003. Rapport sur les crustacés et les échinodermes du site de Béniguet. In Le site préhistorique de Béniguet 3 (île de Béniguet, Le Conquet, Finistère). Dupont, Catherine *et al*. eds. *Bulletin de l' l'A.M.A.R.A.I.* 16: 5-24.

Gruet, Yves, and Catherine Dupont. 2009. I- Les mollusques ou la malacofaune, B- Matériel et méthodes. C- Un ensemble malacologique de référence: Ponthezières. *In Des premiers paysans aux premiers métallurgistes sur les côtes charentaises, Chapitre 7- Économie de subsistance: La part des ressources littorales et continentales. Laporte Luc, ed.* Pp.555-581. Chauvigny: Association des publications chauvinoises.

Gutierrez Zugasti, Igor. 2008. *L'explotación de moluscos y otros recursos litorales en la región cantábrica durante el pleistoceno final y holoceno inicial*. PhD Dissertation, Universidad de Cantabria, Santander.

——. 2011. The use of echinoids and crustaceans as food during the Pleistocene-Holocene transition in Northern Spain: methodological contribution and dietary assessment. *Journal of Island and Coastal Archaeology* 6: 115-133.

Hawkins, Stephen J., and Hugh D. Jones. 1992. *Marine field course guide 1. Rocky shores*. London: Immel Publishing.

Hayward, Peter Joseph and John Stanley Ryland. 1995. *Handbook of the marine fauna of North-West Europe*. Pp.678-680. Oxford: University Press.

Hayward, Peter Joseph, Tony Nelson-Smith, and Chris Schields. 1998. *Guide des bords de mer. Mer du Nord, Manche, Atlantique, Méditerranée.* Lausanne: Delachaux et Niestlé.

Heinrich, Dirk. 1994. Some remarks on the term « thanatocoensis » especially « anthropogenic thanatocoenosis » with particular reference to fish remains. *Archaeofauna* 3: 93-97.

Henri-Martin, 1932. Différents modes de perforation de la coquille chez les mollusques. In *Procès verbaux de la Société Linnéenne de Bordeaux*, séance du 26 juin 1932, Pp. 1-5.

Jammes, Léon. 1904. *Zoologie pratique basée sur la dissection des animaux les plus répandus*. Paris: Masson Editeurs.

Lawrence, John Miller. 2007. Edible sea urchins: use and life-history strategies. In *Edible Sea Urchins Biology and Ecology*. Lawrence, John Miller, ed. Pp. 1-9. Elsevier: Elsevier Science.

Levasseur, Olivier. 1999. Pêches à pied, de nombreuses techniques. In *Pêche à pied et usages de l'estran*. Aubrée, Maurice and Guy Prigent, eds. Pp. 26-29. Rennes: Editions Apogée.

Lewis, John Robert. 1964. *The ecology of rocky shores*. London: the English universities press.

Light, Jan. 2005. Marine mussel shells – wear is the evidence. In *Proceedings of the 9th ICAZ Conference, Durham 2002 Archaeomalacolgy: Molluscs in former environnements of human behaviour*. Bar-Yosef Mayer, Daniella ed. Pp. 56-62. Oxford: Oxbow books.

Lopez-Romero, Elias, and Marie-Yvane Daire. 2008. Des sites archéologiques en danger sur le littoral et les îles de Bretagne: chronique 2007-2008. *Bulletin de l'A.M.A.R.A.I.* 21: 91-104.

MALLORY, JAMES, and FINBAR MC CORMICK. 1988. Excavations at Ballymulholland, Magilligan Foreland, Co. Londonderry. *Ulster Journal of Archaeology* 51: 103-114.

MC CARTHY, AMANDA, NYREE FINLAY and O. MC CLEAN. 1999. Marine Molluscan Remains. In *Excavations at Ferriter's Cove, 1983-95: last foragers, first farmers in the Dingle Peninsula.* Woodman, Peter C., Liz Anderson and Nyree Finlay, eds. Pp. 93-102. Bray: Wordweel.

MC CORMICK, FINBAR, MICHAEL GIBBONS, FRANCIS GERARD MC CORMAC AND JANINE MOORE. 1996. BRONZE AGE TO MEDIEVAL coastal shell middens near Ballyconneely, co. Galway. *The Journal of Irish Archaeology* 7: 77-84.

MÉNIEL, PATRICE. 2008. Cartographie des restes osseux et conservation différentielle dans l'amas coquillier d'Ouessant "Mez Notariou" (du Bronze moyen à la période romaine). *Annales de Paléontologie* 94: 267-285.

MORTENSEN, THEODOR. 1943. *A Monograph of the Echinoidea III.3: Camarodonta II. Echinidae, Strongylocentrotidae, Parasaleniidae, Echinometridae.* Copenhague: Reitzel.

MOUGNE, CAROLINE, and CATHERINE DUPONT. 2011. Etude diachronique des invertébrés marins : accumulation naturelle et dépôts anthropiques. In *Sur les rivages de la Manche... Le site de l'Âge du Fer de Dossen Rouz à Locquémeau-tredrez (Côtes d'Armor).* Etude pluridisciplinaire. Marie-Yvane Daire ed. Alet : Centre Régional d'Archéologie, Pp 93-115.

POPPE, GUIDO T., and YOSHIHIRO GOTO. 1991. *European Seashells: Polyplacophora, Caudofoveata, Solenogastra, Gasteropoda.* vol 1. Germany: Verlag Christa Hemmen.

——. 1993. *European Seashells: Scaphopoda, Bivalvia, Cephalopoda.* vol 2. Germany: Verlag Christa Hemmen.

QUÉRO, JEAN-CLAUDE, and JEAN-JACQUES VAYNE. 1998. *Les fruits de la mer et plantes marines des pêches françaises.* Lausanne: Delachaux et Niestlé.

SOUTHWARD, EVE C., and ANDREW C. CAMPBELL. 2006. *Echinoderms. Keys and Notes for Identification of British Species.* Shrewsbury: Field studies Council.

THEODOROPOULOU, TATIANA. 2008. From sea to land: the exploitation of molluscs in the Geometric artisan site of Skala Oropou, Attica. In *Proceedings of the 4th Symposium of the Hellenic Society for Archaeometry.* Facorellis, Yorgos, Nikos Zacharias and Kiki Polikreti, eds. BAR International Series 1746. Pp.205-214. Athens: Oxford BAR Publishing.

19 - DYNAMICS OF PALAEOENVIRONMENTAL CONDITIONS OVER THE LAST MILLENNIA BY ARCHAEOMALACOLOGICAL DATA (ON EXAMPLE OF ADK-009 SHELL MIDDEN, ADAK ISLAND, ALEUTIAN ISLANDS)

Zhanna ANTIPUSHINA
Laboratory of historical ecology of A.N. Severtsov's Institute of Ecology and Evolution RAS, Leninsky pr, 33, Moscow, Russia
zh.antipushina@gmail.com

Abstract: *This paper endeavours to reconstruct the dynamics of palaeoenvironmental conditions over the last millennia, based on a comprehensive analysis of invertebrate remains from archaeological site ADK-009 on Adak Island, Aleutian Islands. Radiocarbon determinations show that the cultural layers of the site ADK-009 were formed between the 6th to the 17th centuries AD. Analysis of the invertebrate faunal remains from the shell midden and morphometrical analysis of the gastropod shells allow a reconstruction of the sediment type and the wave action from the 6th century until the present. Additionally, zoogeographical analysis and oxygen stable isotope analysis offer insights into the different climatic conditions that prevailed during the occupation of the ancient settlement.*

Keywords: *Shell midden, Bering Sea, Holocene, Palaeoenvironment, Stable isotope.*

Introduction

Within the framework of climate change studies, in-depth study of the Holocene dynamics of ecosystems is necessary to ascertain the mechanisms of those changes and comprehend present dynamics. In order to study the Holocene history of coastal and marine ecosystems, for the last several years our work has focused on the Bering Sea region, namely with Kamchatka, Chukchi peninsula and Aleutian Islands. This unique area has well preserved coastal deposits, peat bog deposits as well as many archaeological sites which inform on the history of ecosystems. Different approaches for palaeoenvironmental reconstructions have been used, including radiocarbon dating, palaeozoological analysis (analysis of vertebrates' bones and remains of marine invertebrates), palaeobotanical analysis (analysis of plant remains, charcoal, spore-pollen, phytolith and diatom analysis), palaeopedological analysis and geomorphological analysis.

Invertebrate remains from archaeological deposits can yield important data for palaeoenvironmental reconstructions and provide excellent material with which to reconstruct past ecological dynamics (Antipushina 2010a). We have attempted to reconstruct the dynamics of palaeoenvironmental conditions, based on comprehensive analysis of invertebrate remains from archaeological site on Adak Island, Aleutian Islands, and the results of this study are presented here.

Study area

The Bering Sea is separated from the Pacific ocean by the Aleutian chain of islands which stretch over about 1800 km. The chain consists of about 150 islands (ca. 37,840 km^2), separated into six groups of islands. Adak Island is located in the central part of the Aleutian chain within the Andreanov Islands. The islands experience a cool, wet and windy maritime climate. The seasonal changes in surface air temperature across the Aleutian Islands are relatively uniform, from 5 to 10°C in summer to near freezing temperatures in winter. Precipitation varies widely, from 530 mm up to 2000 mm (Rodionov *et al.* 2005). The Bering Sea is ice free from July through September. Ice begins moving southward during October and reaches its southern limit near 60° N by February or March. After April the ice pack moves progressively northward and by July it is generally north of Bering Strait. Thus entire water area of Aleutian Islands stays ice free, even in severe winters (Overland 1981). According to zoogeographic zonation of the continental shelf, this area is located in the boreal native zone.

The colonization of the Aleutian Islands began at least 8700 14C years BP (Laughlin 1975). The Aleuts retained ancient traditions of hunting and gathering until the 18th-19th centuries. A characteristic feature of the ancient Aleuts was their dependence on the resources offered by the sea: they relied for subsistence on fishing, the hunting of sea mammals and birds, gathering bird eggs and shellfish in the littoral zone at low tide (McCartney 1975; Veniaminov 1984). As a result, deep cultural layers abundant in mollusc shells, remains of sea urchins, etc., were formed in Aleut settlements over many years.

The first large-scale excavations on the Aleutian Islands were performed by William Dall (1877), who identified invertebrate remains from shell middens in several islands. In spite of the great number of archaeological sites and more than one hundred years of research, some of these shell mounds have not yet been analyzed.

The archaeological site ADK-009 (fig. 19-1) is situated on the mountainous coast of narrow rocky Sweeper Cove (51°51'10" N, 176°38'18" E). It was excavated in 1999 by members of the Western Aleutian Archaeological and Paleobiological Project (WAAPP), leaded by Dr. Dixie West (University of Kansas).

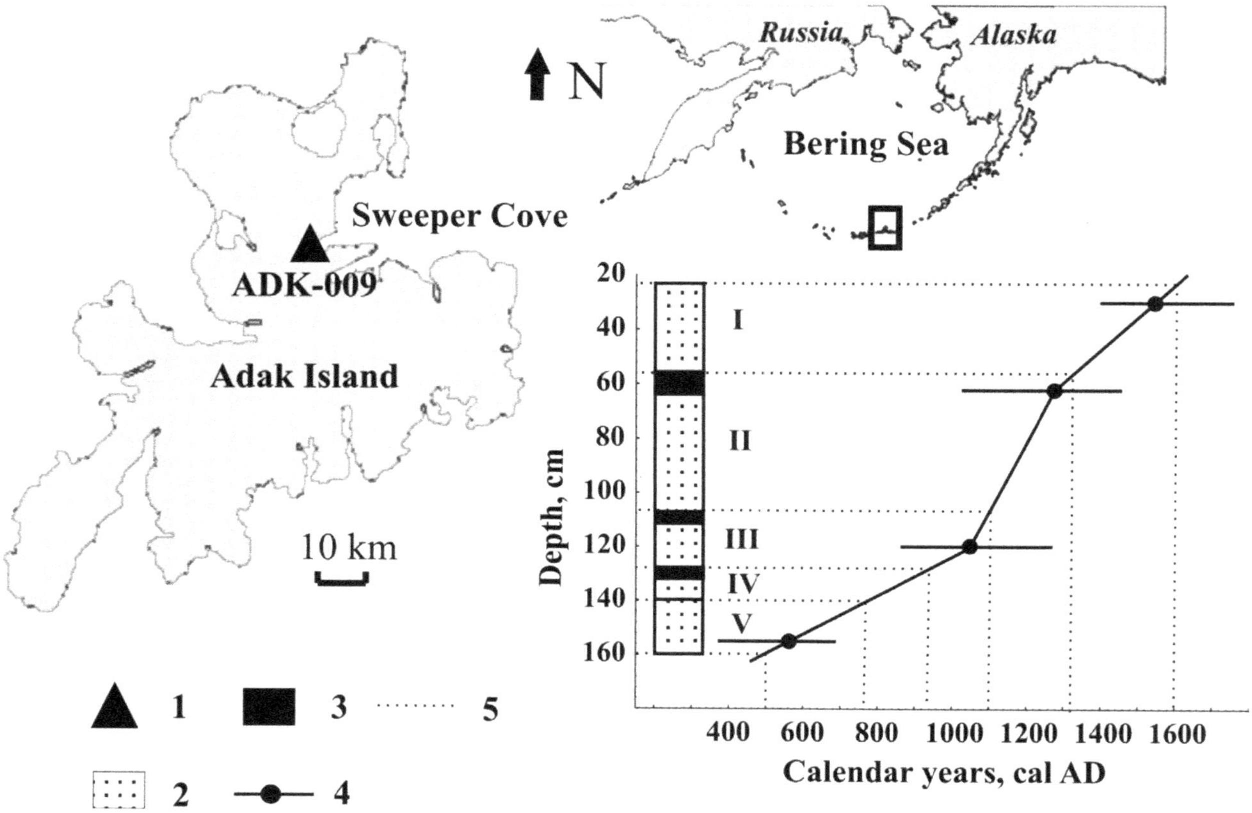

Figure 19-1. The study area and the profile of the ADK-009 shell midden: 1 – Archaeological site ADK-009; 2 – Cultural layers I-V; 3 – Humus-ash interlayers; 4 – Radiocarbon dates and their statistical errors; 5 – Boundaries of horizons within the profile of cultural layers and chronological limits of their formation.

Materials and methods

The shell midden was excavated according to the standard procedure with material collected by recognizable stratigraphic layers no thicker than 10 cm (Causey *et al.* 2005, Antipushina *et al.* 2009). Invertebrate remains were identified with reference to collections of the Zoological Museum of Moscow State University, the Severtsov's Institute of Ecology and Evolution of Russian Academy of Sciences (RAS), and the Zoological Institute of RAS. We used multiple quantification methods: the number of identified specimens (NISP) for quantification of brachiopods, chitons, barnacles, decapods; the total number of umbones for quantification of bivalves; the total number of elements of Aristotle's lantern, madreporits and genital plates for quantification of sea urchins (Antipushina *et al.* 2009; Antipushina 2010b).

We used conventional radiocarbon dating of fish bone collagen extracted by standard methods (Longin 1971). Radiocarbon ages were calibrated to calendrical ages with OxCal 4.10 and the calibration curve Marine 04 with the regional correction ΔR, 220 ± 50, proposed for the southern Bering Sea (Stuiver and Braziunas 1993). In our previous studies we did not correct the dates for carbon isotopic fractionation, but this has subsequently been done. The $\delta^{13}C$ value is −12.8 ppm. The upper, lower and some intermediate layers of the deposit were dated. The plot of the deposit accumulation rate was constructed by the method of linear interpolation (Telford *et al.* 2004). This rate was calculated as the ratio of layer depth to accumulation time, cm/100 years.

For comparison of the zoogeographical compositions of subfossil invertebrates and invertebrates from the modern littoral zone we collected samples of modern macrozoobenthos from Adak's littoral. We collected only those groups whose remains were preserved in cultural layers. For zoogeographical analysis the nomenclature of zoogeographical groups by O. Skarlato (Skarlato 1981) was used.

The sensitivity of marine mollusks to the environment often produces identifiable effects in the shell form. Morphometric analysis of the shells can therefore provide information on present and past environmental conditions. For example, in exposed conditions limpet shells become very flat to help them withstand wave stress (Cabral and Silva 2003). Low *Littorina* ('perwinkle') shell height/aperture height ratios are usually associated with exposed coasts (Janson and Sundberg 1983; Boulding and Van Alstyne 1993). We selected two hundred and thirty shells of limpets and periwinkles for morphometric analysis. Shell characters used for the morphometric analyses were shell height, shell width, aperture height, aperture width for periwinkles and shell height and shell length for limpets.

The isotopic composition of skeletal carbonate of different calcifying organisms is a function of the organism's environment. Shells have been used in environmental reconstruction, predicting seawater paleotemperatures (by analysis of the shell $\delta^{18}O$). Oxygen isotopes undergo a temperature-dependent fractionation during calcification. Increasing temperature results in a decreased fractionation between ^{16}O and ^{18}O, which produces a lower ^{18}O value (Epstein *et al.* 1951). Oxygen isotope analysis was undertaken using the plates of the barnacle Semibalanus cariosus (Savinetsky *et al.* 2011). This particular barnacle was chosen because its plates contain only calcite and because this barnacle is restricted to the littoral zone and does not inhabit the deeper, and subsequently colder, sublittoral zone like other barnacles and some mollusks (Killingley and Newnan 1982).

Oxygen isotope analysis was performed using standard techniques (Jones and Quitmyer 1996; Hu *et al.* 2001). From each layer of ADK-009 we selected several unbroken plates of barnacles. The plates were cleaned under distilled water and then treated with a 5% solution of HCl for three minutes. After rinsing in distilled water and drying, the plates were crushed into powder. The average spacing of samples was between 150 and 200 µg. Samples were reacted at 50°C with concentrated H_3PO_4. The isotopic composition of the evolved CO_2 was analyzed with a Thermo-Finnigan DELTA-V Plus mass spectrometer (IEE RAS, Moscow). The total number of analyzed samples was eighty-four and the total number of laboratory standard replicates was twenty-two with a standard deviation of 0.045‰ for $\delta^{18}O$. All isotope data are expressed in the conventional delta notation, where the isotope ratios of $^{18}O/^{16}O$ are reported relative to the SMOW standard.

Results and discussion

The shell midden ADK-009 has clear stratigraphy. It includes five layers separated by humic-ash layers. Cultural deposits were 140 cm deep (Antipushina 2010b). Four radiometric dates indicate the cultural layers were formed between the 6th and 17th centuries AD (Table 19-1). These radiometric dates further indicate that the formation of the ADK-009 shell midden was uneven. Between the 6th and the middle of the 10th centuries (layers IV and V) the midden accumulated slowly, but formed more rapidly between the 12th and the 14th centuries (layer II).

Depth, cm	Lab. code, IEMAE-	^{14}C age, AD	Correction to ^{13}C, ($\delta^{13}C$, −12.8)	Calendar years, cal AD, (2σ)
23-38	1264	755±72	970±80	1400-1760
45-79	1242	1126±103	1360±110	1020-1450
115-127	1288	1360±74	1560±80	860-1270
150-160	1265	1888±50	2080±55	370-690

Table 19-1 : Radiocarbon dating of ADK-009.

Remains of about 35000 invertebrates, representing sixty-one taxa, were identified in the ADK-009 shell midden. Ethnographic records indicate that eighteen species were important Aleut food resources. Among gathered species the following dominated: foolish mussels *Mytilus trossulus,* Nuttall's cockles *Clinocardium nuttallii,* limpets *Lottia* spp., periwinkles *Littorina* spp., chitons *Katharina tunicata* and *Cryptochiton stelleri* and also barnacles *Semibalanus cariosus* and sea urchins *Strongylocentrotus polyacanthus* (Antipushina 2010b).

The faunal composition of the ADK-009 shell midden does not significantly differ from that of the modern faunal macroinvertebrate community living in the littoral zone at Sweeper Cove. To demonstrate this, we divided all shellfish species in the shell midden into two groups: 1) epifaunal types that prefer rocky substrates and, 2) infaunal types that bury themselves in loose sediments. The remains of epifaunal mollusks dominated in all layers of the ADK-009 shell midden, representing more than 90% of all mollusk remains. This percentage does not significantly change from layer to layer (Antipushina *et al.* 2009). This suggests that the rocky intertidal zone at Sweeper Cove has not dramatically changed from the 6th century to the present.

Moreover, the results of the morphometrical analysis of gastropod shells indicate that changes in wave action and intensity were absent in mentioned time. The limpets and periwinkles' configurations were almost constant (Table 19-2).

	Morphometric ratio	V	IV	III	II	I
Lottia pelta	Shell height to shell length ratio	0	0.34	0.40	0.40	0.38
	Standard deviation		0.08	0.05	0.05	0.05
	N	0	5	10	21	16
Littorina sitkana	Aperture height to shell height ratio	0.74	0.73	0.70	0.69	0.72
	Standard deviation	0.06	0.04	0.03	0.03	0.10
	Aperture width to shell with ratio	0.53	0.57	0.60	0.59	0.58
	Standard deviation	0.08	0.03	0.03	0.03	0.04
	N	11	7	43	79	38

Table 19-2: Results of morphometric analysis of gastropod shells.

The zoogeographical analysis shows that invertebrate's zoogeographical structure of Bering Sea littoral zone did not change significantly over the last millennium. Adak's littoral fauna is zoogeographically heterogeneous (O'Clair and Chew 1971; Antipushina 2010c). The fauna from the archaeological deposits, as well as the modern littoral fauna, contain elements from several geographical provinces, with the Pacific widespread boreal and Arctic fauna dominant (Table 19-3). Boreal Asian and American species in the deposit and noted in modern littoral samples were present at approximately similar rates. This indicates that Aleutian chain and particularly Adak Island occupies an intermediate position between Asia and the Americas. We also discovered several shells of low-boreal species in deposit.

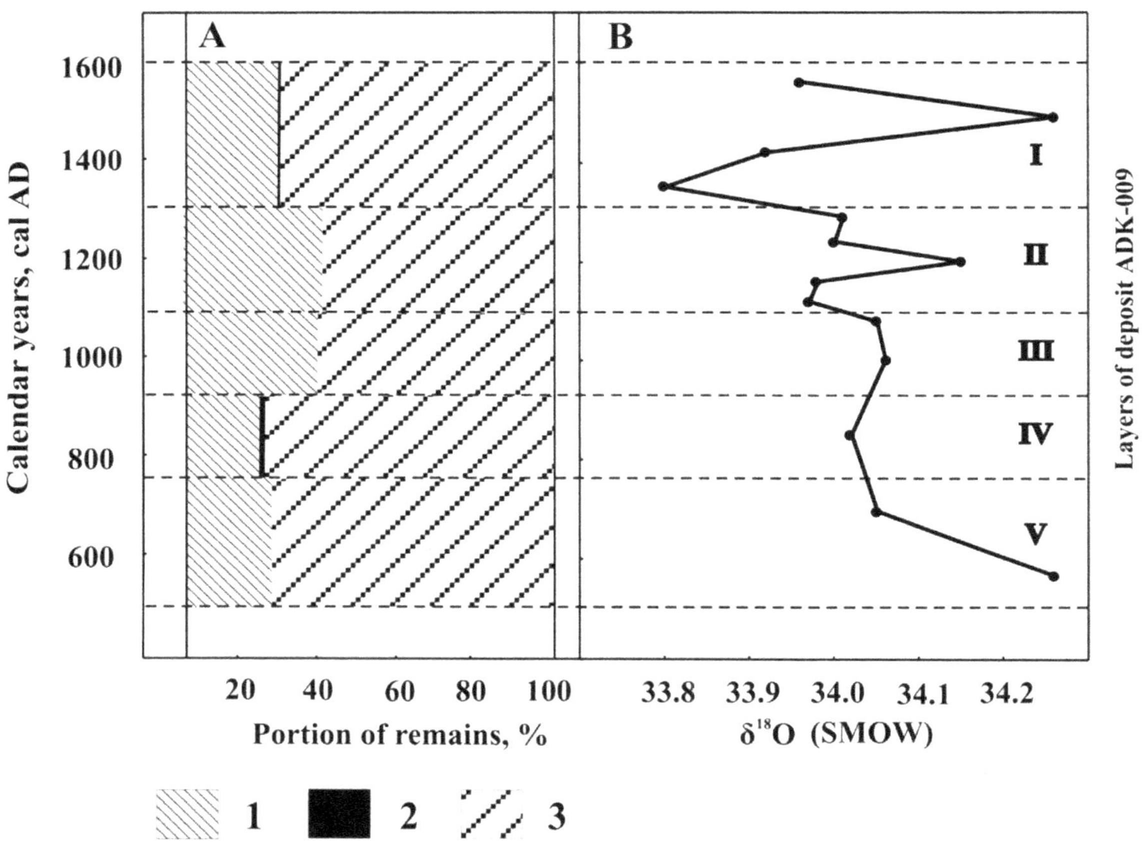

Figure 19-2. Dynamics of the molluscan zoogeographical groups through time (A) and $\delta^{18}O$ value (B) over the last millennia: 1 - Arctic-boreal and high-boreal species; 2 - Low-boreal species; 3 - Boreal species

Zoogeographic group of species	ADK-009 shell midden		Modern Adak's littoral zone	
	Number of species	Proportion, %	Number of species	Proportion, %
Pacific widespread boreal	15	36.6	18	42.9
Arctic-boreal	9	22.0	8	19.0
Pacific widespread high-boreal	5	12.2	2	4.8
Bipolar	1	2.4	2	4.8
Amphiboreal	2	4.9	4	9.5
Pacific Asian boreal	4	9.8	4	9.5
Pacific Asian high-boreal	1	2.4	0	0
Pacific Asian low-boreal	1	2.4	0	0
Pacific American boreal	3	7.3	4	9.5
Total	41	100	42	100

Table 19-3. Zoogeographic composition of the ADK-009 shell midden and the Adak modern littoral zone.

Comparison of the zoogeographic patterns demonstrates that the invertebrate fauna excavated from Holocene archaeological deposits is very similar to the modern littoral macrozoobenthos of the Aleutian Islands. However, the species structure and their population structure changed depending on the dynamics of climatic conditions. This is demonstrated by the dynamics of the zoogeographical composition of shell midden over the time of its accumulation (fig. 19-2, A). The percentages of arctic-boreal and high-boreal species were lower in archaeological layers IV and I. Moreover, remains of the low-boreal gastropod *Nucella heyseana,* atypical in the Bering Sea, were found in the same layers. On the other hand, the percentage of arctic-boreal and high-boreal species increased in layers III and II. This change is probably due to changing temperature over time. The period from the 6th to the middle of the 8th centuries (layer V) was relatively cold. The warmer period from the middle of the 8th to the middle of the 10th centuries (layer IV) encouraged the migration of the low-boreal gastropod *Nucella heyseana* into the higher latitudes. An increase of arctic-boreal and high-boreal species from the middle of the 10th to the 14th centuries (layers II and III) suggests a colder temperature at this time.

Similar and more detailed results were obtained by oxygen stable isotope analysis of barnacle calcite (fig. 19-2, B). As mentioned above, the dynamics of the $\delta^{18}O$ value in organic carbonates is evidence of relative temperature changes; an increase of the $\delta^{18}O$ value is evidence of relative temperature decrease. The coldest time was recorded in the mid-6th, 13th and 16th centuries and the warmer time was marked from the 8th till the 12th, in the mid-15th and the 17th centuries. Our results agree well with dendrochronological data from the Russian Arctic (Hantemirov and Shiyatov 2002) and with the results of ostracod analysis from Alaska's lake sediments (Hu *et al.* 2001). Therefore the marked climate changes were not local and characterized the entire region.

Conclusions

The zoogeographical analysis shows that the general composition of invertebrate fauna now is the same as a millennium ago. The littoral zone of Sweeper Cove was mainly stony, with a certain proportion of sandy ground, and its characteristics have not changed significantly from the 6th to the 17th century. Moreover, results of morphometrical analysis of gastropod shells indicate that changes in wave action and intensity were insignificant over this time period. The zoogeographical and oxygen stable isotope analyses allowed us to reconstruct the dynamics of climatic conditions. The colder time was marked in the mid-6th, the 13th and the 16th centuries, and the warmer time was marked in the 10th, the 12th, the mid-15th and the 17th centuries. The marked climate changes were not local and characterized entire Pacific arctic.

Acknowledgements

I am indebted to Dr. D. West (University of Kansas), Dr. A.B. Savinetsky and all colleagues from Lab. of historical ecology (IEE RAS) for their help. The work was carried out with the financial support of the Russian Foundation for Basic research (№ 12-04-31010 and № 12-04-00655), National Science Foundation (OPP-0353065) and Programs of RAS "Fundamentals of Biological Resources Management", "Origin and Evolution of Biosphere" and "Biodiversity".

References

Antipushina, Zhanna A., Nina K. Kiseleva, Bulat F. Khassanov, Aleksey V. Pakhnevich, Olga A. Krylovich, Boris I. Sirenko, and Arkady B. Savinetsky. 2009. Holocene dynamics of invertebrate fauna in the littoral zone of Adak, the Aleutian Islands, according to archaeozoological data. *Russian Journal of Ecology* 40 (2):128–136.

Antipushina, Zhanna A. 2010a. Using of marine invertebrates from archaeological sites for palaeoenvironmental reconstructions. In *The Holocene dynamics of ecosystems: proceedings of II Russian conference.* N. Smirnov, ed. Pp. 15–20. Ekaterinburg-Chelyabinsk: Riphey. [In Russian.]

—–. 2010b. Analysis of malacofauna remains from archaeological sites on Adak Island (Aleutian Islands, USA). MUNIBE S*uplementost– Gehigarria* 31:120–127.

—–. 2010c. Zoogeographical analysis of the subfossil macrozoobentos of the Commander-Aleutian Islands in the Middle and the Late Holocene. *Biology Bulletin* 37 (9):885–893.

Boulding, Elizabeth G., and Kathryn L. Van Alstyne. 1993. Mechanisms of differential survival and growth of two species of Littorina on wave-exposed and on protected shores. *Journal of Experimental Marine Biology and Ecology* 169:139–166.

Cabral, Joao Paulo, and Armando Coelho F. da Silva. 2003. Morphometric analysis of limpets from an Iron-Age shell midden found in northwest Portugal. *Journal of Archaeological Science* 30 (7):817–829.

Causey, Douglas, Debra G. Corbett, Christine Lefevre, Dixie L. West, Arkady B. Savinetsky, Nina K. Kiseleva, and Bulat F. Khassanov. 2005. The palaeoenvironment of humans and marine birds of the Aleutian Islands: three millennia of change. *Fisheries Oceanography* 14 (1):259–276.

Dall, William. 1877. On succession in the shell-heaps of the Aleutian Islands. In *Contributions to North American Ethnology.* 1:41–91. Washington: Government Printing Office.

Epstein, Samuel, Ralph Buchsbaum, Heinz A. Lowenstam, and Harold C. Urey. 1951. Carbonate-water isotopic temperature scale. *Bulletin of the Geological Society of America* 62 (4):417–426.

Hantemirov, Rashid M., and Stepan G. Shiyatov. 2002. A continuos multimillennial ring-width chronology in Yamal, northwestern Siberia. *The Holocene* 12 (6):717–726.

Hu, Feng Sheng, Emi Ito, Thomas A. Brown, Brandon B. Curry, and Daniel R. Engstrom. 2001. Pronounced climatic variations in Alaska during the last two millennia. *PNAS* 98 (19):10552–10556.

Janson, Kerstin, and Per Sundberg. 1983. Multivariate morphometric analysis of two varieties of *Littorina saxatilis* from the Swedish west coast. *Marine Biology* 74 (1):49–53

Jones, Douglas S., and Irvy R. Quitmyer. 1996. Marking time with bivalve shells: oxygen isotopes and season of annual increment formation. *Palaios* 11 (4):340–346.

Killingley, John S., William A. Newnan. 1982. O18 fractionation in barnacle calcite: a barnacle paleotemperature equation. *Journal of Marine Research* 40:893-902.

Laughlin, William S. 1975. Aleuts: Ecosystem, Holocene History, and Siberian Origin. *Science* 189 (4202):507–515.

Longin, Robert. 1971. New method of collagen extraction for radiocarbon dating. *Nature* 230:241- 242.

McCartney, Allen P. 1975. Maritime Adaptations in Cold Archipelagoes: An Analysis of Environment and Culture in the Aleutian and Other Island Chains. In *Prehistoric Maritime Adaptations of the Circumpolar Zone,* Pp. 281–338. Fitzhugh, W., Ed., The Hague: Mouton, Chicago.

Savinetsky, Arkady B., West, Dixie L., Zhanna A. Antipushina, Bulat F. Khassanov, Nina K. Kiseleva, Olga A. Krylovich, and Andrei M. Pereladov. 2011. The reconstruction of ecosystems history of Adak Island (Aleutian Islands) during the Holocene. In *The People Before: The Geology, Paleoecology and Archaeology of Adak Island, Alaska.* Oxford, British Archaeological Reports International Series. Pp. 77–107.

Skarlato, Orest A. 1981. *Dvystvorchatye molluski umerennyh morei zapadnoi chasti Tihogo okeana* [Bivalves of the temperate seas of the western part of Pacific ocean]. Leningrad: Nauka. [In Russian.]

Stuiver, Minze, and Thomas F. Braziunas. 1993. Modeling Atmospheric 14C Influences and 14C Ages of Marine Samples to 10 000 BC. *Radiocarbon* 35 (1):137–189.

O'Clair, Charles. E., and Kenneth K. Chew. 1971. Transect studies of littoral macrofauna, Amchitka Island, Alaska. *BioScience* 21 (12):661–665.

Overland, James E. 1981. Marine Climatology of the Bering Sea. In: *The Eastern Bering Sea Shelf: Oceanography and Resources.* 1:15–22. D.W. Hood and J.A.

Calder, eds. Seattle: University of Washington Press.

Rodionov, Sergei N., James E. Overland, and Nicholas A. Bond. 2005. Spatial and temporal variability of the Aleutian climate. Fisheries Oceanography 14 (1):3–21.

Telford, Richard J., Einar Heegaard, and H. J.B. Birks. 2004. All age-depth models are wrong: but how badly? *Quaternary Science Reviews. Rapid Communication* 23(1-2):1 – 5.

Veniaminov, Ivan E. 1984. *Notes on Islands of the Unalaska Range*. The Limestone Press: Kingston

20 - BIOSTRATIGRAPHY OF SHELLS AND CLIMATE CHANGES IN THE CANTABRIAN REGION (NORTHERN SPAIN) DURING THE PLEISTOCENE-HOLOCENE TRANSITION

Igor GUTIÉRREZ-ZUGASTI,
Instituto Internacional de Investigaciones Prehistóricas de Cantabria, Ed. Interfacultativo Universidad de Cantabria, Avda. de los Castros s/n, 39005 Santander (Cantabria, Spain),
igorgutierrez.zug@gmail.com

David CUENCA-SOLANA
Centre de Recherche en Archéologie Archéosciences Histoire, UMR 6566 CNRS CReAAH. Université Rennes 1. Campus Beaulieu - Bât 24 – 25. 263 avenue du Général Leclerc - CS 74 205. 35042 Rennes Cedex (France),
david.cuencasolana@gmail.com

Abstract: *In many cases, mollusc species have certain climatic preferences and are, therefore, sensitive to climatic change. Thus, sometimes changes in the composition of archaeomalacological assemblages through time can reflect variations in climatic conditions, which make molluscs an important climatic indicator. In the Cantabrian region, the analysis of 48 stratigraphic units with evidence of molluscan exploitation, ranging from Late Magdalenian (16,000 cal BP) to the Neolithic (5,000 cal BP) have shown changes in the composition of shell assemblages related to climatic changes during the Pleistocene-Holocene transition. The most common species during the colder phases of the Upper Palaeolithic* (Littorina littorea *and* Patella vulgata) *decreased dramatically at the start of the Holocene, while other more temperate species* (Phorcus lineatus *and* Patella depressa) *increased in numbers. Similarly, an increase in the number of exploited species has been documented, which is probably due to the introduction of bivalve species resulting from climatic improvement, sea level rise, and the formation of estuaries.*

Keywords: *Archaeomalacology, Shell midden, Upper Palaeolithic, Mesolithic, Neolithic.*

Introduction

The recent increased interest in global climate change and its implication for marine fauna has encouraged a series of studies on the climate preferences of molluscs and the changes in their modern geographical distributions (Hawkins, Southward, and Genner 2003; Hawthorne 2010; Mieszkowska *et al.* 2007). In archaeology, marine molluscs have sometimes been used as climate proxies, owing to the preferences exhibited by some taxa, although terrestrial gastropods have been employed most often for environmental reconstructions (Evans 1972; Limondin 1995; Sparks 1980). As in the case of actualistic studies, their use as climatic or environmental proxies has generally been based on the determination of the preferences of each taxon through the observation of its modern geographical distribution or biotope conditions. However, this type of inference, based on climate adaptation, can only be made when the molluscs display certain preferences, since some species are eurythermal and can live in a wide range of climatic conditions. The more restricted the preferences of a species are, the more useful it will be as a climate proxy.

In Cantabrian Spain, coastal resources were used by Upper Palaeolithic and Mesolithic hunter-gatherers, and also throughout the Neolithic (Álvarez-Fernández 2011; Bailey and Craighead 2003; Gutiérrez-Zugasti 2011a; Straus and Clark 1986). The presence or absence of some mollusc species has been used to assign archaeological levels to certain periods according to their climate parameters. The most commonly used environmental inference has been to assign a cold climate and Palaeolithic chronology to a level containing the species *Littorina littorea*, and a temperate climate and post-Palaeolithic chronology to levels with *Phorcus lineatus* (Clark 1976; Madariaga 1963; Vega del Sella 1923). The first researcher to see chronological differences in the presence of these two species was Vega del Sella (1923) in the course of his studies of Mesolithic shell middens. He realised that *Littorina littorea* was the species that was found in Upper Palaeolithic levels and was always absent from Mesolithic shell middens, whereas *Phorcus lineatus* behaved inversely. At that time, these differences were conceived in terms of a total substitution of one species by the other, which gave them the value of fossil-guides, indicating the chronology of the levels in which they were found. This view of total substitution was sustained by other authors in the following years (Madariaga 1964) and even into the 1990s (Straus 1992). However, Clark's work (1976) in the Asturian area provided enough data to show that both species were present (albeit in small numbers) in the periods in which the other predominated. It was possible to find *Phorcus lineatus* in Upper Palaeolithic levels (at least in the Late Magdalenian) and *Littorina littorea* in the

Mesolithic. González-Morales (1982) also pointed this out, and proposed that the inversion in the proportions of *Phorcus lineatus* and *Littorina littorea* was a gradual, but relatively rapid, process. In the following years Ortea's studies at La Riera (1986) and Moreno's work at El Perro (1994), with sequences covering the Late Magdalenian, Azilian and Mesolithic, clearly showed that both species had lived together throughout this whole period, but in different proportions depending on the specific time. *Littorina littorea* was more abundant in Late Magdalenian and Azilian levels and *Phorcus lineatus* was predominant during the Mesolithic. An example of a similar change, though less studied in archaeomalacological literature in Cantabria Spain (Clark 1976), is that of the *Patella* genus. In the Upper Palaeolithic the most common species is *Patella vulgata*, whereas at Mesolithic sites, a marked increase is seen in the presence of the species *Patella depressa* and *Patella ulyssiponensis*.

Despite this long history of research, no detailed studies have been undertaken on the evolution of all these species throughout time, nor has this evolution been compared with the available climate information. Similarly, changes in other species, which might also reflect changes in the climate, have not been considered. In this paper we reappraise the existing hypotheses based on the detailed study of the numbers of different species in each stratigraphic unit at multiple sites and compare this data with the evolution in the climate. We also examine the changes in several bivalve species to determine whether their presence in archaeological deposits is related to changes in climatic conditions.

Materials and Methods

To carry out this study data have been used from 48 stratigraphic levels belonging to 18 sites along the northern Spanish coast, dated between the late Upper Palaeolithic and Neolithic (16,000 – 5,000 cal BP) (fig. 20-1, Table 20-1). The data from 24 of these stratigraphic levels at nine sites (La Llana, Mazaculos II – Sector 3, La Pila, La Fragua, La Chora, La Trecha, Arenillas, Kobaederra and Santimamiñe) were obtained through a direct study of the malacological material (Gutiérrez-Zugasti 2009). The information from the remaining 24 stratigraphic levels at 10 sites (Les Pedroses, La Lloseta, La Riera, Poza l'Egua, Mazaculos II – Sector 1, La Garma A, El Perro, Pico Ramos, Laminak II and Ekain) comes from the scientific literature (Álvarez-Fernández 2012; Arias *et al.* 2007; Clark 1976; González-Morales *et al.* 1980; Imaz, 1994; Leoz and Labadía 1984; Moreno 1994, 1995; Ortea 1986).

The mollusc species chosen for the study are the most abundant ones in regional shell middens between the end of the Upper Palaeolithic and the Neolithic (Late Magdalenian, Azilian, Mesolithic and Neolithic). The nomenclature of CLEMAM (Muséum d'Histoire Naturelle Paris, France) has been used in the taxonomic identification. Gastropods are identified as *Littorina littorea* (Linné, 1758: *Turbo*), *Phorcus lineatus* (da Costa, 1778: *Trochus*), *Patella vulgata* Linné, 1758, *Patella depressa* Pennant,177, and *Patella ulyssiponensis* Gmelin, 1791, and bivalves are identified as *Mytilus* sp., *Ostrea edulis* Linné, 1758, *Scrobicularia plana* (da Costa, 1778: *Trigonella*), *Ruditapes decussatus* (Linné, 1758: *Venus*) and *Solen marginatus* Pulteney, 1799. To determine the abundance of each species in the deposits studied by the authors of the present paper, the MNI (Minimum Number of Individuals) was calculated with a method based on fragmentation categories (Gutiérrez-Zugasti 2009, 2011b). Then the importance of the different species in each level was determined with their relative frequencies (%MNI). Data about the MNI, and in most cases about the relative frequencies, were also available for the sites selected from the literature. In all sites, except Les Pedroses and La Lloseta, where 1/8 inches (3.7 mm) mesh screen was used (Clark, 1976), sediments were sieved at least with 2 mm mesh screens, allowing the comparison between assemblages.

Given the spatial and temporal variability observed in the Marine Reservoir Effect in both Atlantic Europe (Russell *et al.* 2010; Soares and Martins 2009) and on the Bay of Biscay (synthesized in Fano 2007), the dates obtained

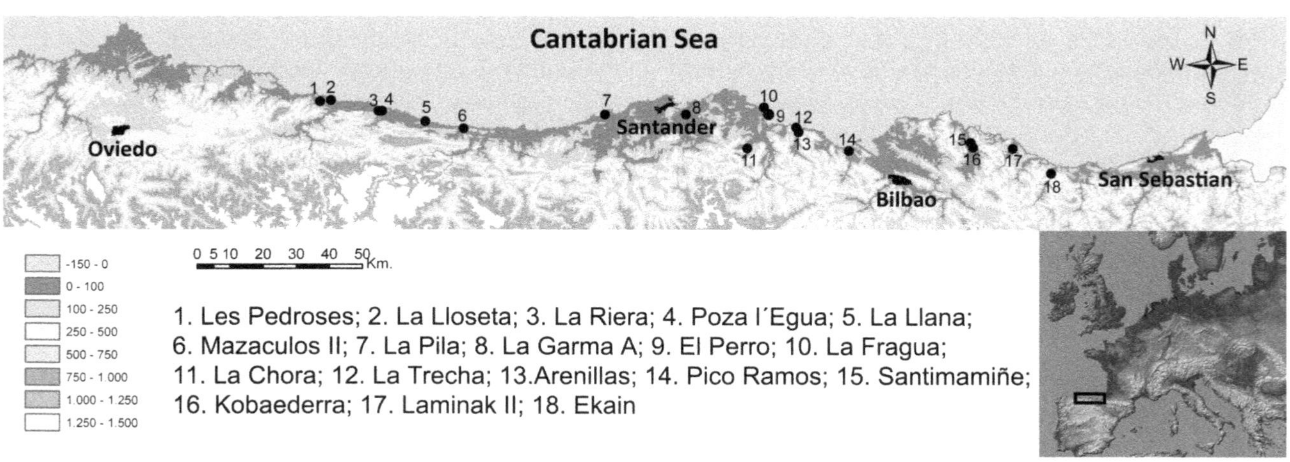

Figure 20-1. Location and distribution of sites included in this study.

using shells have not been calibrated. To assign the stratigraphic levels to a certain climatic period, the data from the Greenland Ice Cores GISP2 and GICC05 have been used (Andersen *et al.* 2006; Rasmussen *et al.* 2006, 2007; Stuiver, Grootes, and Braziunas 1995).

Results and Discussion

The Littorina littorea – Phorcus lineatus *change*

According to the modern geographic distribution of each species, *Littorina littorea* is a cold-loving species whereas *Phorcus lineatus* is adapted to temperate climates (Poppe and Goto 1991). By observing the numbers of each species in each stratigraphic level in the studied sequences (fig. 20-2), *Littorina littorea* is most abundant in the Late Magdalenian and Azilian, whereas *Phorcus lineatus* predominates in the Mesolithic and Neolithic.

The data from the Late Magdalenian at La Garma A, especially from Level O in which the proportions of each species are similar (although favourable to *Littorina littorea*), have led Álvarez-Fernández (2012) to propose that the substitution of *Littorina littorea* by *Phorcus*

Site	Level	BP	cal BP 1σ	Chronoculture	Material	Lab Ref	Climate
La Lloseta	B	4594±680	5150±840	Neolithic?	Charcoal	GaK-2551	Holocene
Mazaculos II	A2	5050±120	5800±120	Neolithic	Charcoal	GaK-15221	Holocene
Kobaederra	II	5200±100	5980±160	Neolithic	Charcoal	UBAR-472	Holocene
Arenillas	Shell midden	5580±80	6380±70	Neolithic	Charcoal	GrN-19596	Holocene
Les Pedroses	Sampling	5760±180	6590±200	Neolithic	Charcoal	GaK-2547	Holocene
Kobaederra	III	5820±240	6670±270	Neolithic	Charcoal	UBAR-471	Holocene
Kobaederra	IV	5630±100	6440±110	Neolithic	Charcoal	UBAR-470	Holocene
Pico Ramos	4	5860±65	6670±80	Neolithic?	Charcoal	Ua-3051	Holocene
La Chora	Shell midden	6360±80	7300±90	Mesolithic	Charcoal	GrN-20961	Holocene
La Fragua	1Up	6650±120	7540±90	Mesolithic	Charcoal	GrN-20963	Holocene
La Fragua	1Mid	6860±60	7710±60	Mesolithic	Charcoal	GrN-20964	Holocene
La Riera	30	6500±200	7370±190	Mesolithic	Charcoal	Gak-3046	Holocene
Mazaculos II	A3	7030±120	7850±110	Mesolithic	Charcoal	GaK-15222	Holocene
Mazaculos II	1,1	7280±220	8110±210	Mesolithic	Charcoal	GaK-8162	Holocene
La Trecha	1	7500±70		Mesolithic	Shell	URU-0038	Holocene
La Fragua	1Low	7530±70	8320±80	Mesolithic	Charcoal	GrN-20665	Holocene
Poza l'Egua	2	8550±80	9550±70	Mesolithic	Bone	TO-10222	Holocene
La Riera	29	8650±300	9720±380	Mesolithic	Charcoal	Gak-2909	Holocene
El Perro	1	9260±110	10450±140	Mesolithic	Charcoal	GrN-18116	Holocene
Mazaculos II	3,3	9290±440	10560±610	Mesolithic	Charcoal	Gak-6884	Holocene
La Riera	28	9230±90		Azilian	Shell	Q-2933	Holocene
Ekain	II	9540±210	10850±280	Azilian	Bone	I-11666	Holocene
La Fragua	3	9600±140	10930±200	Azilian	Charcoal	GrN-20966	Holocene
El Perro	2a/b	10160±110	11790±250	Azilian	Charcoal	GrN-18115	GS1
Laminak II	1	10380±140	12250±260	Azilian	Bone	Ua-2361	GS1
La Riera	27up	10630±120	12540±160	Azilian	Bone	BM-1494	GS1
La Riera	27low	11390±70		Azilian	Shell	Q-2935	GI1/GS1
La Pila	III.3	11710±120	13590±140	Azilian	Bone	Gif-8148	GI1c
Laminak II	II	11700±140	13580±150	Late Magdalenian	Bone	Ua-2362	GI1c
La Riera	26	11850±85		Azilian/Late Magd.	Shell	Q-2925	GI1b/c
La Riera	24	11880±75		Late Magdalenian	Shell	Q-2926	GI1b/c
La Garma A	O	12070±100	14100±230	Late Magdalenian	Bone	OxA-7203	GI1c/d/e
El Perro	2c	12140±180	14260±340	Late Magdalenian	Charcoal	GrN-20962	GI1c/d/e
La Pila	IV.2	12160±130	14270±280	Late Magdalenian	Bone	Gif-8147	GI1c/d/e
La Garma A	N	12420±180	14650±390	Late Magdalenian	Bone	AA-45558	GS2/GI1e
La Riera	23	12620±300	14970±550	Late Magdalenian	Bone	UCR-1274D	GS2/GI1e
La Fragua	4	12960±50	15530±60	Late Magdalenian	Charcoal	GrN-29440	GS2
La Riera	21-23	13200±110		Late Magdalenian	Shell	Q-2932	GS2

Table 20-1. Radiocarbon dates from stratigraphic levels included in this study and reference to the climate. CalCurve: CalPal2007_HULU (Weninger and Jöris 2008; Weninger *et al.* 2008). GI: Greenland Interstadial; GS: Greenland Stadial. Sources: synthesized in Gutiérrez-Zugasti (2009).

lineatus took place from the Late Magdalenian onwards. Around the Last Glacial Maximum, during the Solutrean and Lower Magdalenian, there is hardly any evidence of the gathering of *Phorcus lineatus*. It is, thus, likely that its population was very small or even non-existent at times due to the severe climate conditions. During the Late Magdalenian this species begins to be commonly found in malacological assemblages, but always after 15,000 – 14,700 cal BP, at the end of GS2 and mainly in the GI1. The available date for Level O at La Garma situates it in the GI1e/d/c (Bölling/Older Dryas/Allerød), so the temperate conditions at that time may have led to an increase in the numbers of *Phorcus lineatus*.

However, at other sites with a similar chronology, such as El Perro (Level 2c) and La Pila (Level IV) this increase is not seen. Although *Phorcus lineatus* is found, it is much less abundant than *Littorina littorea*, and therefore these levels may have formed during the colder GI1d period (Older Dryas). The currently available data does not support the hypothesis that the substitution began during the Late Magdalenian, although it is possible that the effects of the GI1 caused an increase in *Phorcus lineatus* populations at certain times.

The levels that formed during the Magdalenian – Azilian transition, belonging to the GI1c temperate period (Allerød), exhibit a predominance of *Littorina littorea*. By contrast, a decrease in its abundance can be seen during this period at La Riera (Levels 24 and 26) and especially at La Pila. Here, the proportions of *Littorina littorea* decrease over time and the *Phorcus lineatus* frequencies increase, possibly resulting from milder climate conditions. However, during the cold GS1 period (Younger Dryas) at La Riera (Levels 27 low and up) there is a decrease in the numbers of *Littorina littorea*, whereas at El Perro (Level 2a/b) this species makes up over 50% of the midden around 11,700 cal BP, probably favoured by the colder conditions in the GS1. In contrast, the data from Laminak II (Level 1, dated to the GS1) do not fit into this pattern as there is a greater proportion of *Phorcus lineatus*, although the sample size is too small to be conclusive (MNI= three

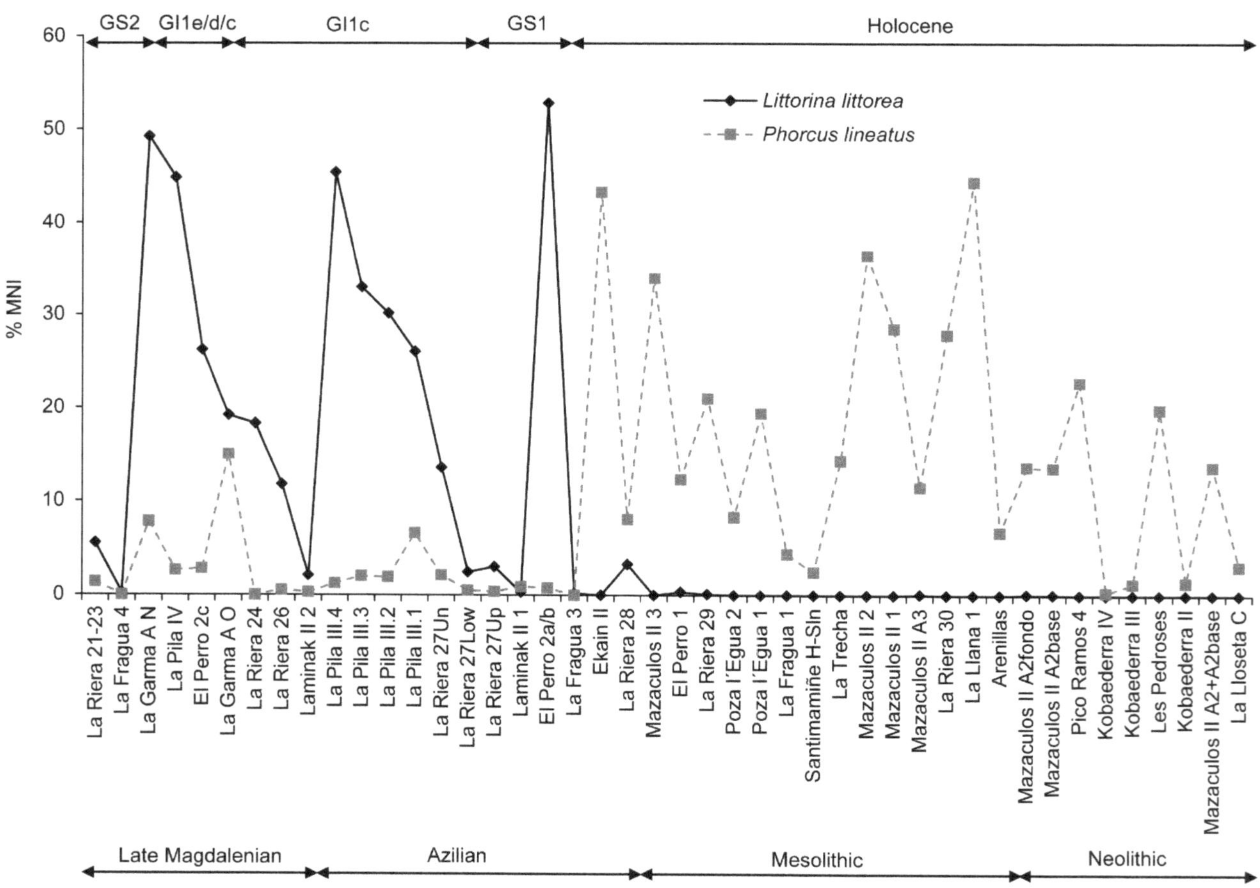

Figure 20-2. Changes in *Littorina littorea* and *Phorcus lineatus* %MNI from the Late Magdalenian to the Neolithic. GI: Greenland Interstadial; GS: Greenland Stadial. Levels dated from shells have been placed according to uncalibrated dates.

Phorcus lineatus and one *Littorina littorea*).

Based on the chronology obtained at El Perro, it was possible to determine that the change in the proportions of the two species took place in the early Holocene, between 11,500 and 10,200 cal BP (González Morales *et al.* 1999). Thus, the only Azilian site with absolute dates within that range where *Littorina littorea* is the most abundant or sole species is La Fragua (Level 3), although the non-quantitative data from Santa Catalina (Level 1) seem to reflect the same pattern (Berganza *et al.* 2012). In contrast, sites where *Phorcus lineatus* is the predominant or only species include La Riera (Level 28) and Ekain (Level II), both attributed to the Azilian, and Mazaculos II (Level 3.3) and El Perro (Level 1), which are assigned to the Mesolithic.

Taking into account all this data, with the exception of the date for Level 28 at La Riera which seems too recent, all the other levels show that the change took place about 10,900 – 10,800 cal BP. However, the Azilian ended about 10,700 cal BP and the Mesolithic was established by about 10,500 cal BP. Therefore the predominance of *Phorcus lineatus* at several sites during the late Azilian, such as Ekain and La Riera, indicates that the change between the two species happened before the end of the period, and it was therefore not due to a shift in subsistence strategy. Instead, it was likely due to the improvement in the climate after the GS1, at the end of the Preboreal oscillation, after 11,200 cal BP. If we examine the percentage of levels in each period in which the two species are found (Gutiérrez-Zugasti 2009), this trend seems to be confirmed. During the Azilian the percentage of levels in which *Phorcus lineatus* has been found is larger than the percentage with *Littorina littorea*. Therefore, the causes of this inversion in proportions are likely climatic and are not related to cultural factors or exploitation strategies.

Changes in the distribution of the genus Patella *species*

The modern geographical distributions indicate that *Patella vulgata* is a species that prefers a cold climate, whereas *Patella depressa* and *Patella ulyssiponensis* are warm-loving species (Poppe and Goto 1991). Despite the bias introduced into studies by the preservation of the shells, which on many occasions only allows the individuals to be identified at genus level, the results of archaeomalacological research has succeeded in defining the evolution of these species. By examining the changes in the relative frequencies of each species, it can be seen that in the Late Magdalenian and Azilian the predominant species is *Patella vulgata*, whereas in the Mesolithic and Neolithic the numbers of *Patella depressa*, and to a slightly lesser extent of *Patella ulyssiponensis*, increase considerably (fig. 20-3).

In the Solutrean and Lower Magdalenian, *Patella vulgata* is practically the only species found in regional archaeomalacological assemblages, while the presence of *Patella depressa* and *Patella ulyssiponensis* is insignificant, similar to *Phorcus lineatus*. The predominance of *Patella vulgata* is maintained during the Late Magdalenian, although *Patella depressa* begins to be found with greater regularity after 15,500 cal BP, albeit still in small numbers. Its percentages in levels formed in the GS2/GI1e (Bölling) transition at La Riera (Level 21-23) and La Garma A (Level N) are over 5%. By contrast, in Level O at La Garma A, dated to the GI1e/d/c (Bölling/Older Dryas/Allerød), *Patella depressa* is more frequent than *Patella vulgata*, probably because of the milder climate conditions. In the GI1c (Allerød), it also appears in similar numbers to *Patella vulgata* in Level 2 at Laminak II. However, as occurred with *Phorcus lineatus*, at other sites with levels formed during the GI1, such as La Pila (Level IV.2) and La Riera (Levels 24 and 26), its percentages are still low. Throughout this time, *Patella ulyssiponensis* has also been documented, but it is only found in significant amounts during the GIc (Allerød) in Level 2 at Laminak II, where it is the predominant species. This might be due to intensified exploitation of low shores, or to mistakes in the identification of this species.

During the Azilian, *Patella vulgata* is still the main species, although a small increase is seen in the percentages of *Patella depressa* in the sequence at La Pila (GI1c - Allerød) and in Level 27un at La Riera, which probably corresponds to the end of GI1. However, at the end of this period and the start of GS1 (Younger Dryas) at La Riera (Levels 27 low and up), the numbers of the latter species remain very low. In the GS1, however, it becomes much more abundant at Laminak II (Level 1). The changes in *Patella ulyssiponensis* are similar, and it is also very abundant at Laminak II in the GS1. The abundance of these temperate species during the GS1 at Laminak II is contradictory with the rest of the available archaeological information, probably for the same reasons proposed for Level 2. Therefore, apart from this exception, in the remaining levels a slight increase in the numbers of temperate climate species coincides with interstadial periods.

In the early Holocene, the proportions of the different species tend to equal out. About 10,900 cal BP at La Fragua (Level 3) *Patella vulgata* is still the most abundant species. However, after 10,800 cal BP *Patella ulyssiponensis* and *Patella depressa* predominate at Ekain (Level II) and La Riera (Level 28), respectively. Hence it is in the Mesolithic, after 10,500 cal BP, when the populations of *Patella depressa* attain greater abundance, though it cannot be said that it substitutes *Patella vulgata* as the main species. Equally, the predominance of *Patella vulgata* in Levels 2 and 1 at Mazaculos II could be connected with the cooling identified at 8,200 cal BP (Von Grafenstein 1998), although the information about the event in this region does not allow this hypothesis to be confirmed. For example, no increase has been documented in the numbers of *Littorina littorea*. In the Early Neolithic, the

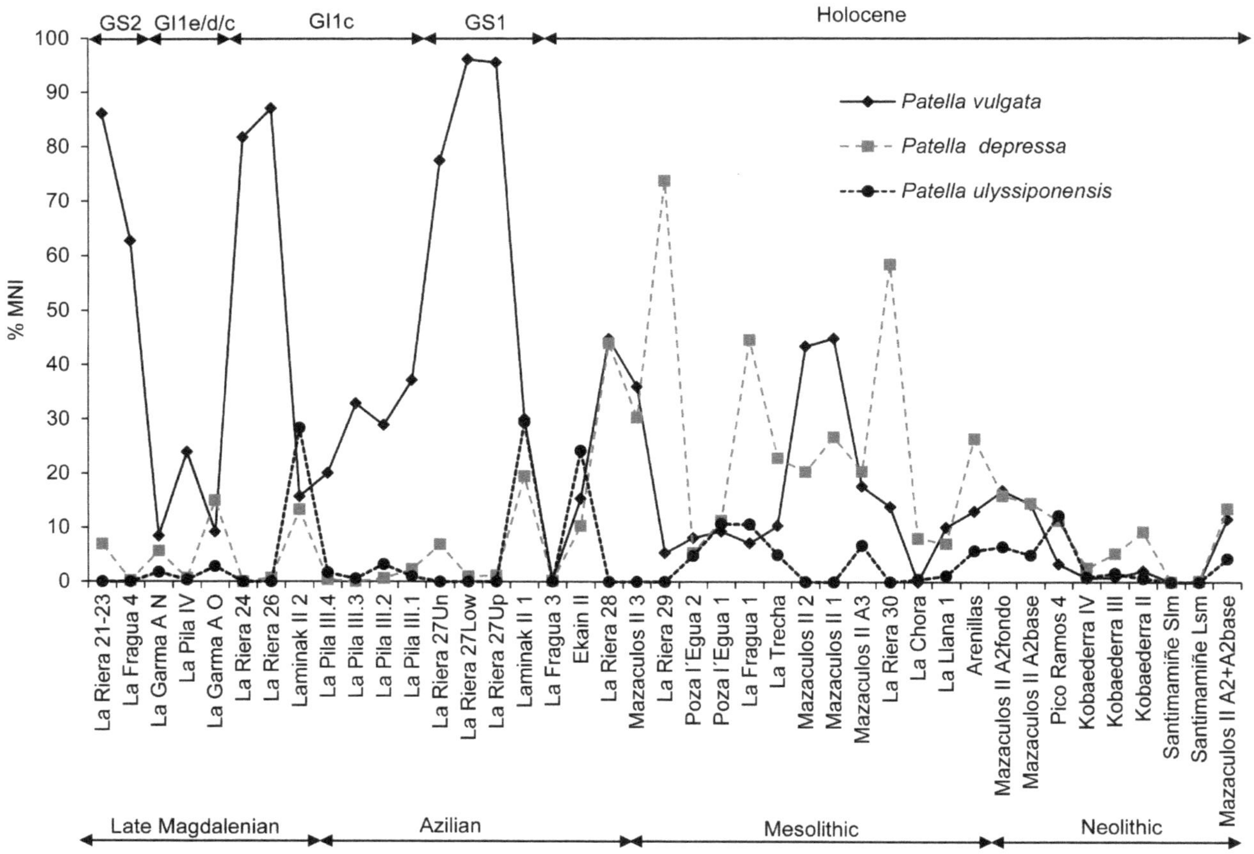

Figure 20-3. Changes in *Patella vulgata, Patella depressa* and *Patella ulyssiponensis* %MNI from the Late Magdalenian to the Neolithic. GI: Greenland Interstadial; GS: Greenland Stadial. Levels dated from shells have been placed according to uncalibrated dates.

greater importance of *Patella depressa* appears to become consolidated, although the small numbers of deposits with appropriate data prevents a more precise assessment of its real importance.

Therefore, as in the case of *Phorcus lineatus*, the value of this species for hunter-gatherer groups in the region begins to attain significance during the Holocene, about 10,800 cal BP. This is supported by the data from Ekain (Level 2) and also probably from La Riera (Level 28), which appear to correspond to the end of the Azilian. It thus seems certain that the changes in *Patella depressa* in the region, at least until the Mesolithic, are determined by climate conditions. The changes in *Patella ulyssiponensis* are very similar to *Patella depressa*, although it generally appears in smaller numbers than the latter species, probably because it lives in low zones that are more difficult for humans to access. The changes in its abundance may therefore also be regarded as influenced by the predominant climate conditions during the period of study.

Bivalves, climate and the formation of estuaries

A very distinct exploitation pattern existed in the Late Magdalenian and Azilian based on two main species, *Patella vulgata* and *Littorina littorea*, which always account for over 80% of the total number of molluscs in the deposits. Consequently, the presence of bivalves at the end of the Upper Palaeolithic is insignificant, even in more temperate phases such as the GI1. The first relevant increase in the numbers of bivalves occurs in the Holocene, in the early Mesolithic, in Level 1 at El Perro (ca. 10,500 cal BP), slightly later than the increase in temperate gastropod species. After this time, the presence of bivalves at archaeological sites in the region becomes more common throughout the Mesolithic, and even more so in the Neolithic when a trend can be seen towards a larger number of species being gathered, especially those living in sandy/muddy shores. This process is especially evident in the eastern part of the region, at sites such as El Perro, La Chora, La Trecha, Arenillas, Santimamiñe and Kobaederra (fig. 20-4).

Climate and environmental conditions (geomorphology, plankton production, salinity and temperature) may

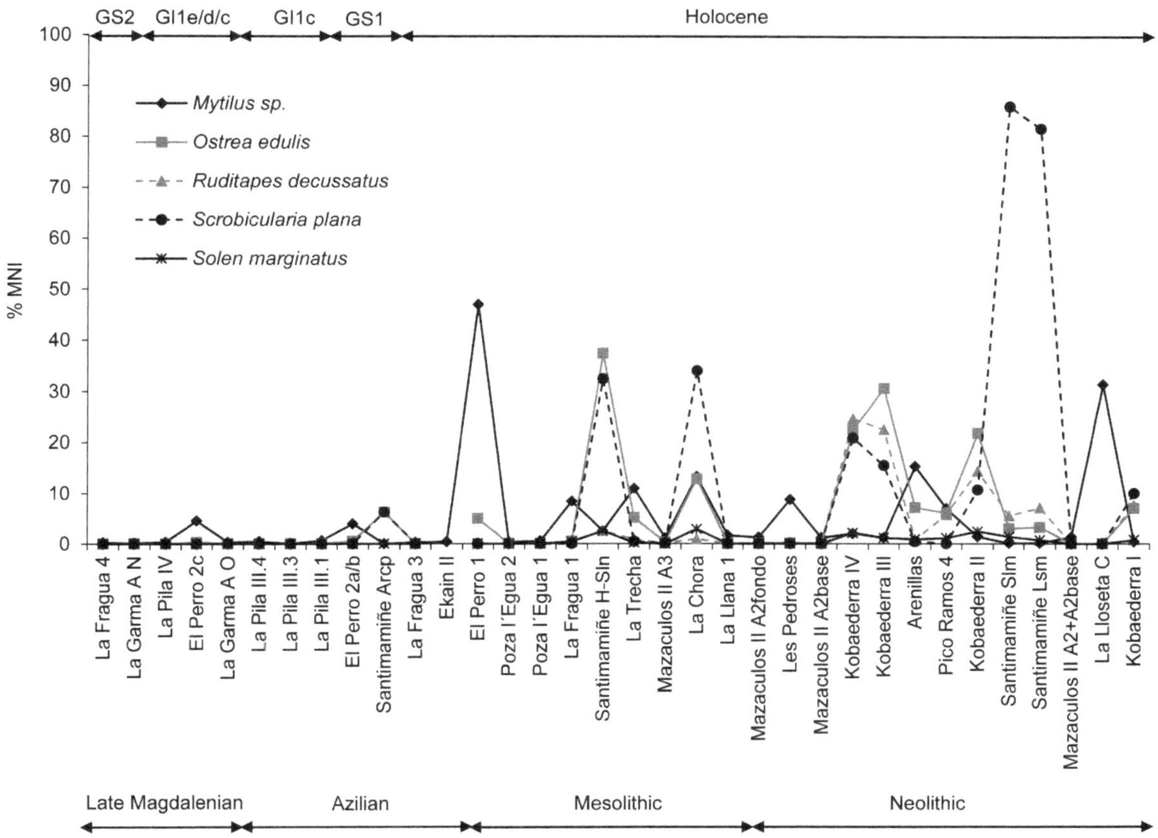

Figure 20-4. Changes in bivalves %MNI from the Late Magdalenian to the Neolithic. GI: Greenland Interstadial; GS: Greenland Stadial. Levels dated from shells have been placed according to uncalibrated dates.

have played an important role in the relative presence or absence of species such as *Mytilus galloprovincialis*, *Ostrea edulis*, and others typical of sandy and/or muddy shores (*Ruditapes decussatus*, *Scrobicularia plana* and *Solen marginatus*) in the Late Pleistocene. The modern geographical distributions of most of these species indicate they can be classed as preferring temperate conditions, except *Solen marginatus*, which has a wider distribution and may therefore be regarded as eurythermal (Poppe and Goto 1993). Since the Gulf Stream did not reach the shores of the Bay of Biscay until the early Holocene, sea surface temperatures during the Late Glacial must have been substantially colder, inhibiting the development of all these species. In addition, the largest populations of a species are located in the centre of their geographical distribution area. Therefore, although these species may all have been present on the shores of northern Spain in the Late Pleistocene, it is unlikely that they would have been very abundant as they would have been near their northern limit. It has already been seen how *Phorcus lineatus* and *Patella depressa* were present from the Late Glacial onwards but did not reach their greatest abundance until the Holocene.

One bivalve, however, does not fit this pattern, since it could have been present in the region since the Late Glacial according to its climatic preferences. *Mytilus edulis*, a cold climate species adapted to even extreme Arctic conditions has a present southern limit in the Ria of Bidasoa (Basque Country). The *Mytilus* genus has been identified at many Late Magdalenian and Azilian sites but its scarcity and degree of fragmentation do not allow the remains to be identified at species level in most cases. However, considering that the northern limit of the modern geographical distribution of *Mytilus galloprovincialis* is the Bidasoa, it is unlikely that this was the species present in the Pleistocene. In any case, if *Mytilus edulis*, because of its preferences, could have lived on the shores of Cantabrian Spain in the Upper Palaeolithic, why it was not exploited more as food? Since the argument of climate improvement cannot be used to explain its absence, perhaps an explanation connected with subsistence strategies is more appropriate. For example, perhaps food poisoning due to a lack of understanding of the effects of red tides on this type of bivalve may have restricted the gathering and consumption of this species, whereas a greater understanding of this natural process in the Mesolithic and Neolithic would have encouraged its consumption.

On the other hand, the increase in the foraging of bivalves in the Holocene is probably not due only to the influence of climate on their geographical distribution areas. It is likely also a result of the rise in sea level and the formation of the modern estuaries, which enlarged the range of environments where these species could live and therefore be gathered by human groups. The gradual sea level rise throughout the Late Glacial period shaped a coastline with certain characteristics. Some authors (Shackleton 1988) have suggested that there would have been few shores with a soft substrate in the Late Glacial, as coastlines in the course of developing tend to be rocky. Although this is applicable to species living in sandy or muddy shores, it does not explain the absence of *Mytilus* and *Ostrea edulis*, unless there were no areas with suitable salinity conditions for these species.

In addition, insufficient geomorphological information is available about Pleistocene coasts in Cantabrian Spain to rule out the possibility that estuaries existed at that time. They may, however, have been different from modern estuaries, with a smaller fluvial contribution in terms of flow and sediment and possibly with less marine sand, as the continental shelf was smaller in size (Alejandro Cearreta, personal conversation). Whether these characteristics could have affected the presence of species preferring both rocky and sandy/muddy shores cannot be reliably determined at present, although it does appear possible. It is clear though that the formation of the modern estuaries about 9,000 – 8,000 cal BP produced the ideal conditions for the growth of bivalve molluscs in the region.

In short, the most probable causes for the changes in the presence of bivalves are environmental, including both climatic and geomorphological factors or a combination of both. This holds true especially when other examples of the sensitivity of molluscs to climate change are known in the region. An explanation connected to subsistence strategies only becomes reasonable in the case of the absence of *Mytilus edulis* from Upper Palaeolithic sites.

Conclusions

In the Late Glacial period, malacological assemblages in Cantabrian Spain are based on species that are typical of a cold climate, in accordance with the climate conditions existing at that time. With the arrival of the Holocene, after the Preboreal oscillation, this pattern underwent a significant change. About 10,800 cal BP the mollusc species present in the region reorganised, with a decrease in the abundance of cold climate species. This transition was especially abrupt in the case of *Littorina littorea*, and more gradual for *Patella vulgata*. Parallel to this decrease, species adapted to a temperate climate like *Phorcus lineatus*, *Patella depressa* and *Patella ulyssiponensis*, which had been represented more scarcely in earlier periods, began to appear in abundance. At the same time, geomorphological changes caused by the rise in sea level associated with the warmer climate led to the formation of estuaries and the proliferation of biotopes suitable for bivalve species living in soft substrates and preferring a temperate climate, such as *Ruditapes decussatus*, *Scrobicularia plana* and *Solen marginatus*. Bivalves living on harder substrates, such as *Ostrea edulis* and *Mytilus galloprovincialis* also thrived. The expansion of all these species enabled the range of species consumed by hunter-gatherer groups to widen during the Mesolithic and especially in the early Neolithic, when the regional estuaries had completed their formation.

Acknowledgements

We would like to thank the University of Cantabria and the Spanish Minister of Science and Innovation (Project HUM 2006–13729) for funding this research, M. R. González Morales, Alejandro García Moreno and Eduardo Palacio for their comments and help with the maps and figures, and Arturo Morales and one anonymous referee for their comments. IGZ is currently funded by the Juan de la Cierva programme and DCS is funded by the Fyssen foundation..

References

Álvarez-Fernández, Esteban, 2011. Humans and marine resource interaction reappraised: Archaeofauna remains during the late Pleistocene and Holocene in Cantabrian Spain. *Journal of Anthropological Archaeology* 30(3):327-343.

——. 2012. Investigaciones arqueomalacológicas en La Garma A (Omoño, Cantabria): los moluscos marinos de los niveles N y O (Magdaleniense Superior). In *El Paleolítico superior Cantábrico. Actas de la Primera Meso Redonda. San Román de Candamo (Asturias) 26-28 de Abril de 2007., Series Monografías 3.* Pablo Arias Cabal, María Soledad Corchón, Mario Menéndez, and Adolfo Rodríguez Asensio (eds.), pp. 145-155. Santander: PUbliCan, Ediciones de la Universidad de Cantabria.

Andersen, Katrine K., Anders Svensson, Sigfus J. Johnsen, Sune O. Rasmussen, Matthias Bigler, Regine Röthlisberger, Urs Ruth, Marie-Louise Siggaard-Andersen, Jørgen Peder Steffensen, Dorthe Dahl-Jensen, Bo M. Vinther, and Henrik B. Clausen. 2006. The Greenland Ice Core Chronology 2005, 15-42 ka. Part 1: constructing the time scale. *Quaternary Science Reviews* 25:3246-3257.

Arias, Pablo, Juan Antonio Fernández -Tresguerres, Esteban Álvarez Fernández, Ángel Armendariz, Marián Cueto Rapado, Miguel Ángel Fano Martínez, Raquel. Fernández García, María Dolores. Garralda, Carmen Mensua, and Luis C. Teira. 2007. Excavación arqueológica de urgencia en la cueva de La Poza l'Egua (Lledías, Llanes). In *Excavaciones Arqueológicas en Asturias 1999-2002.* Pp. 227-239.

Oviedo: Consejería de Cultura, Comunicación Social y Turismo del Principado de Asturias.

Bailey, Geoffrey N., and Alan S. Craighead. 2003. Late Pleistocene and Holocene coastal paleoeconomies: a reconsideration of the molluscan evidence from Northern Spain. *Geoarchaeology: An International Journal* 18 (2):175-204.

Berganza, Eduardo, José Luis Arribas, Pedro Castaños, Mikelo Elorza, Jesús E. González Urquijo, Juan J. Ibáñez, Mª José Iriarte, Arturo Morales, Eduardo Pemán, Eufrasia Roselló, Ana Uriz, Paloma Uzquiano, Víctor Vásquez, and Lidia Zapata. 2012. La transición tardiglaciar en la costa oriental de Bizkaia: el yacimiento de Santa Catalina. In *El Paleolítico Superior Cantábrico, actas de la Primera Mesa Redonda: San Román de Candamo (Asturias) 26-28 de abril de 2007.* Pablo Arias, María Soledad Corchón, Adolfo Rodríguez Asensio, and Mario Menéndez (eds.). Pp. 171-182. Santander: PUbliCan, Ediciones de la Universidad de Cantabria.

Clark, Geoffrey. A. 1976. *El Asturiense Cantábrico.* Biblioteca Prehistórica Hispana Vol. XIII. Madrid: CSIC.

Evans, John G. 1972. *Land Snails in Archaeology.* London: Seminar Press.

Fano, Miguel Ángel. 2007. Un nuevo tiempo: el Mesolítico en la región cantábrica. In *Las Sociedades del Paleolítico en la Región Cantábrica.* Miguel Ángel Fano Martínez, ed. Pp. 337-401. Bilbao: Kobie Anejo 8.

González Morales, Manuel R. 1982. *El Asturiense y otras culturas locales. La explotación de las áreas litorales de la región cantábrica en los tiempos epipaleolíticos.* Santander: Monografía Nº 7, Centro de Investigación y Museo de Altamira.

González Morales, Manuel R., Mª. Carmen Márquez Uría, Tomás E. Díez González, Jose Ángel. Ortea, and Kathleen C. Volman. 1980. *El conchero asturiense de la Cueva de Mazaculos II (La Franca, Asturias): Campañas de 1976 a 1978.* Vol. 9. Noticiario Arqueológico Hispano. Madrid: Ministerio de Cultura

González Morales, Manuel R., Lawrence G. Straus, Agustín Díez Castillo, and Jesús Ruiz Cobo. 1999. Postglacial coast & inland: the Epipaleolithic-Mesolithic-Neolithic transitions in the vasco-cantabrian region. *Munibe* 56:61-78.

Gutiérrez-Zugasti, F. Igor. 2009. *La explotación de moluscos y otros recursos litorales en la región cantábrica durante el Pleistoceno final y el Holoceno inicial.* Santander: PUbliCan, Ediciones de la Universidad de Cantabria.

——. 2011a. Coastal resource intensification across the Pleistocene-Holocene transition in Northern Spain: evidence from shell size and age distributions of marine gastropods. *Quaternary International* 244(1):54-66.

——. 2011b. Shell fragmentation as tool for quantification and identification of taphonomic processes in archaeomalacological analysis: the case of the Cantabrian region (Northern Spain). *Archaeometry* 53 (3): 614-630.

Hawkins, Stephen J., Alan J. Southward, and Martin J. Genner. 2003. Detection of environmental change in a marine ecosystem--evidence from the western English Channel. *The Science of The Total Environment* 310:245-256.

Hawthorne, J. B. 2010. The increasing populations of Osilinus lineatus on East Dorset shores (south-west England) in 2008. *Marine Biodiversity Records* 3:e86 doi:10.1017/S1755267210000825.

Imaz, Margarita. 1994. Los moluscos marinos de Laminak II. *Kobie (Serie Paleoantropología)* XXI:221-224.

Leoz, Ignacio, and Consuelo Labadía. 1984. Malacología marina de Ekain. In *El yacimiento prehistórico de la Cueva de Ekain (Deba, Guipúzcoa).* Jesús Altuna and José M. Merino, eds. Pp. 287-296. Colección Barandiarán 1. San Sebastián: Sociedad de Estudios Vascos y Sociedad de Ciencias Aranzadi.

Limondin, Nicole. 1995. Late-Glacial and Holocene malacofaunas from archaeological sites in the Somme Valley (North France). *Journal of Archaeological Science* 22:683-698.

Madariaga, Benito. 1963. Análisis Paleontológico de la fauna terrestre y marina de la Cueva de La Chora. In *Cueva de La Chora.* Joaquín González Echegaray, Miguel Ángel García Guinea, and A. Begines Ramírez, eds. Pp. 51-74. Excavaciones Arqueológicas en España 26. Madrid: Dirección General de Bellas Artes.

——. 1964. El mar y el hombre prehistórico. *Zephyrus* 15:37-45.

Mieszkowska, Nova, Steve J. Hawkins, Mike T. Burrows, and Michael A. Kendall. 2007. Long-term changes in the geographic distribution and population structures of Osilinus lineatus (Gastropoda: Trochidae) in Britain and Ireland. *Journal of the Marine Biological Association UK* 87:537-545.

MORENO, RUTH. 1994. *Análisis arqueomalacológicos en la Península Ibérica. Contribución metodológica y biocultural*. PhD dissertation, Universidad Autónoma de Madrid, Madrid.

——. 1995. Catálogo de malacofaunas de la Península Ibérica. *Archaeofauna* 4:143-272.

ORTEA, JOSÉ ÁNGEL. 1986. The malacology of La Riera cave. In *La Riera cave. Stone Age hunter-gatherer adaptations in northern Spain*. Lawrence G. Straus and Geoffrey A. Clark, eds. Pp. 289-313. Tempe: Arizona State University.

POPPE, GUIDO T., and YOSIHIRO GOTO. 1991. *European Seashells. Vol. I (Polyplacophora, Caudofoveata, Solenogastra, Gastropoda)*. Germany: Verlag Christa Hemmen.

——. 1993. *European Seashells. Vol. II (Scaphopoda, Bivalvia, Cephalopoda)*. Germany: Verlag Christa Hemmen.

RASMUSSEN, SUNE OLANDER, KATRINE K. ANDERSEN, ANDERS M. SVENSSON, JORGEN P. STEFFENSEN, BO M. VINTHER, HENRIK B. CLAUSEN, MARIE-LOUISE SIGGARD-ANDERSEN, SIGFUS JOHNSEN, LARS BERG LARSEN, DORTHE DAHL-JENSEN, MATTHIAS BIGLER, REGINE RÖTHLISBERGER, H. FISCHER, KUMIKO GOTO-AZUMA, M.E. HANSSON, and URS RUTH. 2006. A new Greenland ice core chronology for the last glacial termination. *Journal of Geophysical Research* 111. D06102, doi:10.1029/2005JD006079.

RASMUSSEN, SUNE OLANDER, BO M. VINTHER, HENRIK B. CLAUSEN, and KATRINE K. ANDERSEN. 2007. Early Holocene climate oscillations recorded in three Greenland ice cores. *Quaternary Science Reviews* 26:1907-1914.

RUSSELL, NICOLA, GORDON T. COOK, PHILIPPA L. ASCOUGH, ANDREW J. DUGMORE. 2010. Spatial variation in the MRE throughout the Scottish Post-Roman to Late Medieval period: North Sea values (500 – 1350BP). *Radiocarbon* 52 (3):1166-1181.

SHACKLETON, JUDITH C. 1988. *Marine molluscan remains from Franchthi Cave*. Bloomington & Indianapolis: Indiana University Press.

SOARES, ANTÓNIO M. MONGE, and JOSÉ M. MATOS MARTINS. 2009. Radiocarbon dating of marine shell samples. The marine radiocarbon reservoir effect of coastal waters off Atlantic Iberia during Late Neolithic and Chalcolithic Periods. *Journal of Archaeological Science* 36:2875-2881.

SPARKS, BRUCE WILFRED 1980. *Los moluscos no marinos en la arqueología.* Ciencia en arqueología. Don Brothwell and Eric Higgs, eds. Pp. 406-417. México: Fondo de Cultura Económica.

STRAUS, LAWRENCE G. 1992. *Iberia before the iberians. The Stone Age Prehistory of Cantabrian Spain.* Albuquerque: University of New Mexico Press.

STRAUS, LAWRENCE G., and GEOFFREY A. CLARK. 1986. *La Riera Cave. Stone Age hunter-gatherer adaptations in northern Spain.* Tempe: Arizona State University.

STUIVER, MINZE, PIETER M. GROOTES, and THOMAS F. BRAZIUNAS. 1995. The GISP2 δ18O Climate Record of the Past 16,500 Years and the Role of the Sun, Ocean, and Volcanoes. *Quaternary Research* 44:341-354.

VEGA DEL SELLA, RICARDO E. CONDE DE LA. 1923. *El Asturiense. Nueva industria preneolítica.* Comisión de Investigaciones Paleontológicas y Prehistóricas, Serie Prehistórica 27. Madrid: Museo Nacional de Ciencias Naturales.

VON GRAFENSTEIN, ULRICH, HELMUT ERLENKEUSER, J. MÜLLER, JEAN JOUZEL, and SIGFUS JOHNSEN, 1998. The cold event 8200 years ago documented in oxygen isotope records of precipitation in Europe and Greenland. *Climate Dynamics* 14(2):73-81.

WENINGER, BERNHARD, and OLAF JÖRIS. 2008. A 14C age calibration curve for the last 60 ka: the Greenland-Hulu U/Th timescale and its impact on understanding the Middle to Upper Paleolithic transition in Western Eurasia. *Journal of Human Evolution* 55:772-781.

WENINGER, BERNHARD, OLAF. JÖRIS, and UWE. DANZAGLOCKE. 2008. *CalPal-2007*. Cologne Radiocarbon Calibration & Paleoclimate Research Package http:/www.calpal.de/, 2010-8-8.

21 - DEPOSITS OF TERRESTRIAL SNAILS: NATURAL OR ANTHROPOGENIC PROCESSES?

Eloisa BERNÁLDEZ-SÁNCHEZ
Laboratorio de Paleobiología, Instituto Andaluz del Patrimonio Histórico, Avda, de los Descubrimientos nº1, 41092 Sevilla (Spain),
eloisa.bernaldez@juntadeandalucia.es
Departamento de Sistemas Físicos, Químicos y Naturales, Universidad Pablo de Olavide, Carretera de Utrera Km1, 41013 Sevilla (Spain),
ebersan@upo.es

Esteban GARCÍA-IÑAS
Laboratorio de Paleobiología, Instituto Andaluz del Patrimonio Histórico, Avda, de los Descubrimientos nº1, 41092 Sevilla (Spain),
esteban.garcia@andaluciajunta.es

Abstract: *In the archaeological sites of the southwestern Iberian Peninsula small dumps of land snails are frequently found. It is likely that these deposits are the traces of human activity. However, it could also be possible that these dumps were produced by a natural process at a period of time when these molluscs were inactive. We will try to explain the origin of these snail deposits by undertaking a statistical analysis to test hypotheses comparing biometric analyses from different groups of snails: those collected by people who usually consume snails in 2009, those found in plants where these molluscs accumulate at the end of the summer of 2009 and those registered in several archaeological sites: "La Gallega" (2,500 BC; Valencina de la Concepción, Seville), "El Carambolo" (ninth-sixth century BC; Camas, Seville), "Cerro de la Albina" (seventh century BC; Puebla del Río, Seville), "C/ San Felipe Neri" (second-third centuries BC; Carmona, Seville), "Hospital de las Cinco Llagas" (twelfth century AC; Seville), "La Almagra" (twelfth century AC; Huelva) and "C/ San Fernando" (twelfth century AC; Seville). The biometric analysis and the biostratinomic observations have provided us a mathematical model to ascertain the origin of deposits of land snails in archaeological sites.*

Keywords: *Biostratinomy, Taphonomy, Biometry, Terrestrial snails, Iberian Peninsula.*

Introduction

Taphonomy is the science which helps us to interpret the origin of the organic record associated with archaeological sites, and is supported on biostratinomical analysis through which we can discern some of the characteristics of the current *post-mortem* processes that underpin taphonomic interpretations (Bernáldez 2009; Davies, Powell and Stanton 1989; Domínguez-Rodrigo 1998). Based on this theoretical framework it is essential to closely study the contents of deposits in order to describe their origin and outline the processes that acted upon them before they were buried. Here, we carry out a biometrical and biostratinomical study of land snails from current deposits from southern Spain to infer these results and distinguish between natural and cultural deposits at archaeological sites. Several studies have already proved the importance of biometrical data (Bernáldez and García-Viñas 2010; Fradkin 2008; Gutiérrez 2009) and biostratinomical analysis in a large number of paleobiological and archaeozoological analysis. In some cases the application of biometrical analysis can help to distinguish between natural and cultural deposits (Bernáldez *et al.* 2010; Hughes and Lampert 1977), but in other cases these studies are not enough (Rowland 1994) and we must support these results with other features of the shell middens. Although we must not forget that uniformitarian research can help us to interpret ancient deposits through inference, we cannot ensure that ancient deposits have experienced the same events as the current ones. With regards to post-depositional processes, different results may be obtained from within the same deposit, or two separate deposits can yield the same results. In taphonomy, this is known as equifinality (Domínguez-Rodrigo, Egeland, and Pickering 2007; Lyman 2004; von Bertalanffy 1956). Thus, we must be careful with our statements about the interpretation of faunal deposits, but without a framework of comparison, our interpretations would be mere speculation (Bernáldez 2009).

Our research team has conducted similar studies with shells of genus *Glycymeris* as well as land snails. In the first case we interpreted the origin of the shells recovered from the floor of the Phoenician sanctuary of "El Carambolo" (Camas, Seville) (Bernáldez *et al.* 2010) and in the second, we deduced the human origin of different samples of land snails at "El Cerro de la Albina" archaeological site (La Puebla del Río, Seville) (Bernáldez and Bernáldez 2001). Archaeozoologists have found evidence of human consumption of land snails in many archaeological sites

(Aparicio 2001; Bernáldez and Bernáldez 2001; Lubell 2004a; Ruiz Cobo, Muñoz Fernández and Smith 1999), some of which are over 20,000 years old (Lubell 2004b). Also, nowadays these molluscs are on the menu of the most prestigious restaurants. Fernández-Armesto (2002) suggests in his book "History of Food" that land snails could be the first domesticated animals and that their importance is underestimated by archaeologists. While not pursuing the issue of domestication, it is true that most archaeologists only focus their efforts on studying the use of marine species for food and/or ornaments (Bar-Yosef 2005; González *et al.* 2010; Szabó and Quitmyer 2008) and that land molluscs are usually studied to characterize palaeoecosystems and environmental factors (Allen 2005; Davies 2008). However, archaeologists usually do not consider land snails as part of the economy of human communities. Nowadays the most commonly consumed land snail in the southwest of the Iberian Peninsula is *Theba pisana pisana* (Arrébola 1999; Arrébola and Álvarez 2001), being also the most common in ecosystems (Mayoral *et al.* 2007) and archaeological sites in this area.

Morphology and ecology of *Theba pisana pisana* (Muller 1774)

Theba pisana pisana is a land snail that belongs to the subfamily Helicinae. It has a moderately strong, opaque and bright shell with a uniform colour. Sometimes its shell has colour bands of varying thickness and regularity. Cowie (1983) has detected changes in the colour of the mantle, caused by differences in the age of individuals and by the amount of irradiation of the ecosystem where they live. Size ranges from between 12-22 mm in diameter and 9-19 mm in height and no significant changes are detected in relation to climatic variations (Cowie 1984b) as have been described for other species of land snails (Goodfriend 1986). This homogeneity of size in this species, not related to variations in temperature, can be explained by the characteristics of their preferred habitats. They often live in coastal areas (Kadmon and Heller 1998) and river valleys up to 600 m altitude, and the presence of water buffers these habitats from sudden temperature changes. *Theba pisana pisana* is a pioneer species which occupies different ecotones (e.g. agricultural fields, dune areas, scrublands, riparian vegetation, roadsides) and avoids hot and dry areas (Cowie 1985). It lives throughout the south of the Iberian Peninsula (Mayoral *et al.* 2007; Ruiz *et al.* 2006), as well as throughout the Mediterranean basin, Eastern Europe and the British Isles (Deisler and Stange 2001). This snail has become an invasive species in many parts of America, Australia or Africa because of its wide range of tolerance (Cowie *et al.* 2009; Odendaal, Haupt and Griffiths 2008; Rumi, Sánchez and Ferrando 2010). *Theba pisana pisana* can accumulate in dense groups and aestivate in diverse places (branches of plants, fence posts).

When identifying *Theba pisana pisana* by the features of its shell, it can be confused with *Cernuella* (*Cernuella*) *virgata* (Da Costa 1778) or *Xerosecta* (*Xeromagna*) *promissa* (Westerlund 1893) which are of similar size and morphology (although we know that more than 95% of individuals in the study area are *Theba pisana pisana*). Nevertheless, in this analysis we have included all exoskeletons of these species (*Theba, Cernuella* and *Xerosecta*) because we are interested in studying the range of size preferred by collectors (who may not discriminate between these species).

Study area

The biostratinomical and biometrical study of the land snails was carried out in the southwest of the Iberian Peninsula (Europe), near the city of Seville (fig. 21-1). We selected different areas near the archaeological sites studied at the Laboratory of Paleobiology of the Instituto Andaluz del Patrimonio Histórico (IAPH) to collect samples of current land snails.

Current ecosystems

We selected two ecosystems: "Loma del Acebuchal" (Alcalá de Guadaira, Seville) and an area near the Guadalquivir River (between Camas and Santiponce, Seville). Both ecosystems have permanent water sources: the first has an artificial lake and at the second the Guadalquivir River runs parallel to the sampling area. The flora includes typical Mediterranean species (e.g. *Pistacia lentiscus, Olea europea*), trees from riparian forests (e.g. *Populus alba, Populus nigra, Ulmus minor*) and ruderal vascular plants.

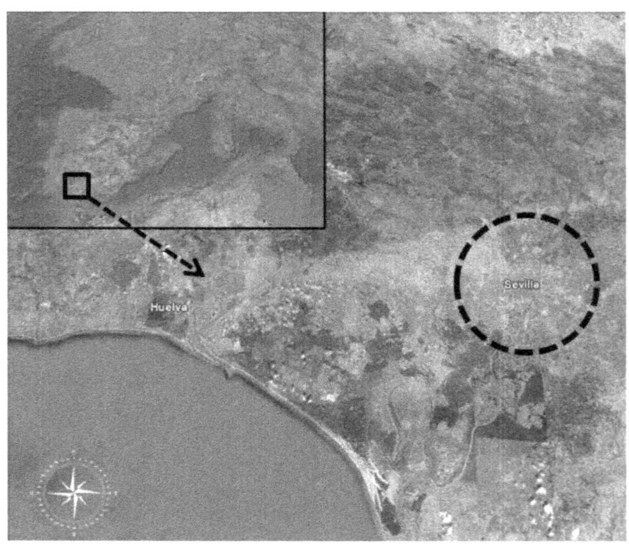

Figure 21-1. The study area is located near Seville, in the southwest Iberian Peninsula (Europe). Source: Google Earth.

Archaeological sites

Seven archaeological sites from southern Spain have been selected to compare with the results obtained in the current biometrical analysis:

- "La Gallega" (2,500 BC; Valencina de la Concepción, Seville). At this site there are different subsurface structures from the Copper Age. This archaeological site is located on a hill on the right bank of the Guadalquivir River and belongs to the larger archaeological site of Valencina de la Concepción-Castilleja de Guzmán. This site is classified as BIC ("Bien de Interés Cultural"), which means it is registered as of cultural interest by the Junta de Andalucía for being considered one of the most important Chalcolithic settlements in the Southern Peninsula. This recognition is based on the fact that the site consists not only of a huge populated area, but also contains an important megalithic area with numerous dolmens (the most notable are La Pastora, Montelirio, Matarrubilla and Ontiveros (Vargas 2004).

- "El Carambolo" (ninth-sixth centuries BC; Camas, Seville). This site is a landmark for specialists studying the legendary Tartessos when in the mid-1970s an important treasure related to this culture was found (Álvarez 2010; Carriazo 1978). Between the years 2002-2005 several excavations were carried out which clarified site funtion and the culture associated with it (Phoenician). In the 1970s this archaeological site was treated as a cabin, but recently new research defined this site as a pit. It is assumed that this pit was used to dispose of the remains of animal sacrifices to the gods *Baal* and *Astarte* made in a near Phoenician sanctuary (fig. 21-2) (Escacena Carrasco, Fernández Flores and Rodríguez Azogue 2007; Fernández and Rodríguez 2006, 2007).

- "Cerro de la Albina" (seventh century BC; La Puebla del Río, Seville). At this site only one structure was found, probably of Tartessian origin, which was possibly used temporarily on several occasions; Escacena and Henares 1999). The archaeologists recorded several deposits of shells with approximately 1,500 land snails in each one (Bernáldez and Bernáldez 2001) buried deliberately in holes. The origin of these deposits of snails could be related to metallurgical activity (Escacena Carrasco, Feliu Ortega and Izquierdo de Montes 2010).

- "C/ San Felipe Neri" (second-third centuries BC; Carmona, Seville). In 2000 archaeologists found a Roman temple in "La Ladera del Corcho", which is currently located 8 m below the surface. The building boasts several galleries filled with a matrix of earthy bricks, pottery fragments, bones and millions of land snails.

- "Hospital de las Cinco Llagas" (twelfth century AD; Seville). When the current Parliament of Andalusia was being built (1998) an archaeological excavation revealed some Roman and Islamic remains of great interest (Tabales *et al.* 2003). The medieval deposits had accumulations of land snails whose average size was less than 10 cm (Bernáldez and Bernáldez 2003a).

- "La Almagra" (twelfth century AD; Huelva). In 2002 an archaeological site was excavated in the "Campus del Carmen" (University of Huelva). There were structures spanning from the Roman period to the present (Campos Carrasco, de la O Teruel and Gómez Rodríguez 2005). We focus on a series of holes excavated in an Islamic kitchen dated from the 12th century. Bernáldez and Bernáldez (2005) described these features as clambakes, in light of the palaeobiological and taphonomic studies. In this paper we will only include groups of land snails recovered from inside these holes.

- "C/ San Fernando" (twelfth century AD; Seville). This archaeological excavation was carried out in the old quarters of Seville. Part of the Islamic wall (twelfth-thirteenth centuries) of the city and other structures dating from the Roman period to the present were discovered due to the tram works on this street (2004).

Both the conservation status of the material and the methodology used during the archaeological excavation ensure that the samples are significant (with small element loss prevented by screening).

Figure 21-2. Aerial view of the Phoenician sanctuary of "El Carambolo". Throughout the entire Phoenician period the building changed its structure and distribution four times. The final state of the building is shown in this image (provided by Álvaro Fernández and Araceli Rodríguez).

Objectives

The main objective of this study is to detect and quantify significant differences between cultural and natural deposits of land snails. We will compare the sizes of the shells in both types of deposits and we will try to define a set of features that allows other researchers to distinguish between the two deposits quickly and easily.

Methodology

To carry out this study different groups of land snails were selected: modern snails collected for consumption, snails from aestivation areas and snails from archaeological sites. The identification of different species was carried out using dedicated literature (Arrébola 2002; Ruiz *et al.* 2006) and the malacological collection of the Laboratory of Paleobiology (Instituto Andaluz del Patrimonio Histórico). All snails were measured by height (LM) and width (AM) in millimeters (fig. 21-3).

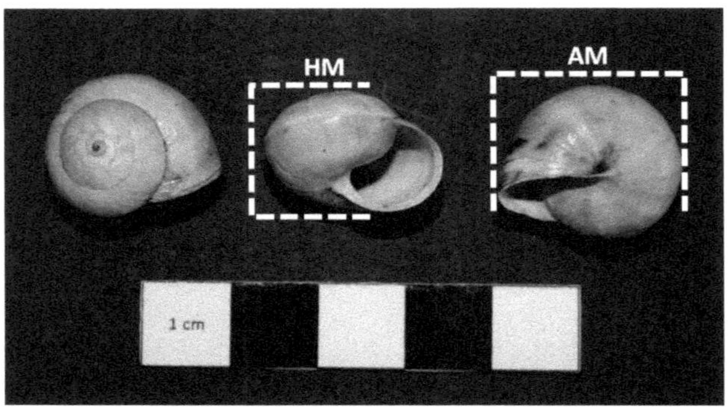

Figure 21-3. Ventral and dorsal view of the shell of *Theba pisana pisana*. We measured width (AM) and height (HM) in mm.

Modern snails gathered for consumption

Each sample of snails was collected by a person or group of people who usually use these molluscs for food. This allowed us to analyze the preferences of the collectors with specific reference to the size of the snails. We studied samples from different collecting seasons in the southern section of the Iberian Peninsula (from April to July) and at different places (see study area) to check if there were changes over time and/or within the collection area. We obtained three samples: two from the "Loma del Acebuchal" (Alcalá de Guadaira, Seville) in April and May of 2009 and one collected near the Guadalquivir river in Camas (Seville) in May of 2009. Once the snails were consumed, they were washed and dried. Afterwards we identified the snails and undertook the biometrical analysis.

Deposits of snails from areas of aestivation

The snails were collected at the "Loma del Acebuchal" (Alcalá de Guadaira, Seville) at the end of July 2009. Most were measured *in situ* and then released into the environment. We collected data from seven groups of this type located in the branches of different trees (*Olea europea*, *Pistacia lentiscus*, *Populus nigra*) and fence posts (fig. 21-4).

Figure 21-4. A group of snails during aestivation on top of a fence post at "La Loma del Acebuchal" (Alcalá del Río, Seville).

Archaeological deposits of land snails

We have deposits of shells from different archaeological sites in the southwest of the Iberian Peninsula, ranging from the Bronze Age to the twelfth century: "La Gallega", "El Carambolo", "Cerro de la Abina", "C/ San Felipe Neri", "Hospital de las Cinco Llagas", "La Almagra" and "C/San Fernando". These snails were cleaned using a combined system of ultrasound, air and water to remove the sediments inside the shells without causing damage to them (fig. 21-5). Once cleaned and dried, the shells were identified and measured with a digital caliper.

Figure 21-5. The snails recovered from archaeological sites were cleaned using a technique that combined air, water and ultrasound to remove all traces of sediment from inside the shells.

For data analysis we used SPSS 14 and we studied normality and homoscedasticity of the different groups. Then we performed mean and variance contrast studies (ANOVA and Scheffé) as parametric contrasts and Kruskal-Wallis and Tamhane as nonparametric. After this, we cross-checked the differences between groups using a distribution analysis and we obtained a formula of classification from the current samples. In order to validate the generated model the group of snails from archaeological sites was included in the analysis.

Results

We studied 7,049 land snails: 2,836 (40.23%) of these derived from the biostratinomical study and 4,213 (59.76%) were selected from archaeological sites.

First we checked whether the differences in size between the groups collected in the actualistic study were statistically significant (three samples were gathered for current consumption and seven samples were selected in areas of aestivation) (fig. 21-6). We analyzed normality using a Shapiro-Wilk test, but it was negative; therefore we applied a non-parametric contrast (Kruskal-Wallis) to test whether these biometrical differences were significant. This one concluded that at least one of the 10 groups analyzed showed significant differences with the others (HM: χ^2 = 1329.538, sig.=.00; AM: χ^2 = 1285.461, sig.=.00). To find out which of the groups showed these differences, we decided to accept normality (depending on Central Limit Theorem which states that samples always follow a normal distribution which adjusts better as the size of the dataset increases until infinity (Milton 2007,

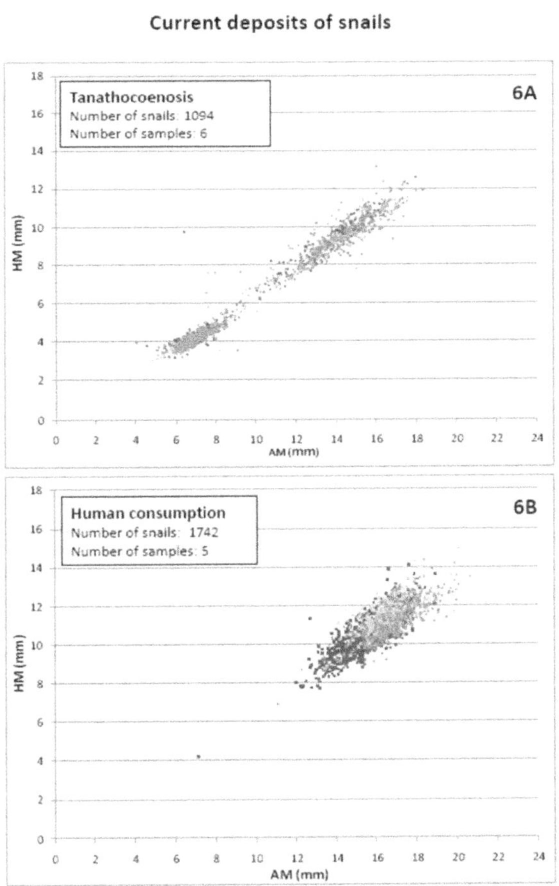

Figure 21-6. In this figure we show the biometrical measures (height and width) of all the land snails which were collected in the actualistic study. We highlight the studied groups: modern consumption and areas of aestivation.

Ramos 2006) and applied the Levene test to check the homoscedasticity. The homocedasticity test was not able to be successfully applied, probably because there were large differences in the sizes of the datasets, and therefore we had to apply a Tamhane test. We concluded that the differences observed between groups (current consumption *vs*. areas of aestivation) of land snails were significant.

Although the differences between the two groups were evidenced by a contrasting average, we cross-checked this through a discriminant analysis. This procedure also identified two groups of snails (Wilks' lambda = 0.481; sig.= .00 / M Box = 1591.700; sig.= .00) and we obtained the following classification formulas:

Natural deposits= -2.096HM + 3.155AM - 10.599
Cultural deposits= -1.545HM + 3.597AM – 20.94

After analyzing the differences between the groups of modern land snails, the 3 samples gathered for consumption were placed into a single group, and the same with the seven samples from areas of aestivation. In order to confirm these differences between the two new groups, we repeated the previous analysis (Kruskal-Wallis and Tamhane tests). These two new groups were compared with the land snails from archaeological sites.

A graph (fig. 21-7) of the biometrical data of the snails found in archaeological sites was similar to figure 21-6 (which represented the groups of current land snails). We thought that this result could be related to three potential origins:

- waste from human consumption,
- natural thanatocoenosis
- and mixed groups of the previous two.

Parametric tests should be applied in order to undertake statistical comparison of samples, but the normality was not proved and neither was the homocedasticity, probably due to the difference in the number of elements of each group: 147 snails from "La Gallega", 312 from "Cerro of the Albina", 606 from "San Felipe Neri", 481 from "La Almagra", 1739 from "C/San Fernando", 472 from "Hospital de las Cinco Llagas" and 457 from "El Carambolo".

Due to these results we decided to implement the Kruscal-Wallis test, which confirmed that at least one of the groups showed differences with the rest in terms of height and width. The pair comparison test was conducted using the Tamhane test with which we obtained four groups according to the size of the snails:
- Group a: Composed of samples from "La Gallega", it had the smallest average size (about 3 mm in HM and 6 mm in AM).
- Group b: Composed of current samples of aestivation and samples of "C / San Felipe Neri".
- Group c: Composed of snails from "Cerro de la Abina", "La Almagra", "C/San Fernando", "El Carambolo" and the current samples of snails collected by humans.
- Group d: Composed of the sample of "Hospital de las cinco Llagas".

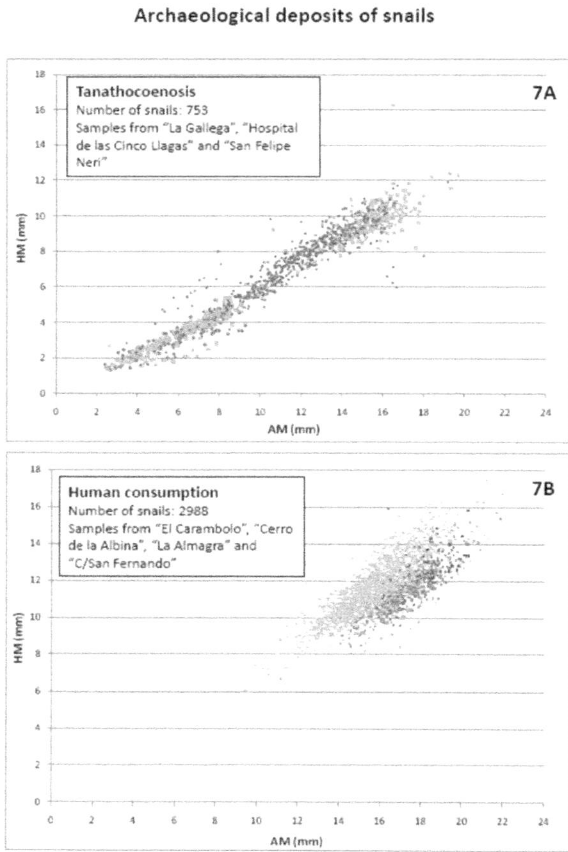

Figure 21-7. In the analysis of size-ranges of the snails from archaeological sites two groups were identified which is a similar result to that obtained in the actualistic study (fig. 21-6).

For this study we were only interested in distinguishing between natural and cultural deposits, therefore those clusters whose average height was less than 10.64 mm, which is the minimum value of the sets associated with human consumption, were assigned as a natural thanatocoenosis. After processing the width data the results obtained were similar to those above, establishing 14.95 mm as the lower limit to consider a group of snails as having a cultural origin.

The average size of snail at "Hospital de las Cinco Llagas" archaeological site (HM=7.35 mm; AM=11.75 mm) is not included in any of the previous deposits (cultural or natural) and it could be an example of a mixed deposit. Nevertheless, these land snails were buried deliberately in a hole, so we think that they were probably collected for human consumption (which could be an exceptional case).

Discussion and conclusions

After the statistical analysis of the biometry of 7,049 shells of *Theba pisana pisana*, we found significant differences between cultural and natural deposits of land snails (fig. 21-9). Therefore, the first conclusion is that we can determine the origin of an accumulation of shells in the southwest Iberian Peninsula using a combined analysis of biometrical and biostratinomical data.

The deposits of land snails were defined by their size (average width and height). A group of snails which measure over 10 mm in height and 15 mm in width could be interpreted as a cultural deposit (waste of human consumption), and those which have an average of about 7 mm in height and 10 mm in width could be related to natural thanatocoenosis (dead snails in aestivation zones). Bernáldez and Bernáldez (2001) suggested that snails over 11 mm in diameter can be selected for consumption, being the minimum and maximum sizes between 11 and 20 mm. These data provide a mean of about 15 mm, a value similar to that obtained in our case. In addition, we have defined two formulae to classify both groups (natural *vs.* cultural deposits) through a discriminant analysis. These are easy to use; simply replace the values of AM and HM and solve it. The highest result will be the solution which will help us to define the deposit. It is important to note that the formulae have been validated with the mean size of the snails collected in the biostratinomical study and at the archaeological sites (but we did not include the sample of the "Hospital de las Cinco Llagas" because this is a special case). In 100% of the cases the result obtained with this method is similar to that already obtained by the Tamhane test. Therefore, we believe that this tool can be a simple way to interpret deposits of land snails which present doubts as to their origin. The homogeneity of the data group should be checked if the group has biometrical characteristics associated with human consumption: a large number of individuals, a mean similar to the median and a low variance. In the case of natural populations these features are not expected, because the bimodality of the size of the snails from these groups (fig. 21-8), related to their biennial cycle of life (Cowie 1984a), prevents the values of mean, median and variance being as adjusted as is required (in fact, the bimodality is a feature which could be used to classify a deposit of snails as a natural thanatocenosis).

Four deposits of land snails from archaeological sites have been statistically considered as a natural thanatocoenosis, something which has already been mentioned in previous studies (Bernáldez and Bernáldez 2002, 2003b). The rest have been interpreted as accumulations produced by human consumption, also classified in this way in previous analysis (Bernáldez and Bernáldez 2001, 2005). "Hospital de las Cinco Llagas" is an exceptional case, the taphonomic interpretation of this site (Bernáldez and Bernáldez 2003a) reveals that the average size of the snails is less than the size of the snails from a cultural deposit, but we are reluctant to identify the origin. Although the statistical analysis classifies this deposit as a natural thanathocenosis, these snails were buried deliberately. Therefore, we could be facing a case of human consumption of *Theba pisana pisana* at a non-optimal time or, perhaps, these molluscs were used in an unknown manner (Bernáldez and Bernáldez 2001).

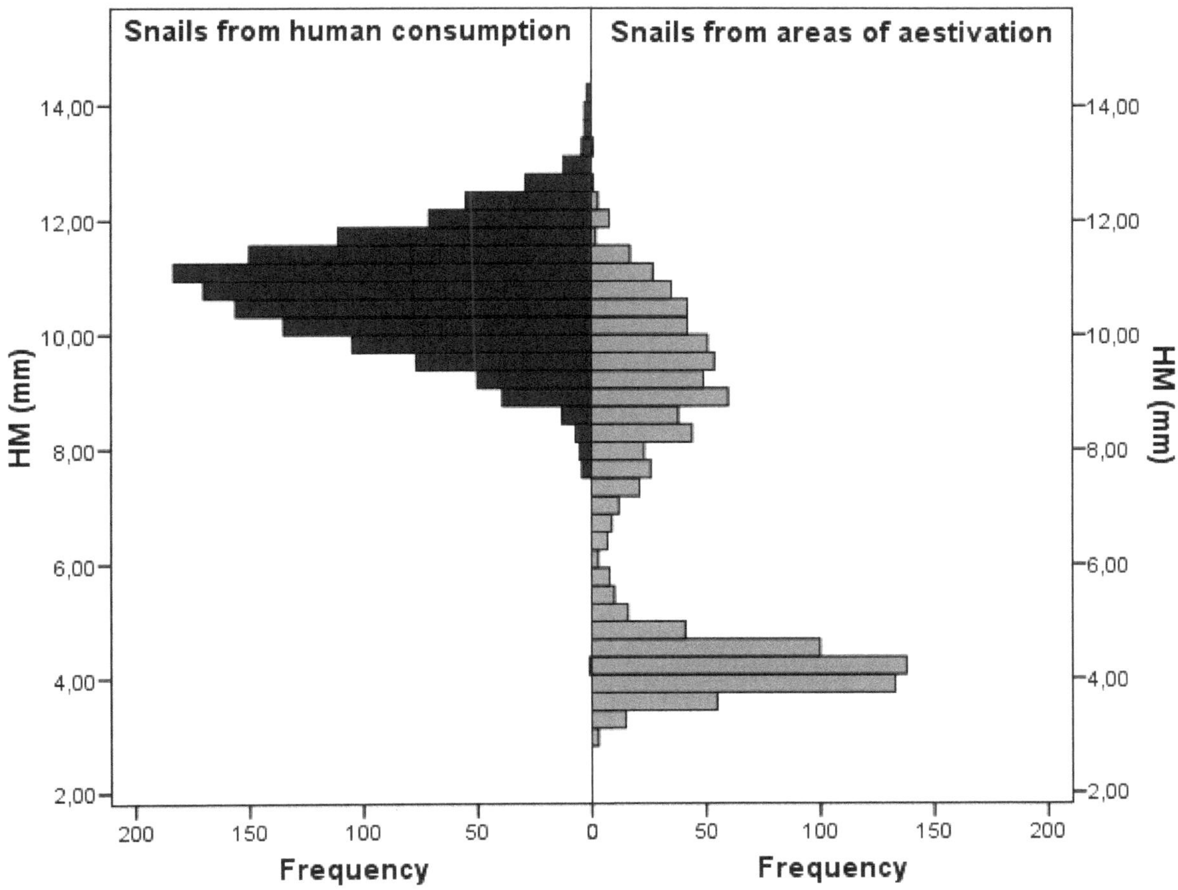

Figure 21-8. The bimodality in the size of the snails from areas of aestivation is explained by the two-year cycle of growth (Baker and Vogelzang 1988; Cowie 1984a).

Although the biostratinomical research could be a science-based tool to discriminate different types of deposits, researchers should be careful when applying the actualistic results in the interpretation of a taphocoenosis or oryctocenosis. A large number of agents may influence both the formation of the deposits (e.g. water, mudslides, animals) and the preservation of their components (e.g. trampling, erosion, dissolutions). In addition, the methodology applied by archaeologists to obtain samples may cause data loss, due to the fragmentation of the elements or the non-collection of some of them. For example, the smaller snails are harder to detect if the material is not sieved, therefore the results of the biometric study could be wrong (because the arithmetical mean could be overestimated). Consequently to complete this research we should study more samples and design biostratinomic experiments to quantify the snail loss during the pre-fossil diagenetic phase.

Figure 21-9. Snails in an aestivation zone. During burial and the action of diagenetic processes several elements were lost through physical, chemical and biological vectors.

Acknowledgements

We thank Marisol Viñas, Santiago, María Cañadas, Belén Cañadas and Antonio Cañadas who gave us samples of snails which they ate. Antonio Augusto and Aurora Ocaña who helped us with biometrical analysis and Joe White for the translation.

References

ALLEN, M.J. 2005. Beaker settlement and environment on the chalk downs of southern England, *Proceedings of the Prehistoric Society* 71:219-45.

ÁLVAREZ, M. 2010. Carriazo y su interpretación de los hallazgos de El Carambolo en el contexto de los estudios sobre Tartesos. In *El Carambolo. 50 años de un tesoro*. M. Luisa de la Bandera and Eduardo Ferrer, coord. Pp. 53-98. Sevilla: Universidad de Sevilla.

APARICIO, M.T. 2001: Malacofauna terrestre del Yacimiento de Cubío Redondo (Matienzo, Cantabria). *Munibe* 53: 61-66.

ARRÉBOLA BURGOS, J.R. 1999. El consumo de caracoles en Andalucía: entre la tradición y la biología. *Noticiario SEM* 31:53-55.

ARRÉBOLA, J.R. 2002. *Caracoles terrestres de Andalucía. Manuales de conservación de la naturaleza, nº 1*. Sevilla: Consejería de Medio Ambiente de la Junta de Andalucía

ARRÉBOLA BURGOS, J.R. and R.M. ÁLVAREZ HALCÓN. 2001. La explotación de los caracoles terrestres en España: aspectos ecológicos y socioculturales. *Temas de Antropología Aragonesa* 11:139-172

BAKER, G.H. and B.K. VOGELZANG. 1988. Life history, population dynamics and polymorphism of Theba pisana (Mollusca: Helicidae) in Australia. *Journal of Applied Ecology* 25(3):867-887.

BAR-YOSEF, D. eds. 2005. *Archaeomalacology. Molluscs in former environments of human behaviour*. Oxford: Oxbow books.

BERNÁLDEZ SÁNCHEZ, E. 2009. *Bioestratinomía de macromamíferos terrestres de Doñana. Inferencias ecológicas en los yacimientos arqueológicos del S.O. de Andalucía*. BAR International Series 1978. Oxford: BAR Publishing.

BERNÁLDEZ SÁNCHEZ, E. and M. BERNÁLDEZ SÁNCHEZ. 2001. "El éxito evolutivo de los recolectores de caracoles de La Puebla del Río". *Boletín del Instituto Andaluz del Patrimonio Histórico,* 35:78-86.

—. 2002. *Estudio malacológico del yacimiento romano de la calle San Felipe en Carmona, Sevilla*. Research report. Laboratorio de Paleobiología: Instituto Andaluz del Patrimonio Histórico.

—. 2003a. "El vertedero islámico del Hospital de las Cinco Llagas. Historias orgánicas en la basura". In *Arqueología y Rehabilitación en el Parlamento de Andalucía. Secretaría General del Parlamento de Andalucía*. Vázquez, A., coord. Pp. 288-318. Sevilla: Secretaría General del Parlamento de Andalucía.

—. 2003b. La basura prehistórica en los últimos metros del subsuelo de Valencina de la Concepción (Sevilla). Paper presented at the *XIX Jornadas de la Sociedad Española de Paleontología en Morella (Castellón),* Morella, October 16-18.

—. 2005. ¿Clambakes en la antigua villa romana de la Almagra (Huelva)? Tafonomía de basureros islámicos. *Boletín del Instituto Andaluz del Patrimonio Histórico,* 53:35-43

BERNÁLDEZ SÁNCHEZ, E. and E. GARCÍA-VIÑAS. 2010. Indirect detection of changes in Seville population studying size changes in oysters? *Munibe* 31:208-215.

BERNÁLDEZ SÁNCHEZ, E., E. GARCÍA-VIÑAS, E. ONTIVEROS, A. GÓMEZ and A. OCAÑA. 2010. Del mar al basurero: una historia de costumbres. In *El Carambolo. 50 años de un tesoro*. M. Luisa de la Bandera and Eduardo Ferrer, coord. Pp. 345-386. Sevilla: Universidad de Sevilla.

BERNÁLDEZ SÁNCHEZ, E. E. GARCÍA-VIÑAS, C. POZO, M.C. LOZANO-FRANCISCO and J.L. VERA-PELÁEZ. 2010. Inferencias biostratinómicas en la interpretación arqueológica de yacimientos del Holoceno. *Férvedes* 6:85-94.

CAMPOS CARRASCO, J.M., N. DE LA O TERUEL and A. GÓMEZ RODRÍGUEZ. 2005. Intervención arqueológica en el cabezo de la Almagra. Avenida de Andalucía/Campus de El Carmen (Huelva). *Anuario Arqueológico de Andalucía 2002. Tomo III Actividades de urgencia*: 513-529.

CARRIAZO, J. DE MATA. 1978. *El Carambolo*. Sevilla: Universidad de Sevilla.

COWIE, R.H. 1983. Variation in mantle collar colour in the land snail *Theba pisana*: Evidence of climatic selection? *Proceedings of the Academy of Natural Sciences of Philadelphia* 135:154-162.

—. 1984a. The Life-Cycle and productivity of the land snail *Theba pisana* (Mollusca: Helicidae). *Journal of Animal Ecology* 53(1):311-325.

—. 1984b. Shell thickness in the land snail *Theba pisana* (Pulmonata: Helicidae). *Biological Journal of the Linnean Society* 21:423-429.

—. 1985. Microhabitat choice and high temperature tolerance in the land snail *Theba pisana* (Mollusca: Gastropoda). *Journal of Zoology* 207:201-211.

Cowie, R.H., R.T. Dillon, D.G. Robinson, and J.W. Smith. 2009. Alien non-marine snails and slugs of priority quarantine importance in the United States: A preliminary risk assessment. *Amer. Malac. Bull.* 27: 113-132.

Davies, D.J., E.N. Powell and R.J. Stanton. 1989. Taphonomic signature as a function of environmental process: Shells and shell beds in a hurricane-influenced inlet on the Texas coast. *Palaeogeography, Palaeoclimatoloogy, Palaeoecology* 72:317-356.

Davies, P. 2008. *Snails. Archaeology and landscape change.* Oxford: Oxbow books.

Deisler J.S. and L.A. Stange. 2001. White garden snail, *Theba pisana* (Mueller) (Gastropoda: Helicidae), University of Florida. http://www.deis.ifas.ufl.edu/in354.

Domínguez-Rodrigo, M. 1998. Tafonomía y ciencia ficción: algunos casos prácticos. *Quaderns de Prehistòria i Arqueología de Castelló* 19:7-25.

Domínguez-Rodrigo, M., C.P. Egeland, and T.R. Pickering. 2007. Equifinality in carnivore tooth mark frequencies and the extended concept of archaeological palimpsests: implications for models of passive scavenging in early hominids. In *Breathing Life into Fossils: Taphonomic Studies in Honor of C.K. "Bob" Brain.* T.R. Pickering, N. Toth and K.D. Schick, eds. Pp. 255-267. Bloomington: Stone Age Institute Press.

Escacena Carrasco, J.L., M.J. Feliu Ortega, and R. Izquierdo de Montes. 2010. El Cerro de la Albina y la metalurgia de la plata en Tartessos. *De Re Metallica* 14:35-51.

Escacena Carrasco, J. L., A. Fernández Flores and A. Rodríguez Azogue. 2007. Sobre El Carambolo: Un *Híppos* sagrado del santuario IV y su contexto arqueológico. *Archivo Español de Arqueología* 80: 5-28.

Escacena Carrasco, J.L. and M.T. Henares. 1999. Un fondo de cabaña de época tartésica en la Puebla del Río (Sevilla). Intervención arqueológica de urgencia. *Anuario Arqueológico de Andalucía. Tomo III: Actividades de urgencia*: 504-510.

Fernández-Armesto, F. 2002. *Near a thousand tables: a History of Food.* New York: Free Press.

Fernández Flores, A. and A. Rodríguez Azogue. 2006. El complejo monumental del Carambolo Alto, Camas (Sevilla). Un santuario orientalizante en la paleodesembocadura del Guadalquivir. *Trabajos de Prehistoria* 62 (1):111-138.

——. 2007. *Tartessos desvelado.* España: Almuzara.

Fradkin, A. 2008. Applying a measure for determining season of oyster harvesting: A case of study of a pre-Columbian Florida site, USA. *Archaeofauna* 17:147-156.

González Gómez de Aguero, E., V. Bejea García, C. Fernández Rodríguez and N. Fuertes Prieto, eds.. 2010. *I Reunión de arqueomalacología de la Península Ibérica. Férvedes* 6. Villalba, Lugo: Museo de Prehistoria y Arqueoloxía de Villalba.

Goodfriend, G.A. 1986. Variation in land-snail shell form and size and its causes: A review. *Systematic Zoology* 35(2):204-223.

Gutiérrez Zugasti, F.I. 2009. Explotación de recursos costeros durante al Neolítico en el área de la Reserva de Urdaibai: los moluscos de la cueva de Kobaeberra (Kortezubi, Bizkaia). *Munibe* 60:175-186.

Hughes, P.J. and R.J. Lampert. 1977. Occupational disturbance and types of archaeological deposits. *Journal of Archaeological Science* 4:135-140.

Kadmon, R. and J. Heller. 1998. Modelling faunal responses to climatic gradients with GIS: Land snails as case study. *Journal of Biogeography* 25(3):527-539.

Lubell, D. 2004a. Are land snails a signature for the Mesolithic-Neolithic transition. *Documenta Praehistorica* XXXI:1-24.

——. 2004b. Prehistoric edible land snails in the circum-Mediterranean: the archaeological evidence. In *Petits animaux et sociétés humaines. Du complément alimentaire aux ressources utilitaires. XXIVe rencontres internationales d'archéologie et d'histoire d'Antibes.* J.P. Brugal and J. Desse, eds. Pp. 77-98. Antibes: Editions APDCA.

Lyman, R.L. 2004. The concept of Equifinality in Taphonomy. *Journal of Taphonomy* 2(1): 15-16.

Mayoral, A.G., J. Perea, R. Martín, R. Acero, E. Camacho, G. Gómez and A. García. 2007. Diversidad malacológica en Andalucía. Recurso renovable. *Arch. Zootec.* 56(1):765-770.

Milton, J.S. 2007. *Estadística para Biología y Ciencias de la Salud.* 3° Edición. Madrid: McGraw Hill.

Odendaal, L.J., T.M. Haupt and C.L. Griffiths. 2008: The alien invasive land snail *Theba pisana* in the West Coast National Park: Is there cause for concern? *Koedoe* 50(1):93-98.

RAMOS ROMERO, H. 2006. Modelos de distribuciones. In *Fundamentos de estadística.* Moreno Rebollo, J.L and A. Pascual Acosta, A. eds. Pp. 29-39. Jaen: Universidad de Jaen.

ROWLAND, M.J. 1994. Size isn't everything. Shells in mounds, middens and natural deposits. *Australian Archaeology* 39:118-124.

RUIZ COBO, J., E. MUÑOZ FERNÁNDEZ, and P. SMITH. 1999. Los concheros de caracoles (Cepaea) en el sector oriental de Catabria. *Altamira* 55:7-27.

RUIZ RUIZ, A., A. CARCABA POZO, A.I. PORRAS CREVILLENT and J.R. ARRÉBOLA BURGOS. 2006. *Guía de los Caracoles Terrestres de Andalucía.* Sevilla: Consejería de Medio Ambiente de la Junta de Andalucía.

RUMI, A., J. SÁNCHEZ and N.S. FERRANDO. 2010: *Theba pisana* (Müller, 1774) (Gastropoda, Helicidae) and other alien land molluscs species in Argentina. *Biol Invasions,* 12:2985–2990.

SZABÓ, K. and I. QUITMYER, eds. 2008. Thematic issue, *Molluscs and other marine resources, Archaeofauna 17.*

TABALES, M.A., P. OLIVA, A. JIMÉNEZ and R. HUARTE. 2003. La intervención arqueológica. In *Arqueología y Rehabilitación en el Parlamento de Andalucía. Secretaría General del Parlamento de Andalucía.* A. Vázquez coord. Pp. 33-100. Sevilla: Secretaría General del Parlamento de Andalucía.

VARGAS, J.M. 2004. *Carta arqueológica de Valencina de la Concepción.* Sevilla: Consejería de Cultura de la Junta de Andalucía.

VON BERTALANFFY, L. 1956. General system theory. *General Systems* 1:1-10.

22 - MICRO-FRESHWATER GASTROPOD REMAINS FROM ÇATALHOYUK, TURKEY: PRELIMINARY ENVIRONMENTAL OBSERVATIONS

Burçin Aşkım GÜMÜŞ
University of Gazi, Faculty of Science, Department of Biology (Zoology), Ankara, Türkiye,
burcinaskim@gazi.edu.tr, burcinaskim@gmail.com

Daniella E. BAR-YOSEF MAYER
Department of Zoology, Tel Aviv University, Tel Aviv, Israel
Peabody Museum, Harvard University, Cambridge MA, U.S.A.,
baryosef@post.tau.ac.il

Abstract: *Çatalhöyük is one of the largest Neolithic sites in Turkey, dated to 7,400 – 6,000 cal BC. It is located east of the Çarşamba River in Çumra (Konya) on the Anatolian plateau. The site contains thousands of molluscan remains originating from marine, paleontological, freshwater and terrestrial contexts. This study is concerned only with the micro-freshwater gastropods that form 43% of all molluscs at the site, and measure under 1 cm. About 4,200 specimens of 12 freshwater gastropod species belonging to 8 families were studied. They derive from several zoogeographic origins, including Palaearctic, Holarctic, and Endemic shells, which in turn represent both lotic and lentic habitats. These environments were noted in other faunal and floral remains from Çatalhöyük and suggest that during the Neolithic Çatalhöyük was surrounded by a lake, riverine habitats including channels and ponds, a marsh, and freshwater springs. At the site the freshwater micro-shells were embedded within mudbricks and mortar that formed the houses of Çatalhöyük. The shells thus represent the environments from which sediments were collected for construction purposes.*

Keywords: *Neolithic, Turkey, Mollusc, Freshwater habitats.*

Introduction

Çatalhöyük is one of the largest Neolithic sites in Turkey, dated to 7,400 – 6,000 cal BC (e.g., Bronk Ramsey *et al.* 2009). It is located east of the Çarşamba River in Çumra (Konya) on the Anatolian plateau, 60km South-East of the modern city of Konya. The site was excavated in the 1960's (Mellaart 1967), and again over the past 17 years as a multidisciplinary project (Hodder 1996, 2000, 2005a, 2005b, 2007). The site consists of over 14 occupational levels and contains very rich and diverse archaeological remains including houses, burials, lithics, ceramics, floral and faunal remains. Mollusc shells at Çatalhöyük are represented by gastropods (snails) that naturally inhabit either marine, freshwater or land environments; bivalves that naturally inhabit marine and freshwater environments; and scaphopods (tusk shells) that inhabit only marine environments. As with the vertebrate fauna, shells were used by the Neolithic inhabitants of Çatalhöyük as food, as raw materials for making ornaments and other artifacts, and, additionally, may indicate the nature of the environment of the site during its occupation. Most of the ornamental shells have been described previously (Bar-Yosef Mayer, Gümüş and İslamoğlu 2010; Reese 2005) but many are still under study. Here we are concerned only with one group of shells: micro-shells under 1 cm in height, that were mostly discovered at the site during the floating and sifting of heavy residue (construction) samples. We assume that these shells were collected neither as food nor as artifacts, but were inadvertently brought into the site with mud or plants that were utilized for construction, and that they therefore represent the natural environment around the site. Thus, the ecological information associated with these species enhances the understanding of the environment in the vicinity of the site during its occupation.

Materials and methods

The shells presented in this study were collected through flotation of the excavated sediments in 1–4 mm mesh screens. Species identification in the field was based on comparisons to various publications (Glöer and Meier-Brook 2003; Schütt 1965; Schütt and Şeşen 1992; Yıldırım 1999; Yıldırım and Schütt 1996; Yıldırım, Gümüş and Kebapçı 2006; Yıldırım, Koca and Kebapçı 2006a, 2006b; Zhadin 1965) as well as a modern reference collection assembled by the first author. The height and length of the complete adult gastropod shells were measured with a digital caliper whenever possible. It is worth noting that some specimens included in this study are juvenile shells of larger gastropods that are within the "micro-shell" size range (e.g., *Viviparus, Lymnaea*, and *Planorbarius*), but they outgrow this category as adults. It is also important to mention that the preliminary report presented here is mostly qualitative in nature, and frequencies of the various species are limited. Quantities, whenever mentioned, represent the Minimum Number of Individuals (MNI) counts, as due to the nature of the heavy residue sampling it was impossible to count tiny fragments that would represent the Number of Identified Specimens (NISP).

Results

The freshwater shells described here were discovered mostly from the debris of mudbricks and mortar that were used to build the houses of Çatalhöyük, but a few were also encountered as inclusions in potsherds (Nurcan Yalman, personal communication). In total 6 prosobranch and 6 pulmonate (basommatophoran) freshwater gastropod species were identified (fig. 22-1). They belong to eight families: Neritidae (1), Viviparidae (1), Hydrobiidae (1), Bithyniidae (1), Valvatidae (1), Melanopsidae (1), Lymnaeidae (3), Planorbidae (3). The abundance of each micro-freshwater species within the sample is presented in table 22-1. The most abundant species at the site is *Valvata piscinalis* (2243), and the least abundant species is *Lymnaea stagnalis* (1). The systematic status, habitat,

Fresh water gastropods	**NISP**	
Bythinella cf. *turca*	27	0.65%
Fagotia esperi	64	1.53%
Galba truncatula	14	0.33%
Gyraulus albus	524	12.52%
Lymnaea stagnalis	1	0.02%
Planorbarius corneus	219	5.23%
Planorbis carinatus	62	1.48%
Radix auricularia	13	0.31%
Stagnicola palustris	70	1.67%
Theodoxus cf. *heldreichi heldreichi*	14	0.33%
Valvata piscinalis	2243	53.60%
Viviparus viviparus	934	22.32%
Total	**4185**	**100.00%**

Table 22-1. Abundance of micro-freshwater gastropod species at Çatalhöyük, presented in alphabetical order. For taxonomic order see table 22-2.

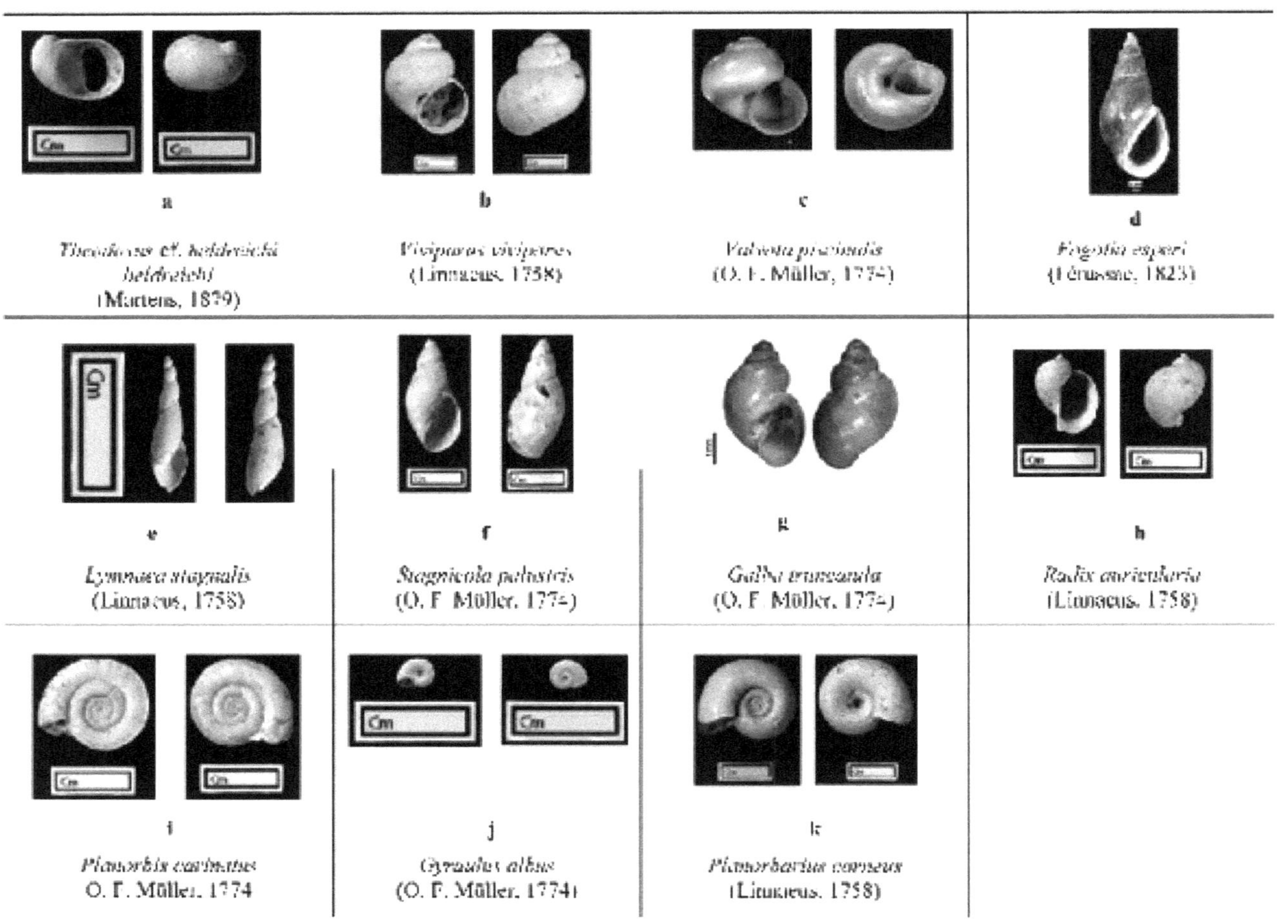

Figure 22-1. The micro-freshwater gastropod species identified at Çatalhöyük except *Bythinella* cf. *turca* (a, b, e, f, h, i, j, k: photographed by Jason Quinlan (2009), Çatalhöyük research project; c: photographed by Peter Glöer (Hamburg–2011); d: after Glöer 2002 (Danube River); g: photographed by Gümüş and De Cupere (2010), Sagalassos research project).

Taxonomy and status	Preferred habitat	Zoogeographic origin
Gastropoda		
Prosobranchia, Archaeogastropoda		
Family: Neritidae		
Theodoxus cf. *heldreichi heldreichi* (Martens, 1879)	Freshwater lakes, the channels and ponds of springs, stony bottom (Littoral)	Endemic
Prosobranchia, Mesogastropoda		
Family: Viviparidae		
Viviparus viviparus (Linnaeus, 1758)	Freshwater river systems, river channels, floodplain lakes and ponds	Palaearctic
Family: Hydrobiidae		
Bythinella cf. *turca* Radoman, 1976	Freshwater river systems, springs, unpolluted streams, with relatively low temperatures	Endemic
Family: Bithyniidae		
Family: Valvatidae		
Valvata piscinalis (O. F. Müller, 1774)	Freshwater river systems, lakes, floodplain ponds; Waters with abundant macrophytes.	Palaearctic
Family: Melanopsidae		
Fagotia esperi (Férussac, 1823)	Freshwater river systems	Palaearctic
Pulmonata, Basommatophora		
Family: Lymnaeidae		
Lymnaea stagnalis (Linnaeus, 1758)	Freshwater river systems, lakes (Littoral); Waters with abundant macrophytes.	Holarctic
Stagnicola palustris (O. F. Müller, 1774)	Freshwater river systems, lakes, shallow water bodies, swamp pools, brooks, ponds (Littoral); Shallow waters with abundant macrophytes	Holarctic
Galba truncatula (O. F. Müller, 1774)	Freshwater river systems, springs, swamps, silty biotopes, NO_3 rich biotopes, living in an amphibian way; Waters with abundant macrophytes	Holarctic
Radix auricularia (Linnaeus, 1758)	Freshwater river systems, brackish lakes, stony biotopes (Littoral); Waters with abundant macrophytes	Palaearctic
Family: Planorbiidae		
Planorbis carinatus O. F. Müller, 1774	Lakes, springs; Waters with abundant macrophytes	Palaearctic
Gyraulus albus (O. F. Müller, 1774)	Freshwater river systems, lakes (Littoral); Waters with abundant macrophytes	Palaearctic
Planorbarius corneus (Linnaeus, 1758)	Freshwater river systems, lakes (Littoral); Waters with abundant macrophytes	Palaearctic

Table 22-2. Micro-freshwater gastropods of Çatalhöyük, their systematic status, habitat and zoogeographic origin.

and zoogeographic origin of the various species is presented in Table 22-2. The zoogeographic origin of the various species are 58% Palaearctic, 25% Holarctic, and 17% endemic. Tables 22-1 and 22-2 only present the gastropod microshells, however, it should be noted that we also identified several freshwater bivalves: *Dreissena* sp. (267), *Pisidium amnicum* (O. F. Müller, 1774) (41), and several juvenile *Unio* sp.

Discussion

Palaeogeography and Environment

The reconstruction of the environment in the vicinity of Çatalhöyük, based on the ecology of the micro freshwater shells, revealed that the site was surrounded by a variety of freshwater bodies including a lake, riverine habitats including channels and ponds, a marsh, and freshwater springs. Geological information tells us that these water bodies were created as a consequence of the Alpine orogenesis in the Late Paleogene, and the Paratethyan basins (Alpine, Pannonian, Dacian, Euxian, Caspian) in the Neogene, which were separated from the Mediterranean (Gray 1988). The evolution of each basin played an important role in the impoverishment of the fauna and the appearance of endemic organisms, as well as the development of each depositional environment, from fully marine through brackish to fluvial. All of the subsequent isolation events took place from this "Tethyan" realm (Müller, Geary and Magyar 1999). An inner freshwater lake was created in the early Eocene by the uplifting of the mountains in Anatolia, and was bordered by the Northern Anatolian and Southern Taurus mountain chains. It ceased to exist in the upper Miocene. The initial occupations of freshwater fauna of Anatolia had started simultaneously in the upper Miocene. The lakes of the Western Taurus Lake District experienced a loss of salinity beginning in the middle Miocene. During the Pliocene, lakes of various origins became connected, creating the ancient Central Anatolian Lake System. The system disintegrated again during the late Pliocene/early Pleistocene, creating several lakes (Wilke 2007). Thus, the molluscan fauna and especially prosobranch and bivalve species living in fresh and brackish water systems in Anatolia, evolved due to the paleo-geomorphological construction of the inner seas and the freshwater lakes (Demirsoy 1999). The malacofauna found at Çatalhöyük reflects the palaeogeography and environmental preferences of this fauna, indicating that the region around Çatalhöyük is ecologically patterned by the inner freshwater lake system (the continuation of the Tethys).

Some of the species present at the site, and especially endemic ones, can provide more detailed ecological information. *Theodoxus heldreichi heldreichi* inhabits the littoral zone of freshwater lakes and is known from Beyşehir, Eğirdir, and Hoyran lakes, the Quaternary sediments of Acıgöl (between Dazkırı and Başmakçı, Afyon), Ereğli Lake (near Çumra, Konya), seed-beds of Hotamış Lake, and channels and ponds of Büyük Gökçeli Spring (Eğirdir-Isparta) (Karaşahin and Yıldırım 1997; Schütt 1991; Schütt and Şeşen 1992; Yıldırım 1998). The freshwater snails of the genus *Bythinella* are widely distributed throughout southern, central, eastern and western Europe and western Asia, from the uplands of Germany and Poland in the north, to the Mediterranean and Northern Africa in the south, and from the Iberian Peninsula in the southwest to the Ukraine in the east and Turkey in the southeast (Glöer and Pešić 2010; Radoman 1976). *Bythinella turca*, an endemic species of Turkish malacofauna, is described from several perennial springs, including Cire (Balkırı); Eğirdir-Isparta and Kuşbaba Spring, Başgöz Spring, Aziziye Köyü, Can Trout Springs; Burdur (Radoman 1976; Yıldırım 1998; Yıldırım, Karaşahin and Kalyoncu 2001). The presence of these endemic species, *Theodoxus* cf. *heldreichi heldreichi* and *Bythinella* cf. *turca* in Çatalhöyük further supports the paleo-geomorphological construction of this region as being the members of genera *Theodoxus* and *Bythinella* that evolved due to the paleo-geomorphological construction of the inner seas and the freshwater lakes in Anatolia (Demirsoy 1999).

Çatalhöyük's environment was reconstructed recently by Roberts and Rosen (2009) and by Roberts, Boyer and Merrick (2007), based on various sedimentological and botanical studies. De Ridder's (1965) study is the only one to date that also addressed malacological aspects. He studied the molluscan fauna in the lake sediments and sand spits of the Konya basin. In his study he mentioned that the Konya basin must have contained a huge lake due to the cooler and moister conditions during the Pleistocene. The relict shore line, sand ridges in the forms of spits and bars, massive deltas at the mouth of stream valleys (Çarşamba, May, Meram, Sille rivers), and the gravel deposits rich in shells form the physical evidence for this former lake. He concluded that the lake, lake-deposited sediments, sand ridges, and abandoned shoreline dated to the Pleistocene, and probably the last glacial period (Würm glacial in his terms). In addition, comparing the recent and the Quaternary malacofauna, he indicated that the climate in the Konya basin has not changed much since the Pleistocene. The lake dried up completely at the end of the last glacial period because of the warmer and drier conditions. We determined that the majority of the micro fresh water gastropods of Çatalhöyük (*Theodoxus heldreichi heldreichi, Viviparus viviparus, Valvata piscinalis, Lymnaea stagnalis, Stagnicola palustris, Radix auricularia, Planorbis carinatus, Planorbarius corneus*) as well as the freshwater bivalves that we identified (*Dreissena* sp., *Pisidium amnicum* and *Unio* sp.) show a great similarity with the fauna obtained from the sediments of Konya basin by De Ridder (1965). Thus this Pleistocene lake, or relics of it, may have continued to exist in the early Holocene during the occupation of Çatalhöyük.

Mollusc habitats and the archaeological context

Most of the freshwater prosobranchs, with some exceptions, are stenoec organisms, meaning that they have a limited range of tolerance to environmental changes in their ecosystem, and they prefer fast running, highly oxygenated waters: a lotic environment. However, most of the freshwater pulmonates, also with some exceptions, are euryoec organisms, meaning that they have a wide tolerance range to environmental changes, and prefer slow running waters with variable levels of dissolved oxygen: a lentic habitat. Much of the remainder of the freshwater gastropod species can derive from either lentic or lotic habitats (Boycott 1936; Fretter 1968; Fretter and Alastair 1962 Gittenberger and Janssen 1998; Karaşahin and Yıldırım 1997; Pennak 1989; Schütt 1991, Schütt and Şeşen 1992; Şahin 2008; Yıldırım 1998; Yıldırım and Schütt 1996; Yıldırım and Şeşen 1994).

The freshwater shells were collected from the debris of mudbricks and mortar that were used to build the houses of Çatalhöyük. While inspecting some of the bricks and mortar in the field we identified the molluscs as inclusions in them, and we assume that they were inadvertently introduced into the site. They thus represent the site surroundings: Fast running water of rivers and a highly oxygenated lake, and slow running water possibly of a marsh. Several of these species, namely, *Valvata piscinalis*, *Lymnaea stagnalis*, *Stagnicola palustris*, *Galba truncatula*, *Radix auricularia*, *Planorbis carinatus*, *Gyraulus albus* and *Planorbarius corneus* are known to live in waters that contain macrophytes (Boycott 1936; Fretter 1968; Fretter and Alastair 1962; Pennak 1989). Since they climb, hide, attach themselves and feed on these plants, they could have been collected, not necessarily only with the mud that was used for the mudbrick and mortar, but also when plants, such as phragmites and Cyperaceae (sedges), were collected. Such plants were used for the production of mats, and were probably also used in construction in addition to the bricks (Rosen 2005, P. Ryan, personal communication).

We also tried to investigate what types of snails are available today in the various environments around the site that are suitable for mudbrick construction. However, the changes in the environment between the Neolithic and today, and especially channeling of water for irrigation over the last few decades, do not allow a reliable analogy. Yet it is worthwhile mentioning that the area around the West and the East Mounds was surveyed, and specimens of *Radix auricularia* and *Gyraulus albus* were collected from the old river bed of the Çarsamba river. Shells sampled from three locations were collected and sifted but the samples were too small to reach any definite conclusions.

A comparison of the environmental information compiled from the shells to that of the environments reflected by the wood remains (see Asouti 2005) indicates similar local environments of lakes and rivers. The site of Çatalhöyük comprises at least 14 levels that span the 1400 years of the site's occupation from 7400 to 6000 cal BC (Bronk Ramsey *et al*. 2009). A preliminary analysis shows that when comparing the frequencies of species by level, only slight differences are observed (Gümüş and Bar-Yosef Mayer, in press). It is important to stress with regard to the sources of freshwater shells, that the changes in proportion between the different shell groups does not necessarily reflect the presence or absence of these environments. Rather, it suggests in what types of water bodies the inhabitants of the site chose to collect the mud and plants.

Conclusions

While the freshwater gastropod fauna discovered within construction materials at Çatalhöyük represent the surroundings of site, we cannot be certain that it reflects all of the possible surrounding environments. It may well be that certain geological or sedimentological settings around the site are not reflected in the malacofauna. But the malacofauna presented here does clearly demonstrate the areas from which construction materials were collected: fast running waters in rivers and a highly oxygenated lake, and slow running water from an environment such as a marsh. The abundance of the identified species and their distribution throughout the site will be evaluated in the near future to examine their origin, and to establish the palaeoenvironment and diversity within the individual excavation units of Çatalhöyük with particular attention to the possibility of change through time.

Acknowledgements

We thank Ian Hodder and Shahina Farid for enabling the study of the Çatalhöyük molluscs. Many of the Çatalhöyük team members were consulted and assisted in various aspects of this study. We would like to thank in particular Slobodan Mitrović and Milena Vasić for obtaining the heavy residue samples, and Serena Love, Arlene Rosen, Philippa Ryan, Nurcan Yalman, and Chris Doherty for their valuable advice. We thank Peter Glöer for the permission in using his marvellous photos of the shells. We are most grateful to Henk K. Mienis (National collections of Natural History, Tel Aviv University), and two anonymous reviewers for their valuable comments on a previous draft.

References

Asouti, E. 2005. Woodland Vegetation and the Exploitation of Fuel and Timber at Neolithic Çatalhöyük: Report on the Wood Charcoal Macro-remains. *In Inhabiting Çatalhöyük: reports from the 1995–1999 seasons,* Edited by Ian Hodder. Çatalhöyük Research Project Volume 4, McDonald Institute Monographs/British Institute of Archaeology at Ankara. Pp. 213–258.

Bar-Yosef Mayer, D. E., B. A. Gümüş, and Y. İslamoğlu. 2010. Fossil Hunting in the Neolithic: Shells from the Taurus Mountains at Çatalhöyük, Turkey. *Geoarchaeology* 25(3):375–392.

Boycott, A. E. 1936. The Habitatas of Freshwater Mollusca in Britain. *Journal of Animal Ecology* 5:116–186.

Bronk Ramsey, C., T. F. G. Higham, F. Brock, D. Baker, and P. Ditchfield. 2009. Radiocarbon dates from the Oxford AMS system: Archaeometry datelist 33. *Archaeometry* 51:323–49.

Demirsoy, A. 1999. *Genel ve Türkiye zoocoğrafyası "Hayvan zoocoğrafyası"*. A.Ş Meteksan (ed), Üçüncü Baskı. Maltepe, Ankara, Pp. 965.

FRETTER, V., G. ALASTAIR. 1962. *British Prosobranch Molluscs, Their Functional Anatomy and Ecology.* The Ray Society, London, 755 p.

FRETTER, V. 1968. *Studies in the Structure, Physiology, and Ecology of Molluscs.* Academic Press New York.

GITTENBERGER, E., A. W. JANSSEN. 1998. De Nederlandse Zoetwatermollusken. Recente en fossiele weekdieren uit zoet en brak water. Nederlandse Fauna, 2: 288 pp. Nationaal Natuurhistorisch Museum Naturalis, KNNV Uitgeverij, European Invertebrate Survey – Nederland. Glöer, P. 2002. *Die Süßwassergastropoden Nord- und Mitteleuropas.* Bestimmungsschlüssel, Lebensweise, Verbreitung. — Die Tierwelt Deutschlands, 73. Teil. 327 S. ConchBooks, Hackenheim.

GLÖER, P., C. MEIER-BROOK. 2003. *Süsswassermollusken.* Deutscher Jugendbund für Naturbeobachtung DJN, ISBN: 3-923376-02-2, 134 pp., 13. überarbeitete Auflage, Hamburg.

GLÖER, P., V. PEŠIĆ. 2010. The Freshwater Snails of the genus *Bythinella* Moquın-Tandon (Gastropoda: Rissooidea: Hydrobiidae) from Montenegro. *Arch. Biol. Sci., Belgrade,* 62 (2):441-447.

GRAY, J. 1988. Evolution of the freshwater ecosystem: the fossil record. *Paleogeography, Paleoclimatology, Paleoecology,* 62:1-214.

GÜMÜŞ, B. A., DE CUPERE, B. 2010. A Revision of the Mollusc Fauna from the Antique Site of Sagalassos (Turkey, Roman–early Byzantine period). *ICAZ–2010, International Council for Archaeozoology, 11th International Conference: 271.*

GÜMÜŞ, B. A., BAR-YOSEF MAYER, D. E., in press, Micro-Freshwater Gastropods at Çatalhöyük as Environmental Indicators. in: I. Hodder (ed.) *Humans and landscapes of Çatalhöyük: reports from the 2000-2008 seasons.* Çatalhöyük research project Volume 8. Cotsen Institute of Archaeology Press, Los Angeles, CA.

HODDER, I. (Ed.) (1996). *On the surface: Çatalhöyük 1993–95.* MacDonald Institute for Archaeological Research, Cambridge.

——. (Ed.). 2000. *Towards reflexive method in archaeology: the example at Çatalhöyük / by members of the Çatalhöyük teams.* MacDonald Institute for Archaeological Research, Cambridge.

——. (Ed.). 2005a. *Changing materialities at Çatalhöyük: reports from the 1995–99 seasons by members of the Çatalhöyük teams.* British Institute at Ankara, London.

——. (Ed.). 2005b. *Inhabiting Çatalhöyük: reports from the 1995–99 seasons / by members of the Çatalhöyük teams.* MacDonald Insititute for Archaeological Research, Cambridge.

——. (Ed.). 2007. *Excavating Çatalhöyük: South, North and Kopal area reports from the 1995–99 Seasons / by members of the Çatalhöyük teams* Vol. 37, British Institute of Archaeology, Ankara.

KARAŞAHIN, S., M. Z.YILDIRIM. 1997. *Eğirdir Civarındaki Bazı Tatlısuların Bentik Faunası Üzerine Bir Araştırma.* III. Ulusal Ekoloji ve Çevre Kongresi, 3-5 Eylül 1997, Kırşehir.

MELLAART, J. 1967. *Çatal Hüyük, a Neolithic town in Anatolia: New aspects of antiquity.* London: Thames and Hudson.

MÜLLER, P., DANA H. GEARY, and IMRE MAGYAR. 1999. The endemic molluscs of the Late Miocene Lake Pannon: their origin, evolution, and family-level taxonomy. *Lethaia,* 32,Pp.:47–60.

PENNAK, R. W. 1989. *Fresh-Water Invertebrates of the United States.* Third Edition. John Wiley & Sons, Inc. 628 p. New York.

RADOMAN, P., 1976. *Speciation within the family Bythinellidae on the Balkans and Asia Minor.* Sonderdruck aus Z. F. Zool. Systematik u. Evolutionsforschung Bd. 14 (1976), H. 2, S. 130–152, Belgrade.

REESE, D. S. 2005. The Çatalhöyük shells. In: I. Hodder (ed), *Inhabiting Çatalhöyük: Reports from the 1995–99 Seasons.* McDonald Institute for Archaeological Research, Cambridge, Pp. 123–127.

DE RIDDER, N. A. 1965 Sediments of Konya Basin, Central Anatolia, Turkey. *Paleogeography, Paleoclimatology, Paleoecology* 1 :225–254.

ROBERTS, N., P. BOYER, and J. MERRICK. 2007. The KOPAL On-site and Off-site Excavations and Sampling, In *Excavating Çatalhöyük: South, North and KOPAL Area reports from the 1995–1999 seasons.* Edited by I. Hodder, 553–72. Cambridge: MacDonald Insititute for Archaeological Research.

ROSEN, A. 2005. Phytolith indicators of plant and land use at Çatalhöyük. In: *Inhabiting Çatalhöyük: reports from the 1995–99 seasons / by members of the Çatalhöyük teams.* Edited by I. Hodder, 203–212. Cambridge: MacDonald Insititute for Archaeological Research.

ROBERTS, N., A. ROSEN. 2009. Diversity and complexity in early farming communities of southwest Asia: new insights into the economic and environmental basis of Neolithic Çatalhöyük. *Current Anthropology* 50:393-402.

SCHÜTT, H., 1965. Zur Systematic und Ökologie Türkischer Süßwasserprosobranchier. *Zoologische Medeedelingen* 41:43–72

——. 1991. Fossile Mollusken dreier anatolischer Ovas. *Archiv fur Molluskenkunde* 120:131–147.

SCHÜTT, H. and R., ŞEŞEN, 1992. The Genus *Theodoxus* in South-western Anatolia, Turkey (Gastropoda, Prosobranchia, Neritidae). *Zoology in the Middle East* 6:63–67.

ŞAHIN, SERAP K. 2008. Mollusc Fauna of Büyükçekmece Lake, and Ecological Parameters that Effect them. 2nd National Malacology Congress. 8-10th of October, Adana, Turkey. Proceedings, Pp:167–176.

WILKE, T., CHRISTIAN ALBRECHT, VITALIY V. ANISTRATENKO, SERAP KOŞAL ŞAHIN and MEHMET Z. YILDIRIM. 2007. Testing biogeographical hypotheses in space and time: faunal relationships of the putative ancient Lake Eğirdir in Asia Minor. *Journal of Biogeography* 34 (10):1807–1821.

YILDIRIM, MEHMET Z. 1998. Isparta İli ve Çevresinin Malakolojik Özellikleri. S. D. Ü. Isparta'nın Dünü, Bugünü ve Yarını Sempozyumu II, 16–17 Mayıs 1998, *Isparta* S: 59–73.

——. 1999. Türkiye Prosobranchia (Gastropoda: Mollusca) Türleri ve Zoocoğrafik Yayılışları, 1. Tatlı ve Acı Sular. *Turkish Journal of Zoology*, 23, Ek Sayı 3:877–900, Ankara

YILDIRIM, M. Z., R. ŞEŞEN. 1994. *Burdur ve Isparta Civarındaki Bazı Tatlı Sulardan Toplanan Mollusca (Yumuşakça) Türleri Üzerinde Zoocoğrafik ve Taksonomik Araştırmalar. XII.* Ulusal Biyoloji Kongresi, 6–8 Temmuz 1994, Edirne, 263–267.

YILDIRIM, M. Z., HARTWIG SCHÜTT. 1996. *Beyşehir Gölü Mollüskleri. XIII.* Ulusal Biyoloji Kongresi 17–20 Eylül 1996, Ankara.

YILDIRIM, M. Z., S. KARAŞAHIN, H. KALYONCU. 2001. Burdur ve Civarı Tatlı Sularında Yayılış Gösteren Gastropoda (Salyangoz) Türleri. *S. D. Ü. Fen Bilimleri Enstitüsü Dergis*, 5 (3):237–256.

YILDIRIM, M. Z., B. A. GÜMÜŞ, ÜMIT KEBAPÇI. 2006. The Basommatophoran Pulmonate Species (Mollusca: Gastropoda) of Turkey. *Turkish Journal of Zoology*, 30(4):445–458.

YILDIRIM, MEHMET Z., SEVAL B. KOCA, ÜMIT KEBAPÇI. 2006a. Supplement to the Prosobranchia (Mollusca, Gastropoda) Fauna of Fresh and Brackish Waters of Turkey. *Turkish J. Of Zoology*, 30:197–204.

——. 2006b. Isparta İli Tatlısularında Yayılış Gösteren Hydrobioidea (Gastropoda: Prosobranchia) Süperfamilyası Türlerinin Bazı Taksonomik Karakterleri. Ege University, *Journal of Fisheries & Aquatic Sciences*, 23 Supplement (1/1):173–177.

ZHADIN, V. I. 1965. *Mollusks of Fresh and Brackish Waters of the U. S. S. R.* Israel Programme for Scientific Translations, Jerusalem.

23 - MOLLUSC SHELLS FROM ARCHAEOLOGICAL BUILDING MATERIALS

Matt LAW
School of History and Archaeology, Cardiff University, Colum Drive, Cardiff, CF10 3EU, United Kingdom,
LawMJ@cf.ac.uk

Abstract : *Mollusc shells can enter the fabric of a building through numerous means. They can be ornamental, structural inserts, deliberately broken up as temper for brick and cement, or incidental inclusions in the sedimentary component of mortar, brick or daub.*
The potential of shell deliberately or accidentally incorporated into building materials to indicate the source of raw materials used in construction is considered. Modern and archaeological examples are given, along with a review of work to date and some recommendations for future research.

Keywords : *Shells, Buildings, Walls, Mortar*

Introduction

Analyses of mollusc remains from archaeological contexts are often employed to address numerous research questions. These include, but are not limited to, questions of palaeoecology and land-use histories, palaeoeconomy, and dating, both relative and absolute. The mollusc remains studied usually derive from death assemblages present within the locality, or may represent shells deliberately brought into the site for economic or other reasons, for example as food waste, currency, or as decorative objects (Claassen 1998:1). Shells may also be present within building materials such as mortars, daubs and plasters, where they derive from the sediment used in the manufacture of the material. The potential of shell deliberately or accidentally incorporated into building materials, and thus imported to a site, to indicate the source of raw materials used in construction is considered here, together with a modern and archaeological example and a review of work to date.

Assigning molluscs to place

Thomas (1985:135) notes that most of the species of land snails present in archaeological contexts in Britain are still extant, and that therefore we can hope to reconstruct past environmental conditions using modern ecological data. The same could also be said of the native marine and brackish or freshwater molluscs (e.g. Davies 2010; Lowe and Walker 1984:195). In the United Kingdom, we are fortunate to have ever-increasing data on modern distributions at our disposal thanks to a number of conservation initiatives, such as the recording carried out by organisations like the Conchological Society of Great Britain and Ireland or the National Biodiversity Network (e.g. Kerney 1999). Using such data in an attempt at reconstruction assumes, however, that the tolerances of the individual species contained within an archaeological assemblage have not changed (Thomas 1985:137). These assumptions are justified in most cases, as 'associations of sub-fossil species resemble present-day associations' (Thomas 1985:137), and Claassen (1998:133) lists a number of projects in which a cultural assemblage has been found to closely match a modern, natural, assemblage in the same locality. Davies (2008:51-66) presents a clear outline of many of the issues inherent in matching sub-fossil Mollusca to particular environments.

Attempts to establish an origin for clearly allochthonous molluscan taxa at a site are relatively commonplace, and often invoke modern species distributions, although Claassen (1998:212 – 218) outlines a number of cases in which chemical sourcing has been used to suggest an origin for archaeological shells.

Mollusc shells in mortar

Mortar can be defined as 'a mixture of lime or more recently cement with sand and water used for bonding stones or bricks' (RCHME 1996:13). It is a relatively common find in British archaeological excavations, and is often sampled on site for laboratory analysis. Hale *et al.* (2003:133-7) list numerous examples where radiocarbon dating has been applied to mortar samples. When encountered on site, it should be recorded as if it were a deposit (MoLAS 1994:Section 3.3.1), which includes making a note of inclusions. The MoLAS Archaeological Site Manual, a standard reference in British field archaeology, specifically mentions the need to look for small shells within mortar (MoLAS 1994:Section 3.3.2).

Mollusc shells are particularly likely to be recovered from within a mortar matrix, or in close association with mortar, as within the organic shell matrix of conchiolin are deposited crystals of calcium carbonate (Morton 1979:29). These survive well in situations where the pH is within the neutral or alkaline range (Evans 1969:171).

The presence of shell within mortar is often recorded in archaeological reports. For example, following a building survey and watching brief carried out between 1998 and 2001, the 16th Century Flodden Wall, an extension to the defensive wall of Edinburgh, Scotland, was described as 'bonded together by a buff coloured, lime-based mortar which included frequent inclusions of grit, shell fragments, pebbles and ceramics' (Lawson and Reed 2003:7). Örstan and Mienis (2010:3-5) describe a modern wall in Turkey whose mortar contains numerous shells thought to derive from the Sea of Marmara.

Mollusc shells in other building materials

Identifiable shells or shell fragments may be also found in daub and lime, and potentially in brick and other ceramic building material. Murphy (2001:22) indicates that the use of alluvial clay to construct an oven was established at St. Martin at Palace-Plain, Norwich, UK from the number of obligate freshwater Mollusca recovered. Murphy (2001:25) also reports whole shells and fragments of cockle in a red sand matrix at a medieval limekiln site at Gilberd School, Colchester, Essex, UK, and suggests that mollusc shells in medieval and later bricks might help to locate clay sources (Murphy 2003:139). The shells of the brackish water snails *Ventrosia* (= *Hydrobia*) *ventrosa* and *Peringia* (= *Hydrobia*) *ulvae* were recorded from the construction trench of a wall at Brean Down, Somerset, most likely from deliberately imported clay (Bell and Johnson 1990:249).

Some shells, particularly oyster valves, have also been used within buildings as structural inserts, for example as compacted foundations or as inserts into masonry or for decoration, such as at the Shell Cottage at the Carton Estate in County Kildare, Ireland (Moorkens 2009:17). Ceci (1984:65) lists a number of uses of shell from middens in construction, and notes that as a bulk material shell compacts and drains well, and can prove very stable, while when broken down it becomes a useful temper for brick, cement and an adobe-like material called 'tabby' used in Puerto Rico and the southern United States (Ceci 1984:67-8). Law and Winder (2009:2) note that oyster valves may be reused after the meat has been consumed, and their use as structural inserts may account for some of the loss of valves they reported from food waste assemblages. The MoLAS Archaeological Site Manual mentions the use of oyster shells to level up courses or fill cracks (MoLAS 1994:Section 3.3.2).

Modern construction sand, Stoke Gifford, South Gloucestershire, UK

As an initial experiment to examine a likely source of shells incorporated in building materials, a spread of modern construction sand of unknown origin on land at Stoke Gifford, South Gloucestershire, was sampled. 1 kg of the sand was wet sieved through a 500 μm mesh, and shells extracted under a low power binocular microscope (table 23-1). A British origin for the material was assumed, and information about habitat was obtained from Hayward *et al.* (1995) and about range in British waters of the taxa recovered from the Marine Life Information Network website (http://www.marlin.ac.uk). Although it was not possible to pinpoint an exact location for the origin of the sand, the species present were generally indicative of lower tidal levels. *Ocenebra erinacea* is especially found on western and south-western coasts of Britain from LWST (Low water spring tide) to 150m depth (Skewes 2005; Hayward *et al.* 1995:534), while *Nassarius incrassatus* and *Emarginula fissura* are not found in seas off East Anglia, or the eastern English Channel and southern North Sea respectively (Peckett 2002, Neal 2004), suggesting a likely south-western inter-tidal or barely sublittoral origin for the material.

GASTROPODA	MNI
Emarginula fissura (Linnaeus, 1758)	1
Gibbula cineraria (Linnaeus, 1758)	9
Littorina saxatilis (Olivi, 1792)	5
Lacuna vincta (Montagu, 1803)	2
Ocenebra erinacea (Linnaeus, 1758)	7
Trophonopsis muricata (Montagu, 1803)	1
Nucella lapillus (Linnaeus, 1758)	1
Buccinum undatum Linnaeus, 1758	12
Nassarius incrassatus (Ström, 1768)	1
BIVALVIA	
Nucula nitidosa Winkworth, 1930	4
Ostrea edulis Linnaeus, 1758	1

Table 23-1. Minimum number of individuals (MNI) of Mollusca recovered from modern construction sand at Stoke Gifford

St. Mary-le-Port, Bristol, UK

In 2006, archaeological evaluations were carried out in Castle Park, Bristol on land adjacent to the ruined church of St Mary-le-Port at the west of the park. Bulk sampling of the site for palaeoenvironmental evidence and smaller artefacts produced a number of residues. One of these, from context 1915, was found to consist of large quantities of lime mortar, one kilogram of which was analysed for mollusc shells incorporated within the soft mortar matrix. Shells were removed from the mortar under a low magnification microscope using fine tweezers and a dissecting needle. The sample yielded very few specimens in general (n=9), however the species present were all found to be freshwater species (table 23-2). Identifications were compared to the reference collection at Cardiff University. Context 1915 filled a pit, 1914, (BaRAS, unpublished data), which was dated to the 17th/ 18th century AD by the pottery types present, although it also included a single sherd of

earlier (c1400 –1450) Tudor Green pot (Reg Jackson, unpublished data). An undated clay pipe stem, the foot of a wine glass, and fragments of window glass were also recovered from the fill (Reg Jackson, unpublished data). A further shell of the freshwater snail *Valvata piscinalis* was recovered from a second sample, of context 1910, the fill of a linear cut, 1909, (BaRAS, unpublished data) which also contained significant quantities of the lime mortar.
All of the species found derive from a freshwater environment, and are recorded on the National Biodiversity Network Gateway (http://data.nbn.org.uk) as present in Bristol. As the River Avon runs immediately to the south of the site, it is likely that the sediment used in the mortar was taken from the closest convenient source, the river.

GASTROPODA	MNI
Theodoxus fluviatilis fluviatilis (Linnaeus)	2
Bithynia tentaculata (Linnaeus, 1758)	2
Valvata piscinalis (O.F. Müller, 1774)	3
Radix balthica (Linnaeus, 1758)	1
Ancylus fluviatilis O.F. Müller, 1774	1

Table 23-2. List of molluscan taxa recovered from mortar at St Mary-le-Port, Bristol

Discussion

The two cases above both present examples in which it is appropriate to attempt to identify a source location for the shells using modern data, as they are both of relatively recent date. Supplementary methods such as chemical sourcing may also be used to help determine a likely source location for shells. Peacock *et al.* (2007) used Laser Ablation-Inductively Coupled Plasma-Mass Spectrometry (LA-ICMP-MS) to chemically identify the waters from which shells used as temper in pottery were being sourced at Lyon's Bluff, Mississippi and such a technique should also be applicable here. LA-ICMP-MS has the additional benefit of causing very little harm to the shell (Peacock *et al.* 2007:320).

Building material samples will not always contain mollusc shell. Carò, Riccardi and Mazzilli Savini (2008:92) recorded the presence in their mortar samples examples of siliceous sponge spicules and radiolarians, but not only did they not record any mollusc shell, they also noted an absence of carbonate rocks among the aggregates. A number of sedimentary deposits in Britain are decalcified, for example many of the brickearths of the Midlands (Murphy 2001:17), and so should not be expected to yield shells when used in building materials.

In some cases, it may be necessary to clean the shells recovered of any encrusted calcareous deposits to aid identification. Allen (1986:52-53) describes a recommended methodology for this that will result in minimal damage to the shells which uses a very weak (0.1-0.5 %) solution of sulphuric acid (H_2SO_4), as well as an alternative method using an ultrasonic bath.

Conclusions

Where shells are present within a building material, it would be useful to analyse the assemblage, with the aim of providing information on the origins of the sedimentary component. In addition to mortars, this may also be true of daubs, plasters, bricks and some artefacts such as loomweights and kiln spacers. This is only likely to yield an accurate result in combination with chemical evidence from the shell, or geological evidence from the aggregate itself however, although the molluscan assemblage should be able to suggest the kind of locality from which the raw material was sourced, and therefore pinpoint possible source locations (subject to the extent of palaeoenvironmental knowledge about the locations in question). The presence of mollusc remains within a building material could also provide an explanation for the presence of allochthonous species in an assemblage derived from a sediment sample. Conversely, the reuse of discarded shell in building material provides a possible destination for missing shell from domestic assemblages, for example where there are unmatched valves of bivalve species.

Acknowledgements

My thanks to Kat Szabó, Paul Davies, Greg Campbell and an anonymous reviewer, who all suggested important improvements to this paper. Any mistakes or omissions are entirely my own however. Ben Rowson confirmed identifications of some of the taxa from the Stoke Gifford sample, and Gareth Dickinson and Tim Longman assisted with sampling at Stoke Gifford. The St Mary-le-Port sample was provided by Bristol and Region Archaeological Services. My attendance at the Paris conference was generously supported by a Cardiff University Graduate Schools Research Support Grant.

References

ALLEN, M. J. 1986. A cleaning technique for land molluscs from archaeological contexts. *Circaea* 4 (1):51-53.

BELL, MARTIN, and SU JOHNSON. 1990. Non-marine Mollusca, Martin Bell, ed, *Brean Down Excavations 1983-1987*.London: English Heritage. Pp 246-250.

CLAASSEN, CHERYL. 1998. *Shells*. Cambridge Manuals in Archaeology. Cambridge: Cambridge University Press.

CARÒ, F., M. P. RICCARDI, and M. T. MAZZILLI SAVINI. 2008. Characterization of plasters and mortars as a tool in archaeological studies: the case of Lardirago Castle in Pavia, Northern Italy. *Archaeometry* 50 (1):85-100.

CECI, LYNN. 1984. Shell midden deposits as coastal resources. *World Archaeology* 16:62-74.

DAVIES, PAUL. 2008 *Snails: archaeology and landscape change.* Oxford: Oxbow.

———. 2010. Land and freshwater molluscs. Terry O'Connor and Naomi Sykes, eds. *Extinctions and Invasions: a social history of the British fauna.* Oxford: Windgather. Pp 175-180

EVANS, J. G. 1969. Land and freshwater molluscs in archaeology: chronological aspects. *World Archaeology* 1:170 –183.

HALE, JOHN, JAN HEINEMEYER, J., LYNNE LANCASTER, ALF LINDROOS, and ÅSA RINGBOM, A. 2003. Dating ancient mortar. *American Scientist* 91(2):130-138.

HAYWARD, P. J., G. D. WIGHAM, and N. YONOW. 1995. Molluscs (Phylum Mollusca), P. J. Hayward., and J. S. Ryland, eds. *Handbook of the Marine Fauna of North-West Europe* Oxford: Oxford University Press, Pp 484-629.

KERNEY, MICHAEL. 1999. *Atlas of the land and freshwater molluscs of Britain and Ireland.* Colchester, UK: Harley.

LAW, MATT, and JESSICA WINDER. 2009. Different rates of survival of right and left valves of European oyster (*Ostrea edulis* L.) from archaeological sites in Britain, *Archaeo+Malacology Newsletter* 16:1-3.

LAWSON, JOHN A., and DAVID REED. 2003. *Conservation and Change on Edinburgh's Defences: archaeological investigation and building recording of the Flodden Wall, Grassmarket 1998-2001*. Scottish Archaeological Internet Report 10.

LOWE, J.J., and M.J.C. WALKER. 1984. *Reconstructing Quaternary Environments.* Harlow, UK: Longman.

MOLAS (Museum of London Archaeology Services), 1994. *Archaeological Site Manual.* 3rd Edition. London: Museum of London.

MOORKENS, ELAINE. 2009. The Shell Cottage, Carton Estate, County Kildare, Ireland *Mollusc World* 19:17.

MORTON , JOHN E. 1979. *Molluscs.* 5th Edition. London: Hutchinson.

MURPHY, PETER. 2001. *Review of Molluscs and other non-insect invertebrates from archaeological sites in the West and East Midlands, and the East of England.* Centre for Archaeology Report 68/2001. London: English Heritage.

MURPHY, PETER. 2003. Charred mollusc shells as indicators of industrial activities, Peter Murphy and Patricia E. J. Wiltshire, eds, *The Environmental Archaeology of Industry*, Symposia of the Association for Environmental Archaeology No. 20. Oxford: Oxbow. Pp 135-140.

NEAL, KEN J. 2004. *Emarginula fissura.* Slit limpet. *Marine Life Information Network: Biology and Sensitivity Key Information Sub-programme* Plymouth: Marine Biological Association of the United Kingdom. http://www.marlin.ac.uk/species/ Emarginulafissura.htm (accessed December 17, 2010).

ÖRSTAN, AYDIN, and HENK K. MIENIS, 2010. Archaeomalacology in the making: a wall of shells in Istanbul. *Archaeo+Malacology Newsletter* 17, 3-5.

PEACOCK, EVAN, HECTOR NEFF, JANE RAFFERTY, and THOMAS MEAKER, 2007. Using Laser Ablation-Inductively Coupled Plasma-Mass Spectrometry (LA-ICP-MS) to source shell in shell-tempered pottery: A pilot study from North Mississippi. *Southeastern Archaeology* 26(2):319-329.

PECKETT, FRANCES. 2002. *Nassarius incrassatus.* Thick-lipped dog whelk. *Marine Life Information Network: Biology and Sensitivity Key Information Sub-programme.* Plymouth: Marine Biological Association of the United Kingdom. http://www.marlin.ac.uk/species/ Nassarius incrassatus.htm (accessed December 17, 2010).

RCHME (Royal Commission on the Historical Monuments of England) 1996, *Thesaurus of Building Materials.* London: RCHME.

SKEWES, M. 2005. *Ocenebra erinacea.* Oyster drill. *Marine Life Information Network: Biology and Sensitivity Key Information Sub-programme.* Plymouth: Marine Biological Association of the United Kingdom, http://www.marlin.ac.uk/species/ Ocenebraerinacea.htm (accessed December 17 2010).

THOMAS, KENNETH D. 1985. Land snail analysis in archaeology: theory and practice, N. R. J. Fieller, D. D Gilbertson and N. G. A. Ralph, eds, *Palaeobiological Investigations: research design, methods and data analysis*. Symposia of the Association for Environmental Archaeology No 5B, Oxford: BAR International Series 266, Pp131-154.

www.ingramcontent.com/pod-product-compliance
Ingram Content Group UK Ltd.
Pitfield, Milton Keynes, MK11 3LW, UK
UKHW061846041026
15591UKWH00015B/941

9 781407 313085